CAMBRIDGE STUDIES IN
ADVANCED MATHEMATICS 38

AN INTRODUCTION TO HOMOLOGICAL ALGEBRA

Already published

AN INTRODUCTION TO HOMOLOGICAL ALGEBRA

CHARLES A. WEIBEL

Department of Mathematics
Rutgers University

CAMBRIDGE
UNIVERSITY PRESS

PUBLISHED BY THE PRESS SYNDICATE OF THE UNIVERSITY OF CAMBRIDGE
The Pitt Building, Trumpington Street, Cambridge CB2 1RP, United Kingdom

CAMBRIDGE UNIVERSITY PRESS
The Edinburgh Building, Cambridge CB2 2RU, United Kingdom
40 West 20th Street, New York, NY 10011-4211, USA
10 Stamford Road, Oakleigh, Melbourne 3166, Australia

First published 1994
First paperback edition 1995
Reprinted 1997

Typeset in Times

A catalogue record for this book is available from the British Library

Library of Congress Cataloguing-in-Publication Data is available

ISBN 0-521-43500-5 hardback
ISBN 0-521-55987-1 paperback

Transferred to digital printing 2003

To my wife, Laurel Van Leer, whose support is invaluable, and to my children, Chad and Aubrey, without whom this book would have been completed much sooner.

Acknowledgments

I wish to express my appreciation to several people for their help in the formation of this book. My viewpoint on the subject comes from S. MacLane, S. Eilenberg, J. Moore, and R. Swan. The notes for the 1985 course were taken by John Lowell, and many topics were suggested by W. Vasconcelos. L. Roberts and W. Vasconcelos used early versions in courses they taught; they helped improve the exposition. Useful suggestions were also made by L. Alfonso, G. Cortiñas, R. Fairman, J.-L. Loday, J. P. May, R. McCarthy, S. Morey, R. Thomason, M. Vigué, R. Wilson, and the referees. Much of the typing was done by A. Boullé and L. Magretto.

Contents

Introduction

Homological algebra is a tool used to prove nonconstructive existence theorems in algebra (and in algebraic topology). It also provides obstructions to carrying out various kinds of constructions; when the obstructions are zero, the construction is possible. Finally, it is detailed enough so that actual calculations may be performed in important cases. The following simple question (taken from Chapter 3) illustrates these points: Given a subgroup A of an abelian group B and an integer n, when is nA the intersection of A and nB? Since the cyclic group $\mathbb{Z}/n$ is not flat, this is not always the case. The obstruction is the group $\mathrm{Tor}(B/A, \mathbb{Z}/n)$, which explicitly is $\{x \in B/A : nx = 0\}$.

This book intends to paint a portrait of the landscape of homological algebra in broad brushstrokes. In addition to the "canons" of the subject (Ext, Tor, cohomology of groups, and spectral sequences), the reader will find introductions to several other subjects: sheaves, $\lim^1$, local cohomology, hypercohomology, profinite groups, the classifying space of a group, Affine Lie algebras, the Dold-Kan correspondence with simplicial modules, triple cohomology, Hochschild and cyclic homology, and the derived category. The historical connections with topology, regular local rings, and semisimple Lie algebras are also described.

After a lengthy gestation period (1890–1940), the birth of homological algebra might be said to have taken place at the beginning of World War II with the crystallization of the notions of homology and cohomology of a topological space. As people (primarily Eilenberg) realized that the same formalism could be applied to algebraic systems, the subject exploded outward, touching almost every area of algebra. This phase of development reached maturity in 1956 with the publication of Cartan and Eilenberg's book [CE] and with the emergence of the central notions of derived functors, projective modules, and injective modules.

Until 1970, almost every mathematician learned the subject from Cartan-Eilenberg [CE]. The canonical list of subjects (Ext, Tor, etc.) came from this book. As the subject gained in popularity, other books gradually appeared on the subject: MacLane's 1963 book [MacH], Hilton and Stammbach's 1971 book [HS], Rotman's 1970 notes, later expanded into the book [Rot], and Bourbaki's 1980 monograph [BX] come to mind. All these books covered the canonical list of subjects, but each had its own special emphasis.

In the meantime, homological algebra continued to evolve. In the period 1955–1975, the subject received another major impetus, borrowing topological ideas. The Dold-Kan correspondence allowed the introduction of simplicial methods, $\lim^1$ appeared in the cohomology of classifying spaces, spectral sequences assumed a central role in calculations, sheaf cohomology became part of the foundations of algebraic geometry, and the derived category emerged as the formal analogue of the topologists' homotopy category.

Largely due to the influence of Grothendieck, homological algebra became increasingly dependent on the central notions of abelian category and derived functor. The cohomology of sheaves, the Grothendieck spectral sequence, local cohomology, and the derived category all owe their existence to these notions. Other topics, such as Galois cohomology, were profoundly influenced.

Unfortunately, many of these later developments are not easily found by students needing homological algebra as a tool. The effect is a technological barrier between casual users and experts at homological algebra. This book is an attempt to break down that barrier by providing an introduction to homological algebra as it exists today.

This book is aimed at a second- or third-year graduate student. Based on the notes from a course I taught at Rutgers University in 1985, parts of it were used in 1990–92 in courses taught at Rutgers and Queens' University (the latter by L. Roberts). After Chapter 2, the teacher may pick and choose topics according to interest and time constraints (as was done in the above courses).

As prerequisites, I have assumed only an introductory graduate algebra course, based on a text such as Jacobson's *Basic Algebra* I [BAI]. This means some familiarity with the basic notions of category theory (category, functor, natural transformation), a working knowledge of the category **Ab** of abelian groups, and some familiarity with the category R–**mod** (resp. **mod**–R) of left (resp. right) modules over an associative ring R. The notions of abelian category (section 1.2), adjoint functor (section 2.3) and limits (section 2.6) are introduced in the text as they arise, and all the category theory introduced in this book is summarized in the Appendix. Several of the motivating examples assume an introductory graduate course in algebraic topology but may be skipped over by the reader willing to accept that such a motivation exists. An exception is the last section (section 10.9), which requires some familiarity with point-set topology.

Many of the modern applications of homological algebra are to algebraic geometry. Inasmuch as I have not assumed any familiarity with schemes or algebraic geometry, the reader will find a discussion of sheaves of abelian groups, but no mention of sheaves of $\mathcal{O}_X$-modules. To include it would have destroyed the flow of the subject; the interested reader may find this material in [Hart].

Chapter 1 introduces chain complexes and the basic operations one can make on them. We follow the indexing and sign conventions of Bourbaki [BX], except that we introduce two total complexes for a double complex: the algebraists' direct sum total complex and the topologists' product total complex. We also generalize complexes to abelian categories in order to facilitate the presentation of Chapter 2, and also in order to accommodate chain complexes of sheaves.

Chapter 2 introduces derived functors via projective modules, injective modules, and δ-functors, following [Tohoku]. In addition to Tor and Ext, this allows us to define sheaf cohomology (section 2.5). Our use of the acyclic assembly lemma in section 2.7 to balance Tor and Ext is new.

Chapter 3 covers the canonical material on Tor and Ext. In addition, we discuss the derived functor $\lim^1$ of the inverse limit of modules (section 3.5), the Künneth Formulas (section 3.6), and their applications to algebraic topology.

Chapter 4 covers the basic homological developments in ring theory. Our discussion of global dimension (leading to commutative regular local rings) follows [KapCR] and [Rot]. Our material on Koszul complexes follows [BX], and of course the material on local cohomology is distilled from [GLC].

Spectral sequences are introduced in Chapter 5, early enough to be able to utilize this fundamental tool in the rest of the book. (A common problem with learning homological algebra from other textbooks is that spectral sequences are often ignored until the last chapter and so are not used in the textbook itself.) Our basic construction follows [CE]. The motivational section 5.3 on the Leray-Serre spectral sequence in topology follows [MacH] very closely. (I first learned about spectral sequences from discussions with MacLane and this section of his book.) Our discussion of convergence covers several results not in the standard literature but widely used by topologists, and is based on unpublished notes of M. Boardman.

In Chapter 6 we finally get around to the homology and cohomology of groups. The material in this chapter is taken from [Brown], [MacH], and [Rot].

We use the Lyndon/Hochschild-Serre spectral sequence to do calculations in section 6.8, and introduce the classifying space BG in section 6.10. The material on universal central extensions (section 6.9) is based on [Milnor] and [Suz]. The material on Galois cohomology (and the Brauer group) comes from [BAII], [Serre], and [Shatz].

Chapter 7 concerns the homology and cohomology of Lie algebras. As Lie algebras aren't part of our prerequisites, the first few sections review the subject, following [JLA] and [Humph]. Most of our material comes from the 1948 Chevalley-Eilenberg paper [ChE] and from [CE], although the emphasis, and our discussion of universal central extensions and Affine Lie algebras, comes from discussions with R. Wilson and [Wil].

Chapter 8 introduces simplicial methods, which have long been a vital part of the homology toolkit of algebraic topologists. The key result is the Dold-Kan theorem, which identifies simplicial modules and positive chain complexes of modules. Applied to adjoint functors, simplicial methods give rise to a host of canonical resolutions (section 8.6), such as the bar resolution, the Godement resolution of a sheaf [Gode], and the triple cohomology resolutions [BB]. Our discussion in section 8.7 of relative Tor and Ext groups parallels that of [MacH], and our short foray into André-Quillen homology comes from [Q] and [Barr].

Chapter 9 discusses Hochschild and cyclic homology of k-algebras. Although part of the discussion is ancient and is taken from [MacH], most is new. The material on differentials and smooth algebras comes from [EGA, IV] and [Mat]. The development of cyclic homology is rather new, and textbooks on it ([Loday],[HK]) are just now appearing. Much of this material is based on the articles [LQ], [Connes], and [Gw].

Chapter 10 is devoted to the derived category of an abelian category. The development here is based upon [Verd] and [HartRD]. The material on the topologists' stable homotopy in section 10.9 is based on [A] and [LMS].

Paris, February 1993

1

Chain Complexes

1.1 Complexes of R-Modules

Homological algebra is a tool used in several branches of mathematics: algebraic topology, group theory, commutative ring theory, and algebraic geometry come to mind. It arose in the late 1800s in the following manner. Let f and g be matrices whose product is zero. If $g \cdot v = 0$ for some column vector v, say, of length n, we cannot always write $v = f \cdot u$. This failure is measured by the *defect*

$$d = n - \text{rank}(f) - \text{rank}(g).$$

In modern language, f and g represent linear maps

$$U \xrightarrow{f} V \xrightarrow{g} W$$

with $gf = 0$, and d is the dimension of the *homology module*

$$H = \ker(g)/f(U).$$

In the first part of this century, Poincaré and other algebraic topologists utilized these concepts in their attempts to describe "n-dimensional holes" in simplicial complexes. Gradually people noticed that "vector space" could be replaced by "R-module" for any ring R.

This being said, we fix an associative ring R and begin again in the category **mod**–R of right R-modules. Given an R-module homomorphism $f: A \to B$, one is immediately led to study the kernel $\ker(f)$, cokernel $\text{coker}(f)$, and image $\text{im}(f)$ of f. Given another map $g: B \to C$, we can form the sequence

(∗) $$A \xrightarrow{f} B \xrightarrow{g} C.$$

1

We say that such a sequence is *exact* (at B) if $\ker(g) = \mathrm{im}(f)$. This implies in particular that the composite $gf: A \to C$ is zero, and finally brings our attention to sequences $(*)$ such that $gf = 0$.

Definition 1.1.1 A *chain complex* C of R-modules is a family $\{C_n\}_{n\in\mathbb{Z}}$ of R-modules, together with R-module maps $d = d_n: C_n \to C_{n-1}$ such that each composite $d \circ d: C_n \to C_{n-2}$ is zero. The maps d_n are called the *differentials* of C. The kernel of d_n is the module of *n-cycles* of C, denoted $Z_n = Z_n(C)$. The image of $d_{n+1}: C_{n+1} \to C_n$ is the module of *n-boundaries* of C, denoted $B_n = B_n(C)$. Because $d \circ d = 0$, we have

$$0 \subseteq B_n \subseteq Z_n \subseteq C_n$$

for all n. The nth *homology module* of C is the subquotient $H_n(C) = Z_n/B_n$ of C_n. Because the dot in C is annoying, we will often write C for C.

Exercise 1.1.1 Set $C_n = \mathbb{Z}/8$ for $n \geq 0$ and $C_n = 0$ for $n < 0$; for $n > 0$ let d_n send $x(\mathrm{mod}\ 8)$ to $4x(\mathrm{mod}\ 8)$. Show that C is a chain complex of $\mathbb{Z}/8$-modules and compute its homology modules.

There is a category **Ch(mod–R)** of chain complexes of (right) R-modules. The objects are, of course, chain complexes. A *morphism* $u: C \to D$ is a chain complex map, that is, a family of R-module homomorphisms $u_n: C_n \to D_n$ commuting with d in the sense that $u_{n-1}d_n = d_{n-1}u_n$. That is, such that the following diagram commutes

$$\cdots \xrightarrow{d} C_{n+1} \xrightarrow{d} C_n \xrightarrow{d} C_{n-1} \xrightarrow{d} \cdots$$
$$\downarrow{u} \qquad \downarrow{u} \qquad \downarrow{u}$$
$$\cdots \xrightarrow{d} D_{n+1} \xrightarrow{d} D_n \xrightarrow{d} D_{n-1} \xrightarrow{d} \cdots.$$

Exercise 1.1.2 Show that a morphism $u: C \to D$ of chain complexes sends boundaries to boundaries and cycles to cycles, hence maps $H_n(C) \to H_n(D)$. Prove that each H_n is a functor from **Ch(mod–R)** to **mod–R**.

Exercise 1.1.3 (Split exact sequences of vector spaces) Choose vector spaces $\{B_n, H_n\}_{n\in\mathbb{Z}}$ over a field, and set $C_n = B_n \oplus H_n \oplus B_{n-1}$. Show that the projection-inclusions $C_n \to B_{n-1} \subset C_{n-1}$ make $\{C_n\}$ into a chain complex, and that every chain complex of vector spaces is isomorphic to a complex of this form.

Exercise 1.1.4 Show that $\{\mathrm{Hom}_R(A, C_n)\}$ forms a chain complex of abelian groups for every R-module A and every R-module chain complex C. Taking $A = Z_n$, show that if $H_n(\mathrm{Hom}_R(Z_n, C)) = 0$, then $H_n(C) = 0$. Is the converse true?

Definition 1.1.2 A morphism $C \to D$ of chain complexes is called a *quasi-isomorphism* (Bourbaki uses *homologism*) if the maps $H_n(C) \to H_n(D)$ are all isomorphisms.

Exercise 1.1.5 Show that the following are equivalent for every C:

1. C is *exact*, that is, exact at every C_n.
2. C is *acyclic*, that is, $H_n(C) = 0$ for all n.
3. The map $0 \to C$ is a quasi-isomorphism, where "0" is the complex of zero modules and zero maps.

The following variant notation is obtained by reindexing with superscripts: $C^n = C_{-n}$. A *cochain complex* $C^{\cdot}$ of R-modules is a family $\{C^n\}$ of R-modules, together with maps $d^n \colon C^n \to C^{n+1}$ such that $d \circ d = 0$. $Z^n(C^{\cdot}) = \ker(d^n)$ is the module of *n-cocycles*, $B^n(C^{\cdot}) = \mathrm{im}(d^{n-1}) \subseteq C^n$ is the module of *n-coboundaries*, and the subquotient $H^n(C^{\cdot}) = Z^n/B^n$ of C^n is the nth *cohomology module* of $C^{\cdot}$. Morphisms and quasi-isomorphisms of cochain complexes are defined exactly as for chain complexes.

A chain complex C is called *bounded* if almost all the C_n are zero; if $C_n = 0$ unless $a \le n \le b$, we say that the complex has *amplitude* in $[a, b]$. A complex C is *bounded above* (resp. *bounded below*) if there is a bound b (resp. a) such that $C_n = 0$ for all $n > b$ (resp. $n < a$). The bounded (resp. bounded above, resp. bounded below) chain complexes form full subcategories of **Ch** $=$ **Ch**(R–**mod**) that are denoted **Ch**$_b$, **Ch**$_-$ and **Ch**$_+$, respectively. The subcategory **Ch**$_{\ge 0}$ of non-negative complexes C ($C_n = 0$ for all $n < 0$) will be important in Chapter 8.

Similarly, a cochain complex $C^{\cdot}$ is called *bounded above* if the chain complex C ($C_n = C^{-n}$) is bounded below, that is, if $C^n = 0$ for all large n; $C^{\cdot}$ is *bounded below* if C is bounded above, and *bounded* if C is bounded. The categories of bounded (resp. bounded above, resp. bounded below, resp. non-negative) cochain complexes are denoted **Ch**^{b}, **Ch**$^-$, **Ch**$^+$, and **Ch**$^{\ge 0}$, respectively.

Exercise 1.1.6 (Homology of a graph) Let Γ be a finite graph with V vertices $(v_1, \cdots, v_V)$ and E edges $(e_1, \cdots, e_E)$. If we orient the edges, we can form the *incidence matrix* of the graph. This is a $V \times E$ matrix whose (ij) entry is $+1$

if the edge e_j starts at v_i, -1 if e_j ends at v_i, and 0 otherwise. Let C_0 be the free R−module on the vertices, C_1 the free R−module on the edges, $C_n = 0$ if $n \neq 0, 1$, and $d: C_1 \to C_0$ be the incidence matrix. If Γ is connected (i.e., we can get from v_0 to every other vertex by tracing a path with edges), show that $H_0(C)$ and $H_1(C)$ are free R−modules of dimensions 1 and $V - E - 1$ respectively. (The number $V - E - 1$ is the number of *circuits* of the graph.) *Hint:* Choose basis $\{v_0, v_1 - v_0, \cdots, v_V - v_0\}$ for C_0, and use a path from v_0 to v_i to find an element of C_1 mapping to $v_i - v_0$.

Application 1.1.3 (Simplicial homology) Here is a topological application we shall discuss more in Chapter 8. Let K be a geometric simplicial complex, such as a triangulated polyhedron, and let K_k $(0 \le k \le n)$ denote the set of k-dimensional simplices of K. Each k-simplex has $k + 1$ faces, which are ordered if the set K_0 of vertices is ordered (do so!), so we obtain $k + 1$ set maps $\partial_i: K_k \to K_{k-1}(0 \le i \le k)$. The *simplicial chain complex* of K with coefficients in R is the chain complex $C_.$, formed as follows. Let C_k be the free R-module on the set K_k; set $C_k = 0$ unless $0 \le k \le n$. The set maps ∂_i yield $k + 1$ module maps $C_k \to C_{k-1}$, which we also call ∂_i; their alternating sum $d = \sum (-1)^i \partial_i$ is the map $C_k \to C_{k-1}$ in the chain complex $C_.$. To see that $C_.$ is a chain complex, we need to prove the algebraic assertion that $d \circ d = 0$. This translates into the geometric fact that each $(k - 2)$-dimensional simplex contained in a fixed k-simplex σ of K lies on exactly two faces of σ. The homology of the chain complex $C_.$ is called the *simplicial homology* of K with coefficients in R. This simplicial approach to homology was used in the first part of this century, before the advent of singular homology.

Exercise 1.1.7 (Tetrahedron) The tetrahedron T is a surface with 4 vertices, 6 edges, and 4 2-dimensional faces. Thus its homology is the homology of a chain complex $0 \to R^4 \to R^6 \to R^4 \to 0$. Write down the matrices in this complex and verify computationally that $H_2(T) \cong H_0(T) \cong R$ and $H_1(T) = 0$.

Application 1.1.4 (Singular homology) Let X be a topological space, and let $S_k = S_k(X)$ be the free R-module on the set of continuous maps from the standard k-simplex Δ_k to X. Restriction to the ith face of Δ_k $(0 \le i \le k)$ transforms a map $\Delta_k \to X$ into a map $\Delta_{k-1} \to X$, and induces an R-module homomorphism ∂_i from S_k to S_{k-1}. The alternating sums $d = \sum (-1)^i \partial_i$ (from S_k to S_{k-1}) assemble to form a chain complex

$$\cdots \xrightarrow{d} S_2 \xrightarrow{d} S_1 \xrightarrow{d} S_0 \longrightarrow 0,$$

called the *singular chain complex* of X. The nth homology module of $S.(X)$ is called the nth *singular homology* of X (with coefficients in R) and is written $H_n(X; R)$. If X is a geometric simplicial complex, then the obvious inclusion $C.(X) \rightarrow S.(X)$ is a quasi-isomorphism, so the simplicial and singular homology modules of X are isomorphic. The interested reader may find details in any standard book on algebraic topology.

1.2 Operations on Chain Complexes

The main point of this section will be that chain complexes form an abelian category. First we need to recall what an abelian category is. A reference for these definitions is [MacCW].

A category $\mathcal{A}$ is called an **Ab**-*category* if every hom-set $\text{Hom}_{\mathcal{A}}(A, B)$ in $\mathcal{A}$ is given the structure of an abelian group in such a way that composition distributes over addition. In particular, given a diagram in $\mathcal{A}$ of the form

$$A \xrightarrow{f} B \underset{g}{\overset{g'}{\rightrightarrows}} C \xrightarrow{h} D$$

we have $h(g + g')f = hgf + hg'f$ in $\text{Hom}(A, D)$. The category **Ch** is an **Ab**-category because we can add chain maps degreewise; if $\{f_n\}$ and $\{g_n\}$ are chain maps from $C.$ to $D.$, their sum is the family of maps $\{f_n + g_n\}$.

An *additive functor* $F: \mathcal{B} \rightarrow \mathcal{A}$ between **Ab**-categories $\mathcal{B}$ and $\mathcal{A}$ is a functor such that each $\text{Hom}_{\mathcal{B}}(B', B) \rightarrow \text{Hom}_{\mathcal{A}}(FB', FB)$ is a group homomorphism.

An *additive category* is an **Ab**-category $\mathcal{A}$ with a zero object (i.e., an object that is initial and terminal) and a product $A \times B$ for every pair A, B of objects in $\mathcal{A}$. This structure is enough to make finite products the same as finite coproducts. The zero object in **Ch** is the complex "0" of zero modules and maps. Given a family $\{A_\alpha\}$ of complexes of R-modules, the product ΠA_α and coproduct (direct sum) $\oplus A_\alpha$ exist in **Ch** and are defined degreewise: the differentials are the maps

$$\prod_\alpha d_\alpha : \prod_\alpha A_{\alpha,n} \rightarrow \prod_\alpha A_{\alpha,n-1} \quad \text{and} \quad \oplus d_\alpha : \oplus_\alpha A_{\alpha,n} \rightarrow \oplus_\alpha A_{\alpha,n-1},$$

respectively. These suffice to make **Ch** into an additive category.

Exercise 1.2.1 Show that direct sum and direct product commute with homology, that is, that $\oplus H_n(A_\alpha) \cong H_n(\oplus A_\alpha)$ and $\Pi H_n(A_\alpha) \cong H_n(\Pi A_\alpha)$ for all n.

Here are some important constructions on chain complexes. A chain complex B is called a *subcomplex* of C if each B_n is a submodule of C_n and the differential on B is the restriction of the differential on C, that is, when the inclusions $i_n : B_n \subseteq C_n$ constitute a chain map $B \to C$. In this case we can assemble the quotient modules C_n/B_n into a chain complex

$$\cdots \to C_{n+1}/B_{n+1} \xrightarrow{d} C_n/B_n \xrightarrow{d} C_{n-1}/B_{n-1} \xrightarrow{d} \cdots$$

denoted C/B and called the *quotient complex*. If $f : B \to C$ is a chain map, the kernels $\{\ker(f_n)\}$ assemble to form a subcomplex of B denoted $\ker(f)$, and the cokernels $\{\mathrm{coker}(f_n)\}$ assemble to form a quotient complex of C denoted $\mathrm{coker}(f)$.

Definition 1.2.1 In any additive category $\mathcal{A}$, a *kernel* of a morphism $f : B \to C$ is defined to be a map $i : A \to B$ such that $fi = 0$ and that is universal with respect to this property. Dually, a *cokernel* of f is a map $e : C \to D$, which is universal with respect to having $ef = 0$. In $\mathcal{A}$, a map $i : A \to B$ is *monic* if $ig = 0$ implies $g = 0$ for every map $g : A' \to A$, and a map $e : C \to D$ is an *epi* if $he = 0$ implies $h = 0$ for every map $h : D \to D'$. (The definition of monic and epi in a non-abelian category is slightly different; see A.1 in the Appendix.) It is easy to see that every kernel is monic and that every cokernel is an epi (exercise!).

Exercise 1.2.2 In the additive category $\mathcal{A} = R\text{–}\mathbf{mod}$, show that:

1. The notions of kernels, monics, and monomorphisms are the same.
2. The notions of cokernels, epis, and epimorphisms are also the same.

Exercise 1.2.3 Suppose that $\mathcal{A} = \mathbf{Ch}$ and f is a chain map. Show that the complex $\ker(f)$ is a kernel of f and that $\mathrm{coker}(f)$ is a cokernel of f.

Definition 1.2.2 An *abelian category* is an additive category $\mathcal{A}$ such that

1. every map in $\mathcal{A}$ has a kernel and cokernel.
2. every monic in $\mathcal{A}$ is the kernel of its cokernel.
3. every epi in $\mathcal{A}$ is the cokernel of its kernel.

The prototype abelian category is the category $\mathbf{mod}\text{–}R$ of R-modules. In any abelian category the *image* $\mathrm{im}(f)$ of a map $f : B \to C$ is the subobject $\ker(\mathrm{coker}\ f)$ of C; in the category of R-modules, $\mathrm{im}(f) = \{f(b) : b \in B\}$. Every map f factors as

$$B \xrightarrow{\ e\ } \operatorname{im}(f) \xrightarrow{\ m\ } C$$

with e an epimorphism and m a monomorphism. A sequence

$$A \xrightarrow{\ f\ } B \xrightarrow{\ g\ } C$$

of maps in $\mathcal{A}$ is called *exact* (at B) if $\ker(g) = \operatorname{im}(f)$.

A subcategory $\mathcal{B}$ of $\mathcal{A}$ is called an *abelian subcategory* if it is abelian, and an exact sequence in $\mathcal{B}$ is also exact in $\mathcal{A}$.

If $\mathcal{A}$ is any abelian category, we can repeat the discussion of section 1.1 to define chain complexes and chain maps in $\mathcal{A}$—just replace **mod**–R by $\mathcal{A}$! These form an additive category **Ch**($\mathcal{A}$), and homology becomes a functor from this category to $\mathcal{A}$. In the sequel we will merely write **Ch** for **Ch**($\mathcal{A}$) when $\mathcal{A}$ is understood.

Theorem 1.2.3 *The category* **Ch** $=$ **Ch**($\mathcal{A}$) *of chain complexes is an abelian category.*

Proof Condition 1 was exercise 1.2.3 above. If $f : B \to C$ is a chain map, I claim that f is monic iff each $B_n \to C_n$ is monic, that is, B is isomorphic to a subcomplex of C. This follows from the fact that the composite $\ker(f) \to C$ is zero, so if f is monic, then $\ker(f) = 0$. So if f is monic, it is isomorphic to the kernel of $C \to C/B$. Similarly, f is an epi iff each $B_n \to C_n$ is an epi, that is, C is isomorphic to the cokernel of the chain map $\ker(f) \to B$. $\diamond$

Exercise 1.2.4 Show that a sequence $0 \to A_. \to B_. \to C_. \to 0$ of chain complexes is exact in **Ch** just in case each sequence $0 \to A_n \to B_n \to C_n \to 0$ is exact in $\mathcal{A}$.

Clearly we can iterate this construction and talk about chain complexes of chain complexes; these are usually called double complexes.

Example 1.2.4 A *double complex* (or *bicomplex*) in $\mathcal{A}$ is a family $\{C_{p,q}\}$ of objects of $\mathcal{A}$, together with maps

$$d^h : C_{p,q} \to C_{p-1,q} \quad \text{and} \quad d^v : C_{p,q} \to C_{p,q-1}$$

such that $d^h \circ d^h = d^v \circ d^v = d^v d^h + d^h d^v = 0$. It is useful to picture the bicomplex $C_{..}$ as a lattice

$$\cdots \quad \cdots \quad \cdots$$

$$\downarrow \qquad\qquad \downarrow \qquad\qquad \downarrow$$

$$\cdots \longleftarrow C_{p-1,q+1} \overset{d^h}{\longleftarrow} C_{p,q+1} \overset{d^h}{\longleftarrow} C_{p+1,p+1} \longleftarrow \cdots$$

$$d^v \downarrow \qquad\qquad d^v \downarrow \qquad\qquad d^v \downarrow$$

$$\cdots \longleftarrow C_{p-1,q} \overset{d^h}{\longleftarrow} C_{p,q} \overset{d^h}{\longleftarrow} C_{p+1,q} \longleftarrow \cdots$$

$$d^v \downarrow \qquad\qquad d^v \downarrow \qquad\qquad d^v \downarrow$$

$$\cdots \longleftarrow C_{p-1,q-1} \overset{d^h}{\longleftarrow} C_{p,q-1} \overset{d^h}{\longleftarrow} C_{p+1,q-1} \longleftarrow \cdots$$

$$\downarrow \qquad\qquad \downarrow \qquad\qquad \downarrow$$

$$\cdots \quad \cdots \quad \cdots$$

in which the maps d^h go horizontally, the maps d^v go vertically, and each square anticommutes. Each row C_{*q} and each column C_{p*} is a chain complex.

We say that a double complex C is *bounded* if C has only finitely many nonzero terms along each diagonal line $p + q = n$, for example, if C is concentrated in the first quadrant of the plane (a *first quadrant double complex*).

Sign Trick 1.2.5 Because of the anticommutivity, the maps d^v are not maps in **Ch**, but chain maps f_{*q} from C_{*q} to $C_{*,q-1}$ can be defined by introducing $\pm$ signs:

$$f_{p,q} = (-1)^p d^v_{p,q}: C_{p,q} \to C_{p,q-1}.$$

Using this sign trick, we can identify the category of double complexes with the category **Ch(Ch)** of chain complexes in the abelian category **Ch**.

Total Complexes 1.2.6 To see why the anticommutative condition $d^v d^h + d^h d^v = 0$ is useful, define the *total complexes* $\mathrm{Tot}(C) = \mathrm{Tot}^{\Pi}(C)$ and $\mathrm{Tot}^{\oplus}(C)$ by

$$\mathrm{Tot}^{\Pi}(C)_n = \prod_{p+q=n} C_{p,q} \quad \text{and} \quad \mathrm{Tot}^{\oplus}(C)_n = \bigoplus_{p+q=n} C_{p,q}.$$

The formula $d = d^h + d^v$ defines maps (check this!)

$$d: \mathrm{Tot}^{\Pi}(C)_n \to \mathrm{Tot}^{\Pi}(C)_{n-1} \quad \text{and} \quad d: \mathrm{Tot}^{\oplus}(C)_n \to \mathrm{Tot}^{\oplus}(C)_{n-1}$$

such that $d \circ d = 0$, making $\mathrm{Tot}^{\Pi}(C)$ and $\mathrm{Tot}^{\oplus}(C)$ into chain complexes. Note that $\mathrm{Tot}^{\oplus}(C) = \mathrm{Tot}^{\Pi}(C)$ if C is bounded, and especially if C is a first quadrant double complex. The difference between $\mathrm{Tot}^{\Pi}(C)$ and $\mathrm{Tot}^{\oplus}(C)$ will become apparent in Chapter 5 when we discuss spectral sequences.

Remark $\text{Tot}^{\Pi}(C)$ and $\text{Tot}^{\oplus}(C)$ do not exist in all abelian categories; they don't exist when $\mathcal{A}$ is the category of all finite abelian groups. We say that an abelian category is *complete* if all infinite direct products exist (and so Tot^{Π} exists) and that it is *cocomplete* if all infinite direct sums exist (and so $\text{Tot}^{\oplus}$ exists). Both these axioms hold in R–**mod** and in the category of chain complexes of R-modules.

Exercise 1.2.5 Give an elementary proof that $\text{Tot}(C)$ is acyclic whenever C is a bounded double complex with exact rows (or exact columns). We will see later that this result follows from the Acyclic Assembly Lemma 2.7.3. It also follows from a spectral sequence argument (see Definition 5.6.2 and exercise 5.6.4).

Exercise 1.2.6 Give examples of (1) a second quadrant double complex C with exact columns such that $\text{Tot}^{\Pi}(C)$ is acyclic but $\text{Tot}^{\oplus}(C)$ is not; (2) a second quadrant double complex C with exact rows such that $\text{Tot}^{\oplus}(C)$ is acyclic but $\text{Tot}^{\Pi}(C)$ is not; and (3) a double complex (in the entire plane) for which every row and every column is exact, yet neither $\text{Tot}^{\Pi}(C)$ nor $\text{Tot}^{\oplus}(C)$ is acyclic.

Truncations 1.2.7 If C is a chain complex and n is an integer, we let $\tau_{\geq n}C$ denote the subcomplex of C defined by

$$(\tau_{\geq n}C)_i = \begin{cases} 0 & \text{if } i < n \\ Z_n & \text{if } i = n \\ C_i & \text{if } i > n. \end{cases}$$

Clearly $H_i(\tau_{\geq n}C) = 0$ for $i < n$ and $H_i(\tau_{\geq n}C) = H_i(C)$ for $i \geq n$. The complex $\tau_{\geq n}C$ is called the (good) *truncation* of C below n, and the quotient complex $\tau_{<n}C = C/(\tau_{\geq n}C)$ is called the (good) truncation of C above n; $H_i(\tau_{<n}C)$ is $H_i(C)$ for $i < n$ and 0 for $i \geq n$.

Some less useful variants are the *brutal truncations* $\sigma_{<n}C$ and $\sigma_{\geq n}C = C/(\sigma_{<n}C)$. By definition, $(\sigma_{<n}C)_i$ is C_i if $i < n$ and 0 if $i \geq n$. These have the advantage of being easier to describe but the disadvantage of introducing the homology group $H_n(\sigma_{\geq n}C) = C_n/B_n$.

Translation 1.2.8 Shifting indices, or translation, is another useful operation we can perform on chain and cochain complexes. If C is a complex and p an integer, we form a new complex $C[p]$ as follows:

$$C[p]_n = C_{n+p} \quad (\text{resp. } C[p]^n = C^{n-p})$$

with differential $(-1)^p d$. We call $C[p]$ the pth *translate* of C. The way to remember the shift is that the degree 0 part of $C[p]$ is C_p. The sign convention is designed to simplify notation later on. Note that translation shifts homology:

$$H_n(C[p]) = H_{n+p}(C) \quad (\text{resp. } H^n(C[p]) = H^{n-p}(C)).$$

We make translation into a functor by shifting indices on chain maps. That is, if $f: C \to D$ is a chain map, then $f[p]$ is the chain map given by the formula

$$f[p]_n = f_{n+p} \quad (\text{resp. } f[p]^n = f^{n-p}).$$

Exercise 1.2.7 If C is a complex, show that there are exact sequences of complexes:

$$0 \longrightarrow Z(C) \longrightarrow C \xrightarrow{d} B(C)[-1] \longrightarrow 0;$$

$$0 \longrightarrow H(C) \longrightarrow C/B(C) \xrightarrow{d} Z(C)[-1] \longrightarrow H(C)[-1] \longrightarrow 0.$$

Exercise 1.2.8 (Mapping cone) Let $f: B \to C$ be a morphism of chain complexes. Form a double chain complex D out of f by thinking of f as a chain complex in **Ch** and using the sign trick, putting $B[-1]$ in the row $q = 1$ and C in the row $q = 0$. Thinking of C and $B[-1]$ as double complexes in the obvious way, show that there is a short exact sequence of double complexes

$$0 \longrightarrow C \longrightarrow D \xrightarrow{\delta} B[-1] \longrightarrow 0.$$

The total complex of D is cone(f'), the mapping cone (see section 1.5) of a map f', which differs from f only by some $\pm$ signs and is isomorphic to f.

1.3 Long Exact Sequences

It is time to unveil the feature that makes chain complexes so special from a computational viewpoint: the existence of long exact sequences.

Theorem 1.3.1 *Let* $0 \to A. \xrightarrow{f} B. \xrightarrow{g} C. \to 0$ *be a short exact sequence of chain complexes. Then there are natural maps* $\partial: H_n(C) \to H_{n-1}(A)$, *called connecting homomorphisms, such that*

$$\cdots \xrightarrow{g} H_{n+1}(C) \xrightarrow{\partial} H_n(A) \xrightarrow{f} H_n(B) \xrightarrow{g} H_n(C) \xrightarrow{\partial} H_{n-1}(A) \xrightarrow{f} \cdots$$

is an exact sequence.

Similarly, if $0 \to A^{\cdot} \xrightarrow{f} B^{\cdot} \xrightarrow{g} C^{\cdot} \to 0$ *is a short exact sequence of cochain complexes, there are natural maps* $\partial \colon H^n(C) \to H^{n+1}(A)$ *and a long exact sequence*

$$\cdots \xrightarrow{g} H^{n-1}(C) \xrightarrow{\partial} H^n(A) \xrightarrow{f} H^n(B) \xrightarrow{g} H^n(C) \xrightarrow{\partial} H^{n+1}(A) \xrightarrow{f} \cdots .$$

Exercise 1.3.1 Let $0 \to A \to B \to C \to 0$ be a short exact sequence of complexes. Show that if two of the three complexes A, B, C are exact, then so is the third.

Exercise 1.3.2 (3×3 lemma) Suppose given a commutative diagram

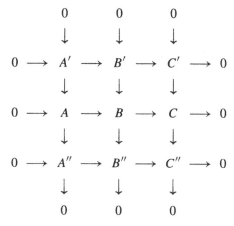

in an abelian category, such that every column is exact. Show the following:

1. If the bottom two rows are exact, so is the top row.
2. If the top two rows are exact, so is the bottom row.
3. If the top and bottom rows are exact, and the composite $A \to C$ is zero, the middle row is also exact.

Hint: Show the remaining row is a complex, and apply exercise 1.3.1.

The key tool in constructing the connecting homomorphism ∂ is our next result, the *Snake Lemma*. We will not print the proof in these notes, because it is best done visually. In fact, a clear proof is given by Jill Clayburgh at the beginning of the movie *It's My Turn* (Rastar-Martin Elfand Studios, 1980). As an exercise in "diagram chasing" of elements, the student should find a proof (but privately—keep the proof to yourself!).

Snake Lemma 1.3.2 *Consider a commutative diagram of R-modules of the form*

$$A' \longrightarrow B' \longrightarrow C' \longrightarrow 0$$
$$f\downarrow \qquad g\downarrow \qquad h\downarrow$$
$$0 \longrightarrow A \xrightarrow{\ i\ } B \longrightarrow C.$$

If the rows are exact, there is an exact sequence

$$\ker(f) \to \ker(g) \to \ker(h) \xrightarrow{\ \partial\ } \operatorname{coker}(f) \to \operatorname{coker}(g) \to \operatorname{coker}(h)$$

with ∂ defined by the formula

$$\partial(c') = i^{-1}gp^{-1}(c'), \quad c' \in \ker(h).$$

Moreover, if $A' \to B'$ is monic, then so is $\ker(f) \to \ker(g)$, and if $B \to C$ is onto, then so is $\operatorname{coker}(f) \to \operatorname{coker}(g)$.

Etymology The term *snake* comes from the following visual mnemonic:

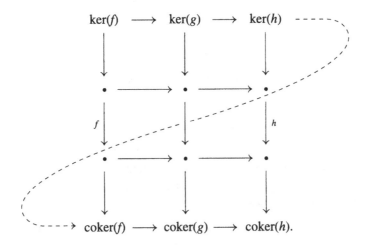

Remark The Snake Lemma also holds in an arbitrary abelian category $\mathcal{C}$. To see this, let $\mathcal{A}$ be the smallest abelian subcategory of $\mathcal{C}$ containing the objects and morphisms of the diagram. Since $\mathcal{A}$ has a set of objects, the Freyd-Mitchell Embedding Theorem (see 1.6.1) gives an exact, fully faithful embedding of $\mathcal{A}$ into R–**mod** for some ring R. Since ∂ exists in R–**mod**, it exists in $\mathcal{A}$ and hence in $\mathcal{C}$. Similarly, exactness in R–**mod** implies exactness in $\mathcal{A}$ and hence in $\mathcal{C}$.

Exercise 1.3.3 (5–Lemma) In any commutative diagram

$$
\begin{array}{ccccccccc}
A' & \longrightarrow & B' & \longrightarrow & C' & \longrightarrow & D' & \longrightarrow & E' \\
a\downarrow\cong & & b\downarrow\cong & & c\downarrow & & d\downarrow\cong & & e\downarrow\cong \\
A & \longrightarrow & B & \longrightarrow & C & \longrightarrow & D & \longrightarrow & E
\end{array}
$$

with exact rows in any abelian category, show that if a, b, d, and e are isomorphisms, then c is also an isomorphism. More precisely, show that if b and d are monic and a is an epi, then c is monic. Dually, show that if b and d are epis and e is monic, then c is an epi.

We now proceed to the construction of the connecting homomorphism ∂ of Theorem 1.3.1 associated to a short exact sequence

$$0 \to A \to B \to C \to 0$$

of chain complexes. From the Snake Lemma and the diagram

$$
\begin{array}{ccccccccc}
& & 0 & & 0 & & 0 & & \\
& & \downarrow & & \downarrow & & \downarrow & & \\
0 & \longrightarrow & Z_nA & \longrightarrow & Z_nB & \longrightarrow & Z_nC & & \\
& & \downarrow & & \downarrow & & \downarrow & & \\
0 & \longrightarrow & A_n & \longrightarrow & B_n & \longrightarrow & C_n & \longrightarrow & 0 \\
& & d\downarrow & & d\downarrow & & d\downarrow & & \\
0 & \longrightarrow & A_{n-1} & \longrightarrow & B_{n-1} & \longrightarrow & C_{n-1} & \longrightarrow & 0 \\
& & \downarrow & & \downarrow & & \downarrow & & \\
& & \dfrac{A_{n-1}}{dA_n} & \longrightarrow & \dfrac{B_{n-1}}{dB_n} & \longrightarrow & \dfrac{C_{n-1}}{dC_n} & \longrightarrow & 0 \\
& & \downarrow & & \downarrow & & \downarrow & & \\
& & 0 & & 0 & & 0 & &
\end{array}
$$

we see that the rows are exact in the commutative diagram

$$
\begin{array}{ccccccc}
\dfrac{A_n}{dA_{n+1}} & \longrightarrow & \dfrac{B_n}{dB_{n+1}} & \longrightarrow & \dfrac{C_n}{dC_{n+1}} & \longrightarrow & 0 \\
d\downarrow & & d\downarrow & & d\downarrow & & \\
0 \longrightarrow Z_{n-1}(A) & \overset{f}{\longrightarrow} & Z_{n-1}(b) & \overset{g}{\longrightarrow} & Z_{n-1}(C). & &
\end{array}
$$

The kernel of the left vertical is $H_n(A)$, and its cokernel is $H_{n-1}(A)$. Therefore the Snake Lemma yields an exact sequence

$$H_n(A) \xrightarrow{f} H_n(B) \xrightarrow{g} H_n(C) \xrightarrow{\partial} H_{n-1}(A) \to H_{n-1}(B) \to H_{n-1}(C).$$

The long exact sequence 1.3.1 is obtained by pasting these sequences together.

Addendum 1.3.3 When one computes with modules, it is useful to be able to push elements around. By decoding the above proof, we obtain the following formula for the connecting homomorphism: Let $z \in H_n(C)$, and represent it by a cycle $c \in C_n$. Lift the cycle to $b \in B_n$ and apply d. The element db of B_{n-1} actually belongs to the submodule $Z_{n-1}(A)$ and represents $\partial(z) \in H_{n-1}(A)$.

We shall now explain what we mean by the naturality of ∂. There is a category S whose objects are short exact sequences of chain complexes (say, in an abelian category C). Commutative diagrams

$$
\begin{array}{ccccccccc}
0 & \longrightarrow & A & \longrightarrow & B & \longrightarrow & C & \longrightarrow & 0 \\
& & \downarrow & & \downarrow & & \downarrow & & \\
0 & \longrightarrow & A' & \longrightarrow & B' & \longrightarrow & C' & \longrightarrow & 0
\end{array}
$$

(*)

give the morphisms in S (from the top row to the bottom row). Similarly, there is a category L of long exact sequences in C.

Proposition 1.3.4 *The long exact sequence is a functor from S to L. That is, for every short exact sequence there is a long exact sequence, and for every map (*) of short exact sequences there is a commutative ladder diagram*

$$
\begin{array}{ccccccccc}
\cdots \xrightarrow{\partial} & H_n(A) & \longrightarrow & H_n(B) & \longrightarrow & H_n(C) & \xrightarrow{\partial} & H_{n-1}(A) \to & \cdots \\
& \downarrow & & \downarrow & & \downarrow & & \downarrow & \\
\cdots \xrightarrow{\partial} & H_n(A') & \longrightarrow & H_n(B') & \longrightarrow & H_n(C') & \xrightarrow{\partial} & H_{n-1}(A') \to & \cdots.
\end{array}
$$

Proof All we have to do is establish the ladder diagram. Since each H_n is a functor, the left two squares commute. Using the Embedding Theorem 1.6.1, we may assume $C = \mathbf{mod}\text{-}R$ in order to prove that the right square commutes. Given $z \in H_n(C)$, represented by $c \in C_n$, its image $z' \in H_n(C')$ is represented by the image of c. If $b \in B_n$ lifts c, its image in B'_n lifts c'. Therefore by 1.3.3 $\partial(z') \in H_{n-1}(A')$ is represented by the image of db, that is, by the image of a representative of $\partial(z)$, so $\partial(z')$ is the image of $\partial(z)$. $\diamond$

Remark 1.3.5 The data of the long exact sequence is sometimes organized into the mnemonic shape

$$H_*(A) \quad \longrightarrow \quad H_*(B)$$
$$\partial \nwarrow \qquad \qquad \swarrow$$
$$H_*(C)$$

This is called an *exact triangle* for obvious reasons. This mnemonic shape is responsible for the term "triangulated category," which we will discuss in Chapter 10. The category **K** of chain equivalence classes of complexes and maps (see exercise 1.4.5 in the next section) is an example of a triangulated category.

Exercise 1.3.4 Consider the boundaries-cycles exact sequence $0 \to Z \to C \to B(-1) \to 0$ associated to a chain complex C (exercise 1.2.7). Show that the corresponding long exact sequence of homology breaks up into short exact sequences.

Exercise 1.3.5 Let f be a morphism of chain complexes. Show that if $\ker(f)$ and $\mathrm{coker}(f)$ are acyclic, then f is a quasi-isomorphism. Is the converse true?

Exercise 1.3.6 Let $0 \to A \to B \to C \to 0$ be a short exact sequence of double complexes of modules. Show that there is a short exact sequence of total complexes, and conclude that if $\mathrm{Tot}(C)$ is acyclic, then $\mathrm{Tot}(A) \to \mathrm{Tot}(B)$ is a quasi-isomorphism.

1.4 Chain Homotopies

The ideas in this section and the next are motivated by homotopy theory in topology. We begin with a discussion of a special case of historical importance. If C is any chain complex of vector spaces over a field, we can always choose vector space decompositions:

$$C_n = Z_n \oplus B'_n, \qquad B'_n \cong C_n/Z_n = d(C_n) = B_{n-1};$$
$$Z_n = B_n \oplus H'_n, \qquad H'_n \cong Z_n/B_n = H_n(C).$$

Therefore we can form the compositions

$$C_n \to Z_n \to B_n \cong B'_{n+1} \subseteq C_{n+1}$$

to get splitting maps $s_n: C_n \to C_{n+1}$, such that $d = dsd$. The compositions ds and sd are projections from C_n onto B_n and B'_n, respectively, so the sum $ds + sd$ is an endomorphism of C_n whose kernel H'_n is isomorphic to the homology $H_n(C)$. The kernel (and cokernel!) of $ds + sd$ is the trivial homology complex $H_*(C)$. Evidently both chain maps $H_*(C) \to C$ and $C \to H_*(C)$ are quasi-isomorphisms. Moreover, C is an exact sequence if and only if $ds + sd$ is the identity map.

Over an arbitrary ring R, it is not always possible to split chain complexes like this, so we give a name to this notion.

Definition 1.4.1 *A complex C is called* split *if there are maps $s_n: C_n \to C_{n+1}$ such that $d = dsd$. The maps s_n are called the* splitting *maps. If in addition C is acyclic (exact as a sequence), we say that C is* split exact.

Example 1.4.2 Let $R = \mathbb{Z}$ or $\mathbb{Z}/4$, and let C be the complex

$$\cdots \xrightarrow{2} \mathbb{Z}/4 \xrightarrow{2} \mathbb{Z}/4 \xrightarrow{2} \mathbb{Z}/4 \xrightarrow{2} \cdots.$$

This complex is acyclic but not split exact. There is no map s such that $ds + sd$ is the identity map, nor is there any direct sum decomposition $C_n \cong Z_n \oplus B'_n$.

Exercise 1.4.1 The previous example shows that even an acyclic chain complex of free R-modules need not be split exact.

1. Show that acyclic *bounded below* chain complexes of free R-modules are always split exact.
2. Show that an acyclic chain complex of finitely generated free abelian groups is always split exact, even when it is not bounded below.

Exercise 1.4.2 Let C be a chain complex, with boundaries B_n and cycles Z_n in C_n. Show that C is split if and only if there are R-module decompositions $C_n \cong Z_n \oplus B'_n$ and $Z_n = B_n \oplus H'_n$. Show that C is split exact iff $H'_n = 0$.

Now suppose that we are given two chain complexes C and D, together with randomly chosen maps $s_n: C_n \to D_{n+1}$. Let f_n be the map from C_n to D_n defined by the formula $f_n = d_{n+1}s_n + s_{n-1}d_n$.

$$
\begin{array}{ccccc}
C_{n+1} & \xrightarrow{d} & C_n & \xrightarrow{d} & C_{n-1} \\
& {\scriptstyle s}\swarrow & {\scriptstyle f}\downarrow & {\scriptstyle s}\swarrow & \\
D_{n+1} & \xrightarrow{d} & D_n & \xrightarrow{d} & D_{n-1}
\end{array}
$$

Dropping the subscripts for clarity, we compute

$$df = d(ds + sd) = dsd = (ds + sd)d = fd.$$

Thus $f = ds + sd$ is a chain map from C to D.

Definition 1.4.3 We say that a chain map $f: C \to D$ is *null homotopic* if there are maps $s_n: C_n \to D_{n+1}$ such that $f = ds + sd$. The maps $\{s_n\}$ are called a *chain contraction* of f.

Exercise 1.4.3 Show that C is a split exact chain complex if and only if the identity map on C is null homotopic.

The chain contraction construction gives us an easy way to proliferate chain maps: if $g: C \to D$ is any chain map, so is $g + (sd + ds)$ for *any* choice of maps s_n. However, $g + (sd + ds)$ is not very different from g, in a sense that we shall now explain.

Definition 1.4.4 We say that two chain maps f and g from C to D are *chain homotopic* if their difference $f - g$ is null homotopic, that is, if

$$f - g = sd + ds.$$

The maps $\{s_n\}$ are called a *chain homotopy* from f to g. Finally, we say that $f: C \to D$ is a *chain homotopy equivalence* (Bourbaki uses *homotopism*) if there is a map $g: D \to C$ such that gf and fg are chain homotopic to the respective identity maps of C and D.

Remark This terminology comes from topology via the following observation. A map f between two topological spaces X and Y induces a map $f_*: S(X) \to S(Y)$ between the corresponding singular chain complexes. It turns out that if f is topologically null homotopic (resp. a homotopy equivalence), then the chain map f_* is null homotopic (resp. a chain homotopy equivalence), and if two maps f and g are topologically homotopic, then f_* and g_* are chain homotopic.

Lemma 1.4.5 *If $f: C \to D$ is null homotopic, then every map $f_*: H_n(C) \to H_n(D)$ is zero. If f and g are chain homotopic, then they induce the same maps $H_n(C) \to H_n(D)$.*

Proof It is enough to prove the first assertion, so suppose that $f = ds + sd$. Every element of $H_n(C)$ is represented by an n-cycle x. But then $f(x) = d(sx)$. That is, $f(x)$ is an n-boundary in D. As such, $f(x)$ represents 0 in $H_n(D)$. ◇

Exercise 1.4.4 Consider the homology $H_*(C)$ of C as a chain complex with zero differentials. Show that if the complex C is split, then there is a chain homotopy equivalence between C and $H_*(C)$. Give an example in which the converse fails.

Exercise 1.4.5 In this exercise we shall show that the chain homotopy classes of maps form a quotient category $\mathbf{K}$ of the category $\mathbf{Ch}$ of all chain complexes. The homology functors H_n on $\mathbf{Ch}$ will factor through the quotient functor $\mathbf{Ch} \to \mathbf{K}$.

1. Show that chain homotopy equivalence is an equivalence relation on the set of all chain maps from C to D. Let $\mathrm{Hom}_{\mathbf{K}}(C, D)$ denote the equivalence classes of such maps. Show that $\mathrm{Hom}_{\mathbf{K}}(C, D)$ is an abelian group.
2. Let f and g be chain homotopic maps from C to D. If $u: B \to C$ and $v: D \to E$ are chain maps, show that vfu and vgu are chain homotopic. Deduce that there is a category $\mathbf{K}$ whose objects are chain complexes and whose morphisms are given in (1).
3. Let f_0, f_1, g_0, and g_1 be chain maps from C to D such that f_i is chain homotopic to g_i ($i = 1, 2$). Show that $f_0 + f_1$ is chain homotopic to $g_0 + g_1$. Deduce that $\mathbf{K}$ is an additive category, and that $\mathbf{Ch} \to \mathbf{K}$ is an additive functor.
4. Is $\mathbf{K}$ an abelian category? Explain.

1.5 Mapping Cones and Cylinders

1.5.1 Let $f: B \to C$ be a map of chain complexes. The *mapping cone* of f is the chain complex $\mathrm{cone}(f)$ whose degree n part is $B_{n-1} \oplus C_n$. In order to match other sign conventions, the differential in $\mathrm{cone}(f)$ is given by the formula

$$d(b, c) = (-d(b), d(c) - f(b)), \quad (b \in B_{n-1}, c \in C_n).$$

That is, the differential is given by the matrix

$$\begin{bmatrix} -d_B & 0 \\ -f & +d_C \end{bmatrix} : \quad \begin{matrix} B_{n-1} & \xrightarrow{\;-\;} & B_{n-2} \\ \oplus & \searrow^{-} & \oplus \\ C_n & \xrightarrow[+]{} & C_{n-1} \end{matrix} .$$

Here is the dual notion for a map $f: B^. \to C^.$ of cochain complexes. The mapping cone, cone(f), is a cochain complex whose degree n part is $B^{n+1} \oplus C^n$. The differential is given by the same formula as above with the same signs.

Exercise 1.5.1 Let cone(C) denote the mapping cone of the identity map id_C of C; it has $C_{n-1} \oplus C_n$ in degree n. Show that cone(C) is split exact, with $s(b, c) = (-c, 0)$ defining the splitting map.

Exercise 1.5.2 Let $f: C \to D$ be a map of complexes. Show that f is null homotopic if and only if f extends to a map $(-s, f): \mathrm{cone}(C) \to D$.

1.5.2 Any map $f_*: H_*(B) \to H_*(C)$ can be fit into a long exact sequence of homology groups by use of the following device. There is a short exact sequence

$$0 \to C \to \mathrm{cone}(f) \xrightarrow{\delta} B[-1] \to 0$$

of chain complexes, where the left map sends c to $(0, c)$, and the right map sends (b, c) to $-b$. Recalling (1.2.8) that $H_{n+1}(B[-1]) \cong H_n(B)$, the homology long exact sequence (with connecting homomorphism ∂) becomes

$$\cdots \to H_{n+1}(\mathrm{cone}(f)) \xrightarrow{\delta_*} H_n(B) \xrightarrow{\partial} H_n(C) \to H_n(\mathrm{cone}(f)) \xrightarrow{\delta_*} H_{n-1}(B) \xrightarrow{\partial} \cdots.$$

The following lemma shows that $\partial = f_*$, fitting f_* into a long exact sequence.

Lemma 1.5.3 *The map ∂ in the above sequence is f_*.*

Proof If $b \in B_n$ is a cycle, the element $(-b, 0)$ in the cone complex lifts b via δ. Applying the differential we get $(db, fb) = (0, fb)$. This shows that

$$\partial[b] = [fb] = f_*[b]. \qquad \Diamond$$

Corollary 1.5.4 *A map $f: B \to C$ is a quasi-isomorphism if and only if the mapping cone complex cone(f) is exact. This device reduces questions about quasi-isomorphisms to the study of split complexes.*

Topological Remark Let K be a simplicial complex (or more generally a cell complex). The *topological cone* CK of K is obtained by adding a new vertex s to K and "coning off" the simplices (cells) to get a new $(n + 1)$-simplex for every old n-simplex of K. (See Figure 1.1.) The simplicial (cellular) chain complex $C_.(s)$ of the one-point space $\{s\}$ is R in degree 0 and zero elsewhere. $C_.(s)$ is a subcomplex of the simplicial (cellular) chain complex $C_.(CK)$ of

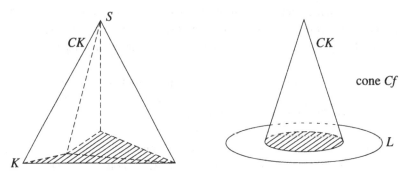

Figure 1.1. The topological cone CK and mapping cone Cf.

the topological cone CK. The quotient $C_.(CK)/C_.(s)$ is the chain complex cone$(C_.K)$ of the identity map of $C_.(K)$. The algebraic fact that cone$(C_.K)$ is split exact (null homotopic) reflects the fact that the topological cone CK is contractible.

More generally, if $f: K \to L$ is a simplicial map (or a cellular map), the *topological mapping cone Cf* of f is obtained by glueing CK and L together, identifying the subcomplex K of CK with its image in L (Figure 1.1). This is a cellular complex, which is simplicial if f is an inclusion of simplicial complexes. Write $C_.(Cf)$ for the cellular chain complex of the topological mapping cone Cf. The quotient chain complex $C_.(Cf)/C_.(s)$ may be identified with cone(f_*), the mapping cone of the chain map $f_*: C_.(K) \to C_.(L)$.

1.5.5 A related construction is that of the *mapping cylinder* cyl(f) of a chain complex map $f: B_. \to C_.$. The degree n part of cyl(f) is $B_n \oplus B_{n-1} \oplus C_n$, and the differential is

$$d(b, b', c) = (d(b) + b', -d(b'), d(c) - f(b')).$$

That is, the differential is given by the matrix

$$
\begin{bmatrix}
d_B & \mathrm{id}_B & 0 \\
0 & -d_B & 0 \\
0 & -f & d_C
\end{bmatrix}
$$

$$
\begin{array}{ccc}
B_n & \xrightarrow{\;+\;} & B_{n-1} \\
\oplus & {}^{+}\nearrow & \oplus \\
B_{n-1} & \xrightarrow{\;-\;} & B_{n-2} \\
\oplus & \searrow^{-} & \oplus \\
C_n & \xrightarrow{\;+\;} & C_{n-1}
\end{array}
$$

The cylinder is a chain complex because

$$d^2 = \begin{bmatrix} d_B^2 & d_B - d_B & 0 \\ 0 & d_B^2 & 0 \\ 0 & f d_B - d_C f & d_C^2 \end{bmatrix} = 0.$$

Exercise 1.5.3 Let cyl(C) denote the mapping cylinder of the identity map id$_C$ of C; it has $C_n \oplus C_{n-1} \oplus C_n$ in degree n. Show that two chain maps $f, g: C \to D$ are chain homotopic if and only if they extend to a map (f, s, g): cyl(C) $\to D$.

Lemma 1.5.6 *The subcomplex of elements $(0, 0, c)$ is isomorphic to C, and the corresponding inclusion $\alpha: C \to$ cyl(f) is a quasi-isomorphism.*

Proof The quotient cyl(f)/$\alpha(C)$ is the mapping cone of $-$id$_B$, so it is null-homotopic (exercise 1.5.1). The lemma now follows from the long exact homology sequence for

$$0 \longrightarrow C \xrightarrow{\;\alpha\;} \text{cyl}(f) \longrightarrow \text{cone}(-\text{id}_B) \longrightarrow 0. \qquad \Diamond$$

Exercise 1.5.4 Show that $\beta(b, b', c) = f(b) + c$ defines a chain map from cyl(f) to C such that $\beta\alpha = $ id$_C$. Then show that the formula $s(b, b', c) = (0, b, 0)$ defines a chain homotopy from the identity of cyl(f) to $\alpha\beta$. Conclude that α is in fact a chain homotopy equivalence between C and cyl(f).

Topological Remark Let X be a cellular complex and let I denote the interval $[0,1]$. The space $I \times X$ is the topological cylinder of X. It is also a cell complex; every n-cell e^n in X gives rise to three cells in $I \times X$: the two n-cells, $0 \times e^n$ and $1 \times e^n$, and the $(n + 1)$-cell $(0, 1) \times e^n$. If $C.(X)$ is the cellular chain complex of X, then the cellular chain complex $C.(I \times X)$ of $I \times X$ may be identified with cyl(id$_{C.X}$), the mapping cylinder chain complex of the identity map on $C.(X)$.

More generally, if $f: X \to Y$ is a cellular map, then the topological mapping cylinder cyl(f) is obtained by glueing $I \times X$ and Y together, identifying $0 \times X$ with the image of X under f (see Figure 1.2). This is also a cellular complex, whose cellular chain complex $C.(\text{cyl}(f))$ may be identified with the mapping cylinder of the chain map $C.(X) \to C.(Y)$.

The constructions in this section are the algebraic analogues of the usual topological constructions $I \times X \simeq X$, cyl(f) $\simeq Y$, and so forth which were used by Dold and Puppe to get long exact sequences for any generalized homology theory on topological spaces.

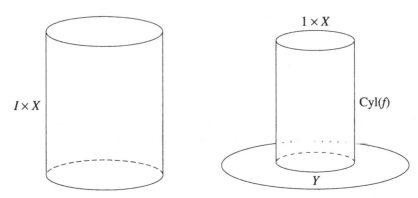

Figure 1.2.   The topological cylinder of X and mapping cylinder $\mathrm{cyl}(f)$.

Here is how to use mapping cylinders to fit f_* into a long exact sequence of homology groups. The subcomplex of elements $(b, 0, 0)$ in $\mathrm{cyl}(f)$ is isomorphic to B, and the quotient $\mathrm{cyl}(f)/B$ is the mapping cone of f. The composite $B \to \mathrm{cyl}(f) \xrightarrow{\beta} C$ is the map f, where β is the equivalence of exercise 1.5.4, so on homology $f_*: H(B) \to H(C)$ factors through $H(B) \to H(\mathrm{cyl}(f))$. Therefore we may construct a commutative diagram of chain complexes with exact rows:

$$
\begin{array}{ccccccccc}
& & & & C & & & & \\
& & f\nearrow & & \uparrow\beta & & & & \\
0 & \longrightarrow & B & \longrightarrow & \mathrm{cyl}(f) & \longrightarrow & \mathrm{cone}(f) & \longrightarrow & 0 \\
& & & & \uparrow\alpha & & \| & & \\
0 & \longrightarrow & C & \longrightarrow & \mathrm{cone}(f) & \xrightarrow{\delta} & B[-1] & \longrightarrow & 0.
\end{array}
$$

The homology long exact sequences fit into the following diagram:

$$
\begin{array}{ccccccccc}
\cdots \xrightarrow{-\partial} & H_n(B) & \to & H_n(\mathrm{cyl}(f)) & \to & H_n(\mathrm{cone}(f)) & \xrightarrow{-\partial} & H_{n-1}(B) & \to \cdots \\
& \| \wr \quad f \searrow & & \| \wr & & \| & & \| \wr & \\
\cdots \to & H_{n+1}(B[-1]) & \xrightarrow{\partial} & H_n(C) & \to & H_n(\mathrm{cone}(f)) & \xrightarrow{\delta} & H_n(B[-1]) & \xrightarrow{\partial} \cdots
\end{array}
$$

Lemma 1.5.7 *This diagram is commutative, with exact rows.*

Proof It suffices to show that the right square (with $-\partial$ and δ) commutes.

Let (b, c) be an n-cycle in cone(f), so $d(b) = 0$ and $f(b) = d(c)$. Lift it to $(0, b, c)$ in cyl(f) and apply the differential:

$$d(0, b, c) = (0 + b, -db, dc - fb) = (b, 0, 0).$$

Therefore ∂ maps the class of (b, c) to the class of $b = -\delta(b, c)$ in $H_{n-1}(B)$.

$\diamond$

1.5.8 The cone and cylinder constructions provide a natural way to fit the homology of *every* chain map $f: B \to C$ into *some* long exact sequence (see 1.5.2 and 1.5.7). To show that the long exact sequence is well defined, we need to show that the usual long exact homology sequence attached to any short exact sequence of complexes

$$0 \to B \xrightarrow{f} C \xrightarrow{g} D \to 0$$

agrees both with the long exact sequence attached to f and with the long exact sequence attached to g.

We first consider the map f. There is a chain map $\varphi: \text{cone}(f) \to D$ defined by the formula $\varphi(b, c) = g(c)$. It fits into a commutative diagram with exact rows:

$$
\begin{array}{ccccccccc}
0 & \longrightarrow & C & \longrightarrow & \text{cone}(f) & \xrightarrow{\delta} & B[-1] & \longrightarrow & 0 \\
& & \downarrow{\alpha} & & \| & & & & \\
0 & \longrightarrow & B & \longrightarrow & \text{cyl}(f) & \longrightarrow & \text{cone}(f) & \longrightarrow & 0 \\
& & \| & & \downarrow{\beta} & & \downarrow{\varphi} & & \\
0 & \longrightarrow & B & \xrightarrow{f} & C & \xrightarrow{g} & D & \longrightarrow & 0.
\end{array}
$$

Since β is a quasi-isomorphism, it follows from the 5-lemma and 1.3.4 that φ is a quasi-isomorphism as well. The following exercise shows that φ need not be a chain homotopy equivalence.

Exercise 1.5.5 Suppose that the B and C of 1.5.8 are modules, considered as chain complexes concentrated in degree zero. Then cone(f) is the complex $0 \to B \xrightarrow{-f} C \to 0$. Show that φ is a chain homotopy equivalence iff $f : B \subset C$ is a split injection.

To continue, the naturality of the connecting homomorphism ∂ provides us with a natural isomorphism of long exact sequences:

$$\cdots \xrightarrow{\partial} H_n(B) \longrightarrow H_n(\mathrm{cyl}(f)) \longrightarrow H_n(\mathrm{cone}(f)) \xrightarrow{\partial} H_{n-1}(B) \longrightarrow \cdots$$

$$\Big\| \qquad\qquad \Big\downarrow \cong \qquad\qquad \Big\downarrow \cong \qquad\qquad \Big\| \wr$$

$$\cdots \xrightarrow{\partial} H_n(B) \longrightarrow H_n(C) \longrightarrow H_n(D) \xrightarrow{\partial} H_{n-1}(B) \longrightarrow \cdots.$$

Exercise 1.5.6 Show that the composite

$$H_n(D) \cong H_n(\mathrm{cone}(f)) \xrightarrow{-\delta_*} H_n(B[-1]) \cong H_{n-1}(B)$$

is the connecting homomorphism ∂ in the homology long exact sequence for

$$0 \to B \to C \to D \to 0.$$

Exercise 1.5.7 Show that there is a quasi-isomorphism $B[-1] \to \mathrm{cone}(g)$ dual to φ. Then dualize the preceding exercise, by showing that the composite

$$H_n(D) \xrightarrow{\partial} H_{n-1}(B) \xrightarrow{\simeq} H_n(\mathrm{cone}(g))$$

is the usual map induced by the inclusion of D in $\mathrm{cone}(g)$.

Exercise 1.5.8 Given a map $f: B \to C$ of complexes, let v denote the inclusion of C into $\mathrm{cone}(f)$. Show that there is a chain homotopy equivalence $\mathrm{cone}(v) \to B[-1]$. This equivalence is the algebraic analogue of the topological fact that for any map $f: K \to L$ of (topological) cell complexes the cone of the inclusion $L \subset Cf$ is homotopy equivalent to the suspension of K.

Exercise 1.5.9 Let $f: B \to C$ be a morphism of chain complexes. Show that the natural maps $\ker(f)[-1] \xrightarrow{\partial} \mathrm{cone}(f) \xrightarrow{\beta} \mathrm{coker}(f)$ give rise to a long exact sequence:

$$\cdots \xrightarrow{\partial} H_{n-1}(\ker(f)) \xrightarrow{\alpha} H_n(\mathrm{cone}(f)) \xrightarrow{\beta} H_n(\mathrm{coker}(f)) \xrightarrow{\partial} H_{n-2}(\ker(f)) \cdots.$$

Exercise 1.5.10 Let C and C' be split complexes, with splitting maps s, s'. If $f: C \to C'$ is a morphism, show that $\sigma(c, c') = (-s(c), s'(c') - s'fs(c))$ defines a splitting of $\mathrm{cone}(f)$ if and only if the map $f_*: H_*(C) \to H_*(C')$ is zero.

1.6 More on Abelian Categories

We have already seen that R–**mod** is an abelian category for every associative ring R. In this section we expand our repertoire of abelian categories to include functor categories and sheaves. We also introduce the notions of left exact and right exact functors, which will form the heart of the next chapter. We give the Yoneda embedding of an additive category, which is exact and fully faithful, and use it to sketch a proof of the following result, which has already been used. Recall that a category is called *small* if its class of objects is in fact a set.

Freyd-Mitchell Embedding Theorem 1.6.1 (1964) *If $\mathcal{A}$ is a small abelian category, then there is a ring R and an exact, fully faithful functor from $\mathcal{A}$ into R–**mod**, which embeds $\mathcal{A}$ as a full subcategory in the sense that $\text{Hom}_{\mathcal{A}}(M, N) \cong \text{Hom}_R(M, N)$.*

We begin to prepare for this result by introducing some examples of abelian categories. The following criterion, whose proof we leave to the reader, is frequently useful:

Lemma 1.6.2 *Let $\mathcal{C} \subset \mathcal{A}$ be a full subcategory of an abelian category $\mathcal{A}$.*

1. *$\mathcal{C}$ is additive $\Leftrightarrow$ $0 \in \mathcal{C}$, and $\mathcal{C}$ is closed under $\oplus$.*
2. *$\mathcal{C}$ is abelian and $\mathcal{C} \subset \mathcal{A}$ is exact $\Leftrightarrow$ $\mathcal{C}$ is additive, and $\mathcal{C}$ is closed under* ker *and* coker.

Examples 1.6.3

1. Inside R–**mod**, the finitely generated R-modules form an additive category, which is abelian if and only if R is noetherian.
2. Inside **Ab**, the torsionfree groups form an additive category, while the p-groups form an abelian category. (A is a *p-group* if $(\forall a \in A)$ some $p^n a = 0$.) Finite p-groups also form an abelian category. The category $(\mathbb{Z}/p)$–**mod** of vector spaces over the field $\mathbb{Z}/p$ is also a full subcategory of **Ab**.

Functor Categories 1.6.4 Let C be any category, $\mathcal{A}$ an abelian category. The *functor category* $\mathcal{A}^C$ is the abelian category whose objects are functors $F: C \to \mathcal{A}$. The maps in $\mathcal{A}^C$ are natural transformations. Here are some relevant examples:

1. If C is the discrete category of integers, **Ab**^{C} contains the abelian category of *graded abelian groups* as a full subcategory.

2. If C is the poset category of integers $(\cdots \to n \to (n+1) \to \cdots)$ then the abelian category $\mathbf{Ch}(\mathcal{A})$ of cochain complexes is a full subcategory of $\mathcal{A}^C$.

3. If R is a ring considered as a one-object category, then R–**mod** is the full subcategory of all additive functors in $\mathbf{Ab}^R$.

4. Let X be a topological space, and $\mathcal{U}$ the poset of open subsets of X. A contravariant functor F from $\mathcal{U}$ to $\mathcal{A}$ such that $F(\emptyset) = \{0\}$ is called a *presheaf* on X with values in $\mathcal{A}$, and the presheaves are the objects of the abelian category $\mathcal{A}^{\mathcal{U}^{op}} = \text{Presheaves}(X)$.

A typical example of a presheaf with values in $\mathbb{R}$–**mod** is given by $C^0(U) = \{\text{continuous functions } f: U \to \mathbb{R}\}$. If $U \subset V$ the maps $C^0(V) \to C^0(U)$ are given by restricting the domain of a function from V to U. In fact, C^0 is a sheaf:

Definition 1.6.5 (Sheaves) A *sheaf* on X (with values in $\mathcal{A}$) is a presheaf F satisfying the

  Sheaf Axiom. Let $\{U_i\}$ be an open covering of an open subset U of X.
  If $\{f_i \in F(U_i)\}$ are such that each f_i and f_j agree in $F(U_i \cap U_j)$, then
  there is a unique $f \in F(U)$ that maps to every f_i under $F(U) \to F(U_i)$.

Note that the uniqueness of f is equivalent to the assertion that if $f \in F(U)$ vanishes in every $F(U_i)$, then $f = 0$. In fancy (element-free) language, the sheaf axiom states that for every covering $\{U_i\}$ of every open U the following sequence is exact:

$$0 \to F(U) \longrightarrow \prod F(U_i) \xrightarrow{\text{diff}} \prod_{i<j} F(U_i \cap U_j).$$

Exercise 1.6.1 Let M be a smooth manifold. For each open U in M, let $C^\infty(M)$ be the set of smooth functions from U to $\mathbb{R}$. Show that $C^\infty(M)$ is a sheaf on M.

Exercise 1.6.2 (Constant sheaves) Let A be any abelian group. For every open subset U of X, let $A(U)$ denote the set of continuous maps from U to the discrete topological space A. Show that A is a sheaf on X.

The category Sheaves(X) of sheaves forms an abelian category contained in Presheaves(X), but it is not an abelian subcategory; cokernels in Sheaves(X) are different from cokernels in Presheaves(X). This difference gives rise to sheaf cohomology (Chapter 2, section 2.6). The following example lies at the heart of the subject. For any space X, let $\mathcal{O}$ (resp. $\mathcal{O}^*$) be the sheaf such that

$\mathcal{O}(U)$ (resp. $\mathcal{O}^*(U)$) is the group of continuous maps from U into $\mathbb{C}$ (resp. $\mathbb{C}^*$). Then there is a short exact sequence of sheaves:

$$0 \to \mathbb{Z} \xrightarrow{2\pi i} \mathcal{O} \xrightarrow{\exp} \mathcal{O}^* \to 0.$$

When X is the space $\mathbb{C}^*$, this sequence is not exact in Presheaves(X) because the exponential map from $\mathbb{C} = \mathcal{O}(X)$ to $\mathcal{O}^*(X)$ is not onto; the cokernel is $\mathbb{Z} = H^1(X, \mathbb{Z})$, generated by the global unit $1/z$. In effect, there is no global logarithm function on X, and the contour integral $\frac{1}{2\pi i} \oint f(z)\,dz$ gives the image of $f(z)$ in the cokernel.

Definition 1.6.6 Let $F: \mathcal{A} \to \mathcal{B}$ be an additive functor between abelian categories. F is called *left exact* (resp. *right exact*) if for every short exact sequence $0 \to A \to B \to C \to 0$ in $\mathcal{A}$, the sequence $0 \to F(A) \to F(B) \to F(C)$ (resp. $F(A) \to F(B) \to F(C) \to 0$) is exact in $\mathcal{B}$. F is called *exact* if it is both left and right exact, that is, if it preserves exact sequences. A contravariant functor F is called left exact (resp. right exact, resp. exact) if the corresponding covariant functor $F': \mathcal{A}^{op} \to \mathcal{B}$ is left exact (resp. ...).

Example 1.6.7 The inclusion of Sheaves(X) into Presheaves(X) is a left exact functor. There is also an exact functor Presheaves(X) $\to$ Sheaves(X), called "sheafification." (See 2.6.5; the sheafification functor is left adjoint to the inclusion.)

Exercise 1.6.3 Show that the above definitions are equivalent to the following, which are often given as the definitions. (See [Rot], for example.) A (covariant) functor F is left exact (resp. right exact) if exactness of the sequence

$$0 \to A \to B \to C \quad (\text{resp. } A \to B \to C \to 0)$$

implies exactness of the sequence

$$0 \to FA \to FB \to FC \quad (\text{resp. } FA \to FB \to FC \to 0).$$

Proposition 1.6.8 *Let $\mathcal{A}$ be an abelian category. Then $\mathrm{Hom}_{\mathcal{A}}(M, -)$ is a left exact functor from $\mathcal{A}$ to $\mathbf{Ab}$ for every M in $\mathcal{A}$. That is, given an exact sequence $0 \to A \xrightarrow{f} B \xrightarrow{g} C \to 0$ in $\mathcal{A}$, the following sequence of abelian groups is also exact:*

$$0 \to \mathrm{Hom}(M, A) \xrightarrow{f_*} \mathrm{Hom}(M, B) \xrightarrow{g_*} \mathrm{Hom}(M, C).$$

Proof If $\alpha \in \mathrm{Hom}(M, A)$ then $f_*\alpha = f \circ \alpha$; if this is zero, then α must be zero since f is monic. Hence f_* is monic. Since $g \circ f = 0$, we have $g_* f_*(\alpha) = g \circ f \circ \alpha = 0$, so $g_* f_* = 0$. It remains to show that if $\beta \in \mathrm{Hom}(M, B)$ is such that $g_*\beta = g \circ \beta$ is zero, then $\beta = f \circ \alpha$ for some α. But if $g \circ \beta = 0$, then $\beta(M) \subseteq f(A)$, so β factors through A. $\diamond$

Corollary 1.6.9 $\mathrm{Hom}_{\mathcal{A}}(-, M)$ *is a left exact contravariant functor.*

Proof $\mathrm{Hom}_{\mathcal{A}}(A, M) = \mathrm{Hom}_{\mathcal{A}^{op}}(M, A)$. $\diamond$

Yoneda Embedding 1.6.10 Every additive category $\mathcal{A}$ can be embedded in the abelian category $\mathbf{Ab}^{\mathcal{A}^{op}}$ by the functor h sending A to $h_A = \mathrm{Hom}_{\mathcal{A}}(-, A)$. Since each $\mathrm{Hom}_{\mathcal{A}}(M, -)$ is left exact, h is a left exact functor. Since the functors h_A are left exact, the Yoneda embedding actually lands in the abelian subcategory $\mathcal{L}$ of all left exact contravariant functors from $\mathcal{A}$ to $\mathbf{Ab}$ whenever $\mathcal{A}$ is an abelian category.

Yoneda Lemma 1.6.11 *The Yoneda embedding h reflects exactness. That is, a sequence $A \xrightarrow{\alpha} B \xrightarrow{\beta} C$ in $\mathcal{A}$ is exact, provided that for every M in $\mathcal{A}$ the following sequence is exact:*

$$\mathrm{Hom}_{\mathcal{A}}(M, A) \xrightarrow{\alpha*} \mathrm{Hom}_{\mathcal{A}}(M, B) \xrightarrow{\beta*} \mathrm{Hom}_{\mathcal{A}}(M, C).$$

Proof Taking $M = A$, we see that $\beta\alpha = \beta^*\alpha^*(id_A) = 0$. Taking $M = \ker(\beta)$, we see that the inclusion $\iota: \ker(\beta) \to B$ satisfies $\beta^*(\iota) = \beta\iota = 0$. Hence there is a $\sigma \in \mathrm{Hom}(M, A)$ with $\iota = \alpha^*(\sigma) = \alpha\sigma$, so that $\ker(\beta) = \mathrm{im}(\iota) \subseteq \mathrm{im}(\alpha)$. $\diamond$

We now sketch a proof of the Freyd-Mitchell Embedding Theorem 1.6.1; details may be found in [Freyd] or [Swan, pp. 14–22]. Consider the failure of the Yoneda embedding $h: \mathcal{A} \to \mathbf{Ab}^{\mathcal{A}^{op}}$ to be exact: if $0 \to A \to B \to C \to 0$ is exact in $\mathcal{A}$ and $M \in \mathcal{A}$, then define the abelian group $W(M)$ by exactness of

$$0 \to \mathrm{Hom}_{\mathcal{A}}(M, A) \to \mathrm{Hom}_{\mathcal{A}}(M, B) \to \mathrm{Hom}_{\mathcal{A}}(M, C) \to W(M) \to 0.$$

In general $W(M) \neq 0$, and there is a short exact sequence of functors:

$$(*)\qquad\qquad 0 \to h_A \to h_B \to h_C \to W \to 0.$$

W is an example of a *weakly effaceable functor*, that is, a functor such that for all $M \in \mathcal{A}$ and $x \in W(M)$ there is a surjection $P \to M$ in $\mathcal{A}$ so that the map $W(M) \to W(P)$ sends x to zero. (To see this, take P to be the pullback $M \times_C B$, where $M \to C$ represents x, and note that $P \to C$ factors through B.) Next (see *loc. cit.*), one proves:

Proposition 1.6.12 *If $\mathcal{A}$ is small, the subcategory $\mathcal{W}$ of weakly effaceable functors is a localizing subcategory of $\mathbf{Ab}^{\mathcal{A}^{op}}$ whose quotient category is $\mathcal{L}$. That is, there is an exact "reflection" functor R from $\mathbf{Ab}^{\mathcal{A}^{op}}$ to $\mathcal{L}$ such that $R(L) = L$ for every left exact L and $R(W) \cong 0$ iff W is weakly effaceable.*

Remark Cokernels in $\mathcal{L}$ are different from cokernels in $\mathbf{Ab}^{\mathcal{A}^{op}}$, so the inclusion $\mathcal{L} \subset \mathbf{Ab}^{\mathcal{A}^{op}}$ is not exact, merely left exact. To see this, apply the reflection R to (*). Since $R(h_A) = h_A$ and $R(W) \cong 0$, we see that

$$0 \to h_A \to h_B \to h_C \to 0$$

is an exact sequence in $\mathcal{L}$, but not in $\mathbf{Ab}^{\mathcal{A}^{op}}$.

Corollary 1.6.13 *The Yoneda embedding $h \colon \mathcal{A} \to \mathcal{L}$ is exact and fully faithful.*

Finally, one observes that the category $\mathcal{L}$ has arbitrary coproducts and has a faithfully projective object P. By a result of Gabriel and Mitchell [Freyd, p. 106], $\mathcal{L}$ is equivalent to the category R–**mod** of modules over the ring $R = \mathrm{Hom}_{\mathcal{L}}(P, P)$. This finishes the proof of the Embedding Theorem.

Example 1.6.14 The abelian category of graded R-modules may be thought of as the full subcategory of $(\prod_{i \in \mathbb{Z}} R)$-modules of the form $\oplus_{i \in \mathbb{Z}} M_i$. The abelian category of chain complexes of R-modules may be embedded in S–**mod**, where

$$S = (\prod_{i \in \mathbb{Z}} R)[d]/(d^2 = 0, \{dr = rd\}_{r \in R}, \{de_i = e_{i-1}d\}_{i \in \mathbb{Z}}).$$

Here $e_i \colon \prod R \to R \to \prod R$ is the ith coordinate projection.

2
Derived Functors

2.1 δ-Functors

The right context in which to view derived functors, according to Grothendieck [Tohoku], is that of δ-functors between two abelian categories $\mathcal{A}$ and $\mathcal{B}$.

Definition 2.1.1 A (covariant) *homological* (resp. *cohomological*) *δ-functor* between $\mathcal{A}$ and $\mathcal{B}$ is a collection of additive functors $T_n : \mathcal{A} \to \mathcal{B}$ (resp. $T^n : \mathcal{A} \to \mathcal{B}$) for $n \geq 0$, together with morphisms

$$\delta_n : T_n(C) \to T_{n-1}(A)$$

$$(\text{resp. } \delta^n : T^n(C) \to T^{n+1}(A))$$

defined for each short exact sequence $0 \to A \to B \to C \to 0$ in $\mathcal{A}$. Here we make the convention that $T^n = T_n = 0$ for $n < 0$. These two conditions are imposed:

1. For each short exact sequence as above, there is a long exact sequence

$$\cdots \ T_{n+1}(C) \ \xrightarrow{\delta} \ T_n(A) \ \to \ T_n(B) \ \to \ T_n(C) \ \xrightarrow{\delta} \ T_{n-1}(A) \ \cdots$$

(resp.

$$\cdots \ T^{n-1}(C) \ \xrightarrow{\delta} \ T^n(A) \ \to \ T^n(B) \ \to \ T^n(C) \ \xrightarrow{\delta} \ T^{n+1}(A) \ \cdots).$$

In particular, T_0 is right exact, and T^0 is left exact.

2. For each morphism of short exact sequences from $0 \to A' \to B' \to C' \to 0$ to $0 \to A \to B \to C \to 0$, the δ's give a commutative diagram

$$
\begin{array}{ccc}
T_n(C') & \overset{\delta}{\longrightarrow} & T_{n-1}(A') \\
\downarrow & & \downarrow \\
T_n(C) & \overset{\delta}{\longrightarrow} & T_{n-1}(A)
\end{array}
\qquad \text{resp.} \qquad
\begin{array}{ccc}
T^n(C') & \overset{\delta}{\longrightarrow} & T^{n+1}(A') \\
\downarrow & & \downarrow \\
T^n(C) & \overset{\delta}{\longrightarrow} & T^{n+1}(A).
\end{array}
$$

Example 2.1.2 Homology gives a homological δ-functor H_* from $\mathbf{Ch}_{\geq 0}(\mathcal{A})$ to $\mathcal{A}$; cohomology gives a cohomological δ-functor H^* from $\mathbf{Ch}^{\geq 0}(\mathcal{A})$ to $\mathcal{A}$.

Exercise 2.1.1 Let S be the category of short exact sequences

$$(*) \qquad\qquad 0 \to A \to B \to C \to 0$$

in $\mathcal{A}$. Show that δ_i is a natural transformation from the functor sending $(*)$ to $T_i(C)$ to the functor sending $(*)$ to $T_{i-1}(A)$.

Example 2.1.3 (p-torsion) If p is an integer, the functors $T_0(A) = A/pA$ and

$$T_1(A) = {}_pA \equiv \{a \in A : pa = 0\}$$

fit together to form a homological δ-functor, or a cohomological δ-functor (with $T^0 = T_1$ and $T^1 = T_0$) from **Ab** to **Ab**. To see this, apply the Snake Lemma to

$$
\begin{array}{ccccccccc}
0 & \longrightarrow & A & \longrightarrow & B & \longrightarrow & C & \longrightarrow & 0 \\
 & & p\downarrow & & p\downarrow & & p\downarrow & & \\
0 & \longrightarrow & A & \longrightarrow & B & \longrightarrow & C & \longrightarrow & 0
\end{array}
$$

to get the exact sequence

$$0 \to {}_pA \to {}_pB \to {}_pC \overset{\delta}{\longrightarrow} A/pA \to B/pB \to C/pC \to 0.$$

Generalization The same proof shows that if r is any element in a ring R, then $T_0(M) = M/rM$ and $T_1(M) = {}_rM$ fit together to form a homological δ-functor (or cohomological δ-functor, if that is one's taste) from R–**mod** to **Ab**.

Vista We will see in 2.6.3 that $T_n(M) = \operatorname{Tor}_n^R(R/r, M)$ is also a homological δ-functor with $T_0(M) = M/rM$. If r is a left nonzerodivisor (meaning that $_rR = \{s \in R : rs = 0\}$ is zero), then in fact $\operatorname{Tor}_1^R(R/r, M) = {}_rM$ and $\operatorname{Tor}_n^R(R/r, M) = 0$ for $n \geq 2$; see 3.1.7. However, in general $_rR \neq 0$, while $\operatorname{Tor}_1^R(R/r, R) = 0$, so they aren't the same; $\operatorname{Tor}_1^R(M, R/r)$ is the quotient of $_rM$ by the submodule $(_rR)M$ generated by $\{sm : rs = 0, s \in R, m \in M\}$. The Tor_n will be *universal* δ-functors in a sense that we shall now make precise.

Definition 2.1.4 A *morphism* $S \to T$ of δ-functors is a system of natural transformations $S_n \to T_n$ (resp. $S^n \to T^n$) that commute with δ. This is fancy language for the assertion that there is a commutative ladder diagram connecting the long exact sequences for S and T associated to any short exact sequence in $\mathcal{A}$.

A homological δ-functor T is *universal* if, given any other δ-functor S and a natural transformation $f_0: S_0 \to T_0$, there exists a unique morphism $\{f_n: S_n \to T_n\}$ of δ-functors that extends f_0.

A cohomological δ-functor T is *universal* if, given S and $f^0: T^0 \to S^0$, there exists a unique morphism $T \to S$ of δ-functors extending f^0.

Example 2.1.5 We will see in section 2.4 that homology $H_*: \mathbf{Ch}_{\geq 0}(\mathcal{A}) \to \mathcal{A}$ and cohomology $H^*: \mathbf{Ch}^{\geq 0}(\mathcal{A}) \to \mathcal{A}$ are universal δ-functors.

Exercise 2.1.2 If $F: \mathcal{A} \to \mathcal{B}$ is an exact functor, show that $T_0 = F$ and $T_n = 0$ for $n \neq 0$ defines a universal δ-functor (of both homological and cohomological type).

Remark If $F: \mathcal{A} \to \mathcal{B}$ is an additive functor, then we can ask if there is *any* δ-functor T (universal or not) such that $T_0 = F$ (resp. $T^0 = F$). One obvious obstruction is that T_0 must be right exact (resp. T^0 must be left exact). By definition, however, we see that there is at most one (up to isomorphism) *universal* δ-functor T with $T_0 = F$ (resp. $T^0 = F$). If a universal T exists, the T_n are sometimes called the *left satellite functors* of F (resp. the T^n are called the *right satellite functors* of F). This terminology is due to the pervasive influence of the book [CE].

We will see that derived functors, when they exist, are indeed universal δ-functors. For this we need the concept of projective and injective resolutions.

2.2 Projective Resolutions

An object P in an abelian category $\mathcal{A}$ is *projective* if it satisfies the following universal lifting property: Given a surjection $g: B \to C$ and a map $\gamma: P \to C$, there is at least one map $\beta: P \to B$ such that $\gamma = g \circ \beta$.

$$P$$
$${}^{\exists\beta}\swarrow \quad \downarrow\gamma$$
$$B \longrightarrow C \longrightarrow 0$$

We shall be mostly concerned with the special case of projective modules ($\mathcal{A}$ being the category **mod–R**). The notion of projective module first appeared in the book [CE]. It is easy to see that free R-modules are projective (lift a basis). Clearly, direct summands of free modules are also projective modules.

Proposition 2.2.1 *An R-module is projective iff it is a direct summand of a free R-module.*

Proof Letting $F(A)$ be the free R-module on the set underlying an R-module A, we see that for every R-module A there is a surjection $\pi: F(A) \to A$. If A is a projective R-module, the universal lifting property yields a map $i: A \to F(A)$ so that $\pi i = 1_A$, that is, A is a direct summand of the free module $F(A)$. $\diamond$

Example 2.2.2 Over many nice rings ($\mathbb{Z}$, fields, division rings, $\cdots$) every projective module is in fact a free module. Here are two examples to show that this is not always the case:

1. If $R = R_1 \times R_2$, then $P = R_1 \times 0$ and $0 \times R_2$ are projective because their sum is R. P is not free because $(0, 1)P = 0$. This is true, for example, when R is the ring $\mathbb{Z}/6 = \mathbb{Z}/2 \times \mathbb{Z}/3$.
2. Consider the ring $R = M_n(F)$ of $n \times n$ matrices over a field F, acting on the left on the column vector space $V = F^n$. As a left R-module, R is the direct sum of its columns, each of which is the left R-module V. Hence $R \cong V \oplus \cdots \oplus V$, and V is a projective R-module. Since any free R-module would have dimension dn^2 over F for some cardinal number d, and $\dim_F(V) = n$, V cannot possibly be free over R.

Remark The category $\mathcal{A}$ of finite abelian groups is an example of an abelian category that has *no* projective objects. We say that $\mathcal{A}$ *has enough projectives* if for every object A of $\mathcal{A}$ there is a surjection $P \to A$ with P projective.

Here is another characterization of projective objects in $\mathcal{A}$:

Lemma 2.2.3 *M is projective iff* $\text{Hom}_{\mathcal{A}}(M, -)$ *is an exact functor. That is, iff the sequence of groups*

$$0 \to \text{Hom}(M, A) \to \text{Hom}(M, B) \xrightarrow{g_*} \text{Hom}(M, C) \to 0$$

is exact for every exact sequence $0 \to A \to B \to C \to 0$ *in* $\mathcal{A}$.

Proof Suppose that $\text{Hom}(M, -)$ is exact and that we are given a surjection $g: B \to C$ and a map $\gamma: M \to C$. We can lift $\gamma \in \text{Hom}(M, C)$ to $\beta \in \text{Hom}(M, B)$ such that $\gamma = g_*\beta = g \circ \beta$ because g_* is onto. Thus M has the universal lifting property, that is, it is projective. Conversely, suppose M is projective. In order to show that $\text{Hom}(M, -)$ is exact, it suffices to show that g_* is onto for every short exact sequence as above. Given $\gamma \in \text{Hom}(M, C)$, the universal lifting property of M gives $\beta \in \text{Hom}(M, B)$ so that $\gamma = g \circ \beta = g_*(\beta)$, that is, g_* is onto. $\diamond$

A chain complex P in which each P_n is projective in $\mathcal{A}$ is called a *chain complex of projectives*. It need not be a projective object in **Ch**.

Exercise 2.2.1 Show that a chain complex P is a projective object in **Ch** if and only if it is a split exact complex of projectives. *Hint:* To see that P must be split exact, consider the surjection from $\text{cone}(\text{id}_P)$ to $P[-1]$. To see that split exact complexes are projective objects, consider the special case $0 \to P_1 \cong P_0 \to 0$.

Exercise 2.2.2 Use the previous exercise 2.2.1 to show that if $\mathcal{A}$ has enough projectives, then so does the category $\textbf{Ch}(\mathcal{A})$ of chain complexes over $\mathcal{A}$.

Definition 2.2.4 Let M be an object of $\mathcal{A}$. A *left resolution* of M is a complex P with $P_i = 0$ for $i < 0$, together with a map $\epsilon: P_0 \to M$ so that the augmented complex

$$\cdots \xrightarrow{d} P_2 \xrightarrow{d} P_1 \xrightarrow{d} P_0 \xrightarrow{\epsilon} M \to 0$$

is exact. It is a *projective resolution* if each P_i is projective.

Lemma 2.2.5 *Every R-module M has a projective resolution. More generally, if an abelian category $\mathcal{A}$ has enough projectives, then every object M in $\mathcal{A}$ has a projective resolution.*

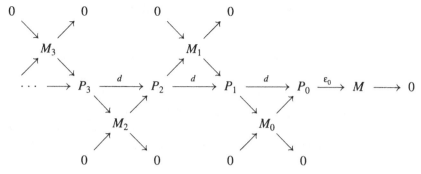

Figure 2.1.   Forming a resolution by splicing.

Proof Choose a projective P_0 and a surjection $\epsilon_0: P_0 \to M$, and set $M_0 =$ $\ker(\epsilon_0)$. Inductively, given a module M_{n-1}, we choose a projective P_n and a surjection $\epsilon_n: P_n \to M_{n-1}$. Set $M_n = \ker(\epsilon_n)$, and let d_n be the composite $P_n \to M_{n-1} \to P_{n-1}$. Since $d_n(P_n) = M_{n-1} = \ker(d_{n-1})$, the chain complex P is a resolution of M. (See Figure 2.1.)   ◇

Exercise 2.2.3 Show that if P is a complex of projectives with $P_i = 0$ for $i < 0$, then a map $\epsilon: P_0 \to M$ giving a resolution for M is the same thing as a chain map $\epsilon: P \to M$, where M is considered as a complex concentrated in degree zero.

Comparison Theorem 2.2.6 *Let* $P \xrightarrow{\epsilon} M$ *be a projective resolution of* M *and* $f': M \to N$ *a map in* $\mathcal{A}$. *Then for every resolution* $Q \xrightarrow{\eta} N$ *of* N *there is a chain map* $f: P \to Q$ *lifting* f' *in the sense that* $\eta \circ f_0 = f' \circ \epsilon$. *The chain map* f *is unique up to chain homotopy equivalence.*

$$
\begin{array}{ccccccccc}
\cdots & \longrightarrow & P_2 & \longrightarrow & P_1 & \longrightarrow & P_0 & \xrightarrow{\epsilon} & M & \longrightarrow & 0 \\
 & & \exists\downarrow & & \exists\downarrow & & \exists\downarrow & & \downarrow f' & & \\
\cdots & \longrightarrow & Q_2 & \longrightarrow & Q_1 & \longrightarrow & Q_0 & \xrightarrow{\eta} & N & \longrightarrow & 0
\end{array}
$$

Porism 2.2.7 The proof will make it clear that the hypothesis that $P \to M$ be a projective resolution is too strong. It suffices to be given a chain complex

$$\cdots \to P_2 \to P_1 \to P_0 \to M \to 0$$

with the P_i projective. Then for every resolution $Q \to N$ of N, every map $M \to N$ lifts to a map $P \to Q$, which is unique up to chain homotopy. This stronger version of the Comparison Theorem will be used in section 2.7 to construct the external product for Tor.

Proof We will construct the f_n and show their uniqueness by induction on n, thinking of f_{-1} as f'. Inductively, suppose f_i has been constructed for $i \leq n$ so that $f_{i-1}d = df_i$. In order to construct f_{n+1} we consider the n-cycles of P and Q. If $n = -1$, we set $Z_{-1}(P) = M$ and $Z_{-1}(Q) = N$; if $n \geq 0$, the fact that $f_{n-1}d = df_n$ means that f_n induces a map f_n' from $Z_n(P)$ to $Z_n(Q)$. Therefore we have two diagrams with exact rows

$$\cdots \xrightarrow{d} P_{n+1} \xrightarrow{d} Z_n(P) \longrightarrow 0 \qquad\qquad 0 \longrightarrow Z_n(P) \longrightarrow P_n \longrightarrow P_{n-1}$$
$$\exists\downarrow \qquad\quad \downarrow f_n' \qquad\qquad \text{and} \qquad\qquad \downarrow f_n' \qquad \downarrow f_n \qquad \downarrow f_{n-1}$$
$$\cdots \longrightarrow Q_{n+1} \xrightarrow{d} Z_n(Q) \longrightarrow 0 \qquad\qquad 0 \longrightarrow Z_n(Q) \longrightarrow Q_n \longrightarrow Q_{n-1}$$

The universal lifting property of the projective P_{n+1} yields a map f_{n+1} from P_{n+1} to Q_{n+1}, so that $df_{n+1} = f_n'd = f_nd$. This finishes the inductive step and proves that the chain map $f: P \to Q$ exists.

To see uniqueness of f up to chain homotopy, suppose that $g: P \to Q$ is another lift of f' and set $h = f - g$; we will construct a chain contraction $\{s_n: P_n \to Q_{n+1}\}$ of h by induction on n. If $n < 0$, then $P_n = 0$, so we set $s_n = 0$. If $n = 0$, note that since $\eta h_0 = \epsilon(f' - f') = 0$, the map h_0 sends P_0 to $Z_0(Q) = d(Q_1)$. We use the lifting property of P_0 to get a map $s_0: P_0 \to Q_1$ so that $h_0 = ds_0 = ds_0 + s_{-1}d$. Inductively, we suppose given maps $s_i (i < n)$ so that $ds_{n-1} = h_{n-1} - s_{n-2}d$ and consider the map $h_n - s_{n-1}d$ from P_n to Q_n. We compute that

$$d(h_n - s_{n-1}d) = dh_n - (h_{n-1} - s_{n-2}d)d = (dh - hd) + s_{n-2}dd = 0.$$

Therefore $h_n - s_{n-1}d$ lands in $Z_n(Q)$, a quotient of Q_{n+1}. The lifting property of P_n yields the desired map $s_n: P_n \to Q_{n+1}$ such that $ds_n = h_n - s_{n-1}d$. $\diamond$

$$\begin{array}{ccc} & P_n & \\ \exists\swarrow & \downarrow h-sd & \\ Q_{n+1} \xrightarrow{d} & Z_n(Q) \longrightarrow 0 & \end{array} \qquad \text{and} \qquad \begin{array}{ccccc} P_n & \xrightarrow{d} & P_{n-1} & \xrightarrow{d} & P_{n-2} \\ \downarrow h & \swarrow s & \downarrow h & \swarrow s & \\ Q_n & \longrightarrow & Q_{n-1} & & \end{array}$$

Here is another way to construct projective resolutions. It is called the Horseshoe Lemma because we are required to fill in the horseshoe-shaped diagram.

Horseshoe Lemma 2.2.8 *Suppose given a commutative diagram*

$$
\begin{array}{ccccccccc}
& & & & & & 0 & & \\
& & & & & & \downarrow & & \\
\cdots & P_2' & \longrightarrow & P_1' & \longrightarrow & P_0' & \overset{\epsilon'}{\longrightarrow} & A' & \longrightarrow & 0 \\
& & & & & & \downarrow^{i_A} & & \\
& & & & & & A & & \\
& & & & & & \downarrow^{\pi_A} & & \\
\cdots & P_2'' & \longrightarrow & P_1'' & \longrightarrow & P_0'' & \overset{\epsilon''}{\longrightarrow} & A'' & \longrightarrow & 0 \\
& & & & & & \downarrow & & \\
& & & & & & 0 & &
\end{array}
$$

where the column is exact and the rows are projective resolutions. Set $P_n = P_n' \oplus P_n''$. Then the P_n assemble to form a projective resolution P of A, and the right-hand column lifts to an exact sequence of complexes

$$0 \to P' \overset{i}{\longrightarrow} P \overset{\pi}{\longrightarrow} P'' \to 0,$$

where $i_n: P_n' \to P_n$ and $\pi_n: P_n \to P_n''$ are the natural inclusion and projection, respectively.

Proof Lift ϵ'' to a map $P_0'' \to A$; the direct sum of this with the map $i_A \epsilon': P_0' \to A$ gives a map $\epsilon: P_0 \to A$. The diagram $(*)$ below commutes.

$$(*)$$

$$
\begin{array}{ccccccccc}
& & 0 & & 0 & & 0 & & \\
& & \downarrow & & \downarrow & & \downarrow & & \\
0 & \longrightarrow & \ker(\epsilon') & \longrightarrow & P_0' & \overset{\epsilon'}{\longrightarrow} & A' & \longrightarrow & 0 \\
& & \downarrow & & \downarrow & & \downarrow & & \\
0 & \longrightarrow & \ker(\epsilon) & \longrightarrow & P_0 & \overset{\epsilon}{\longrightarrow} & A & \longrightarrow & 0 \\
& & \downarrow & & \downarrow & & \downarrow & & \\
0 & \longrightarrow & \ker(\epsilon'') & \longrightarrow & P_0'' & \overset{\epsilon''}{\longrightarrow} & A'' & \longrightarrow & 0 \\
& & \downarrow & & \downarrow & & \downarrow & & \\
& & 0 & & 0 & & 0 & &
\end{array}
$$

The right two columns of (∗) are short exact sequences. The Snake Lemma 1.3.2 shows that the left column is exact and that $\operatorname{coker}(\epsilon) = 0$, so that P_0 maps onto A. This finishes the initial step and brings us to the situation

$$0$$
$$\downarrow$$
$$\cdots \longrightarrow P_1' \xrightarrow{d'} \ker(\epsilon') \longrightarrow 0$$
$$\downarrow$$
$$\ker(\epsilon)$$
$$\downarrow$$
$$\cdots \longrightarrow P_1'' \xrightarrow{d''} \ker(\epsilon'') \longrightarrow 0$$
$$\downarrow$$
$$0.$$

The filling in of the "horseshoe" now proceeds by induction.   ◇

Exercise 2.2.4 Show that there are maps $\lambda_n: P_n'' \to P_{n-1}'$ so that

$$d = \begin{bmatrix} d' & \lambda \\ 0 & d'' \end{bmatrix}, \quad \text{i.e.,} \quad d' \begin{bmatrix} p' \\ p'' \end{bmatrix} = \begin{bmatrix} d'(p') + \lambda(p'') \\ d''(p'') \end{bmatrix}.$$

2.3 Injective Resolutions

An object I in an abelian category $\mathcal{A}$ is *injective* if it satisfies the following universal lifting property: Given an injection $f: A \to B$ and a map $\alpha: A \to I$, there exists at least one map $\beta: B \to I$ such that $\alpha = \beta \circ f$.

$$0 \longrightarrow A \xrightarrow{f} B$$
$$\alpha \downarrow \quad \swarrow \exists \beta$$
$$I$$

We say that $\mathcal{A}$ *has enough injectives* if for every object A in $\mathcal{A}$ there is an injection $A \to I$ with I injective. Note that if $\{I_\alpha\}$ is a family of injectives, then the product $\prod I_\alpha$ is also injective. The notion of injective module was invented by R. Baer in 1940, long before projective modules were thought of.

Baer's Criterion 2.3.1 *A right R-module E is injective if and only if for every right ideal J of R, every map J → E can be extended to a map R → E.*

Proof The "only if" direction is a special case of the definition of injective. Conversely, suppose given an R-module B, a submodule A and a map $\alpha: A \to E$. Let $\mathcal{E}$ be the poset of all extensions $\alpha': A' \to E$ of α to an intermediate submodule $A \subseteq A' \subseteq B$; the partial order is that $\alpha' \leq \alpha''$ if α'' extends α'. By Zorn's lemma there is a maximal extension $\alpha': A' \to E$ in $\mathcal{E}$; we have to show that $A' = B$. Suppose there is some $b \in B$ not in A'. The set $J = \{r \in R : br \in A'\}$ is a right ideal of R. By assumption, the map $J \xrightarrow{b} A' \xrightarrow{\alpha'} E$ extends to a map $f: R \to E$. Let A'' be the submodule $A' + bR$ of B and define $\alpha'': A'' \to E$ by

$$\alpha''(a + br) = \alpha'(a) + f(r), \quad a \in A' \text{ and } r \in R.$$

This is well defined because $\alpha'(br) = f(r)$ for br in $A' \cap bR$, and α'' extends α', contradicting the existence of b. Hence $A' = B$. $\diamond$

Exercise 2.3.1 Let $R = \mathbb{Z}/m$. Use Baer's criterion to show that R is an injective R-module. Then show that $\mathbb{Z}/d$ is *not* an injective R-module when $d|m$ and some prime p divides both d and m/d. (The hypothesis ensures that $\mathbb{Z}/m \not\cong \mathbb{Z}/d \oplus \mathbb{Z}/e$.)

Corollary 2.3.2 *Suppose that $R = \mathbb{Z}$, or more generally that R is a principal ideal domain. An R-module A is injective iff it is* divisible, *that is, for every $r \neq 0$ in R and every $a \in A$, $a = br$ for some $b \in A$.*

Example 2.3.3 The divisible abelian groups $\mathbb{Q}$ and $\mathbb{Z}_{p^\infty} = \mathbb{Z}[\frac{1}{p}]/\mathbb{Z}$ are injective ($\mathbb{Z}[\frac{1}{p}]$ is the group of rational numbers of the form a/p^n, $n \geq 1$). Every injective abelian group is a direct sum of these [KapIAB,section 5]. In particular, the injective abelian group $\mathbb{Q}/\mathbb{Z}$ is isomorphic to $\oplus \mathbb{Z}_{p^\infty}$.

We will now show that **Ab** has enough injectives. If A is an abelian group, let $I(A)$ be the product of copies of the injective group $\mathbb{Q}/\mathbb{Z}$, indexed by the set $\text{Hom}_{\mathbf{Ab}}(A, \mathbb{Q}/\mathbb{Z})$. Then $I(A)$ is injective, being a product of injectives, and there is a canonical map $e_A: A \to I(A)$. This is our desired injection of A into an injective by the following exercise.

Exercise 2.3.2 Show that e_A is an injection. *Hint:* If $a \in A$, find a map $f: a\mathbb{Z} \to \mathbb{Q}/\mathbb{Z}$ with $f(a) \neq 0$ and extend f to a map $f': A \to \mathbb{Q}/\mathbb{Z}$.

Exercise 2.3.3 Show that an abelian group A is zero iff $\mathrm{Hom}_{\mathbf{Ab}}(A, \mathbb{Q}/\mathbb{Z}) = 0$.

Now it is a fact, easily verified, that if $\mathcal{A}$ is an abelian category, then the opposite category $\mathcal{A}^{op}$ is also abelian. The definition of injective is dual to that of projective, so we immediately can deduce the following results (2.3.4–2.3.7) by arguing in $\mathcal{A}^{op}$.

Lemma 2.3.4 *The following are equivalent for an object I in an abelian category $\mathcal{A}$:*

1. *I is injective in $\mathcal{A}$.*
2. *I is projective in $\mathcal{A}^{op}$.*
3. *The contravariant functor $\mathrm{Hom}_{\mathcal{A}}(-, I)$ is exact, that is, it takes short exact sequences in $\mathcal{A}$ to short exact sequences in $\mathbf{Ab}$.*

Definition 2.3.5 Let M be an object of $\mathcal{A}$. A *right resolution* of M is a cochain complex $I^{\cdot}$ with $I^i = 0$ for $i < 0$ and a map $M \to I^0$ such that the augmented complex

$$0 \to M \to I^0 \xrightarrow{d} I^1 \xrightarrow{d} I^2 \xrightarrow{d} \cdots$$

is exact. This is the same as a cochain map $M \to I^{\cdot}$, where M is considered as a complex concentrated in degree 0. It is called an *injective resolution* if each I^i is injective.

Lemma 2.3.6 *If the abelian category $\mathcal{A}$ has enough injectives, then every object in $\mathcal{A}$ has an injective resolution.*

Comparison Theorem 2.3.7 *Let $N \to I^{\cdot}$ be an injective resolution of N and $f': M \to N$ a map in $\mathcal{A}$. Then for every resolution $M \to E^{\cdot}$ there is a cochain map $F: E^{\cdot} \to I^{\cdot}$ lifting f'. The map f is unique up to cochain homotopy equivalence.*

$$
\begin{array}{ccccccccc}
0 & \longrightarrow & M & \longrightarrow & E^0 & \longrightarrow & E^1 & \longrightarrow & E^2 & \longrightarrow & \cdots \\
 & & f'\downarrow & & \exists\downarrow & & \exists\downarrow & & \exists\downarrow & & \\
0 & \longrightarrow & N & \longrightarrow & I^0 & \longrightarrow & I^1 & \xrightarrow{\eta} & I^2 & \longrightarrow & \cdots
\end{array}
$$

Exercise 2.3.4 Show that I is an injective object in the category of chain complexes iff I is a split exact complex of injectives. Then show that if $\mathcal{A}$ has enough injectives, so does the category $\mathbf{Ch}(\mathcal{A})$ of chain complexes over $\mathcal{A}$. *Hint:* $\mathbf{Ch}(\mathcal{A})^{op} \approx \mathbf{Ch}(\mathcal{A}^{op})$.

We now show that there are enough injective R-modules for every ring R. Recall that if A is an abelian group and B is a left R-module, then $\mathrm{Hom}_{\mathbf{Ab}}(B, A)$ is a right R-module via the rule $fr: b \mapsto f(rb)$.

Lemma 2.3.8 *For every right R-module M, the natural map*

$$\tau: \mathrm{Hom}_{\mathbf{Ab}}(M, A) \to \mathrm{Hom}_{\mathbf{mod}-R}(M, \mathrm{Hom}_{\mathbf{Ab}}(R, A))$$

is an isomorphism, where $(\tau f)(m)$ is the map $r \mapsto f(mr)$.

Proof We define a map μ backwards as follows: If $g: M \to \mathrm{Hom}(R, A)$ is an R-module map, μg is the abelian group map sending m to $g(m)(1)$. Since $\tau(\mu g) = g$ and $\mu \tau(f) = f$ (check this!), τ is an isomorphism. $\diamond$

Definition 2.3.9 A pair of functors $L: \mathcal{A} \to \mathcal{B}$ and $R: \mathcal{B} \to \mathcal{A}$ are *adjoint* if there is a natural bijection for all A in $\mathcal{A}$ and B in $\mathcal{B}$:

$$\tau = \tau_{AB} : \mathrm{Hom}_{\mathcal{B}}(L(A), B) \xrightarrow{\cong} \mathrm{Hom}_{\mathcal{A}}(A, R(B)).$$

Here "natural" means that for all $f: A \to A'$ in $\mathcal{A}$ and $g: B \to B'$ in $\mathcal{B}$ the following diagram commutes:

$$
\begin{array}{ccccc}
\mathrm{Hom}_{\mathcal{B}}(L(A'), B) & \xrightarrow{Lf^*} & \mathrm{Hom}_{\mathcal{B}}(L(A), B) & \xrightarrow{g_*} & \mathrm{Hom}_{\mathcal{B}}(L(A), B') \\
\downarrow{\scriptstyle\tau} & & \downarrow{\scriptstyle\tau} & & \downarrow{\scriptstyle\tau} \\
\mathrm{Hom}_{\mathcal{A}}(A', R(B)) & \xrightarrow{f^*} & \mathrm{Hom}_{\mathcal{A}}(A, R(B)) & \xrightarrow{Rg_*} & \mathrm{Hom}_{\mathcal{A}}(A, R(B')).
\end{array}
$$

We call L the *left adjoint* and R the *right adjoint* of this pair. The above lemma states that the forgetful functor from **mod**–R to **Ab** has $\mathrm{Hom}_{\mathbf{Ab}}(R, -)$ as its right adjoint.

Proposition 2.3.10 *If an additive functor $R: \mathcal{B} \to \mathcal{A}$ is right adjoint to an exact functor $L: \mathcal{A} \to \mathcal{B}$ and I is an injective object of $\mathcal{B}$, then $R(I)$ is an injective object of $\mathcal{A}$. (We say that R preserves injectives.)*

Dually, if an additive functor $L: \mathcal{A} \to \mathcal{B}$ is left adjoint to an exact functor $R: \mathcal{B} \to \mathcal{A}$ and P is a projective object of $\mathcal{A}$, then $L(P)$ is a projective object of $\mathcal{B}$. (We say that L preserves projectives.)

Proof We must show that $\mathrm{Hom}_{\mathcal{A}}(-, R(I))$ is exact. Given an injection $f: A \to A'$ in $\mathcal{A}$ the diagram

$$\text{Hom}_B(L(A'), I) \xrightarrow{\ Lf^*\ } \text{Hom}_B(L(A), I)$$

$$\Big\downarrow\cong \qquad\qquad\qquad \Big\downarrow\cong$$

$$\text{Hom}_A(A', R(I)) \xrightarrow{\ f^*\ } \text{Hom}_A(A, R(I))$$

commutes by naturality of τ. Since L is exact and I is injective, the top map Lf^* is onto. Hence the bottom map f^* is onto, proving that $R(I)$ is an injective object in $\mathcal{A}$. $\diamond$

Corollary 2.3.11 *If I is an injective abelian group, then $\text{Hom}_{\mathbf{Ab}}(R, I)$ is an injective R-module.*

Exercise 2.3.5 If M is an R-module, let $I(M)$ be the product of copies of $I_0 = \text{Hom}_{\mathbf{Ab}}(R, \mathbb{Q}/\mathbb{Z})$, indexed by the set $\text{Hom}_R(M, I_0)$. There is a canonical map $e_M : M \to I(M)$; show that e_M is an injection. Being a product of injectives, $I(M)$ is injective, so this will prove that R–**mod** has enough injectives. An important consequence of this is that every R-module has an injective resolution.

Example 2.3.12 The category Sheaves(X) of abelian group sheaves (1.6.5) on a topological space X has enough injectives. To see this, we need two constructions. The *stalk* of a sheaf $\mathcal{F}$ at a point $x \in X$ is the abelian group $\mathcal{F}_x = \varinjlim \{\mathcal{F}(U) : x \in U\}$. "Stalk at x" is an exact functor from Sheaves(X) to **Ab**. If A is any abelian group, the *skyscraper sheaf* x_*A at the point $x \in X$ is defined to be the presheaf

$$(x_*A)(U) = \begin{cases} A & \text{if } x \in U \\ 0 & \text{otherwise.} \end{cases}$$

Exercise 2.3.6 Show that x_*A is a sheaf and that

$$\text{Hom}_{\mathbf{Ab}}(\mathcal{F}_x, A) \cong \text{Hom}_{\text{Sheaves}(X)}(\mathcal{F}, x_*A)$$

for every sheaf $\mathcal{F}$. Use 2.3.10 to conclude that if A_x is an injective abelian group, then $x_*(A_x)$ is an injective object in Sheaves(X) for each x, and that $\prod_{x \in X} x_*(A_x)$ is also injective.

Given a fixed sheaf $\mathcal{F}$, choose an injection $\mathcal{F}_x \to I_x$ with I_x injective in **Ab** for each $x \in X$. Combining the natural maps $\mathcal{F} \to x_*\mathcal{F}_x$ with $x_*\mathcal{F}_x \to x_*I_x$ yields a map from $\mathcal{F}$ to the injective sheaf $\mathcal{I} = \prod_{x \in X} x_*(I_x)$. The map $\mathcal{F} \to \mathcal{I}$ is an injection (see [Gode], for example) showing that Sheaves(X) has enough injectives.

Example 2.3.13 Let I be a small category and $\mathcal{A}$ an abelian category. If the product of any set of objects exists in $\mathcal{A}$ ($\mathcal{A}$ is *complete*) and $\mathcal{A}$ has enough injectives, we will show that the functor category $\mathcal{A}^I$ has enough injectives. For each k in I, the kth coordinate $A \mapsto A(k)$ is an exact functor from $\mathcal{A}^I$ to $\mathcal{A}$. Given A in $\mathcal{A}$, define the functor $k_* A: I \to \mathcal{A}$ by sending $i \in I$ to

$$k_* A(i) = \prod_{\mathrm{Hom}_I(i,k)} A.$$

If $\eta: i \to j$ is a map in I, the map $k_* A(i) \to k_* A(j)$ is determined by the index map $\eta^*: \mathrm{Hom}(j, k) \to \mathrm{Hom}(i, k)$. That is, the coordinate $k_* A(i) \to A$ of this map corresponding to $\varphi \in \mathrm{Hom}(j, k)$ is the projection of $k_* A(i)$ onto the factor corresponding to $\eta^* \varphi = \varphi \eta \in \mathrm{Hom}(i, k)$. If $f: A \to B$ is a map in $\mathcal{A}$, there is a corresponding map $k_* A \to k_* B$ defined slotwise. In this way, k_* becomes an additive functor from $\mathcal{A}$ to $\mathcal{A}^I$, assuming that $\mathcal{A}$ has enough products for $k_* A$ to be defined.

Exercise 2.3.7 Assume that $\mathcal{A}$ is complete and has enough injectives. Show that k_* is right adjoint to the kth coordinate functor, so that k_* preserves injectives by 2.3.10. Given $F \in \mathcal{A}^I$, embed each $F(k)$ in an injective object A_k of $\mathcal{A}$, and let $F \to k_* A_k$ be the corresponding adjoint map. Show that the product $E = \prod_{k \in I} k_* A_k$ exists in $\mathcal{A}^I$, that E is an injective object, and that $F \to E$ is an injection. Conclude that $\mathcal{A}^I$ has enough injectives.

Exercise 2.3.8 Use the isomorphism $(\mathcal{A}^I)^{op} \cong \mathcal{A}^{(I^{op})}$ to dualize the previous exercise. That is, assuming that $\mathcal{A}$ is cocomplete and has enough projectives, show that $\mathcal{A}^I$ has enough projectives.

2.4 Left Derived Functors

Let $F: \mathcal{A} \to \mathcal{B}$ be a right exact functor between two abelian categories. If $\mathcal{A}$ has enough projectives, we can construct the *left derived functors* $L_i F$ ($i \geq 0$) of F as follows. If A is an object of $\mathcal{A}$, choose (once and for all) a projective resolution $P \to A$ and define

$$L_i F(A) = H_i(F(P)).$$

Note that since $F(P_1) \to F(P_0) \to F(A) \to 0$ is exact, we always have $L_0 F(A) \cong F(A)$. The aim of this section is to show that the $L_* F$ form a universal homological δ-functor.

Lemma 2.4.1 *The objects $L_i F(A)$ of B are well defined up to natural isomorphism. That is, if $Q \to A$ is a second projective resolution, then there is a canonical isomorphism:*

$$L_i F(A) = H_i(F(P)) \xrightarrow{\cong} H_i(F(Q)).$$

In particular, a different choice of the projective resolutions would yield new functors $\hat{L}_i F$, which are naturally isomorphic to the functors $L_i F$.

Proof By the Comparison Theorem (2.2.6), there is a chain map $f: P \to Q$ lifting the identity map id_A, yielding a map f_* from $H_i F(P)$ to $H_i F(Q)$. Any other such chain map $f': P \to Q$ is a chain homotopic to f, so $f_* = f'_*$. Therefore, the map f_* is canonical. Similarly, there is a chain map $g: Q \to P$ lifting id_A and a map g_*. Since gf and id_P are both chain maps $P \to P$ lifting id_A, we have

$$g_* f_* = (gf)_* = (\mathrm{id}_P)_* = \text{ identity map on } H_i F(P).$$

Similarly, fg and id_Q both lift id_A, so $f_* g_*$ is the identity. This proves that f_* and g_* are isomorphisms.   ◇

Corollary 2.4.2 *If A is projective, then $L_i F(A) = 0$ for $i \neq 0$.*

F-Acyclic Objects 2.4.3 An object Q is called *F-acyclic* if $L_i F(Q) = 0$ for all $i \neq 0$, that is, if the higher derived functors of F vanish on Q. Clearly, projectives are F-acyclic for every right exact functor F, but there are others; flat modules are acyclic for tensor products, for example. An *F-acyclic resolution* of A is a left resolution $Q \to A$ for which each Q_i is F-acyclic. We will see later (using dimension shifting, exercise 2.4.3 and 3.2.8) that we can also compute left derived functors from F-acyclic resolutions, that is, that $L_i(A) \cong H_i(F(Q))$ for any F-acyclic resolution Q of A.

Lemma 2.4.4 *If $f: A' \to A$ is any map in $\mathcal{A}$, there is a natural map $L_i F(f)$: $L_i F(A') \to L_i F(A)$ for each i.*

Proof Let $P' \to A'$ and $P \to A$ be the chosen projective resolutions. The comparison theorem yields a lift of f to a chain map $\tilde{f}$ from P' to P, hence a map $\tilde{f}_*$ from $H_i F(P')$ to $H_i F(P)$. Any other lift is chain homotopic to $\tilde{f}$, so the map $\tilde{f}_*$ is independent of the choice of $\tilde{f}$. The map $L_i F(f)$ is $\tilde{f}_*$.   ◇

Exercise 2.4.1 Show that $L_0 F(f) = F(f)$ under the identification $L_0 F(A) \cong F(A)$.

Theorem 2.4.5 *Each $L_i F$ is an additive functor from $\mathcal{A}$ to $\mathcal{B}$.*

Proof The identity map on P lifts the identity on A, so $L_i F(id_A)$ is the identity map. Given maps $A' \xrightarrow{f} A \xrightarrow{g} A''$ and chain maps $\tilde{f}, \tilde{g}$ lifting f and g, the composite $\tilde{g}\tilde{f}$ lifts gf. Therefore $g_* f_* = (gf)_*$, proving that $L_i F$ is a functor. If $f_i : A' \to A$ are two maps with lifts $\tilde{f}_i$, the sum $\tilde{f}_1 + \tilde{f}_2$ lifts $f_1 + f_2$. Therefore $f_{1*} + f_{2*} = (f_1 + f_2)_*$, proving that $L_i F$ is additive. ◇

Exercise 2.4.2 (Preserving derived functors) If $U : \mathcal{B} \to \mathcal{C}$ is an exact functor, show that

$$U(L_i F) \cong L_i(UF).$$

Forgetful functors such as **mod**–$R \to$ **Ab** are often exact, and it is often easier to compute the derived functors of UF due to the absence of cluttering restrictions.

Theorem 2.4.6 *The derived functors $L_* F$ form a homological δ-functor.*

Proof Given a short exact sequence

$$0 \to A' \to A \to A'' \to 0,$$

choose projective resolutions $P' \to A'$ and $P'' \to A''$. By the Horseshoe Lemma 2.2.8, there is a projective resolution $P \to A$ fitting into a short exact sequence $0 \to P' \to P \to P'' \to 0$ of projective complexes in $\mathcal{A}$. Since the P_n'' are projective, each sequence $0 \to P_n' \to P_n \to P_n'' \to 0$ is split exact. As F is additive, each sequence

$$0 \to F(P_n') \to F(P_n) \xleftarrow{} F(P_n'') \to 0$$

is split exact in $\mathcal{B}$. Therefore

$$0 \to F(P') \to F(P) \to F(P'') \to 0$$

is a short exact sequence of chain complexes. Writing out the corresponding long exact homology sequence, we get

$$\cdots \xrightarrow{\partial} L_i F(A') \to L_i F(A) \to L_i F(A'') \xrightarrow{\partial} L_{i-1} F(A') \to L_{i-1} F(A) \to L_{i-1} F(A'') \xrightarrow{\partial} \cdots$$

To see the naturality of the ∂_i, assume we are given a commutative diagram

$$
\begin{array}{ccccccccc}
0 & \longrightarrow & A' & \longrightarrow & A & \longrightarrow & A'' & \longrightarrow & 0 \\
& & f'\downarrow & & f\downarrow & & \downarrow f'' & & \\
0 & \longrightarrow & B' & \longrightarrow & B & \longrightarrow & B'' & \longrightarrow & 0 \\
& & & i_B & & \pi_B & & &
\end{array}
$$

in $\mathcal{A}$, and projective resolutions of the corners: $\epsilon' : P' \to A'$, $\epsilon'' : P'' \to A''$, $\eta' : Q' \to B'$ and $\eta'' : Q'' \to B''$. Use the Horseshoe Lemma 2.2.8 to get projective resolutions $\epsilon : P \to A$ and $\eta : Q \to B$. Use the Comparison Theorem 2.2.6 to obtain chain maps $F' : P' \to Q'$ and $F'' : P'' \to Q''$ lifting the maps f' and f'', respectively. We shall show that there is also a chain map $F : P \to Q$ lifting f, and giving a commutative diagram of chain complexes with exact rows:

$$
\begin{array}{ccccccccc}
0 & \longrightarrow & P' & \longrightarrow & P & \longrightarrow & P'' & \longrightarrow & 0 \\
& & F'\downarrow & & F\downarrow & & \downarrow F'' & & \\
0 & \longrightarrow & Q' & \longrightarrow & Q & \longrightarrow & Q'' & \longrightarrow & 0.
\end{array}
$$

The naturality of the connecting homomorphism in the long exact homology sequence now translates into the naturality of the ∂_i. In order to produce F, we will construct maps (*not* chain maps) $\gamma_n : P''_n \to Q'_n$ such that F_n is

$$
F_n = \begin{bmatrix} F'_n & \gamma_n \\ 0 & F''_n \end{bmatrix} : \begin{array}{c} P'_n \\ \oplus \\ P''_n \end{array} \longrightarrow \begin{array}{c} Q'_n \\ \oplus \\ Q''_n \end{array}
$$

$$
F_n(p', p'') = (F'(p') + \gamma(p''), F''(p'')).
$$

Assuming that F is a chain map over f, this choice of F will yield our commutative diagram of chain complexes. In order for F to be a lifting of f, the map $(\eta F_0 - f\epsilon)$ from $P_0 = P'_0 \oplus P''_0$ to B must vanish. On P'_0 this is no problem, so this just requires that

$$
i_B \eta' \gamma_0 = f \lambda_P - \lambda_Q F''_0
$$

as maps from P''_0 to B, where λ_P and λ_Q are the restrictions of ϵ and η to P''_0 and Q''_0, and i_B is the inclusion of B' in B. There is some map $\beta : P''_0 \to B'$ so that $i_B \beta = f\lambda - \lambda F''_0$ because in B'' we have

$$
\pi_B(f\lambda - \lambda F''_0) = f'' \pi_A \lambda_P - \pi_B \lambda F''_0 = f'' \epsilon'' - \eta'' F''_0 = 0.
$$

We may therefore define γ_0 to be any lift of β to Q'_0.

$$P''_0$$

$$\gamma_0 \swarrow \quad \downarrow \beta$$

$$Q'_0 \xrightarrow{\eta'} B' \longrightarrow 0$$

In order for F to be a chain map, we must have

$$dF - Fd = \left[\begin{pmatrix} d' & \lambda \\ 0 & d'' \end{pmatrix}, \begin{pmatrix} F' & \gamma \\ 0 & F'' \end{pmatrix}\right]$$

$$= \begin{pmatrix} d'F' - F'd' & (d'\gamma - \gamma d'' + \lambda F'' - F'\lambda') \\ 0 & d''F'' - F''d'' \end{pmatrix}$$

vanishing. That is, the map $d'\gamma_n: P''_n \to Q'_{n-1}$ must equal

$$g_n = \gamma_{n-1}d'' - \lambda_n F'_n + F''_{n-1}\lambda_n.$$

Inductively, we may suppose γ_i defined for $i < n$, so that g_n exists. A short calculation, using the inductive formula for $d'\gamma_{n-1}$, shows that $d'g_n = 0$. As the complex Q' is exact, the map g_n factors through a map $\beta: P''_n \to d(Q'_n)$. We may therefore define γ_n to be any lift of β to Q'_n. This finishes the construction of the chain map F and the proof. $\diamond$

Exercise 2.4.3 (Dimension shifting) If $0 \to M \to P \to A \to 0$ is exact with P projective (or F-acyclic 2.4.3), show that $L_i F(A) \cong L_{i-1}F(M)$ for $i \geq 2$ and that $L_1 F(A)$ is the kernel of $F(M) \to F(P)$. More generally, show that if

$$0 \to M_m \to P_m \to P_{m-1} \to \cdots \to P_0 \to A \to 0$$

is exact with the P_i projective (or F-acyclic), then $L_i F(A) \cong L_{i-m-1}F(M_m)$ for $i \geq m+2$ and $L_{m+1}F(A)$ is the kernel of $F(M_m) \to F(P_m)$. Conclude that if $P \to A$ is an F-acyclic resolution of A, then $L_i F(A) = H_i(F(P))$.

The object M_m, which obviously depends on the choices made, is called the mth *syzgy* of A. The word "syzygy" comes from astronomy, where it was originally used to describe the alignment of the Sun, Earth, and Moon.

Theorem 2.4.7 *Assume that $\mathcal{A}$ has enough projectives. Then for any right exact functor $F: \mathcal{A} \to \mathcal{B}$, the derived functors $L_n F$ form a universal δ-functor.*

Remark This result was first proven in [CE, III.5], but is commonly attributed to [Tohoku], where the term "universal δ-functor" first appeared.

Proof Suppose that T_* is a homological δ-functor and that $\varphi_0: T_0 \to F$ is given. We need to show that φ_0 admits a unique extension to a morphism $\varphi: T_* \to L_*F$ of δ-functors. Suppose inductively that $\varphi_i: T_i \to L_i F$ are already defined for $0 \leq i < n$, and that they commute with all the appropriate δ_i's. Given A in $\mathcal{A}$, select an exact sequence $0 \to K \to P \to A \to 0$ with P projective. Since $L_n F(P) = 0$, this yields a commutative diagram with exact rows:

$$
\begin{array}{ccccc}
T_n(A) & \xrightarrow{\delta_n} & T_{n-1}(K) & \longrightarrow & T_{n-1}(P) \\
& & \downarrow{\varphi_{n-1}} & & \downarrow{\varphi_{n-1}} \\
0 \longrightarrow L_n F(A) & \xrightarrow{\delta_n} & L_{n-1}F(K) & \longrightarrow & L_{n-1}F(P).
\end{array}
$$

A diagram chase reveals that there exists a *unique* map $\varphi_n(A)$ from $T_n(A)$ to $L_n F(A)$ commuting with the given δ_n's. We need to show that φ_n is a natural transformation commuting with all δ_n's for all short exact sequences.

To see that φ_n is a natural transformation, suppose given $f: A' \to A$ and an exact sequence $0 \to K' \to P' \to A' \to 0$ with P' projective. As P' is projective we can lift f to $g: P' \to P$, which induces a map $h: K' \to K$.

$$
\begin{array}{ccccccccc}
0 & \longrightarrow & K' & \longrightarrow & P' & \longrightarrow & A' & \longrightarrow & 0 \\
& & \downarrow{h} & & \downarrow{g} & & \downarrow{f} & & \\
0 & \longrightarrow & K & \longrightarrow & P & \longrightarrow & A & \longrightarrow & 0
\end{array}
$$

To see that φ_n commutes with f, we note that in the following diagram that each small quadrilateral commutes.

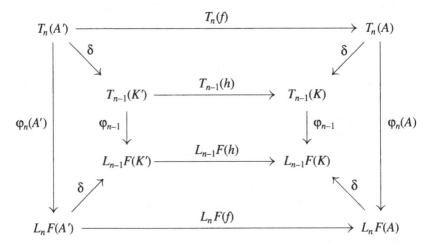

A chase reveals that

$$\delta \circ L_n(f) \circ \varphi_n(A') = \delta \circ \varphi_n(A) \circ T_n(f).$$

Because $\delta: L_n F(A) \to L_{n-1} F(K)$ is monic, we can cancel it from the equation to see that the outer square commutes, that is, that φ_n is a natural transformation. Incidentally, this argument (with $A = A'$ and $f = id_A$) also shows that $\varphi_n(A)$ doesn't depend on the choice of P.

Finally, we need to verify that φ_n commutes with δ_n. Given a short exact sequence $0 \to A' \to A \to A'' \to 0$ and a chosen exact sequence $0 \to K'' \to P'' \to A'' \to 0$ with P'' projective, we can construct maps f and g making the diagram

$$
\begin{array}{ccccccccc}
0 & \longrightarrow & K'' & \longrightarrow & P'' & \longrightarrow & A'' & \longrightarrow & 0 \\
& & \downarrow g & & \downarrow f & & \| & & \\
0 & \longrightarrow & A' & \longrightarrow & A & \longrightarrow & A'' & \longrightarrow & 0
\end{array}
$$

commute. This yields a commutative diagram

$$
\begin{array}{ccccc}
T_n(A'') & \xrightarrow{\delta} & T_{n-1}(K'') & \xrightarrow{T(g)} & T_{n-1}(A') \\
\varphi_n \downarrow & & \downarrow \varphi_{n-1} & & \downarrow \varphi_{n-1} \\
L_n F(A'') & \xrightarrow{\delta} & L_{n-1} F(K'') & \xrightarrow{LF(g)} & L_{n-1} F(A').
\end{array}
$$

Since the horizontal composites are the δ_n maps of the bottom row, this implies the desired commutativity relation. ◇

Exercise 2.4.4 Show that homology $H_*: \mathbf{Ch}_{\geq 0}(\mathcal{A}) \to \mathcal{A}$ and cohomology $H^*: \mathbf{Ch}^{\geq 0}(\mathcal{A}) \to \mathcal{A}$ are universal δ-functors. *Hint:* Copy the proof above, replacing P by the mapping cone cone(A) of exercise 1.5.1.

Exercise 2.4.5 ([Tohoku]) An additive functor $F: \mathcal{A} \to \mathcal{B}$ is called *effaceable* if for each object A of $\mathcal{A}$ there is a monomorphism $u: A \to I$ such that $F(u) = 0$. We call F *coeffaceable* if for every A there is a surjection $u: P \to A$ such that $F(u) = 0$. Modify the above proof to show that if T_* is a homological δ-functor such that each T_n is coeffaceable (except T_0), then T_* is universal. Dually, show that if T^* is a cohomological δ-functor such that each T^n is effaceable (except T^0), then T^* is universal.

2.5 Right Derived Functors

2.5.1 Let $F: \mathcal{A} \to \mathcal{B}$ be a left exact functor between two abelian categories. If $\mathcal{A}$ has enough injectives, we can construct the *right derived functors*

$R^i F(i \geq 0)$ of F as follows. If A is an object of $\mathcal{A}$, choose an injective resolution $A \to I^{\cdot}$ and define

$$R^i F(A) = H^i(F(I)).$$

Note that since $0 \to F(A) \to F(I^0) \to F(I^1)$ is exact, we always have $R^0 F(A) \cong F(A)$.

Since F also defines a right exact functor $F^{op}: \mathcal{A}^{op} \to \mathcal{B}^{op}$, and $\mathcal{A}^{op}$ has enough projectives, we can construct the left derived functors $L_i F^{op}$ as well. Since $I^{\cdot}$ becomes a projective resolution of A in $\mathcal{A}^{op}$, we see that

$$R^i F(A) = (L_i F^{op})^{op}(A).$$

Therefore all the results about right exact functors apply to left exact functors. In particular, the objects $R^i F(A)$ are independent of the choice of injective resolutions, $R^* F$ is a universal cohomological δ-functor, and $R^i F(I) = 0$ for $i \neq 0$ whenever I is injective. Calling an object Q *F-acyclic* if $R^i F(Q) = 0$ $(i \neq 0)$, as in 2.4.3, we see that the right derived functors of F can also be computed from F-acyclic resolutions.

Definition 2.5.2 (Ext functors) For each R-module A, the functor $F(B) = \operatorname{Hom}_R(A, B)$ is left exact. Its right derived functors are called the *Ext* groups:

$$\operatorname{Ext}^i_R(A, B) = R^i \operatorname{Hom}_R(A, -)(B).$$

In particular, $\operatorname{Ext}^0(A, B)$ is $\operatorname{Hom}(A, B)$, and injectives are characterized by Ext via the following exercise.

Exercise 2.5.1 Show that the following are equivalent.

1. B is an *injective* R-module.
2. $\operatorname{Hom}_R(-, B)$ is an exact functor.
3. $\operatorname{Ext}^i_R(A, B)$ vanishes for all $i \neq 0$ and all A (B is $\operatorname{Hom}_R(-, B)$-acyclic for all A).
4. $\operatorname{Ext}^1_R(A, B)$ vanishes for all A.

The behavior of Ext with respect to the variable A characterizes projectives.

Exercise 2.5.2 Show that the following are equivalent.

1. A is a *projective* R-module.
2. $\operatorname{Hom}_R(A, -)$ is an exact functor.
3. $\operatorname{Ext}^i_R(A, B)$ vanishes for all $i \neq 0$ and all B (A is $\operatorname{Hom}_R(-, B)$-acyclic for all B).
4. $\operatorname{Ext}^1_R(A, B)$ vanishes for all B.

The notion of derived functor has obvious variations for contravariant functors. For example, let F be a contravariant left exact functor from $\mathcal{A}$ to $\mathcal{B}$. This is the same as a covariant left exact functor from $\mathcal{A}^{op}$ to $\mathcal{B}$, so if $\mathcal{A}$ has enough projectives (i.e., $\mathcal{A}^{op}$ has enough injectives), we can define the right derived functors $R^*F(A)$ to be the cohomology of $F(P_{\cdot})$, $P_{\cdot} \to A$ being a projective resolution in $\mathcal{A}$. This too is a universal δ-functor with $R^0F(A) = F(A)$, and $R^iF(P) = 0$ for $i \neq 0$ whenever P is projective.

Example 2.5.3 For each R-module B, the functor $G(A) = \operatorname{Hom}_R(A, B)$ is contravariant and left exact. It is therefore entitled to right derived functors $R^*G(A)$. However, we will see in 2.7.6 that these are just the functors $\operatorname{Ext}^*(A, B)$. That is,

$$R^* \operatorname{Hom}(-, B)(A) \cong R^* \operatorname{Hom}(A, -)(B) = \operatorname{Ext}^*(A, B).$$

Application 2.5.4 Let X be a topological space. The *global sections* functor Γ from Sheaves(X) to **Ab** is the functor $\Gamma(\mathcal{F}) = \mathcal{F}(X)$. It turns out (see 2.6.1 and exercise 2.6.3 below) that Γ is right adjoint to the constant sheaves functor, so Γ is left exact. The right derived functors of Γ are the *cohomology functors* on X:

$$H^i(X, \mathcal{F}) = R^i\Gamma(\mathcal{F}).$$

The cohomology of a sheaf is arguably the central notion in modern algebraic geometry. For more details about sheaf cohomology, we refer the reader to [Hart].

Exercise 2.5.3 Let X be a topological space and $\{A_x\}$ any family of abelian groups, parametrized by the points $x \in X$. Show that the skyscraper sheaves $x_*(A_x)$ of 2.3.12 as well as their product $\mathcal{F} = \Pi x_*(A_x)$ are Γ-acyclic, that is, that $H^i(X, \mathcal{F}) = 0$ for $i \neq 0$. This shows that sheaf cohomology can also be computed from resolutions by products of skyscraper sheaves.

2.6 Adjoint Functors and Left/Right Exactness

We begin with a useful trick for constructing left and right exact functors.

Theorem 2.6.1 *Let $L: \mathcal{A} \to \mathcal{B}$ and $R: \mathcal{B} \to \mathcal{A}$ be an adjoint pair of additive functors. That is, there is a natural isomorphism*

$$\tau: \operatorname{Hom}_\mathcal{B}(L(A), B) \xrightarrow{\cong} \operatorname{Hom}_\mathcal{A}(A, R(B)).$$

Then L is right exact, and R is left exact.

Proof Suppose that $0 \to B' \to B \to B'' \to 0$ is exact in $\mathcal{B}$. By naturality of τ there is a commutative diagram for every A in $\mathcal{A}$.

$$
\begin{array}{ccccccc}
0 & \longrightarrow & \operatorname{Hom}_{\mathcal{B}}(L(A), B') & \longrightarrow & \operatorname{Hom}_{\mathcal{B}}(L(A), B) & \longrightarrow & \operatorname{Hom}_{\mathcal{B}}(L(A), B'') \\
& & \downarrow{\cong} & & \downarrow{\cong} & & \downarrow{\cong} \\
0 & \longrightarrow & \operatorname{Hom}_{\mathcal{A}}(A, R(B')) & \longrightarrow & \operatorname{Hom}_{\mathcal{A}}(A, R(B)) & \longrightarrow & \operatorname{Hom}_{\mathcal{A}}(A, R(B''))
\end{array}
$$

The top row is exact because $\operatorname{Hom}(LA, -)$ is left exact, so the bottom row is exact for all A. By the Yoneda Lemma 1.6.11,

$$0 \to R(B') \to R(B) \to R(B'')$$

must be exact. This proves that every right adjoint R is left exact. In particular $L^{op}: \mathcal{A}^{op} \to \mathcal{B}^{op}$ (which is a right adjoint) is left exact, that is, L is right exact.
$\diamond$

Remark Left adjoints have left derived functors, and right adjoints have right derived functors. This of course assumes that $\mathcal{A}$ has enough projectives, and that $\mathcal{B}$ has enough injectives for the derived functors to be defined.

Application 2.6.2 Let R be a ring and B a left R-module. The following standard proposition shows that $\otimes_R B: \mathbf{mod}\text{-}R \to \mathbf{Ab}$ is left adjoint to $\operatorname{Hom}_{\mathbf{Ab}}(B, -)$, so $\otimes_R B$ is right exact. More generally, if S is another ring, and B is an $R\text{-}S$ bimodule, then $\otimes_R B$ takes $\mathbf{mod}\text{-}R$ to $\mathbf{mod}\text{-}S$ and is a left adjoint, so it is right exact.

Proposition 2.6.3 *If B is an $R\text{-}S$ bimodule and C a right S-module, then $\operatorname{Hom}_S(B, C)$ is naturally a right R-module by the rule $(fr)(b) = f(rb)$ for $f \in \operatorname{Hom}(B, C)$, $r \in R$ and $b \in B$. The functor $\operatorname{Hom}_S(B, -)$ from $\mathbf{mod}\text{-}S$ to $\mathbf{mod}\text{-}R$ is right adjoint to $\otimes_R B$. That is, for every R-module A and S-module C there is a natural isomorphism*

$$\tau: \operatorname{Hom}_S(A \otimes_R B, C) \xrightarrow{\cong} \operatorname{Hom}_R(A, \operatorname{Hom}_S(B, C)).$$

Proof Given $f: A \otimes_R B \to C$, we define $(\tau f)(a)$ as the map $b \mapsto f(a \otimes b)$ for each $a \in A$. Given $g: A \to \operatorname{Hom}_S(B, C)$, we define $\tau^{-1}(g)$ to be the map defined by the bilinear form $a \otimes b \mapsto g(a)(b)$. We leave the verification that

$\tau(f)(a)$ is an S-module map, that $\tau(f)$ is an R-module map, $\tau^{-1}(g)$ is an R-module map, τ is an isomorphism with inverse τ^{-1}, and that τ is natural as an exercise for the reader.   ◇

Definition 2.6.4 Let B be a left R-module, so that $T(A) = A \otimes_R B$ is a right exact functor from **mod–R** to **Ab**. We define the abelian groups

$$\operatorname{Tor}_n^R(A, B) = (L_n T)(A).$$

In particular, $\operatorname{Tor}_0^R(A, B) \cong A \otimes_R B$. Recall that these groups are computed by finding a projective resolution $P \to A$ and taking the homology of $P \otimes_R B$. In particular, if A is a projective R-module, then $\operatorname{Tor}_n(A, B) = 0$ for $n \neq 0$.

More generally, if B is an R–S bimodule, we can think of $T(A) = A \otimes_R B$ as a right exact functor landing in **mod–S**, so we can think of the $\operatorname{Tor}_n^R(A, B)$ as S-modules. Since the forgetful functor U from **mod–S** to **Ab** is exact, this generalization does not change the underlying abelian groups, it merely adds an S-module structure, because $U(L_* \otimes B) \cong L_* U(\otimes B)$ as derived functors.

The reader may notice that the functor $A \otimes_R$ is also right exact, so we could also form the derived functors $L_*(A \otimes_R)$. We will see in section 2.7 that this yields nothing new in the sense that $L_*(A \otimes_R)(B) \cong L_*(\otimes_R B)(A)$.

Application 2.6.5 Now we see why the inclusion "incl" of Sheaves(X) into Presheaves(X) is a left exact functor, as claimed in 1.6.7; it is the right adjoint to the sheafification functor. The fact that sheafification is right exact is automatic; it is a theorem that sheafification is exact.

Exercise 2.6.1 Show that the derived functor $R^i(\text{incl})$ sends a sheaf $\mathcal{F}$ to the presheaf $U \mapsto H^i(U, \mathcal{F}|U)$, where $\mathcal{F}|U$ is the restriction of $\mathcal{F}$ to U and H^i is the sheaf cohomology of 2.5.4. *Hint:* Compose $R^i(\text{incl})$ with the exact functors Presheaves$(X) \to$ **Ab** sending $\mathcal{F}$ to $\mathcal{F}(U)$.

Application 2.6.6 Let $f: X \to Y$ be a continuous map of topological spaces. For any sheaf $\mathcal{F}$ on X, we define the *direct image sheaf* $f_*\mathcal{F}$ on Y by $(f_*\mathcal{F})(V) = \mathcal{F}(f^{-1}V)$ for every open V in Y. (*Exercise:* Show that $f_*\mathcal{F}$ is a sheaf!) For any sheaf $\mathcal{G}$ on Y, we define the *inverse image sheaf* $f^{-1}\mathcal{G}$ to be the sheafification of the presheaf sending an open set U in X to the direct limit $\varinjlim \mathcal{G}(V)$ over the poset of all open sets V in Y containing $f(U)$. The following exercise shows that f^{-1} is right exact and that f_* is left exact because they are adjoint. The derived functors $R^i f_*$ are called the *higher direct image sheaf functors* and also play a key role in algebraic geometry. (See [Hart] for more details.)

Exercise 2.6.2 Show that for any sheaf $\mathcal{F}$ on X there is a natural map $f^{-1}f_*\mathcal{F} \to \mathcal{F}$, and that for any sheaf $\mathcal{G}$ on Y there is a natural map $\mathcal{G} \to f_*f^{-1}\mathcal{G}$. Conclude that f^{-1} and f_* are adjoint to each other, that is, that there is a natural isomorphism

$$\mathrm{Hom}_X(f^{-1}\mathcal{G}, \mathcal{F}) \cong \mathrm{Hom}_Y(\mathcal{G}, f_*\mathcal{F}).$$

Exercise 2.6.3 Let $*$ denote the one-point space, so that $\mathrm{Sheaves}(*) \cong \mathbf{Ab}$.

1. If $f: X \to *$ is the collapse map, show that f_* and f^{-1} are the global sections functor Γ and the constant sheaves functor, respectively. This proves that Γ is right adjoint to the constant sheaves functor. By 2.6.1, Γ is left exact, as asserted in 2.5.4.
2. If $x: * \to X$ is the inclusion of a point in X, show that x_* and x^{-1} are the skyscraper sheaf and stalk functors of 2.3.12.

Application 2.6.7 (Colimits) Let I be a fixed category. There is a diagonal functor Δ from every category $\mathcal{A}$ to the functor category $\mathcal{A}^I$; if $A \in \mathcal{A}$, then ΔA is the constant functor: $(\Delta A)_i = A$ for all i. Recall that the *colimit* of a functor $F: I \to \mathcal{A}$ is an object of $\mathcal{A}$, written $\mathrm{colim}_{i \in I} F_i$, together with a natural transformation from F to $\Delta(\mathrm{colim}\ F_i)$, which is universal among natural transformations $F \to \Delta A$ with $A \in \mathcal{A}$. (See the appendix or [MacCW, III.3].) This universal property implies that colim is a functor from $\mathcal{A}^I$ to $\mathcal{A}$, at least when the colimit exists for all $F: I \to \mathcal{A}$.

Exercise 2.6.4 Show that colim is left adjoint to Δ. Conclude that colim is a right exact functor when $\mathcal{A}$ is abelian (and colim exists). Show that pushout (the colimit when I is $\cdot \leftarrow \cdot \to \cdot$) is not an exact functor in $\mathbf{Ab}$.

Proposition 2.6.8 *The following are equivalent for an abelian category $\mathcal{A}$:*

1. *The direct sum $\oplus A_i$ exists in $\mathcal{A}$ for every set $\{A_i\}$ of objects in $\mathcal{A}$.*
2. *$\mathcal{A}$ is cocomplete, that is, $\mathrm{colim}_{i \in I} A_i$ exists in $\mathcal{A}$ for each functor $A: I \to \mathcal{A}$ whose indexing category I has only a set of objects.*

Proof As (1) is a special case of (2), we assume (1) and prove (2). Given $A: I \to \mathcal{A}$, the cokernel C of

$$\bigoplus_{\varphi: i \to j} A_i \longrightarrow \bigoplus_{i \in I} A_i$$

$$a_i[\varphi] \mapsto \varphi(a_i) - a_i$$

solves the universal problem defining the colimit, so $C = \mathrm{colim}_{i \in I} A_i$. $\diamond$

Remark **Ab**, **mod**–*R*, Presheaves(*X*), and Sheaves(*X*) are cocomplete because (1) holds. (If *I* is infinite, the direct sum in Sheaves(*X*) is the sheafification of the direct sum in Presheaves(*X*)). The category of finite abelian groups has only *finite* direct sums, so it is not cocomplete.

Variation 2.6.9 (Limits) The limit of a functor $A: I \to \mathcal{A}$ is the colimit of the corresponding functor $A^{op}: I^{op} \to \mathcal{A}^{op}$, so all the above remarks apply in dual form to limits. In particular, $\lim: \mathcal{A}^I \to \mathcal{A}$ is right adjoint to the diagonal functor Δ, so lim is a left exact functor when it exists. If the product ΠA_i of every set $\{A_i\}$ of objects exists in $\mathcal{A}$, then $\mathcal{A}$ is *complete*, that is, $\lim_{i \in I} A_i$ exists for every $A: I \to \mathcal{A}$ with *I* having only a set of objects. **Ab**, **mod**–*R*, Presheaves(*X*), and Sheaves(*X*) are complete because such products exist.

One of the most useful properties of adjoint functors is the following result, which we quote without proof from [MacCW, V.5].

Adjoints and Limits Theorem 2.6.10 *Let* $L: \mathcal{A} \to \mathcal{B}$ *be left adjoint to a functor* $R: \mathcal{B} \to \mathcal{A}$*, where* $\mathcal{A}$ *and* $\mathcal{B}$ *are arbitrary categories. Then*

1. *L preserves all colimits (coproducts, direct limits, cokernels, etc.). That is, if* $A: I \to \mathcal{A}$ *has a colimit, then so does* $LA: I \to \mathcal{B}$*, and*

$$L(\operatorname*{colim}_{i \in I} A_i) = \operatorname*{colim}_{i \in I} L(A_i).$$

2. *R preserves all limits (products, inverse limits, kernels, etc.). That is, if* $B: I \to \mathcal{B}$ *has a limit, then so does* $RB: I \to \mathcal{A}$*, and*

$$R(\operatorname*{lim}_{i \in I} B_i) = \operatorname*{lim}_{i \in I} R(B_i).$$

Here are two consequences that use the fact that homology commutes with arbitrary direct sums of chain complexes. (Homology does not commute with arbitrary colimits; the derived functors of colim intervene via a spectral sequence.)

Corollary 2.6.11 *If a cocomplete abelian category* $\mathcal{A}$ *has enough projectives, and* $F: \mathcal{A} \to \mathcal{B}$ *is a left adjoint, then for every set* $\{A_i\}$ *of objects in* $\mathcal{A}$:

$$L_* F \left(\bigoplus_{i \in I} A_i \right) \cong \bigoplus_{i \in I} L_* F(A_i).$$

Proof If $P_i \to A_i$ are projective resolutions, then so is $\oplus P_i \to \oplus A_i$. Hence

$$L_* F(\oplus A_i) = H_*(F(\oplus P_i)) \cong H_*(\oplus F(P_i)) \cong \oplus H_*(F(P_i)) = \oplus L_* F(A_i). \quad \diamond$$

Corollary 2.6.12 $\mathrm{Tor}_*(A, \oplus_{i \in I} B_i) = \oplus_{i \in I} \mathrm{Tor}_*(A, B_i)$.

Proof If $P \to A$ is a projective resolution, then

$$\mathrm{Tor}_*(A, \oplus B_i) = H_*(P \otimes (\oplus B_i)) \cong H_*(\oplus(P \otimes B_i)) \cong \oplus H_*(P \otimes B_i)$$
$$= \oplus Tor_*(A, B_i). \qquad \qquad \diamond$$

Definition 2.6.13 A nonempty category I is called *filtered* if

1. For every $i, j \in I$ there are arrows $\begin{smallmatrix} i \\ j \end{smallmatrix} \searrow\!\!\!\!\!\nearrow k$ to some $k \in I$.
2. For every two parallel arrows $u, v: i \rightrightarrows j$ there is an arrow $w: j \to k$ such that $wu = wv$.

A *filtered colimit* in $\mathcal{A}$ is just the colimit of a functor $A: I \to \mathcal{A}$ in which I is a filtered category. We shall use the notation $\underrightarrow{\mathrm{colim}}(A_i)$ for such a filtered colimit.

If I is a partially ordered set (poset), considered as a category, then condition (1) always holds, and (2) just requires that every pair of elements has an upper bound in I. A filtered poset is often called *directed;* filtered colimits over directed posets are often called *direct limits* and are often written $\varinjlim A_i$.

We are going to show that direct limits and filtered colimits of modules are exact. First we obtain a more concrete description of the elements of $\underrightarrow{\mathrm{colim}}(A_i)$.

Lemma 2.6.14 *Let I be a filtered category and $A: I \to$ **mod**–R a functor. Then*

1. *Every element $a \in \underrightarrow{\mathrm{colim}}(A_i)$ is the image of some element $a_i \in A_i$ (for some $i \in I$) under the canonical map $A_i \to \underrightarrow{\mathrm{colim}}(A_i)$.*
2. *For every i, the kernel of the canonical map $A_i \to \underrightarrow{\mathrm{colim}}(A_i)$ is the union of the kernels of the maps $\varphi: A_i \to A_j$ (where $\varphi: i \to j$ is a map in I).*

Proof We shall use the explicit construction of $\underrightarrow{\mathrm{colim}}(A_i)$. Let $\lambda_i: A_i \to \oplus_{i \in I} A_i$ be the canonical maps. Every element a of $\underrightarrow{\mathrm{colim}} A_i$ is the image of

$$\sum_{j \in J} \lambda_j(a_j)$$

for some finite set $J = \{i_1, \cdots, i_n\}$. There is an upper bound i in I for

$\{i_1, \cdots, i_n\}$; using the maps $A_j \to A_i$ we can represent each a_j as an element in A_i and take a_i to be their sum. Evidently, a is the image of a_i, so (1) holds.

Now suppose that $a_i \in A_i$ vanishes in $\underrightarrow{\text{colim}}(A_i)$. Then there are $\varphi_{jk}: j \to k$ in I and $a_{jk} \in A_j$ so that $\lambda_i(a_i) = \sum \lambda_k(\varphi_{jk}(a_j)) - \lambda_j(a_j)$ in $\oplus A_i$. Choose an upper bound t in I for all the i, j, k in this expression. Adding $\lambda_t(\varphi_{it}a_i) - \lambda_i(a_i)$ to both sides we may assume that $i = t$. Adding zero terms of the form

$$[\lambda_t \varphi_{jt}(a_j) - \lambda_k \varphi_{jk}(a_j)] + [\lambda_t \varphi_{jt}(-a_j) - \lambda_k \varphi_{jk}(-a_j)],$$

we can assume that all the k's are t. If any φ_{jt} are parallel arrows in I, then by changing t we can equalize them. Therefore we have

$$\lambda_t(a_t) = \lambda_t(\sum \varphi_{jt}(a_j)) - \sum \lambda_j(a_j)$$

with all the j's distinct and none equal to t. Since the λ_j are injections, all the a_j must be zero. Hence $\varphi_{it}(a_i) = a_t = 0$, that is, $a_i \in \ker(\varphi_{it})$. $\diamond$

Theorem 2.6.15 *Filtered colimits (and direct limits) of R-modules are exact, considered as functors from* (**mod–R**)^I *to* **mod–R**.

Proof Set $\mathcal{A} = $ **mod–R**. We have to show that if I is a filtered category (*e.g.*, a directed poset), then $\underrightarrow{\text{colim}}: \mathcal{A}^I \to \mathcal{A}$ is exact. Exercise 2.6.4 showed that $\underrightarrow{\text{colim}}$ is right exact, so we need only prove that if $t: A \to B$ is monic in $\mathcal{A}^I$(*i.e.*, each t_i is monic), then $\underrightarrow{\text{colim}}(A_i) \to \underrightarrow{\text{colim}}(B_i)$ is monic in $\mathcal{A}$. Let $a \in \underrightarrow{\text{colim}}(A_i)$ be an element that vanishes in $\underrightarrow{\text{colim}}(B_i)$. By the lemma above, a is the image of some $a_i \in A_i$. Therefore $t_i(a_i) \in B_i$ vanishes in $\underrightarrow{\text{colim}}(B_i)$, so there is some $\varphi: i \to j$ so that

$$0 = \varphi(t_i(a_i)) = t_j(\varphi(a_i)) \text{ in } B_j.$$

Since t_j is monic, $\varphi(a_i) = 0$ in A_j. Hence $a = 0$ in $\underrightarrow{\text{colim}}(A_i)$. $\diamond$

Exercise 2.6.5 (AB5) The above theorem does not hold for every cocomplete abelian category $\mathcal{A}$. Show that if $\mathcal{A}$ is the opposite category **Ab**^{op} of abelian groups, then the functor $\underrightarrow{\text{colim}}: \mathcal{A}^I \to \mathcal{A}$ need not be exact when I is filtered.

An abelian category $\mathcal{A}$ is said to satisfy axiom (AB5) if it is cocomplete and filtered colimits are exact. Thus the above theorem states that **mod–R** and **R–mod** satisfy axiom (AB5), and this exercise shows that **Ab**^{op} does not.

Exercise 2.6.6 Let $f: X \to Y$ be a continuous map. Show that the inverse image sheaf functor f^{-1}: Sheaves$(Y) \to$ Sheaves(X) is exact. (See 2.6.6.)

The following consequences are proven in the same manner as their counterparts for direct sum. Note that in categories like R–**mod** for which filtered colimits are exact, homology commutes with filtered colimits.

Corollary 2.6.16 *If $\mathcal{A} = R$–**mod** (or $\mathcal{A}$ is any abelian category with enough projectives, satisfying axiom (AB5)), and $F: \mathcal{A} \to \mathcal{B}$ is a left adjoint, then for every $A: I \to \mathcal{A}$ with I filtered*

$$L_* F(\operatorname*{colim}_{\longrightarrow}(A_i)) \cong \operatorname*{colim}_{\longrightarrow} L_* F(A).$$

Corollary 2.6.17 *For every filtered $B: I \to R$–**mod** and every $A \in$ **mod**–R,*

$$\operatorname{Tor}_*(A, \operatorname*{colim}_{\longrightarrow}(B_i)) \cong \operatorname*{colim}_{\longrightarrow} \operatorname{Tor}_*(A, B_i).$$

2.7 Balancing Tor and Ext

In earlier sections we promised to show that the two left derived functors of $A \otimes_R B$ gave the same result and that the two right derived functors of Hom(A, B) gave the same result. It is time to deliver on these promises.

Tensor Product of Complexes 2.7.1 Suppose that P and Q are chain complexes of right and left R-modules, respectively. Form the double complex $P \otimes_R Q = \{P_p \otimes_R Q_q\}$ using the sign trick, that is, with horizontal differentials $d \otimes 1$ and vertical differentials $(-1)^p \otimes d$. $P \otimes_R Q$ is called the *tensor product double complex*, and Tot$^{\oplus}(P \otimes_R Q)$ is called the *(total) tensor product chain complex* of P and Q.

Theorem 2.7.2 $L_n(A \otimes_R)(B) \cong L_n(\otimes_R B)(A) = \operatorname{Tor}_n^R(A, B)$ *for all* n.

Proof Choose a projective resolution $P \xrightarrow{\epsilon} A$ in **mod**–R and a projective resolution $Q \xrightarrow{\eta} B$ in R–**mod**. Thinking of A and B as complexes concentrated in degree zero, we can form the three tensor product double complexes $P \otimes Q$, $A \otimes Q$, and $P \otimes B$. The augmentations ϵ and η induce maps from $P \otimes Q$ to $A \otimes Q$ and $P \otimes B$.

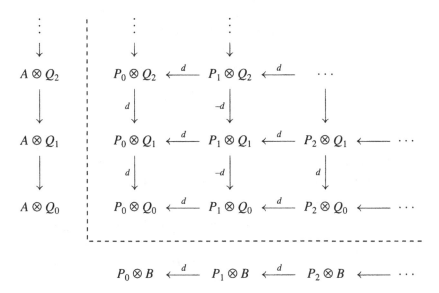

Using the Acyclic Assembly Lemma 2.7.3, we will show that the maps

$$A \otimes Q = \mathrm{Tot}(A \otimes Q) \xleftarrow{\epsilon \otimes Q} \mathrm{Tot}(P \otimes Q) \xrightarrow{P \otimes \eta} \mathrm{Tot}(P \otimes B) = P \otimes B$$

are quasi-isomorphisms, inducing the promised isomorphisms on homology:

$$L_*(A\otimes_R)(B) \xleftarrow{\cong} H_*(\mathrm{Tot}(P \otimes Q)) \xrightarrow{\cong} L_*(\otimes_R B)(A).$$

Consider the double complex C obtained from $P \otimes Q$ by adding $A \otimes Q[-1]$ in the column $p = -1$. The translate $\mathrm{Tot}(C)[1]$ is the mapping cone of the map $\epsilon \otimes Q$ from $\mathrm{Tot}(P \otimes Q)$ to $A \otimes Q$ (see 1.2.8 and 1.5.1), so in order to show that $\epsilon \otimes Q$ is a quasi-isomorphism, it suffices to show that $\mathrm{Tot}(C)$ is acyclic. Since each $\otimes Q_q$ is an exact functor, every row of C is exact, so $\mathrm{Tot}(C)$ is exact by the Acyclic Assembly Lemma.

Similarly, the mapping cone of $P \otimes \eta: \mathrm{Tot}(P \otimes Q) \to P \otimes B$ is the translate $\mathrm{Tot}(D)[1]$, where D is the double complex obtained from $P \otimes Q$ by adding $P \otimes B[-1]$ in the row $q = -1$. Since each $P_p\otimes$ is an exact functor, every column of D is exact, so $\mathrm{Tot}(D)$ is exact by the Acyclic Assembly Lemma 2.7.3. Hence $\mathrm{cone}(P \otimes \eta)$ is acyclic, and $P \otimes \eta$ is also a quasi-isomorphism.

$\diamond$

Acyclic Assembly Lemma 2.7.3 *Let C be a double complex in* **mod–R**. *Then*

- $\text{Tot}^{\Pi}(C)$ *is an acyclic chain complex, assuming either of the following:*
    1. *C is an upper half-plane complex with exact columns.*
    2. *C is a right half-plane complex with exact rows.*
- $\text{Tot}^{\oplus}(C)$ *is an acyclic chain complex, assuming either of the following:*
    3. *C is an upper half-plane complex with exact rows.*
    4. *C is a right half-plane complex with exact columns.*

Remark The proof will show that in (1) and (3) it suffices to have every diagonal bounded on the lower right, and in (2) and (4) it suffices to have every diagonal bounded on the upper left. See 5.5.1 and 5.5.10.

Proof We first show that it suffices to establish case (1). Interchanging rows and columns also interchanges (1) and (2), and (3) and (4), so (1) implies (2) and (4) implies (3). Suppose we are in case (4), and let $\tau_n C$ be the double subcomplex of C obtained by truncating each column at level n:

$$(\tau_n C)_{pq} = \begin{cases} C_{pq} & \text{if } q > n \\ \ker(d^v \colon C_{pn} \to C_{p,n-1}) & \text{if } q = n \\ 0 & \text{if } q < n \end{cases}.$$

Each $\tau_n C$ is, up to vertical translation, a first quadrant double complex with exact columns, so (1) implies that $\text{Tot}^{\oplus}(\tau_n C) = \text{Tot}^{\Pi}(\tau_n C)$ is acyclic. This implies that $\text{Tot}^{\oplus}(C)$ is acyclic, because every cycle of $\text{Tot}^{\oplus}(C)$ is a cycle (hence a boundary) in some subcomplex $\text{Tot}^{\oplus}(\tau_n C)$. Therefore (1) implies (4) as well.

In case (1), translating C left and right, suffices to prove that $H_0(\text{Tot}(C))$ is zero. Let

$$c = (\cdots, c_{-p,p}, \cdots, c_{-2,2}, c_{-1,1}, c_{0,0}) \in \prod C_{-p,p} = \text{Tot}(C)_0$$

be a 0-cycle; we will find elements $b_{-p,p+1}$ by induction on p so that

$$d^v(b_{-p,p+1}) + d^h(b_{-p+1,p}) = c_{-p,p}.$$

Assembling the b's will yield an element b of $\prod C_{-p,p+1}$ such that $d(b) = c$, proving that $H_0(\text{Tot}(C)) = 0$. The following schematic should help give the idea.

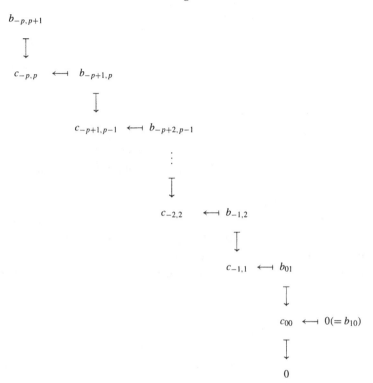

We begin the induction by choosing $b_{10} = 0$ for $p = -1$. Since $C_{0,-1} = 0$, $d^v(c_{00}) = 0$; since the 0th column is exact, there is a $b_{01} \in C_{01}$ so that $d^v(b_{01}) = c_{00}$. Inductively, we compute that

$$d^v(c_{-p,p} - d^h(b_{-p+1,p})) = d^v(c_{-pp}) + d^h d^v(b_{-p+1,p})$$
$$= d^v(c_{-pp}) + d^h(c_{-p+1,p-1}) - d^h d^h(b_{-p+2,p-1})$$
$$= 0.$$

Since the $-p^{th}$ column is exact, there is a $b_{-p,p+1}$ so that

$$d^v(b_{-p,p+1}) = c_{-p,p} - d^h(b_{-p+1,p})$$

as desired. ◇

Exercise 2.7.1 Let C be the periodic upper half-plane complex with $C_{pq} = \mathbb{Z}/4$ for all p and $q \geq 0$, all differentials being multiplication by 2.

$$\downarrow 2 \qquad \downarrow 2 \qquad \downarrow 2$$

$$\cdots \xleftarrow{2} \mathbb{Z}/4 \xleftarrow{2} \mathbb{Z}/4 \xleftarrow{2} \mathbb{Z}/4 \xleftarrow{2} \cdots$$

$$\downarrow 2 \qquad \downarrow 2 \qquad \downarrow 2$$

$$\cdots \xleftarrow{2} \mathbb{Z}/4 \xleftarrow{2} \mathbb{Z}/4 \xleftarrow{2} \mathbb{Z}/4 \xleftarrow{2} \cdots$$

1. Show that $H_0(\mathrm{Tot}^\Pi(C)) \cong \mathbb{Z}/2$ on the cycle $(\ldots, 1, 1, 1) \in \prod C_{-p,p}$ even though the rows of C are exact. *Hint:* First show that the 0-boundaries are $\prod 2\mathbb{Z}/4$.
2. Show that $\mathrm{Tot}^\oplus(C)$ is acyclic.
3. Now extend C downward to form a doubly periodic plane double complex D with $D_{pq} = \mathbb{Z}/4$ for all $p, q \in \mathbb{Z}$. Show that $H_0(\mathrm{Tot}^\Pi(D))$ maps onto $H_0(\mathrm{Tot}^\Pi C) \cong \mathbb{Z}/2$. Hence $\mathrm{Tot}^\Pi(D)$ is not acyclic, even though every row and column of D is exact. Finally, show that $\mathrm{Tot}^\oplus(D)$ is acyclic.

Exercise 2.7.2

1. Give an example of a 2nd quadrant double chain complex C with exact columns for which $\mathrm{Tot}^\oplus(C)$ is not an acyclic chain complex.
2. Give an example of a 4th quadrant double complex C with exact columns for which $\mathrm{Tot}^\Pi(C)$ is not acyclic.

Hom Cochain Complex 2.7.4 Given a chain complex P and a cochain complex I, form the double cochain complex $\mathrm{Hom}(P, I) = \{\mathrm{Hom}(P_p, I^q)\}$ using a variant of the sign trick. That is, if $f: P_p \to I^q$, then $d^h f: P_{p+1} \to I^q$ by $(d^h f)(p) = f(dp)$, while we define $d^v f: P_p \to I^{q+1}$ by

$$(d^v f)(p) = (-1)^{p+q+1} d(fp) \quad \text{for } p \in P_p.$$

$\mathrm{Hom}(P, I)$ is called the *Hom double complex*, and $\mathrm{Tot}^\Pi(\mathrm{Hom}(P, I))$ is called the *(total) Hom cochain complex*. *Warning:* Different conventions abound in the literature. Bourbaki [BX] converts $\mathrm{Hom}(P, I)$ into a double chain complex and obtains a total Hom chain complex. Others convert I into a chain complex Q with $Q_q = I^{-q}$ and form $\mathrm{Hom}(P, Q)$ as a chain complex, and so on.

Morphisms and Hom 2.7.5 To explain our sign convention, suppose that C and D are two chain complexes. If we reindex D as a cochain complex, then an n-cycle f of $\mathrm{Hom}(C, D)$ is a sequence of maps $f_p: C_p \to D^{n-p} = D_{p-n}$

such that $f_p d = (-1)^n d f_{p+1}$, that is, a morphism of chain complexes from C to the translate $D[-n]$ of D. An n-boundary is a morphism f that is null homotopic. Thus $H^n \operatorname{Hom}(C, D)$ is the group of chain homotopy equivalence classes of morphisms $C \to D[-n]$, the morphisms in the quotient category **K** of the category of chain complexes discussed in exercise 1.4.5.

Similarly, if X and Y are cochain complexes, we may form $\operatorname{Hom}(X, Y)$ by reindexing X. Our conventions about reindexing and translation ensure that once again an n-cycle of $\operatorname{Hom}(X, Y)$ is a morphism $X \to Y[-n]$ and that $H^n \operatorname{Hom}(X, Y)$ is the group of chain homotopy equivalence classes of such morphisms. We will return to this point in Chapter 10 when we discuss **RHom** in the derived category $\mathbf{D}(\mathcal{A})$.

Exercise 2.7.3 To see why $\operatorname{Tot}^{\oplus}$ is used for the tensor product $P \otimes_R Q$ of right and left R-module complexes, while Tot^{Π} is used for Hom, let I be a cochain complex of abelian groups. Show that there is a natural isomorphism of double complexes:

$$\operatorname{Hom}_{\mathbf{Ab}}(\operatorname{Tot}^{\oplus}(P \otimes_R Q), I) \cong \operatorname{Hom}_R(P, \operatorname{Tot}^{\Pi}(\operatorname{Hom}_{\mathbf{Ab}}(Q, I))).$$

Theorem 2.7.6 *For every pair of R-modules A and B, and all n,*

$$\operatorname{Ext}_R^n(A, B) = R^n \operatorname{Hom}_R(A, -)(B) \cong R^n \operatorname{Hom}_R(-, B)(A).$$

Proof Choose a projective resolution P of A and an injective resolution I of B. Form the first quadrant double cochain complex $\operatorname{Hom}(P, I)$. The augmentations induce maps from $\operatorname{Hom}(A, I)$ and $\operatorname{Hom}(P, B)$ to $\operatorname{Hom}(P, I)$. As in the proof of 2.7.2, the mapping cones of $\operatorname{Hom}(A, I) \to \operatorname{Tot}(\operatorname{Hom}(P, I))$ and $\operatorname{Hom}(P, B) \to \operatorname{Tot}(\operatorname{Hom}(P, I))$ are translates of the total complexes obtained from $\operatorname{Hom}(P, I)$ by adding $\operatorname{Hom}(A, I)[-1]$ and $\operatorname{Hom}(P, B)[-1]$, respectively. By the Acyclic Assembly Lemma 2.7.3 (or rather its dual), both mapping cones are exact. Therefore the maps

$$\operatorname{Hom}(A, I) \to \operatorname{Tot}(\operatorname{Hom}(P, I)) \leftarrow \operatorname{Hom}(P, B)$$

are quasi-isomorphisms. Taking cohomology yields the result:

$$\begin{aligned} R^* \operatorname{Hom}(A, -)(B) &= H^* \operatorname{Hom}(A, I) \\ &\cong H^* \operatorname{Tot}(\operatorname{Hom}(P, I)) \\ &\cong H^* \operatorname{Hom}(P, B) = R^* \operatorname{Hom}(-, B)(A). \quad \diamond \end{aligned}$$

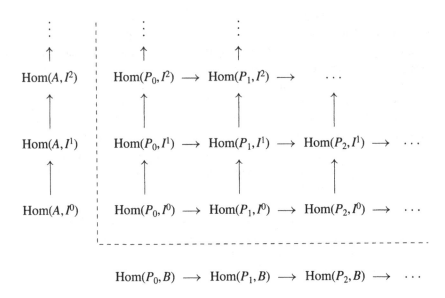

$$\text{Hom}(P_0,B) \ \longrightarrow \ \text{Hom}(P_1,B) \ \longrightarrow \ \text{Hom}(P_2,B) \ \longrightarrow \ \cdots$$

Definition 2.7.7 ([CE]) In view of the two above theorems, the following definition seems natural. Let T be a left exact functor of p "variable" modules, some covariant and some contravariant. T will be called *right balanced* under the following conditions:

1. When any one of the covariant variables of T is replaced by an injective module, T becomes an exact functor in each of the remaining variables.
2. When any one of the contravariant variables of T is replaced by a projective module, T becomes an exact functor in each of the remaining variables. The functor Hom is an example of a right balanced functor, as is $\text{Hom}(A \otimes B, C)$.

Exercise 2.7.4 Show that all p of the right derived functors $R^*T(A_1, \cdots, \hat{A}_i, \cdots, A_p)(A_i)$ of T are naturally isomorphic.

A similar discussion applies to right exact functors T which are *left balanced*. The prototype left balanced functor is $A \otimes B$. In particular, all of the left derived functors associated to a left balanced functor are isomorphic.

Application 2.7.8 (External product for Tor) Suppose that R is a commutative ring and that A, A', B, B' are R-modules. The *external product* is the map

$$\text{Tor}_i(A, B) \otimes_R \text{Tor}_j(A', B') \to \text{Tor}_{i+j}(A \otimes_R A', B \otimes_R B')$$

constructed for every i and j in the following manner. Choose projective resolutions $P \to A$, $P' \to A'$, and $P'' \to A \otimes A'$. The Comparison Theorem 2.2.6 gives a chain map $\mathrm{Tot}(P \otimes P') \to P''$ which is unique up to chain homotopy equivalence. (We saw above that $H_i\,\mathrm{Tot}(P \otimes P') = \mathrm{Tor}_i(A, A')$, so we actually need the version of the Comparison Theorem contained in the porism 2.2.7.) This yields a natural map

$$H_n(P \otimes B \otimes P' \otimes B') \cong H_n(P \otimes P' \otimes B \otimes B') \to H_n(P'' \otimes B \otimes B')$$

$$= \mathrm{Tor}_n(A \otimes A', B \otimes B').$$

On the other hand, there are natural maps $H_i(C) \otimes H_j(C') \to H_{i+j}\,\mathrm{Tot}(C \otimes C')$ for every pair of complexes C, C'; one maps the tensor product $c \otimes c'$ of cycles $c \in C_i$ and $c' \in C'_j$ to $c \otimes c' \in C_i \otimes C'_j$. (Check this!) The external product is obtained by composing the special case $C = P \otimes B$, $C' = P' \otimes B'$:

$$\mathrm{Tor}_i(A, B) \otimes \mathrm{Tor}_j(A', B') = H_i(P \otimes B) \otimes H_j(P' \otimes B') \to H_{i+j}(P \otimes B \otimes P' \otimes B')$$

with the above map.

Exercise 2.7.5

1. Show that the external product is independent of the choices of P, P', P'' and that it is natural in all four modules A, A', B, B'.
2. Show that the product is associative as a map to $\mathrm{Tor}_*(A \otimes A' \otimes A'', B \otimes B' \otimes B'')$.
3. Show that the external product commutes with the connecting homomorphism δ in the long exact Tor sequences associated to $0 \to B_0 \to B \to B_1 \to 0$.
4. (Internal product) Suppose that A and B are R-algebras. Use (1) and (2) to show that $\mathrm{Tor}_*^R(A, B)$ is a graded R-algebra.

3

Tor and Ext

3.1 Tor for Abelian Groups

The first question many people ask about $\mathrm{Tor}_*(A, B)$ is "Why the name 'Tor'?" The results of this section should answer that question. Historically, the first Tor groups to arise were the groups $\mathrm{Tor}_1(\mathbb{Z}/p, B)$ associated to abelian groups. The following simple calculation describes these groups.

Calculation 3.1.1 $\mathrm{Tor}_0^{\mathbb{Z}}(\mathbb{Z}/p, B) = B/pB$, $\mathrm{Tor}_1^{\mathbb{Z}}(\mathbb{Z}/p, B) = {}_pB = \{b \in B : pB = 0\}$ and $\mathrm{Tor}_n^{\mathbb{Z}}(\mathbb{Z}/p, B) = 0$ for $n \geq 2$. To see this, use the resolution

$$0 \to \mathbb{Z} \xrightarrow{p} \mathbb{Z} \to \mathbb{Z}/p \to 0$$

to see that $\mathrm{Tor}_*(\mathbb{Z}/p, B)$ is the homology of the complex $0 \to B \xrightarrow{p} B \to 0$.

Proposition 3.1.2 *For all abelian groups A and B:*

(a) $\mathrm{Tor}_1^{\mathbb{Z}}(A, B)$ *is a torsion abelian group.*
(b) $\mathrm{Tor}_n^{\mathbb{Z}}(A, B) = 0$ *for $n \geq 2$.*

Proof A is the direct limit of its finitely generated subgroups A_α, so by 2.6.17 $\mathrm{Tor}_n(A, B)$ is the direct limit of the $\mathrm{Tor}_n(A_\alpha, B)$. As the direct limit of torsion groups is a torsion group, we may assume that A is finitely generated, that is, $A \cong \mathbb{Z}^m \oplus \mathbb{Z}/p_1 \oplus \mathbb{Z}/p_2 \oplus \cdots \oplus \mathbb{Z}/p_r$ for appropriate integers $m, p_1, \ldots, p_r$. As $\mathbb{Z}^m$ is projective, $\mathrm{Tor}_n(\mathbb{Z}^m, -)$ vanishes for $n \neq 0$, and so we have

$$\mathrm{Tor}_n(A, B) \cong \mathrm{Tor}_n(\mathbb{Z}/p_1, B) \oplus \cdots \oplus \mathrm{Tor}_n(\mathbb{Z}/p_r, B).$$

The proposition holds in this case by calculation 3.1.1 above. $\diamond$

Proposition 3.1.3 $\text{Tor}_1^{\mathbb{Z}}(\mathbb{Q}/\mathbb{Z}, B)$ *is the torsion subgroup of B for every abelian group B.*

Proof As $\mathbb{Q}/\mathbb{Z}$ is the direct limit of its finite subgroups, each of which is isomorphic to $\mathbb{Z}/p$ for some integer p, and Tor commutes with direct limits,

$$\text{Tor}_*^{\mathbb{Z}}(\mathbb{Q}/\mathbb{Z}, B) \cong \varinjlim \text{Tor}_1^{\mathbb{Z}}(\mathbb{Z}/p, B) \cong \varinjlim({}_pB) = \cup_p\{b \in B : pb = 0\},$$

which is the torsion subgroup of B. $\diamond$

Proposition 3.1.4 *If A is a torsionfree abelian group, then $\text{Tor}_n^{\mathbb{Z}}(A, B) = 0$ for $n \neq 0$ and all abelian groups B.*

Proof A is the direct limit of its finitely generated subgroups, each of which is isomorphic to $\mathbb{Z}^m$ for some m. Therefore, $\text{Tor}_n(A, B) \cong \varinjlim \text{Tor}_n(\mathbb{Z}^m, B) = 0$. $\diamond$

Remark (Balancing Tor) If R is any commutative ring, then $\text{Tor}_*^R(A, B) \cong \text{Tor}_*^R(B, A)$. In particular, this is true for $R = \mathbb{Z}$, that is, for abelian groups. This is because for fixed B, both are universal δ-functors over $F(A) = A \otimes B \cong B \otimes A$. Therefore $\text{Tor}_1^{\mathbb{Z}}(A, \mathbb{Q}/\mathbb{Z})$ is the torsion subgroup of A. From this we obtain the following.

Corollary 3.1.5 *For every abelian group A,*

$$\text{Tor}_1^{\mathbb{Z}}(A, -) = 0 \Leftrightarrow A \text{ is torsionfree} \Leftrightarrow \text{Tor}_1^{\mathbb{Z}}(-, A) = 0.$$

Calculation 3.1.6 All this fails if we replace $\mathbb{Z}$ by another ring. For example, if we take $R = \mathbb{Z}/m$ and $A = \mathbb{Z}/d$ with $d|m$, then we can use the periodic free resolution

$$\cdots \xrightarrow{d} \mathbb{Z}/m \xrightarrow{m/d} \mathbb{Z}/m \xrightarrow{d} \mathbb{Z}/m \xrightarrow{\epsilon} \mathbb{Z}/d \to 0$$

to see that for all $\mathbb{Z}/m$-modules B we have

$$\text{Tor}_n^{\mathbb{Z}/m}(\mathbb{Z}/d, B) = \begin{cases} B/dB & \text{if } n = 0 \\ \{b \in B : db = 0\}/(m/d)B & \text{if } n \text{ is odd, } n > 0 \\ \{b \in B : (m/d)b = 0\}/dB & \text{if } n \text{ is even, } n > 0. \end{cases}$$

Example 3.1.7 Suppose that $r \in R$ is a left nonzerodivisor on R, that is, $_rR = \{s \in R : rs = 0\}$ is zero. For every R-module B, set $_rB = \{b \in B : rb = 0\}$. We can repeat the above calculation with R/rR in place of $\mathbb{Z}/p$ to see that $\mathrm{Tor}_0(R/rR, B) = B/rB$, $\mathrm{Tor}_1^R(R/rR, B) = {}_rB$ and $\mathrm{Tor}_n^R(R/rR, B) = 0$ for all B when $n \geq 2$.

Exercise 3.1.1 If $_rR \neq 0$, all we have is the non-projective resolution

$$0 \to {}_rR \to R \xrightarrow{r} R \to R/rR \to 0.$$

Show that there is a short exact sequence

$$0 \longrightarrow \mathrm{Tor}_2^R(R/rR, B) \longrightarrow {}_rR \otimes_R B \xrightarrow{\text{multiply}} {}_rB \longrightarrow \mathrm{Tor}_1^R(R/rR, B) \longrightarrow 0$$

and that $\mathrm{Tor}_n^R(R/rR, B) \cong \mathrm{Tor}_{n-2}^R({}_rR, B)$ for $n \geq 3$.

Exercise 3.1.2 Suppose that R is a commutative domain with field of fractions F. Show that $\mathrm{Tor}_1^R(F/R, B)$ is the torsion submodule $\{b \in B : (\exists r \neq 0)\ rb = 0\}$ of B for every R-module B.

Exercise 3.1.3 Show that $\mathrm{Tor}_1^R(R/I, R/J) \cong \frac{I \cap J}{IJ}$ for every right ideal I and left ideal J of R. In particular, $\mathrm{Tor}_1(R/I, R/I) \cong I/I^2$ for every 2-sided ideal I. *Hint:* Apply the Snake Lemma to

$$
\begin{array}{ccccccccc}
0 & \longrightarrow & IJ & \longrightarrow & I & \longrightarrow & I \otimes R/J & \longrightarrow & 0 \\
& & \downarrow & & \downarrow & & \downarrow & & \\
0 & \longrightarrow & J & \longrightarrow & R & \longrightarrow & R \otimes R/J & \longrightarrow & 0.
\end{array}
$$

3.2 Tor and Flatness

In the last chapter, we saw that if A is a right R-module and B is a left R-module, then $\mathrm{Tor}_*^R(A, B)$ may be computed either as the left derived functors of $A \otimes_R$ evaluated at B or as the left derived functors of $\otimes_R B$ evaluated at A. It follows that if either A or B is projective, then $\mathrm{Tor}_n(A, B) = 0$ for $n \neq 0$.

Definition 3.2.1 A left R-module B is *flat* if the functor $\otimes_R B$ is exact. Similarly, a right R-module A is *flat* if the functor $A \otimes_R$ is exact. The above remarks show that projective modules are flat. The example $R = \mathbb{Z}$, $B = \mathbb{Q}$ shows that flat modules need not be projective.

Theorem 3.2.2 *If S is a central multiplicatively closed set in a ring R, then $S^{-1}R$ is a flat R-module.*

Proof Form the filtered category I whose objects are the elements of S and whose morphisms are $\mathrm{Hom}_I(s_1, s_2) = \{s \in S : s_1 s = s_2\}$. Then $\mathrm{colim}\, F(s) \cong$ $S^{-1}R$ for the functor $F: I \to R\text{–}\mathbf{mod}$ defined by $F(s) = R$, $F(s_1 \xrightarrow{s} s_2)$ being multiplication by s. (*Exercise:* Show that the maps $F(s) \to S^{-1}R$ sending 1 to $1/s$ induce an isomorphism $\mathrm{colim}\, F(s) \cong S^{-1}R$.) Since $S^{-1}R$ is the filtered colimit of the free R-modules $F(s)$, it is flat by 2.6.17. $\diamond$

Exercise 3.2.1 Show that the following are equivalent for every left R-module B.

1. B is flat.
2. $\mathrm{Tor}_n^R(A, B) = 0$ for all $n \neq 0$ and all A.
3. $\mathrm{Tor}_1^R(A, B) = 0$ for all A.

Exercise 3.2.2 Show that if $0 \to A \to B \to C \to 0$ is exact and both B and C are flat, then A is also flat.

Exercise 3.2.3 We saw in the last section that if $R = \mathbb{Z}$ (or more generally, if R is a principal ideal domain), a module B is flat iff B is torsionfree. Here is an example of a torsionfree ideal I that is not a flat R-module. Let k be a field and set $R = k[x, y]$, $I = (x, y)R$. Show that $k = R/I$ has the projective resolution

$$0 \to R \xrightarrow{\begin{bmatrix} -y \\ x \end{bmatrix}} R^2 \xrightarrow{(x\ y)} R \to k \to 0.$$

Then compute that $\mathrm{Tor}_1^R(I, k) \cong \mathrm{Tor}_2^R(k, k) \cong k$, showing that I is not flat.

Definition 3.2.3 The *Pontrjagin dual* B^* of a left R-module B is the right R-module $\mathrm{Hom}_{\mathbf{Ab}}(B, \mathbb{Q}/\mathbb{Z})$; an element r of R acts via $(fr)(b) = f(rb)$.

Proposition 3.2.4 *The following are equivalent for every left R-module B :*

1. *B is a flat R-module.*
2. *B^* is an injective right R-module.*
3. *$I \otimes_R B \cong IB = \{x_1 b_1 + \cdots + x_n b_n \in B : x_i \in I, b_i \in B\} \subset B$ for every right ideal I of R.*
4. *$\mathrm{Tor}_1^R(R/I, B) = 0$ for every right ideal I of R.*

Proof The equivalence of (3) and (4) follows from the exact sequence

$$0 \to \operatorname{Tor}_1(R/I, B) \to I \otimes B \to B \to B/IB \to 0.$$

Now for every inclusion $A' \subset A$ of right modules, the adjoint functors $\otimes B$ and $\operatorname{Hom}(-, B)$ give a commutative diagram

$$\operatorname{Hom}(A, B^*) \qquad \longrightarrow \qquad \operatorname{Hom}(A', B^*)$$

$$\downarrow \cong \qquad\qquad\qquad \cong \downarrow$$

$$(A \otimes B)^* = \operatorname{Hom}(A \otimes B, \mathbb{Q}/\mathbb{Z}) \longrightarrow \operatorname{Hom}(A' \otimes B, \mathbb{Q}/\mathbb{Z}) = (A' \otimes B)^*.$$

Using the lemma below and Baer's criterion 2.3.1, we see that

B^* is injective $\Leftrightarrow (A \otimes B)^* \to (A' \otimes B)^*$ is surjective for all $A' \subset A$.

$\qquad \Leftrightarrow A' \otimes B \to A \otimes B$ is injective for all $A' \subset A \Leftrightarrow B$ is flat.

B^* is injective $\Leftrightarrow (R \otimes B)^* \to (I \otimes B)^*$ is surjective for all $I \subset R$

$\qquad \Leftrightarrow I \otimes B \to R \otimes B$ is injective for all I

$\qquad \Leftrightarrow I \otimes B \cong IB$ for all I. $\qquad\qquad\qquad\qquad \diamond$

Lemma 3.2.5 *A map* $f \colon B \to C$ *is injective iff the dual map* $f^* \colon C^* \to B^*$ *is surjective.*

Proof If A is the kernel of f, then A^* is the cokernel of f^*, because $\operatorname{Hom}(-, \mathbb{Q}/\mathbb{Z})$ is contravariant exact. But we saw in exercise 2.3.3 that $A = 0$ iff $A^* = 0$. $\qquad\qquad\qquad\qquad\qquad\qquad\qquad\qquad\qquad\qquad \diamond$

Exercise 3.2.4 Show that a sequence $A \to B \to C$ is exact iff its dual $C^* \to B^* \to A^*$ is exact.

An R-module M is called *finitely presented* if it can be presented using finitely many generators $(e_1, \ldots, e_n)$ and relations $(\sum \alpha_{ij} e_j = 0,\ j = 1, \ldots, m)$. That is, there is an $m \times n$ matrix α and an exact sequence $R^m \xrightarrow{\alpha} R^n \to M \to 0$. If M is finitely generated, the following exercise shows that the property of being finitely presented is independent of the choice of generators.

Exercise 3.2.5 Suppose that $\varphi \colon F \to M$ is *any* surjection, where F is finitely generated and M is finitely presented. Use the Snake Lemma to show that $\ker(\varphi)$ is finitely generated.

Still letting A^* denote the Pontrjagin dual 3.2.3 of A, there is a natural map $\sigma \colon A^* \otimes_R M \to \operatorname{Hom}_R(M, A)^*$ defined by $\sigma(f \otimes m) \colon h \mapsto f(h(m))$ for $f \in A^*$, $m \in M$ and $h \in \operatorname{Hom}(M, A)$. (*Exercise:* If $M = \oplus_{i=1}^{\infty} R$, show that σ is not an isomorphism.)

Lemma 3.2.6 *The map σ is an isomorphism for every finitely presented M and all A.*

Proof A simple calculation shows that σ is an isomorphism if $M = R$. By additivity, σ is an isomorphism if $M = R^m$ or R^n. Now consider the diagram

$$
\begin{array}{ccccccc}
A^* \otimes R^m & \longrightarrow & A^* \otimes R^n & \longrightarrow & A^* \otimes M & \longrightarrow & 0 \\
\sigma \downarrow \cong & & \sigma \downarrow \cong & & \sigma \downarrow & & \\
\mathrm{Hom}(R^m, A)^* & \xrightarrow{\alpha^*} & \mathrm{Hom}(R^n, A)^* & \longrightarrow & \mathrm{Hom}(M, A)^* & \longrightarrow & 0.
\end{array}
$$

The rows are exact because $\otimes$ is right exact, Hom is left exact, and Pontrjagin dual is exact by 2.3.3. The 5-lemma shows that σ is an isomorphism. $\diamond$

Theorem 3.2.7 *Every finitely presented flat R-module M is projective.*

Proof In order to show that M is projective, we shall show that $\mathrm{Hom}_R(M, -)$ is exact. To this end, suppose that we are given a surjection $B \to C$. Then $C^* \to B^*$ is an injection, so if M is flat, the top arrow of the square

$$
\begin{array}{ccc}
(C^*) \otimes_R M & \longrightarrow & (B^*) \otimes_R M \\
\cong \downarrow & & \cong \downarrow \\
\mathrm{Hom}(M, C)^* & \longrightarrow & \mathrm{Hom}(M, B)^*
\end{array}
$$

is an injection. Hence the bottom arrow is an injection. As we have seen, this implies that $\mathrm{Hom}(M, B) \to \mathrm{Hom}(M, C)$ is a surjection, as required. $\diamond$

Flat Resolution Lemma 3.2.8 *The groups $\mathrm{Tor}_*(A, B)$ may be computed using resolutions by flat modules. That is, if $F \to A$ is a resolution of A with the F_n being flat modules, then $\mathrm{Tor}_*(A, B) \cong H_*(F \otimes B)$. Similarly, if $F' \to B$ is a resolution of B by flat modules, then $\mathrm{Tor}_*(A, B) \cong H_*(A \otimes F')$.*

Proof We use induction and dimension shifting (exercise 2.4.3) to prove that $\mathrm{Tor}_n(A, B) \cong H_n(F \otimes B)$ for all n; the second part follows by arguing over R^{op}. The assertion is true for $n = 0$ because $\otimes B$ is right exact. Let K be such that $0 \to K \to F_0 \to A \to 0$ is exact; if $E = (\cdots \to F_2 \to F_1 \to 0)$, then $E \to K$ is a resolution of K by flat modules. For $n = 1$ we simply compute

$$
\mathrm{Tor}_1(A, B) = \ker(K \otimes B \to F_0 \otimes B)
$$

$$
= \ker \left\{ \frac{F_1 \otimes B}{\mathrm{im}(F_2 \otimes B)} \to F_0 \otimes B \right\} = H_1(F \otimes B).
$$

For $n \geq 2$ we use induction to see that

$$\mathrm{Tor}_n(A, B) \cong \mathrm{Tor}_{n-1}(K, B) \cong H_{n-1}(E \otimes B) = H_n(F \otimes B). \quad \diamond$$

Proposition 3.2.9 (Flat base change for Tor) *Suppose $R \to T$ is a ring map such that T is flat as an R-module. Then for all R-modules A, all T-modules C and all n*

$$\mathrm{Tor}_n^R(A, C) \cong \mathrm{Tor}_n^T(A \otimes_R T, C).$$

Proof Choose an R-module projective resolution $P \to A$. Then $\mathrm{Tor}_*^R(A, C)$ is the homology of $P \otimes_R C$. Since T is R-flat, and each $P_n \otimes_R T$ is a projective T-module, $P \otimes T \to A \otimes T$ is a T-module projective *resolution*. Thus $\mathrm{Tor}_*^T(A \otimes_R T, C)$ is the homology of the complex $(P \otimes_R T) \otimes_T C \cong P \otimes_R C$ as well. $\diamond$

Corollary 3.2.10 *If R is commutative and T is a flat R-algebra, then for all R-modules A and B, and for all n*

$$T \otimes_R \mathrm{Tor}_n^R(A, B) \cong \mathrm{Tor}_n^T(A \otimes_R T, T \otimes_R B).$$

Proof Setting $C = T \otimes_R B$, it is enough to show that $\mathrm{Tor}_*^R(A, T \otimes B) = T \otimes \mathrm{Tor}_*^R(A, B)$. As $T \otimes_R$ is an exact functor, $T \otimes \mathrm{Tor}_*^R(A, B)$ is the homology of $T \otimes_R (P \otimes_R B) \cong P \otimes_R (T \otimes_R B)$, the complex whose homology is $\mathrm{Tor}_*^R(A, T \otimes B)$. $\diamond$

Now we shall suppose that R is a commutative ring, so that the $\mathrm{Tor}_*^R(A, B)$ are actually R-modules in order to show how Tor_* localizes.

Lemma 3.2.11 *If $\mu \colon A \to A$ is multiplication by a central element $r \in R$, so are the induced maps $\mu_* \colon \mathrm{Tor}_n^R(A, B) \to \mathrm{Tor}_n^R(A, B)$ for all n and B.*

Proof Pick a projective resolution $P \to A$. Multiplication by r is an R-module chain map $\tilde{\mu} \colon P \to P$ over μ (this uses the fact that r is central), and $\tilde{\mu} \otimes B$ is multiplication by r on $P \otimes B$. The induced map μ_* on the subquotient $\mathrm{Tor}_n(A, B)$ of $P_n \otimes B$ is therefore also multiplication by r. $\diamond$

Corollary 3.2.12 *If A is an R/r-module, then for every R-module B the R-modules $\mathrm{Tor}_*^R(A, B)$ are actually R/r-modules, that is, annihilated by the ideal rR.*

3.3 Ext for Nice Rings

Corollary 3.2.13 (Localization for Tor) *If R is commutative and A and B are R-modules, then the following are equivalent for each n:*

1. $\text{Tor}_n^R(A, B) = 0$.
2. *For every prime ideal p of R* $\text{Tor}_n^{R_p}(A_p, B_p) = 0$.
3. *For every maximal ideal m of R* $\text{Tor}_n^{R_m}(A_m, B_m) = 0$.

Proof For any R-module M, $M = 0 \Leftrightarrow M_p = 0$ for every prime $p \Leftrightarrow M_m = 0$ for every maximal ideal m. In the case $M = \text{Tor}_m^R(A, B)$ we have

$$M_p = R_p \otimes_R M = \text{Tor}_n^{R_p}(A_p, B_p). \qquad \diamond$$

3.3 Ext for Nice Rings

We first turn to a calculation of $\text{Ext}_{\mathbb{Z}}^*$ groups to get a calculational feel for what these derived functors do to abelian groups.

Lemma 3.3.1 $\text{Ext}_{\mathbb{Z}}^n(A, B) = 0$ *for $n \geq 2$ and all abelian groups A, B.*

Proof Embed B in an injective abelian group I^0; the quotient I^1 is divisible, hence injective. Therefore, $\text{Ext}^*(A, B)$ is the cohomology of

$$0 \to \text{Hom}(A, I^0) \to \text{Hom}(A, I^1) \to 0. \qquad \diamond$$

Calculation 3.3.2 $(A = \mathbb{Z}/p)$ $\text{Ext}_{\mathbb{Z}}^0(\mathbb{Z}/p, B) = {}_pB$, $\text{Ext}_{\mathbb{Z}}^1(\mathbb{Z}/p, B) = B/pB$ and $\text{Ext}_{\mathbb{Z}}^n(\mathbb{Z}/p, B) = 0$ for $n \geq 2$. To see this, use the resolution

$$0 \to \mathbb{Z} \xrightarrow{p} \mathbb{Z} \to \mathbb{Z}/p \to 0 \text{ and the fact that } \text{Hom}(\mathbb{Z}, B) \cong B$$

to see that $\text{Ext}^*(\mathbb{Z}/p, B)$ is the cohomology of $0 \leftarrow B \xleftarrow{p} B \leftarrow 0$.

Since $\mathbb{Z}$ is projective, $\text{Ext}^1(\mathbb{Z}, B) = 0$. Hence we can calculate $\text{Ext}^*(A, B)$ for every finitely generated abelian group $A \cong \mathbb{Z}^m \oplus \mathbb{Z}/p_1 \oplus \cdots \oplus \mathbb{Z}/p_n$ by taking a finite direct sum of $\text{Ext}^*(\mathbb{Z}/p, B)$ groups. For infinitely generated groups, the calculation is much more complicated than it was for Tor.

Example 3.3.3 $(B = \mathbb{Z})$ Let A be a torsion group, and write A^* for its Pontrjagin dual $\text{Hom}(A, \mathbb{Q}/\mathbb{Z})$ as in 3.2.3. Using the injective resolution $0 \to \mathbb{Z} \to \mathbb{Q} \to \mathbb{Q}/\mathbb{Z} \to 0$ to compute $\text{Ext}^*(A, \mathbb{Z})$, we see that $\text{Ext}_{\mathbb{Z}}^0(A, \mathbb{Z}) = 0$ and

$\text{Ext}^1_{\mathbb{Z}}(A, \mathbb{Z}) = A^*$. To get a feel for this, note that because $\mathbb{Z}_{p^\infty}$ is the union (colimit) of its subgroups $\mathbb{Z}/p^n$, the group

$$\text{Ext}^1_{\mathbb{Z}}(\mathbb{Z}_{p^\infty}, \mathbb{Z}) = (\mathbb{Z}_{p^\infty})^*$$

is the torsionfree group of p-adic integers, $\hat{\mathbb{Z}}_p = \varprojlim(\mathbb{Z}/p^n)$. We will calculate $\text{Ext}^1_{\mathbb{Z}}(\mathbb{Z}_{p^\infty}, B)$ more generally in section 3.5, using $\varprojlim^1$.

Exercise 3.3.1 Show that $\text{Ext}^1_{\mathbb{Z}}(\mathbb{Z}[\frac{1}{p}], \mathbb{Z}) \cong \hat{\mathbb{Z}}_p/\mathbb{Z} \cong \mathbb{Z}_{p^\infty}$. This shows that $\text{Ext}^1(-, \mathbb{Z})$ does not vanish on flat abelian groups.

Exercise 3.3.2 When $R = \mathbb{Z}/m$ and $B = \mathbb{Z}/p$ with $p|m$, show that

$$0 \to \mathbb{Z}/p \overset{\iota}{\hookrightarrow} \mathbb{Z}/m \overset{p}{\longrightarrow} \mathbb{Z}/m \overset{m/p}{\longrightarrow} \mathbb{Z}/m \overset{p}{\longrightarrow} \mathbb{Z}/m \overset{m/p}{\longrightarrow} \cdots$$

is an infinite periodic injective resolution of B. Then compute the groups $\text{Ext}^n_{\mathbb{Z}/m}(A, \mathbb{Z}/p)$ in terms of $A^* = \text{Hom}(A, \mathbb{Z}/m)$. In particular, show that if $p^2|m$, then $\text{Ext}^n_{\mathbb{Z}/m}(\mathbb{Z}/p, \mathbb{Z}/p) \cong \mathbb{Z}/p$ for all n.

Proposition 3.3.4 *For all n and all rings R*

1. $\text{Ext}^n_R(\oplus_\alpha A_\alpha, B) \cong \prod_\alpha \text{Ext}^n_R(A_\alpha, B)$.
2. $\text{Ext}^n_R(A, \prod_\beta B_\beta) \cong \prod_\beta \text{Ext}^n_R(A, B_\beta)$.

Proof If $P_\alpha \to A_\alpha$ are projective resolutions, so is $\oplus P_\alpha \to \oplus A_\alpha$. If $B_\beta \to I_\beta$ are injective resolutions, so is $\prod B_\beta \to \prod I_\beta$. Since $\text{Hom}(\oplus P_\alpha, B) = \prod \text{Hom}(P_\alpha, B)$ and $\text{Hom}(A, \prod I_\beta) = \prod \text{Hom}(A, I_\beta)$, the result follows from the fact that for any family C_γ of cochain complexes,

$$H^*(\prod C_\gamma) \cong \prod H^*(C_\gamma). \qquad \diamond$$

Examples 3.3.5

1. If $p^2|m$ and A is a $\mathbb{Z}/p$-vector space of countably infinite dimension, then $\text{Ext}^n_{\mathbb{Z}/m}(A, \mathbb{Z}/p) \cong \prod_{i=1}^{\infty} \mathbb{Z}/p$ is a $\mathbb{Z}/p$-vector space of dimension $2^{\aleph_0}$.
2. If B is the product $\mathbb{Z}/2 \times \mathbb{Z}/3 \times \mathbb{Z}/4 \times \mathbb{Z}/5 \times \cdots$ then B is *not* a torsion group, and

$$\text{Ext}^1(A, B) = \prod_{p=2}^{\infty} A/pA = 0$$

vanishes if and only if A is divisible.

Lemma 3.3.6 *Suppose that R is a commutative ring, so that $\mathrm{Hom}_R(A, B)$ and the $\mathrm{Ext}^*_R(A, B)$ are actually R-modules. If $\mu: A \to A$ and $\nu: B \to B$ are multiplication by $r \in R$, so are the induced endomorphisms μ^* and ν_* of $\mathrm{Ext}^n_R(A, B)$ for all n.*

Proof Pick a projective resolution $P \to A$. Multiplication by r is an R-module chain map $\tilde{\mu}: P \to P$ over μ (as r is central); the map $\mathrm{Hom}(\tilde{\mu}, B)$ on $\mathrm{Hom}(P, B)$ is multiplication by r, because it sends $f \in \mathrm{Hom}(P_n, B)$ to $f\tilde{\mu}$, which takes $p \in P_n$ to $f(rp) = rf(p)$. Hence the map μ^* on the subquotient $\mathrm{Ext}^n(A, B)$ of $\mathrm{Hom}(P_n, B)$ is also multiplication by r. The argument for ν_* is similar, using an injective resolution $B \to I$. $\diamond$

Corollary 3.3.7 *If R is commutative and A is actually an R/r-module, then for every R-module B the R-modules $\mathrm{Ext}^*_R(A, B)$ are actually R/r-modules.*

We would like to conclude, as we did for Tor, that Ext commutes with localization in some sense. Indeed, there is a natural map Φ from $S^{-1}\mathrm{Hom}_R(A, B)$ to $\mathrm{Hom}_{S^{-1}R}(S^{-1}A, S^{-1}B)$, but it need not be an isomorphism. A sufficient condition is that A be finitely presented, that is, some $R^m \xrightarrow{\alpha} R^n \to A \to 0$ is exact.

Lemma 3.3.8 *If A is a finitely presented R-module, then for every central multiplicative set S in R, Φ is an isomorphism:*

$$\Phi : S^{-1}\mathrm{Hom}_R(A, B) \cong \mathrm{Hom}_{S^{-1}R}(S^{-1}A, S^{-1}B).$$

Proof Φ is trivially an isomorphism when $A = R$; as Hom is additive, Φ is also an isomorphism when $A = R^m$. The result now follows from the 5-lemma and the following diagram:

$$
\begin{array}{ccccccc}
0 & \to & S^{-1}\mathrm{Hom}_R(A, B) & \to & S^{-1}\mathrm{Hom}_R(R^n, B) & \xrightarrow{\alpha} & S^{-1}\mathrm{Hom}_R(R^m, B) \\
& & \Phi \downarrow & & \cong \downarrow & & \cong \downarrow \\
0 & \to & \mathrm{Hom}(S^{-1}A, S^{-1}B) & \to & \mathrm{Hom}(S^{-1}R^n, S^{-1}B) & \xrightarrow{\alpha} & \mathrm{Hom}(S^{-1}R^m, S^{-1}B).
\end{array}
$$
$\diamond$

Definition 3.3.9 A ring R is *(right) noetherian* if every (right) ideal is finitely generated, that is, if every module R/I is finitely presented. It is well known that if R is noetherian, then every finitely generated (right) R-module is finitely presented. (See [BAII,§3.2].) It follows that every finitely generated module A has a resolution $F \to A$ in which each F_n is a finitely generated free R-module.

Proposition 3.3.10 *Let A be a finitely generated module over a commutative noetherian ring R. Then for every multiplicative set S, all modules B, and all n*

$$\Phi: S^{-1}\operatorname{Ext}^n_R(A, B) \cong \operatorname{Ext}^n_{S^{-1}R}(S^{-1}A, S^{-1}B).$$

Proof Choose a resolution $F \to A$ by finitely generated free R-modules. Then $S^{-1}F \to S^{-1}A$ is a resolution by finitely generated free $S^{-1}R$-modules. Because S^{-1} is an exact functor from R-modules to $S^{-1}R$-modules,

$$S^{-1}\operatorname{Ext}^*_R(A, B) = S^{-1}(H^*\operatorname{Hom}_R(F, B)) \cong H^*(S^{-1}\operatorname{Hom}_R(F, B))$$

$$\cong H^*\operatorname{Hom}_{S^{-1}R}(S^{-1}F, S^{-1}B) = \operatorname{Ext}^*_{S^{-1}R}(S^{-1}A, S^{-1}B).\diamond$$

Corollary 3.3.11 (Localization for Ext) *If R is commutative noetherian and A is a finitely generated R-module, then the following are equivalent for all modules B and all n:*

1. $\operatorname{Ext}^n_R(A, B) = 0.$
2. *For every prime ideal p of R,* $\operatorname{Ext}^n_{R_p}(A_p, B_p) = 0.$
3. *For every maximal ideal m of R,* $\operatorname{Ext}^n_{R_m}(A_m, B_m) = 0.$

3.4 Ext and Extensions

An *extension* ξ of A by B is an exact sequence $0 \to B \to X \to A \to 0$. Two extensions ξ and ξ' are *equivalent* if there is a commutative diagram

$$
\begin{array}{ccccccccc}
\xi: & 0 & \longrightarrow & B & \longrightarrow & X & \longrightarrow & A & \longrightarrow & 0 \\
     &   &   & \| &   & \downarrow\cong & & \| & & \\
\xi': & 0 & \longrightarrow & B & \longrightarrow & X' & \longrightarrow & A & \longrightarrow & 0
\end{array}
$$

An extension is *split* if it is equivalent to $0 \to B \xrightarrow{(0,1)} A \oplus B \to A \to 0$.

Exercise 3.4.1 Show that if p is prime, there are exactly p equivalence classes of extensions of $\mathbb{Z}/p$ by $\mathbb{Z}/p$ in **Ab**: the split extension and the extensions

$$0 \to \mathbb{Z}/p \xrightarrow{p} \mathbb{Z}/p^2 \xrightarrow{i} \mathbb{Z}/p \to 0 \qquad (i = 1, 2, \cdots, p-1).$$

Lemma 3.4.1 *If* $\mathrm{Ext}^1(A, B) = 0$, *then every extension of A by B is split.*

Proof Given an extension ξ, applying $\mathrm{Ext}^*(A, -)$ yields the exact sequence

$$\mathrm{Hom}(A, X) \to \mathrm{Hom}(A, A) \xrightarrow{\partial} \mathrm{Ext}^1(A, B)$$

so the identity map id_A lifts to a map $\sigma\colon A \to X$ when $\mathrm{Ext}^1(A, B) = 0$. As σ is a section of $X \twoheadrightarrow A$, evidently $X \cong A \oplus B$ and ξ is split. $\diamond$

Porism 3.4.2 Taking the construction of this lemma to heart, we see that the class $\Theta(\xi) = \partial(\mathrm{id}_A)$ in $\mathrm{Ext}^1(A, B)$ is an *obstruction* to ξ being split: ξ is split iff id_A lifts to $\mathrm{Hom}(A, X)$ iff the class $\Theta(\xi) \in \mathrm{Ext}^1(A, B)$ vanishes. Equivalent extensions have the same obstruction by naturality of the map ∂, so the obstruction $\Theta(\xi)$ only depends on the equivalence class of ξ.

Theorem 3.4.3 *Given two R-modules A and B, the mapping* $\Theta\colon \xi \mapsto \partial(\mathrm{id}_A)$ *establishes a 1–1 correspondence*

$$\left\{ \begin{array}{c} \text{equivalence classes of} \\ \text{extensions of } A \text{ by } B \end{array} \right\} \overset{1-1}{\longleftrightarrow} \mathrm{Ext}^1(A, B)$$

in which the split extension corresponds to the element $0 \in \mathrm{Ext}^1(A, B)$.

Proof Fix an exact sequence $0 \to M \xrightarrow{j} P \to A \to 0$ with P projective. Applying $\mathrm{Hom}(-, B)$ yields an exact sequence

$$\mathrm{Hom}(P, B) \to \mathrm{Hom}(M, B) \xrightarrow{\partial} \mathrm{Ext}^1(A, B) \to 0.$$

Given $x \in \mathrm{Ext}^1(A, B)$, choose $\beta \in \mathrm{Hom}(M, B)$ with $\partial(\beta) = x$. Let X be the pushout of j and β, i.e., the cokernel of $M \to P \oplus B$ $(m \mapsto (j(m), -\beta(m)))$. There is a diagram

$$
\begin{array}{ccccccccc}
0 & \longrightarrow & M & \xrightarrow{j} & P & \longrightarrow & A & \longrightarrow & 0 \\
& & \beta\downarrow & & \llcorner\downarrow\sigma & & \| & & \\
\xi: & 0 & \longrightarrow & B & \xrightarrow{i} & X & \longrightarrow & A & \longrightarrow & 0,
\end{array}
$$

where the map $X \to A$ is induced by the maps $B \xrightarrow{0} A$ and $P \to A$. (*Exercise:* Show that the bottom sequence ξ is exact.) By naturality of the connecting map ∂, we see that $\Theta(\xi) = x$, that is, that Θ is a surjection.

In fact, this construction gives a set map Ψ from $\text{Ext}^1(A, B)$ to the set of equivalence classes of extensions. For if $\beta' \in \text{Hom}(M, B)$ is another lift of x, then there is an $f \in \text{Hom}(P, B)$ so that $\beta' = \beta + fj$. If X' is the pushout of j and β', then the maps $i: B \to X$ and $\sigma + if: P \to X$ induce an isomorphism $X' \cong X$ and an equivalence between ξ' and ξ. (Check this!)

Conversely, given an extension ξ of A by B, the lifting property of P gives a map $\tau: P \to X$ and hence a commutative diagram

(∗)
$$
\begin{array}{ccccccccc}
0 & \longrightarrow & M & \xrightarrow{\ j\ } & P & \longrightarrow & A & \longrightarrow & 0 \\
& & \downarrow{\gamma} & & \downarrow{\tau} & & \| & & \\
\xi: \quad 0 & \longrightarrow & B & \xrightarrow{\ i\ } & X & \longrightarrow & A & \longrightarrow & 0.
\end{array}
$$

Now X is the pushout of j and γ. (*Exercise:* Check this!) Hence $\Psi(\Theta(\xi)) = \xi$, showing that Θ is injective.   ◇

Definition 3.4.4 (Baer sum) Let $\xi: 0 \to B \to X \to A \to 0$ and $\xi': 0 \to B \to X' \to A \to 0$ be two extensions of A by B. Let X'' be the pullback $\{(x, x') \in X \times X' : \bar{x} = \bar{x}' \text{ in } A\}$.

$$
\begin{array}{ccc}
X'' & \longrightarrow & X' \\
\downarrow{\ulcorner} & & \downarrow \\
X & \longrightarrow & A
\end{array}
$$

X'' contains three copies of $B : B \times 0, 0 \times B$, and the skew diagonal $\{(-b, b) : b \in B\}$. The copies $B \times 0$ and $0 \times B$ are identified in the quotient Y of X'' by the skew diagonal. Since $X''/0 \times B \cong X$ and $X/B \cong A$, it is immediate that the sequence

$$
\varphi: \quad 0 \to B \to Y \to A \to 0
$$

is also an extension of A by B. The class of φ is called the *Baer sum* of the extensions ξ and ξ', since this construction was introduced by R. Baer in 1934.

Corollary 3.4.5 *The set of (equiv. classes of) extensions is an abelian group under Baer sum, with zero being the class of the split extension. The map Θ is an isomorphism of abelian groups.*

Proof We will show that $\Theta(\varphi) = \Theta(\xi) + \Theta(\xi')$ in $\text{Ext}^1(A, B)$. This will prove that Baer sum is well defined up to equivalence, and the corollary will then follow. We shall adopt the notation used in (∗) in the proof of the above theorem. Let $\tau'': P \to X''$ be the map induced by $\tau: P \to X$ and $\tau': P \to X'$, and let $\bar{\tau}: P \to Y$ be the induced map. The restriction of $\bar{\tau}$ to M is induced by the map $\gamma + \gamma': M \to B$, so

$$
\begin{array}{ccccccccc}
0 & \longrightarrow & M & \longrightarrow & P & \longrightarrow & A & \longrightarrow & 0 \\
 & & {\scriptstyle \gamma+\gamma'}\downarrow & & {\scriptstyle \bar{\tau}}\downarrow & & \| & & \\
\varphi: \quad 0 & \longrightarrow & B & \longrightarrow & Y & \longrightarrow & A & \longrightarrow & 0
\end{array}
$$

commutes. Hence, $\Theta(\varphi) = \partial(\gamma + \gamma')$, where ∂ is the map from $\mathrm{Hom}(M, B)$ to $\mathrm{Ext}^1(A, B)$. But $\partial(\gamma + \gamma') = \partial(\gamma) + \partial(\gamma') = \Theta(\xi) + \Theta(\xi')$. $\quad\diamond$

Vista 3.4.6 (Yoneda Ext groups) We can define $\mathrm{Ext}^1(A, B)$ in *any* abelian category $\mathcal{A}$, even if it has no projectives and no injectives, to be the set of equivalence classes of extensions under Baer sum (if indeed this is a set). The Freyd-Mitchell Embedding Theorem 1.6.1 shows that $\mathrm{Ext}^1(A, B)$ is an abelian group—but one could also prove this fact directly. Similarly, we can recapture the groups $\mathrm{Ext}^n(A, B)$ without mentioning projectives or injectives. This approach is due to Yoneda. An element of the Yoneda $\mathrm{Ext}^n(A, B)$ is an equivalence class of exact sequences of the form

$$\xi: \quad 0 \to B \to X_n \to \cdots \to X_1 \to A \to 0.$$

The equivalence relation is generated by the relation that $\xi' \sim \xi''$ if there is a diagram

$$
\begin{array}{ccccccccccc}
\xi': & 0 & \longrightarrow & B & \longrightarrow & X'_n & \longrightarrow & \cdots & \longrightarrow & X'_1 & \longrightarrow & A & \longrightarrow & 0 \\
 & & & \| & & \downarrow & & & & \downarrow & & \| & & \\
\xi'': & 0 & \longrightarrow & B & \longrightarrow & X''_n & \longrightarrow & \cdots & \longrightarrow & X''_1 & \longrightarrow & A & \longrightarrow & 0.
\end{array}
$$

To "add" ξ and ξ' when $n \geq 2$, let X''_1 be the pullback of X_1 and X'_1 over A, let X''_n be the pushout of X_n and X''_n under B, and let Y_n be the quotient of X''_n by the skew diagonal copy of B. Then $\xi + \xi'$ is the class of the extension

$$0 \to B \to Y_n \to X_{n-1} \oplus X'_{n-1} \to \cdots \to X_2 \oplus X'_2 \to X''_1 \to A \to 0.$$

Now suppose that $\mathcal{A}$ has enough projectives. If $P \to A$ is a projective resolution, the Comparison Theorem 2.2.6 yields a map from P to ξ, hence a diagram

$$0 \longrightarrow M \longrightarrow P_{n-1} \longrightarrow \cdots \longrightarrow P_0 \longrightarrow A \longrightarrow 0$$

$$\qquad\quad \beta\downarrow \qquad\quad \downarrow \gamma_n \qquad\qquad\qquad\qquad \downarrow \qquad\quad \|$$

$$\xi: \quad 0 \longrightarrow B \longrightarrow X_n \longrightarrow \cdots \longrightarrow X_1 \longrightarrow A \longrightarrow 0.$$

By dimension shifting, there is an exact sequence

$$\mathrm{Hom}(P_{n-1}, B) \to \mathrm{Hom}(M, B) \xrightarrow{\partial} \mathrm{Ext}^n(\Lambda, B) \to 0.$$

The association $\Theta(\xi) = \partial(\beta)$ gives the 1–1 correspondence between the Yoneda Ext^n and the derived functor Ext^n. For more details we refer the reader to [BX, §7.5] or [MacH, pp. 82–87].

3.5 Derived Functors of the Inverse Limit

Let I be a small category and $\mathcal{A}$ an abelian category. We saw in Chapter 2 that the functor category $\mathcal{A}^I$ has enough injectives, at least when $\mathcal{A}$ is complete and has enough injectives. (For example, $\mathcal{A}$ could be **Ab**, R–**mod**, or Sheaves(X).) Therefore we can define the right derived functors $R^n \lim_{i \in I}$ from $\mathcal{A}^I$ to $\mathcal{A}$.

We are most interested in the case in which $\mathcal{A}$ is **Ab** and I is the poset $\cdots \to 2 \to 1 \to 0$ of whole numbers in reverse order. We shall call the objects of **Ab**^{I} (countable) *towers* of abelian groups; they have the form

$$\{A_i\}: \quad \cdots \to A_2 \to A_1 \to A_0.$$

In this section we shall give the alternative construction $\underleftarrow{\lim}^1$ of $R^1\underleftarrow{\lim}$ for countable towers due to Eilenberg and prove that $R^n \underleftarrow{\lim} = 0$ for $n \neq 0, 1$. This construction generalizes from **Ab** to other abelian categories that satisfy the following axiom, introduced by Grothendieck in [Tohoku]:

$(AB4^*)$ $\mathcal{A}$ is complete, and the product of any set of surjections is a surjection.

Explanation If I is a discrete set, $\mathcal{A}^I$ is the product category $\Pi_{i \in I}\mathcal{A}$ of indexed families of objects $\{A_i\}$ in $\mathcal{A}$. For $\{A_i\}$ in $\mathcal{A}^I$, $\lim_{i \in I} A_i$ is the product $\prod A_i$. Axiom $(AB4^*)$ states that the left exact functor $\prod$ from $\mathcal{A}^I$ to $\mathcal{A}$ is exact for all discrete I. Axiom $(AB4^*)$ fails $(\prod_{i=1}^\infty$ is not exact) for some important abelian categories, such as Sheaves(X). On the other hand, axiom $(AB4^*)$ is satisfied by many abelian categories in which objects have underlying sets, such as **Ab**, **mod**–R, and **Ch**(**mod**–R).

Definition 3.5.1 Given a tower $\{A_i\}$ in **Ab**, define the map

$$\Delta: \prod_{i=0}^{\infty} A_i \to \prod_{i=0}^{\infty} A_i$$

by the element-theoretic formula

$$\Delta(\cdots, a_i, \cdots, a_0) = (\cdots, a_i - \bar{a}_{i+1}, \cdots, a_1 - \bar{a}_2, a_0 - \bar{a}_1),$$

where $\bar{a}_{i+1}$ denotes the image of $a_{i+1} \in A_{i+1}$ in A_i. The kernel of Δ is $\varprojlim A_i$ (check this!). We define $\varprojlim^1 A_i$ to be the cokernel of Δ, so that $\varprojlim^1$ is a functor from **Ab**^{I} to **Ab**. We also set $\varprojlim^0 A_i = \varprojlim A_i$ and $\varprojlim^n A_i = 0$ for $n \neq 0, 1$.

Lemma 3.5.2 *The functors* $\{\varprojlim^n\}$ *form a cohomological δ-functor.*

Proof If $0 \to \{A_i\} \to \{B_i\} \to \{C_i\} \to 0$ is a short exact sequence of towers, apply the Snake Lemma to

$$
\begin{array}{ccccccccc}
0 & \longrightarrow & \prod A_i & \longrightarrow & \prod B_i & \longrightarrow & \prod C_i & \longrightarrow & 0 \\
& & \downarrow^{\Delta} & & \downarrow^{\Delta} & & \downarrow^{\Delta} & & \\
0 & \longrightarrow & \prod A_i & \longrightarrow & \prod B_i & \longrightarrow & \prod C_i & \longrightarrow & 0
\end{array}
$$

to get the requisite natural long exact sequence. $\diamond$

Lemma 3.5.3 *If all the maps* $A_{i+1} \to A_i$ *are onto, then* $\varprojlim^1 A_i = 0$. *Moreover* $\varprojlim A_i \neq 0$ *(unless every* $A_i = 0$*), because each of the natural projections* $\varprojlim A_i \to A_j$ *are onto.*

Proof Given elements $b_i \in A_i$ ($i = 0, 1, \cdots$), and any $a_0 \in A_0$, inductively choose $a_{i+1} \in A_{i+1}$ to be a lift of $a_i - b_i \in A_i$. The map Δ sends $(\cdots, a_1, a_0)$ to $(\cdots, b_1, b_0)$, so Δ is onto and $\operatorname{coker}(\Delta) = 0$. If all the $b_i = 0$, then $(\cdots, a_1, a_0) \in \varprojlim A_i$. $\diamond$

Corollary 3.5.4 $\varprojlim^1 A_i \cong (R^1 \varprojlim)(A_i)$ *and* $R^n \varprojlim = 0$ *for* $n \neq 0, 1$.

Proof In order to show that the $\varprojlim^n$ forms a universal δ-functor, we only need to see that $\varprojlim^1$ vanishes on enough injectives. In Chapter 2 we constructed enough injectives by taking products of towers

$$k_* E: \quad \cdots = E = E \to 0 \to 0 \cdots \to 0$$

with E injective. All the maps in $k_* E$ (and hence in the product towers) are onto, so $\varprojlim^1$ vanishes on these injective towers. $\diamond$

Remark If we replace **Ab** by $\mathcal{A} = $ **mod**–R, **Ch(mod**–R) or any abelian category $\mathcal{A}$ satisfying Grothendieck's axiom $(AB4^*)$, the above proof goes through to show that $\varprojlim^1 = R^1(\varprojlim)$ and $R^n(\varprojlim) = 0$ for $n \neq 0, 1$ as functors on the category of towers in $\mathcal{A}$. However, the proof breaks down for other abelian categories.

Example 3.5.5 Set $A_0 = \mathbb{Z}$ and let $A_i = p^i \mathbb{Z}$ be the subgroup generated by p^i. Applying $\varprojlim$ to the short exact sequence of towers

$$0 \to \{p^i \mathbb{Z}\} \to \{\mathbb{Z}\} \to \{\mathbb{Z}/p^i \mathbb{Z}\} \to 0$$

with p prime yields the uncountable group

$$\varprojlim{}^1 \{p^i \mathbb{Z}\} \cong \hat{\mathbb{Z}}_p / \mathbb{Z}.$$

Here $\hat{\mathbb{Z}}_p = \varprojlim \mathbb{Z}/p^i \mathbb{Z}$ is the group of p-adic integers.

Exercise 3.5.1 Let $\{A_i\}$ be a tower in which the maps $A_{i+1} \to A_i$ are inclusions. We may regard $A = A_0$ as a topological group in which the sets $a + A_i (a \in A, i \geq 0)$ are the open sets. Show that $\varprojlim A_i = \cap A_i$ is zero iff A is *Hausdorff*. Then show that $\varprojlim^1 A_i = 0$ iff A is *complete* in the sense that every Cauchy sequence has a limit, not necessarily unique. *Hint:* Show that A is complete iff $A \cong \varprojlim(A/A_i)$.

Definition 3.5.6 A tower $\{A_i\}$ of abelian groups satisfies the *Mittag-Leffler condition* if for each k there exists a $j \geq k$ such that the image of $A_i \to A_k$ equals the image of $A_j \to A_k$ for all $i \geq j$. (The images of the A_i in A_k satisfy the *descending chain condition*.) For example, the Mittag-Leffler condition is satisfied if all the maps $A_{i+1} \to A_i$ in the tower $\{A_i\}$ are onto. We say that $\{A_i\}$ satisfies the *trivial* Mittag-Leffler condition if for each k there exists a $j > k$ such that the map $A_j \to A_k$ is zero.

Proposition 3.5.7 *If $\{A_i\}$ satisfies the Mittag-Leffler condition, then*

$$\varprojlim {}^1 A_i = 0.$$

Proof If $\{A_i\}$ satisfies the trivial Mittag-Leffler condition, and $b_i \in A_i$ are given, set $a_k = b_k + \bar{b}_{k+1} + \cdots + \bar{b}_{j-1}$, where $\bar{b}_i$ denotes the image of b_i in A_k. (Note that $\bar{b}_i = 0$ for $i \geq j$.) Then Δ maps $(\cdots, a_1, a_0)$ to $(\cdots, b_1, b_0)$. Thus Δ is onto and $\varprojlim {}^1 A_i = 0$ when $\{A_i\}$ satisfies the trivial Mittag-Leffler condition. In the general case, let $B_k \subseteq A_k$ be the image of $A_i \to A_k$ for large i. The maps $B_{k+1} \to B_k$ are all onto, so $\varprojlim {}^1 B_k = 0$. The tower $\{A_k/B_k\}$ satisfies the trivial Mittag-Leffler condition, so $\varprojlim {}^1 A_k/B_k = 0$. From the short exact sequence

$$0 \to \{B_i\} \to \{A_i\} \to \{A_i/B_i\} \to 0$$

of towers, we see that $\varprojlim {}^1 A_i = 0$ as claimed. $\diamond$

Exercise 3.5.2 Show that $\varprojlim {}^1 A_i = 0$ if $\{A_i\}$ is a tower of finite abelian groups, or a tower of finite-dimensional vector spaces over a field.

The following formula presages the Universal Coefficient theorems of the next section, as well as the spectral sequences of Chapter 5.

Theorem 3.5.8 *Let $\cdots \to C_1 \to C_0$ be a tower of chain complexes of abelian groups satisfying the Mittag-Leffler condition, and set $C = \varprojlim C_i$. Then there is an exact sequence for each q:*

$$0 \to \varprojlim {}^1 H_{q+1}(C_i) \to H_q(C) \to \varprojlim H_q(C_i) \to 0.$$

Proof Let $B_i \subseteq Z_i \subseteq C_i$ be the subcomplexes of boundaries and cycles in the complex C_i, so that Z_i/B_i is the chain complex $H_*(C_i)$ with zero differentials. Applying the left exact functor $\varprojlim$ to $0 \to \{Z_i\} \to \{C_i\} \xrightarrow{d} \{C_i[-1]\}$ shows that in fact $\varprojlim Z_i$ is the subcomplex Z of cycles in C. (The $[-1]$ refers to the surpressed subscript on the chain complexes.) Let B denote the subcomplex $d(C)[1] = (C/Z)[1]$ of boundaries in C, so that Z/B is the chain complex $H_*(C)$ with zero differentials. From the exact sequence of towers

$$0 \to \{Z_i\} \to \{C_i\} \xrightarrow{d} \{B_i[-1]\} \to 0$$

we see that $\lim^1 B_i = (\lim{}^1 B_i[-1])[+1] = 0$ and that

$$0 \to B[-1] \to \lim B_i[-1] \to \lim{}^1 Z_i \to 0$$

is exact. From the exact sequence of towers

$$0 \to \{B_i\} \to \{Z_i\} \to H_*(C_i) \to 0$$

we see that $\lim^1 Z_i \cong \lim^1 H_*(C_i)$ and that

$$0 \to \lim B_i \to Z \to \lim H_*(C_i) \to 0$$

is exact. Hence C has the filtration by subcomplexes

$$0 \subseteq B \subseteq \lim B_i \subseteq Z \subseteq C$$

whose filtration quotients are B, $\lim^1 H_*(C_i)[1]$, $\lim H_*(C_i)$, and C/Z respectively. The theorem follows, since $Z/B = H_*(C)$.   ◇

Variant If $\cdots \to C_1 \to C_0$ is a tower of cochain complexes satisfying the Mittag-Leffler condition, the sequences become

$$0 \to \lim{}^1 H^{q-1}(C_i) \to H^q(C) \to \lim H^q(C_i) \to 0.$$

Application 3.5.9 Let $H^*(X)$ denote the integral cohomology of a topological CW complex X. If $\{X_i\}$ is an increasing sequence of subcomplexes with $X = \cup X_i$, there is an exact sequence

$$(*) \qquad 0 \to \lim{}^1 H^{q-1}(X_i) \to H^q(X) \to \lim H^q(X_i) \to 0$$

for each q. This use of $\lim^1$ to perform calculations in algebraic topology was discovered by Milnor in 1960 [Milnor] and thrust $\lim^1$ into the limelight.

To derive this formula, let C_i denote the chain complex $\mathrm{Hom}(S(X_i), \mathbb{Z})$ used to compute $H^*(X_i)$. Since the inclusion $S(X_i) \subseteq S(X_{i+1})$ splits (because each $S_n(X_{i+1})/S_n(X_i)$ is a free abelian group), the maps $C_{i+1} \to C_i$ are onto, and the tower satisfies the Mittag-Leffler condition. Since X has the weak topology, $S(X)$ is the union of the $S(X_i)$, and therefore $H^*(X)$ is the cohomology of the cochain complex

$$\mathrm{Hom}(\cup S(X_i), \mathbb{Z}) = \lim \mathrm{Hom}(S(X_i), \mathbb{Z}) = \lim C_i.$$

A historical remark: Milnor proved that the sequence $(*)$ is also valid if H^* is replaced by any generalized cohomology theory, such as topological K–theory.

Application 3.5.10 Let A be an R-module that is the union of submodules $\cdots \subseteq A_i \subseteq A_{i+1} \subseteq \cdots$. Then for every R-module B and every q the sequence

$$0 \to \varprojlim{}^1 \operatorname{Ext}_R^{q-1}(A_i, B) \to \operatorname{Ext}_R^q(A, B) \to \varprojlim \operatorname{Ext}_R^q(A_i, B) \to 0$$

is exact. For $\mathbb{Z}_{p^\infty} = \cup \mathbb{Z}/p^i$, this gives a short exact sequence for every B:

$$0 \to \varprojlim{}^1 \operatorname{Hom}(\mathbb{Z}/p^i, B) \to \operatorname{Ext}_{\mathbb{Z}}^1(\mathbb{Z}_{p^\infty}, B) \to \hat{B}_p \to 0,$$

where the group $\hat{B}_p = \varprojlim (B/p^i B)$ is the p-adic completion of B. This generalizes the calculation $\operatorname{Ext}_{\mathbb{Z}}^1(\mathbb{Z}_{p^\infty}, \mathbb{Z}) \cong \hat{\mathbb{Z}}_p$ of 3.3.3. To see this, let E be a fixed injective resolution of B, and consider the tower of cochain complexes

$$\operatorname{Hom}(A_{i+1}, E) \to \operatorname{Hom}(A_i, E) \to \cdots \to \operatorname{Hom}(A_0, E).$$

Each $\operatorname{Hom}(-, E_n)$ is contravariant exact, so each map in the tower is a surjection. The cohomology of $\operatorname{Hom}(A_i, E)$ is $\operatorname{Ext}^*(A_i, B)$, and $\operatorname{Ext}^*(A, B)$ is the cohomology of

$$\operatorname{Hom}(\cup A_i, E) = \varprojlim \operatorname{Hom}(A_i, E).$$

Exercise 3.5.3 Show that $\operatorname{Ext}_{\mathbb{Z}}^1(\mathbb{Z}[\frac{1}{p}], \mathbb{Z}) \cong \hat{\mathbb{Z}}_p/\mathbb{Z}$ using $\mathbb{Z}[\frac{1}{p}] = \cup p^{-i}\mathbb{Z}$; cf. exercise 3.3.1. Then show that $\operatorname{Ext}_{\mathbb{Z}}^1(\mathbb{Q}, B) = (\prod_p \hat{B}_p)/B$ for torsionfree B.

Application 3.5.11 Let $C = C_{**}$ be a double chain complex, viewed as a lattice in the plane, and let $T_n C$ be the quotient double complex obtained by brutally truncating C at the vertical line $p = -n$:

$$(T_n C)_{pq} = \begin{cases} C_{pq} & \text{if } p \geq -n \\ 0 & \text{if } p < -n \end{cases}.$$

Then $\operatorname{Tot}(C)$ is the inverse limit of the tower of surjections

$$\cdots \to \operatorname{Tot}(T_{i+1}C) \to \operatorname{Tot}(T_i C) \to \cdots \to \operatorname{Tot}(T_0 C).$$

Therefore there is a short exact sequence for each q:

$$0 \to \varprojlim{}^1 H_{q+1}(\operatorname{Tot}(T_i C)) \to H_q(\operatorname{Tot}(C)) \to \varprojlim H_q(\operatorname{Tot}(T_i C)) \to 0.$$

This is especially useful when C is a second quadrant double complex, because the truncated complexes have only a finite number of nonzero rows.

Exercise 3.5.4 Let C be a second quadrant double complex with exact rows, and let B_{pq}^h be the image of $d^h\colon C_{pq} \to C_{p-1,q}$. Show that $H_{p+q}\operatorname{Tot}(T_{-p}C) \cong H_q(B_{p*}^h, d^v)$. Then let $b = d^h(a)$ be an element of B_{pq}^h representing a cycle ξ in $H_{p+q}\operatorname{Tot}(T_{-p}C)$ and show that the image of ξ in $H_{p+q}\operatorname{Tot}(T_{-p-1}C)$ is represented by $d^v(a) \in B_{p+1,q-1}^h$. This provides an effective method for calculating $H_*\operatorname{Tot}(C)$.

Vista 3.5.12 Let I be any poset and $\mathcal{A}$ any abelian category satisfying $(AB4^*)$. The following construction of the right derived functors of lim is taken from [Roos] and generalizes the construction of $\underleftarrow{\lim}^1$ in this section.

Given $A\colon I \to \mathcal{A}$, we define C_k to be the product over the set of all chains $i_k < \cdots < i_0$ in I of the objects A_{i_0}. Letting $pr_{i_k \cdots i_1}$ denote the projection of C_k onto the $(i_k < \cdots < i_1)^{st}$ factor and f_0 denote the map $A_{i_1} \to A_{i_0}$ associated to $i_1 < i_0$, we define $d^0\colon C_{k-1} \to C_k$ to be the map whose $(i_k < \cdots < i_0)^{th}$ factor is $f_0(pr_{i_k \cdots i_1})$. For $1 \le p \le k$, we define $d^p\colon C_{k-1} \to C_k$ to be the map whose $(i_k < \cdots < i_0)^{th}$ factor is the projection onto the $(i_k < \cdots < \hat{i}_p < \cdots < i_0)^{th}$ factor. This data defines a cochain complex C_*A whose differential $C_{k-1} \to C_k$ is the alternating sum $\sum_{p=0}^{k}(-1)^p d^p$, and we define $\lim_{i\in I}^n A$ to be $H^n(C_*A)$. (The data actually forms a *cosimplicial object* of $\mathcal{A}$; see Chapter 8.)

It is easy to see that $\lim_{i\in I}^0 A$ is the limit $\lim_{i\in I} A$. An exact sequence $0 \to A \to B \to C \to 0$ in $\mathcal{A}^I$ gives rise to a short exact sequence $0 \to C_*A \to C_*B \to C_*C \to 0$ in $\mathcal{A}$, whence an exact sequence

$$0 \to \lim_{i\in I} A \to \lim_{i\in I} B \to \lim_{i\in I} C \to \lim_{i\in I}^1 A \to \lim_{i\in I}^1 B \to \lim_{i\in I}^1 C \to \lim_{i\in I}^2 A \to \cdots.$$

Therefore the functors $\{\lim_{i\in I}^n\}$ form a cohomological δ-functor. It turns out that they are universal when $\mathcal{A}$ has enough injectives, so in fact $R^n \lim_{i\in I} \cong \lim_{i\in I}^n$.

Remark Let $\aleph_d$ denote the dth infinite cardinal number, $\aleph_0$ being the cardinality of $\{1, 2, \cdots\}$. If I is a directed poset of cardinality $\aleph_d$, or a filtered category with $\aleph_d$ morphisms, Mitchell proved in [Mitch] that $R^n \underleftarrow{\lim}$ vanishes for $n \ge d + 1$.

Exercise 3.5.5 (Pullback) Let $\to\leftarrow$ denote the poset $\{x, y, z\}$, $x < z$ and $y < z$, so that $\lim_{\to\leftarrow} A_i$ is the pullback of A_x and A_y over A_z. Show that $\lim_{\to\leftarrow}^1 A_i$

is the cokernel of the difference map $A_x \times A_y \to A_z$ and that $\lim_{\to \leftarrow}{}^n = 0$ for $n \neq 0, 1$.

3.6 Universal Coefficient Theorems

There is a very useful formula for using the homology of a chain complex P to compute the homology of the complex $P \otimes M$. Here is the most useful general formulation we can give:

Theorem 3.6.1 (Künneth formula) *Let P be a chain complex of flat right R-modules such that each submodule $d(P_n)$ of P_{n-1} is also flat. Then for every n and every left R-module M, there is an exact sequence*

$$0 \to H_n(P) \otimes_R M \to H_n(P \otimes_R M) \to \mathrm{Tor}_1^R(H_{n-1}(P), M) \to 0.$$

Proof The long exact Tor sequence associated to $0 \to Z_n \to P_n \to d(P_n) \to 0$ shows that each Z_n is also flat (exercise 3.2.2). Since $\mathrm{Tor}_1^R(d(P_n), M) = 0$,

$$0 \to Z_n \otimes M \to P_n \otimes M \to d(P_n) \otimes M \to 0$$

is exact for every n. These assemble to give a short exact sequence of chain complexes $0 \to Z \otimes M \to P \otimes M \to d(P) \otimes M \to 0$. Since the differentials in the Z and $d(P)$ complexes are zero, the homology sequence is

$$H_{n+1}(dP \otimes M) \xrightarrow{\partial} H_n(Z \otimes M) \to H_n(P \otimes M) \to H_n(dP \otimes M) \xrightarrow{\partial} H_{n-1}(Z \otimes M)$$

$$\Big\| \wr \qquad\qquad \Big\| \wr \qquad\qquad\qquad \Big\| \wr \qquad\qquad \Big\| \wr$$

$$d(P_{n+1}) \otimes M \qquad Z_n \otimes M \qquad\qquad d(P_n) \otimes M \qquad Z_{n-1} \otimes M.$$

Using the definition of ∂, it is immediate that $\partial = i \otimes M$, where i is the inclusion of $d(P_{n+1})$ in Z_n. On the other hand,

$$0 \to d(P_{n+1}) \xrightarrow{i} Z_n \to H_n(P) \to 0$$

is a flat resolution of $H_n(P)$, so $\mathrm{Tor}_*(H_n(P), M)$ is the homology of

$$0 \to d(P_{n+1}) \otimes M \xrightarrow{\partial} Z_n \otimes M \to 0.$$

$\diamond$

Universal Coefficient Theorem for Homology 3.6.2 *Let P be a chain complex of free abelian groups. Then for every n and every abelian group M the*

Künneth formula 3.6.1 splits noncanonically, yielding a direct sum decomposition

$$H_n(P \otimes M) \cong H_n(P) \otimes M \oplus \operatorname{Tor}_1^{\mathbb{Z}}(H_{n-1}(P), M).$$

Proof We shall use the well-known fact that every subgroup of a free abelian group is free abelian [KapIAB, section 15]. Since $d(P_n)$ is a subgroup of P_{n+1}, it is free abelian. Hence the surjection $P_n \to d(P_n)$ splits, giving a noncanonical decomposition

$$P_n \cong Z_n \oplus d(P_n).$$

Applying $\otimes M$, we see that $Z_n \otimes M$ is a direct summand of $P_n \otimes M$; a fortiori, $Z_n \otimes M$ is a direct summand of the intermediate group

$$\ker(d_n \otimes 1: P_n \otimes M \to P_{n-1} \otimes M).$$

Modding out $Z_n \otimes M$ and $\ker(d_n \otimes 1)$ by the common image of $d_{n+1} \otimes 1$, we see that $H_n(P) \otimes M$ is a direct summand of $H_n(P \otimes M)$. Since P and $d(P)$ are flat, the Künneth formula tells us that the other summand is $\operatorname{Tor}_1(H_{n-1}(P), M)$. $\diamond$

Theorem 3.6.3 (Künneth formula for complexes) *Let P and Q be right and left R-module complexes, respectively. Recall from 2.7.1 that the* tensor product *complex $P \otimes_R Q$ is the complex whose degree n part is $\bigoplus_{p+q=n} P_p \otimes Q_q$ and whose differential is given by $d(a \otimes b) = (da) \otimes b + (-1)^p a \otimes (db)$ for $a \in P_p$, $b \in Q_q$. If P_n and $d(P_n)$ are flat for each n, then there is an exact sequence*

$$0 \to \bigoplus_{p+q=n} H_p(P) \otimes H_q(Q) \to H_n(P \otimes_R Q) \to \bigoplus_{\substack{p+q=\\n-1}} \operatorname{Tor}_1^R(H_p(P), H_q(Q)) \to 0$$

for each n. If $R = \mathbb{Z}$ and P is a complex of free abelian groups, this sequence is noncanonically split.

Proof Modify the proof given in 3.6.1 for $Q = M$. $\diamond$

Application 3.6.4 (Universal Coefficient Theorem in topology) Let $S(X)$ denote the singular chain complex of a topological space X; each $S_n(X)$ is a free abelian group. If M is any abelian group, the homology of X with "coefficients" in M is

$$H_*(X; M) = H_*(S(X) \otimes M).$$

Writing $H_*(X)$ for $H_*(X; \mathbb{Z})$, the formula in this case becomes

$$H_n(X; M) \cong H_n(X) \otimes M \oplus \operatorname{Tor}_1^{\mathbb{Z}}(H_{n-1}(X), M).$$

This formula is often called the Universal Coefficient Theorem in topology.

If Y is another topological space, the Eilenberg-Zilber theorem 8.5.1 (see [MacH, VIII.8]) states that $H_*(X \times Y)$ is the homology of the tensor product complex $S(X) \otimes S(Y)$. Therefore the Künneth formula yields the "Künneth formula for cohomology:"

$$H_n(X \times Y) \cong \left\{ \bigoplus_{p=0}^{n} H_p(X) \otimes H_{n-p}(Y) \right\} \otimes \left\{ \bigoplus_{p=1}^{n} \operatorname{Tor}_1^{\mathbb{Z}}(H_{p-1}(X), H_{n-p}(Y)) \right\}.$$

We now turn to the analogue of the Künneth formula for Hom in place of $\otimes$.

Universal Coefficient Theorem for Cohomology 3.6.5 *Let P be a chain complex of projective R-modules such that each $d(P_n)$ is also projective. Then for every n and every R-module M, there is a (noncanonically) split exact sequence*

$$0 \to \operatorname{Ext}_R^1(H_{n-1}(P), M) \to H^n(\operatorname{Hom}_R(P, M)) \to \operatorname{Hom}_R(H_n(P), M) \to 0.$$

Proof Since $d(P_n)$ is projective, there is a (noncanonical) isomorphism $P_n \cong Z_n \oplus d(P_n)$ for each n. Therefore each sequence

$$0 \to \operatorname{Hom}(d(P_n), M) \to \operatorname{Hom}(P_n, M) \to \operatorname{Hom}(Z_n, M) \to 0$$

is exact. We may now copy the proof of the Künneth formula 3.6.1 for $\otimes$, using $\operatorname{Hom}(-, M)$ instead of $\otimes M$, to see that the sequence is indeed exact. We may copy the proof of the Universal Coefficient Theorem 3.6.2 for $\otimes$ in the same way to see that the sequence is split. ◇

Application 3.6.6 (Universal Coefficient theorem in topology) The cohomology of a topological space X with "coefficients" in M is defined to be

$$H^*(X; M) = H^*(\operatorname{Hom}(S(X), M)).$$

In this case, the Universal Coefficient theorem becomes

$$H^n(X; M) \cong \operatorname{Hom}(H_n(X), M) \oplus \operatorname{Ext}_{\mathbb{Z}}^1(H_{n-1}(X), M).$$

Example 3.6.7 If X is path-connected, then $H_0(X) = \mathbb{Z}$ and $H^1(X; \mathbb{Z}) \cong \text{Hom}(H_1(X), \mathbb{Z})$, which is a torsionfree abelian group.

Exercise 3.6.1 Let P be a chain complex and Q a cochain complex of R-modules. As in 2.7.4, form the Hom double cochain complex $\text{Hom}(P, Q) = \{\text{Hom}_R(P_p, Q^q)\}$, and then write $H^* \text{Hom}(P, Q)$ for the cohomology of $\text{Tot}(\text{Hom}(P, Q))$. Show that if each P_n and $d(P_n)$ is projective, there is an exact sequence

$$0 \to \prod_{\substack{p+q \\ n-1}} \text{Ext}_R^1(H_p(P), H^q(Q)) \to H^n \text{Hom}(P, Q) \to \prod_{\substack{p+q= \\ n}} \text{Hom}_R(H_p(P), H^q(Q)) \to 0.$$

Exercise 3.6.2 A ring R is called *right hereditary* if every submodule of every (right) free module is a projective module. (See 4.2.10 and exercise 4.2.6 below.) Any principal ideal domain (for example, $R = \mathbb{Z}$) is hereditary, as is any commutative Dedekind domain. Show that the universal coefficient theorems of this section remain valid if $\mathbb{Z}$ is replaced by an arbitrary right hereditary ring R.

4

Homological Dimension

4.1 Dimensions

Definitions 4.1.1 Let A be a right R-module.

1. The *projective dimension* $pd(A)$ is the minimum integer n (if it exists) such that there is a resolution of A by projective modules

$$0 \to P_n \to \cdots \to P_1 \to P_0 \to A \to 0.$$

2. The *injective dimension* $id(A)$ is the minimum integer n (if it exists) such that there is a resolution of A by injective modules

$$0 \to A \to E^0 \to E^1 \to \cdots \to E^n \to 0.$$

3. The *flat dimension* $fd(A)$ is the minimum integer n (if it exists) such that there is a resolution of A by flat modules

$$0 \to F_n \to \cdots \to F_1 \to F_0 \to A \to 0.$$

If no finite resolution exists, we set $pd(A)$, $id(A)$, or $fd(A)$ equal to ∞.

We are going to prove the following theorems in this section, which allow us to define the global and Tor dimensions of a ring R.

Global Dimension Theorem 4.1.2 The following numbers are the same for any ring R:

1. $\sup\{id(B) : B \in \mathbf{mod}\text{–}R\}$
2. $\sup\{pd(A) : A \in \mathbf{mod}\text{–}R\}$
3. $\sup\{pd(R/I) : I \text{ is a right ideal of } R\}$
4. $\sup\{d : \operatorname{Ext}_R^d(A, B) \neq 0 \text{ for some right modules } A, B\}$

This common number (possibly ∞) is called the (right) global dimension *of R*, *r.gl.* dim(R). *Bourbaki [BX] calls it the* homological dimension *of R.*

Remark One may define the left global dimension $\ell.gl.$ dim(R) similarly. If R is commutative, we clearly have $\ell.gl.$ dim$(R) = r.gl.$ dim(R). Equality also holds if R is left *and* right noetherian. Osofsky [Osof] proved that if every one-sided ideal can be generated by at most $\aleph_n$ elements, then $|\ell.gl.$ dim$(R) -$ $r.gl.$ dim$(R)| \leq n + 1$. The continuum hypothesis of set theory lurks at the fringe of this subject whenever we encounter non-constructible ideals over uncountable rings.

Tor-dimension Theorem 4.1.3 *The following numbers are the same for any ring R:*

1. sup$\{fd(A) : A$ *is a right R-module*$\}$
2. sup$\{fd(R/J) : J$ *is a right ideal of R*$\}$
3. sup$\{fd(B) : B$ *is a left R-module*$\}$
4. sup$\{fd(R/I) : I$ *is a left ideal of R*$\}$
5. sup$\{d : \operatorname{Tor}_d^R(A, B) \neq 0$ *for some R-modules A, B*$\}$

This common number (possibly ∞) is called the Tor-dimension *of R. Due to the influence of [CE], the less descriptive name* weak dimension *of R is often used.*

Example 4.1.4 Obviously every field has both global and Tor-dimension zero. The Tor and Ext calculations for abelian groups show that $R = \mathbb{Z}$ has global dimension 1 and Tor-dimension 1. The calculations for $R = \mathbb{Z}/m$ show that if some $p^2 | m$ (so R isn't a product of fields), then $\mathbb{Z}/m$ has global dimension ∞ and Tor-dimension ∞.

As projective modules are flat, $fd(A) \leq pd(A)$ for every R-module A. We need not have equality: over $\mathbb{Z}$, $fd(\mathbb{Q}) = 0$, but $pd(\mathbb{Q}) = 1$. Taking the supremum over all A shows that Tor–dim$(R) \leq r.gl.$ dim(R). We will see examples in the next section where Tor–dim$(R) \neq r.gl.$ dim(R). These examples are perforce non-noetherian, as we now prove, assuming the global and Tor-dimension theorems.

Proposition 4.1.5 *If R is right noetherian, then*

1. $fd(A) = pd(A)$ *for every finitely generated R-module A.*
2. Tor–dim$(R) = r.gl.$ dim(R).

Proof Since we can compute Tor–dim(R) and $r.gl.$ dim(R) using the modules R/I, it suffices to prove (1). Since $fd(A) \leq pd(A)$, it suffices to suppose that $fd(A) = n < \infty$ and prove that $pd(A) \le n$. As R is noetherian, there is a resolution

$$0 \to M \to P_{n-1} \to \cdots \to P_1 \to P_0 \to A \to 0$$

in which the P_i are finitely generated free modules and M is finitely presented. The fd lemma 4.1.10 below implies that the syzygy M is a flat R-module, so M must also be projective (3.2.7). This proves that $pd(A) \le n$, as required. $\diamond$

Exercise 4.1.1 Use the Tor-dimension theorem to prove that if R is both left and right noetherian, then $r.gl.\dim(R) = l.gl.\dim(R)$.

The pattern of proof for both theorems will be the same, so we begin with the characterization of projective dimension.

pd Lemma 4.1.6 *The following are equivalent for a right R-module A:*

1. $pd(A) \le d$.
2. $\mathrm{Ext}_R^n(A, B) = 0$ *for all* $n > d$ *and all R-modules B.*
3. $\mathrm{Ext}_R^{d+1}(A, B) = 0$ *for all R-modules B.*
4. *If* $0 \to M_d \to P_{d-1} \to P_{d-2} \to \cdots \to P_1 \to P_0 \to A \to 0$ *is any resolution with the P's projective, then the syzygy M_d is also projective.*

Proof Since $\mathrm{Ext}^*(A, B)$ may be computed using a projective resolution of A, it is clear that $(4) \Rightarrow (1) \Rightarrow (2) \Rightarrow (3)$. If we are given a resolution of A as in (4), then $\mathrm{Ext}^{d+1}(A, B) \cong \mathrm{Ext}^1(M_d, B)$ by dimension shifting. Now M_d is projective iff $\mathrm{Ext}^1(M_d, B) = 0$ for all B (exercise 2.5.2), so (3) implies (4). $\diamond$

Example 4.1.7 In 3.1.6 we produced an infinite projective resolution of $A = \mathbb{Z}/p$ over the ring $R = \mathbb{Z}/p^2$. Each syzygy was $\mathbb{Z}/p$, which is not a projective $\mathbb{Z}/p^2$-module. Therefore by (4) we see that $\mathbb{Z}/p$ has $pd = \infty$ over $R = \mathbb{Z}/p^2$. On the other hand, $\mathbb{Z}/p$ has $pd = 0$ over $R = \mathbb{Z}/p$ and $pd = 1$ over $R = \mathbb{Z}$.

The following two lemmas have the same proof as the preceding lemma.

id Lemma 4.1.8 *The following are equivalent for a right R-module B:*

1. $id(B) \le d$.
2. $\mathrm{Ext}_R^n(A, B) = 0$ *for all* $n > d$ *and all R-modules A.*
3. $\mathrm{Ext}_R^{d+1}(A, B) = 0$ *for all R-modules B.*
4. *If* $0 \to B \to E^0 \to \cdots \to E^{d-1} \to M^d \to 0$ *is a resolution with the E^i injective, then M^d is also injective.*

Example 4.1.9 In 3.1.6 we gave an infinite injective resolution of $B = \mathbb{Z}/p$ over $R = \mathbb{Z}/p^2$ and showed that $\mathrm{Ext}_R^n(\mathbb{Z}/p, \mathbb{Z}/p) \cong \mathbb{Z}/p$ for all n. Therefore $\mathbb{Z}/p$ has $id = \infty$ over $R = \mathbb{Z}/p^2$. On the other hand, it has $id = 0$ over $R = \mathbb{Z}/p$ and $id = 1$ over $\mathbb{Z}$.

fd Lemma 4.1.10 *The following are equivalent for a right R-module A:*

1. *$fd(A) \le d$.*
2. *$\mathrm{Tor}_n^R(A, B) = 0$ for all $n > d$ and all left R-modules B.*
3. *$\mathrm{Tor}_{d+1}^R(A, B) = 0$ for all left R-modules B.*
4. *If $0 \to M_d \to F_{d-1} \to F_{d-2} \to \cdots \to F_0 \to A \to 0$ is a resolution with the F_i all flat, then M_d is also a flat R-module.*

Lemma 4.1.11 *A left R-module B is injective iff $\mathrm{Ext}^1(R/I, B) = 0$ for all left ideals I.*

Proof Applying $\mathrm{Hom}(-, B)$ to $0 \to I \to R \to R/I \to 0$, we see that

$$\mathrm{Hom}(R, B) \to \mathrm{Hom}(I, B) \to \mathrm{Ext}^1(R/I, B) \to 0$$

is exact. By Baer's criterion 2.3.1, B is injective iff the first map is surjective, that is, iff $\mathrm{Ext}^1(R/I, B) = 0$.   ◇

Proof of Global Dimension Theorem The lemmas characterizing $pd(A)$ and $id(A)$ show that $\sup(2) = \sup(4) = \sup(1)$. As $\sup(2) \ge \sup(3)$, we may assume that $d = \sup\{pd(R/I)\}$ is finite and that $id(B) > d$ for some R-module B. For this B, choose a resolution

$$0 \to B \to E^0 \to E^1 \to \cdots \to E^{d-1} \to M \to 0$$

with the E's injective. But then for all ideals I we have

$$0 = \mathrm{Ext}_R^{d+1}(R/I, B) \cong \mathrm{Ext}_R^1(R/I, M).$$

By the preceding lemma 4.1.11, M is injective, a contradiction to $id(B) > d$.
                                                                              ◇

Proof of Tor-dimension theorem The lemma 4.1.10 characterizing $fd(A)$ over R shows that $\sup(5) = \sup(1) \ge \sup(2)$. The same lemma over R^{op} shows that $\sup(5) = \sup(3) \ge \sup(4)$. We may assume that $\sup(2) \le \sup(4)$, that is, that $d = \sup\{fd(R/J) : J \text{ is a right ideal}\}$ is at most the supremum over left ideals.

We are done unless d is finite and $fd(B) > d$ for some left R-module B. For this B, choose a resolution $0 \to M \to F_{d-1} \to \cdots \to F_0 \to B \to 0$ with the F's flat. But then for all ideals J we have

$$0 = \mathrm{Tor}^R_{d+1}(R/J, B) = \mathrm{Tor}^R_1(R/J, M).$$

We saw in 3.2.4 that this implies that M is flat, contradicting $fd(B) > d$. $\diamond$

Exercise 4.1.2 If $0 \to A \to B \to C \to 0$ is an exact sequence, show that

1. $pd(B) \le \max\{pd(A), pd(C)\}$ with equality except when $pd(C) = pd(A) + 1$.
2. $id(B) \le \max\{id(A), id(C)\}$ with equality except when $id(A) = id(C) + 1$.
3. $fd(B) \le \max\{fd(A), fd(C)\}$ with equality except when $fd(C) = fd(A) + 1$.

Exercise 4.1.3

1. Given a (possibly infinite) family $\{A_i\}$ of modules, show that

$$\mathrm{pd}\left(\bigoplus A_i\right) = \sup\{\mathrm{pd}(A_i)\}.$$

2. Conclude that if S is an R-algebra and P is a projective S-module considered as an R-module, the $pd_R(P) \le pd_R(S)$.
3. Show that if $r.gl.\dim(R) = \infty$, there actually is an R-module A with $pd(A) = \infty$.

4.2 Rings of Small Dimension

Definition 4.2.1 A ring R is called *(right) semisimple* if every right ideal is a direct summand of R or, equivalently, if R is the direct sum of its minimal ideals. Wedderburn's theorem (see [Lang]) classifies semisimple rings: they are finite products $R = \prod_{i=1}^r R_i$ of matrix rings $R_i = M_{n_i}(D_i) = \mathrm{End}_{D_i}(V_i)$ ($n_i = \dim(V_i)$) over division rings D_i. It follows that right semisimple is the same as left semisimple, and that every semisimple ring is (both left and right) noetherian. By Maschke's theorem, the group ring $k[G]$ of a finite group G over a field k is semisimple if char(k) doesn't divide the order of G.

Theorem 4.2.2 *The following are equivalent for every ring R, where by "R-module" we mean either left R-module or right R-module.*

   1. R is semisimple.
   2. R has (left and/or right) global dimension 0.
   3. Every R-module is projective.
   4. Every R-module is injective.
   5. R is noetherian, and every R-module is flat.
   6. R is noetherian and has Tor-dimension 0.

Proof We showed in the last section that (2) $\Leftrightarrow$ (3) $\Leftrightarrow$ (4) for left R-modules and also for right R-modules. R is semisimple iff every short exact sequence $0 \to I \to R \to R/I \to 0$ splits, that is, iff $pd(R/I) = 0$ for every (right and/or left) ideal I. This proves that (1)$\Leftrightarrow$ (2). As (1) and (3) imply (5), and (5)$\Leftrightarrow$ (6) by definition, we only have to show that (5) implies (1). If I is an ideal of R, then (5) implies that R/I is finitely presented and flat, hence projective by 3.2.7. Since R/I is projective, $R \to R/I$ splits, and I is a direct summand of R, that is, (1) holds. $\diamond$

Definition 4.2.3 A ring R is *quasi-Frobenius* if it is (left and right) noetherian and R is an injective (left and right) R-module. Our interest in quasi-Frobenius rings stems from the following result of Faith and Faith-Walker, which we quote from [Faith].

Theorem 4.2.4 *The following are equivalent for every ring R:*

   1. R is quasi-Frobenius.
   2. Every projective right R-module is injective.
   3. Every injective right R-module is projective.
   4. Every projective left R-module is injective.
   5. Every injective left R-module is projective.

Exercise 4.2.1 Show that $\mathbb{Z}/m$ is a quasi-Frobenius ring for every integer m.

Exercise 4.2.2 Show that if R is quasi-Frobenius, then either R is semisimple or R has global dimension ∞. *Hint:* Every finite projective resolution is split.

Definition 4.2.5 A *Frobenius algebra* over a field k is a finite-dimensional algebra R such that $R \cong \text{Hom}_k(R, k)$ as (right) R-modules. Frobenius algebras are quasi-Frobenius; more generally, $\text{Hom}_k(R, k)$ is an injective R-module for any algebra R over any field k, since k is an injective k-module and $\text{Hom}_k(R, -)$ preserves injectives (being right adjoint to the forgetful functor **mod**–$R \to$ **mod**–k). Frobenius algebras were introduced in 1937 by Brauer and Nesbitt in order to generalize group algebras $k[G]$ of a finite group, especially when char$(k) = p$ divides the order of G so that $k[G]$ is not semisimple.

Proposition 4.2.6 *If G is a finite group, then k[G] is a Frobenius algebra.*

Proof Set $R = k[G]$ and define $f: R \to k$ by letting $f(r)$ be the coefficient of $g = 1$ in the unique expression $r = \sum_{g \in G} r_g g$ of every element $r \in k[G]$. Let $\alpha: R \to \operatorname{Hom}_k(R, k)$ be the map $\alpha(r): x \mapsto f(rx)$. Since $\alpha(r) = fr$, α is a right R-module map; we claim that α is an isomorphism. If $\alpha(r) = 0$ for $r = \sum r_g g$, then $r = 0$ as each $r_g = f(rg^{-1}) = \alpha(r)(g^{-1}) = 0$. Hence α is an injection. As R and $\operatorname{Hom}_k(R, k)$ have the same finite dimension over k, α must be an isomorphism. $\diamond$

Vista 4.2.7 Let R be a commutative noetherian ring. R is called a *Gorenstein ring* if $id(R)$ is finite; in this case $id(R)$ is the Krull dimension of R, defined in 4.4.1. Therefore a quasi-Frobenius ring is just a Gorenstein ring of Krull dimension zero, and in particular a finite product of 0-dimensional local rings. If R is a 0-dimensional local ring with maximal ideal $\mathfrak{m}$, then R is quasi-Frobenius $\Leftrightarrow \operatorname{ann}_R(\mathfrak{m}) = \{r \in R : r\mathfrak{m} = 0\} \cong R/\mathfrak{m}$. This recognition criterion is at the heart of current research into the Gorenstein rings that arise in algebraic geometry.

Now we shall characterize rings of Tor-dimension zero. A ring R is called *von Neumann regular* if for every $a \in R$ there is an $x \in R$ for which $axa = a$. These rings were introduced by J. von Neumann in 1936 in order to study continuous geometries such as the lattices of projections in "von Neumann algebras" of bounded operators on a Hilbert space. For more information about von Neumann regular rings, see [Good].

Remark A commutative ring R is von Neumann regular iff R has no nilpotent elements and has Krull dimension zero. On the other hand, a commutative ring R is semisimple iff it is a finite product of fields.

Exercise 4.2.3 Show that an infinite product of fields is von Neumann regular. This shows that not every von Neumann regular ring is semisimple.

Exercise 4.2.4 If V is a vector space over a field k (or a division ring k), show that $R = End_k(V)$ is von Neumann regular. Show that R is semisimple iff $\dim_k(V) < \infty$.

Lemma 4.2.8 *If R is von Neumann regular and I is a finitely generated right ideal of R, then there is an* idempotent e *(an element with $e^2 = e$) such that $I = eR$. In particular, I is a projective R-module, because $R \cong I \oplus (1 - e)R$.*

Proof Suppose first that $I = aR$ and that $axa = a$. It follows that $e = ax$ is idempotent and that $I = eR$. By induction on the number of generators of

I, we may suppose that $I = aR + bR$ with $a \in I$ idempotent. Since $bR = abR + (1 - a)bR$, we have $I = aR + cR$ for $c = (1 - a)b$. If $cyc = c$, then $f = cy$ is idempotent and $af = a(1 - a)by = 0$. As fa may not vanish, we consider $e = f(1 - a)$. Then $e \in I$, $ae = 0 = ea$, and e is idempotent:

$$e^2 = f(1 - a)f(1 - a) = f(f - af)(1 - a) = f^2(1 - a) = f(1 - a) = e.$$

Moreover, $eR = cR$ because $c = fc = ffc = f(1 - a)fc = efc$. Finally, we claim that I equals $J = (a + e)R$. Since $a + e \in I$, we have $J \subseteq I$; the reverse inclusion follows from the observation that $a = (a + e)a \in J$ and $e = (a + e)e \in J$.   ◇

Exercise 4.2.5 Show that the converse holds: If every fin. gen. right ideal I of R is generated by an idempotent (*i.e.*, $R \cong I \oplus R/I$), then R is von Neumann regular.

Theorem 4.2.9 *The following are equivalent for every ring R:*

1. *R is von Neumann regular.*
2. *R has Tor-dimension 0.*
3. *Every R-module is flat.*
4. *R/I is projective for every finitely generated ideal I.*

Proof By definition, (2) ⇔ (3). If I is a fin. generated ideal, then R/I is finitely presented. Thus R/I is flat iff it is projective, hence iff $R \cong I \oplus R/I$ as a module. Therefore (3) ⇒ (4) ⇔ (1). Finally, any ideal I is the union of its finitely generated subideals I_α, and we have $R/I = \varinjlim(R/I_\alpha)$. Hence (4) implies that each R/I is flat, that is, that (2) holds.   ◇

Remark Since the Tor-dimension of a ring is at most the global dimension, noetherian von Neumann regular rings must be semisimple (4.1.5). Von Neumann regular rings that are not semisimple show that we can have Tor–dim$(R) <$ gl. dim(R). For example, the global dimension of $\prod_{i=1}^{\infty} \mathbb{C}$ is ≥ 2, with equality iff the Continuum Hypothesis holds.

Definition 4.2.10 A ring R is called (*right*) *hereditary* if every right ideal is projective. A commutative integral domain R is hereditary iff it is a *Dedekind domain* (noetherian, Krull dimension 0 or 1 and every local ring R_m is a discrete valuation ring). Principal ideal domains (*e.g*, $\mathbb{Z}$ or $k[t]$) are Dedekind, and of course every semisimple ring is hereditary.

Theorem 4.2.11 *A ring R is right hereditary iff r.gl. dim$(R) \leq 1$.*

Proof The exact sequences $0 \to I \to R \to R/I \to 0$ show that R is hereditary iff $r.gl. \dim(R) \le 1$.   ◇

Exercise 4.2.6 Show that R is right hereditary iff every submodule of every
free module is projective. This was used in exercise 3.6.2.

4.3 Change of Rings Theorems

General Change of Rings Theorem 4.3.1 *Let $f: R \to S$ be a ring map, and
let A be an S-module. Then as an R-module*

$$pd_R(A) \le pd_S(A) + pd_R(S).$$

Proof There is nothing to prove if $pd_S(A) = \infty$ or $pd_R(S) = \infty$, so assume
that $pd_S(A) = n$ and $pd_R(S) = d$ are finite. Choose an S-module projective
resolution $Q \to A$ of length n. Starting with R-module projective resolutions
of A and of each syzygy in Q, the Horseshoe Lemma 2.2.8 gives us R-module
projective resolutions $\tilde{P}_{*q} \to Q_q$ such that $\tilde{P}_{*q} \to \tilde{P}_{*,q-2}$ is zero. We saw in
section 4.1 that $pd_R(Q_q) \le d$ for each q. The truncated resolutions $P_{*q} \to Q_q$
of length d ($P_{iq} = 0$ for $i > d$ and $P_{dq} = \tilde{P}_{dq}/\operatorname{im}(\tilde{P}_{d+1,q})$, as in 1.2.7) have
the same property. By the sign trick, we have a double complex P_{**} and an
augmentation $P_{0*} \to Q_*$.

$$
\begin{array}{c|ccccc}
0 & 0 & 0 & & 0 \\
\downarrow & \downarrow & \downarrow & & \downarrow \\
Q_n & P_{0n} \leftarrow & P_{1n} \leftarrow & \cdots \leftarrow \cdots \leftarrow & P_{dn} \leftarrow 0 \\
\downarrow & \downarrow & \downarrow & & \downarrow \\
\cdots & \cdots & \cdots & \cdots & \cdots \\
\downarrow & \downarrow & \downarrow & & \downarrow \\
Q_1 & P_{01} \leftarrow & P_{11} \leftarrow & P_{21} \leftarrow \cdots \leftarrow & P_{d1} \leftarrow 0 \\
\downarrow & \downarrow & \downarrow & \downarrow & \downarrow \\
Q_0 & P_{00} \leftarrow & P_{10} \leftarrow & P_{20} \leftarrow \cdots \leftarrow & P_{d0} \leftarrow 0 \\
\downarrow & \downarrow & \downarrow & \downarrow & \downarrow \\
0 & 0 & 0 & 0 & 0 \\
\end{array}
$$

The argument used in 2.7.2 to balance Tor shows that $\operatorname{Tot}(P) \to Q$ is a quasi-
isomorphism, because the rows of the augmented double complex (add $Q[-1]$

in column -1) are exact. Hence $\text{Tot}(P) \to A$ is an R-module projective resolution of A. But then $pd_R(A)$ is at most the length of $\text{Tot}(P)$, that is, $d + n$.

$\diamond$

Example 4.3.2 If R is a field and $pd_S(A) \neq 0$, we have strict inequality.

Remark The above argument presages the use of spectral sequences in getting more explicit information about $\text{Ext}_R^*(A, B)$. An important case in which we have equality is the case $S = R/xR$ when x is a nonzerodivisor, so $pd_R(R/xR) = 1$.

First Change of Rings Theorem 4.3.3 *Let x be a central nonzerodivisor in a ring R. If $A \neq 0$ is a R/x-module with $pd_{R/x}(A)$ finite, then*

$$pd_R(A) = 1 + pd_{R/x}(A).$$

Proof As $xA = 0$, A cannot be a projective R-module, so $pd_R(A) \geq 1$. On the other hand, if A is a projective R/x-module, then evidently $pd_R(A) = pd_R(R/x) = 1$. If $pd_{R/x}(A) \geq 1$, find an exact sequence

$$0 \to M \to P \to A \to 0$$

with P a projective R/x-module, so that $pd_{R/x}(A) = pd_{R/x}(M) + 1$. By induction, $pd_R(M) = 1 + pd_{R/x}(M) = pd_{R/x}(A) \geq 1$. Either $pd_R(A)$ equals $pd_R(M) + 1$ or $1 = pd_R(P) = \sup\{pd_R(M), pd_R(A)\}$. We shall conclude the proof by eliminating the possibility that $pd_R(A) = 1 = pd_{R/x}(A)$.

Map a free R-module F onto A with kernel K. If $pd_R(A) = 1$, then K is a projective R-module. Tensoring with R/xR yields the sequence of R/x-modules:

$$0 \to \text{Tor}_1^R(A, R/x) \to K/xK \to F/xF \to A \to 0.$$

If $pd_{R/x}(A) \leq 2$, then $\text{Tor}_1^R(A, R/x)$ is a projective R/x-module. But

$$\text{Tor}_1^R(A, R/x) \cong \{a \in A : xa = 0\} = A, \quad \text{so } pd_{R/x}(A) = 0. \qquad \diamond$$

Example 4.3.4 The conclusion of this theorem fails if $pd_{R/x}(A) = \infty$ but $pd_R(A) < \infty$. For example, $pd_{\mathbb{Z}/4}(\mathbb{Z}/2) = \infty$ but $pd_{\mathbb{Z}}(\mathbb{Z}/2) = 1$.

Exercise 4.3.1 Let R be the power series ring $k[[x_1, \cdots, x_n]]$ over a field k. R is a noetherian local ring with residue field k. Show that $gl.\dim(R) = pd_R(k) = n$.

Second Change of Rings Theorem 4.3.5 *Let x be a central nonzerodivisor in a ring R. If A is an R-module and x is a nonzerodivisor on A (i.e., $a \neq 0 \Rightarrow xa \neq 0$), then*

$$pd_R(A) \geq pd_{R/x}(A/xA).$$

Proof If $pd_R(A) = \infty$, there is nothing to prove, so we assume $pd_R(A) = n < \infty$ and proceed by induction on n. If A is a projective R-module, then A/xA is a projective R/x-module, so the result is true if $pd_R(A) = 0$. If $pd_R(A) \neq 0$, map a free R-module F onto A with kernel K. As $pd_R(K) = n - 1$, $pd_{R/x}(K/xK) \leq n - 1$ by induction. Tensoring with R/x yields the sequence

$$0 \to \mathrm{Tor}_1^R(A, R/x) \to K/xK \to F/xF \to A/xA \to 0.$$

As x is a nonzerodivisor on A, $\mathrm{Tor}_1(A, R/x) \cong \{a \in A : xa = 0\} = 0$. Hence either A/xA is projective or $pd_{R/x}(A/xA) = 1 + pd_{R/x}(K/xK) \leq 1 + (n - 1) = pd_R(A)$. $\diamond$

Exercise 4.3.2 Use the first Change of Rings Theorem 4.3.3 to find another proof when $pd_{R/x}(A/xA)$ is finite.

Now let $R[x]$ be a polynomial ring in one variable over R. If A is an R-module, write $A[x]$ for the $R[x]$-module $R[x] \otimes_R A$.

Corollary 4.3.6 $pd_{R[x]}(A[x]) = pd_R(A)$ *for every R-module A.*

Proof Writing $T = R[x]$, we note that x is a nonzerodivisor on $A[x] = T \otimes_R A$. Hence $pd_T(A[x]) \geq pd_R(A)$ by the second Change of Rings theorem 4.3.5. On the other hand, if $P \to A$ is an R-module projective resolution, then $T \otimes_R P \to T \otimes_R A$ is a T-module projective resolution (T is flat over R), so $pd_R(A) \geq pd_T(T \otimes A)$. $\diamond$

Theorem 4.3.7 *If $R[x_1, \cdots, x_n]$ denotes a polynomial ring in n variables, then $gl.\dim(R[x_1, \cdots, x_n]) = n + gl.\dim(R)$.*

Proof It suffices to treat the case $T = R[x]$. If $gl.\dim(R) = \infty$, then by the above corollary $gl.\dim(T) = \infty$, so we may assume $gl.\dim(R) = n < \infty$. By the first Change of Rings theorem 4.3.3, $gl.\dim(T) \geq 1 + gl.\dim(R)$. Given a T-module M, write $U(M)$ for the underlying R-module and consider the sequence of T-modules

$(*)$ $$0 \to T \otimes_R U(M) \xrightarrow{\beta} T \otimes_R U(M) \xrightarrow{\mu} M \to 0,$$

where μ is multiplication and β is defined by the bilinear map $\beta(t \otimes m) = t[x \otimes m - 1 \otimes (xm)]$ $(t \in T, m \in M)$. We claim that $(*)$ is exact, which yields the inequality $pd_T(M) \leq 1 + pd_T(T \otimes_R U(M)) = 1 + pd_R(U(M)) \leq 1 + n$. The supremum over all M gives the final inequality $gl.\dim(T) \leq 1 + n$.

To finish the proof, we must establish the claim that $(*)$ is exact. We first observe that, since T is a free R-module on basis $\{1, x, x^2, \cdots\}$, we can write every nonzero element f of $T \otimes U(M)$ as a polynomial with coefficients $m_i \in M$:

$$f = x^k \otimes m_k + \cdots + x^2 \otimes m_2 + x \otimes m_1 + 1 \otimes m_0 \quad (m_k \neq 0).$$

Since the leading term of $\beta(f)$ is $x^{k+1} \otimes m_k$, we see that β is injective. Clearly $\mu\beta = 0$. Finally, we prove by induction on k (the degree of f) that if $f \in \ker(\mu)$, then $f \in im(\beta)$. Since $\mu(1 \otimes m) = m$, the case $k = 0$ is trivial (if $\mu(f) = 0$, then $f = 0$). If $k \neq 0$, then $\mu(f) = \mu(g)$ for the polynomial $g = f - \beta(x^{k-1} \otimes m_k)$ of lower degree. By induction, if $f \in \ker(\mu)$, then $g = \beta(h)$ for some h, and hence $f = \beta(h + x^{k-1} \otimes m_k)$. $\diamond$

Corollary 4.3.8 (Hilbert's theorem on syzygies) *If k is a field, then the polynomial ring $k[x_1, \cdots, x_n]$ has global dimension n. Thus the $(n-1)^{st}$ syzygy of every module is a projective module.* $\diamond$

We now turn to the third Change of Rings theorem. For simplicity we deal with commutative local rings, that is, commutative rings with a unique maximal ideal. Here is the fundamental tool used to study local rings.

Nakayama's Lemma 4.3.9 *Let R be a commutative local ring with unique maximal ideal $\mathfrak{m}$ and let B be a nonzero finitely generated R-module. Then*

 1. $B \neq \mathfrak{m}B$.
 2. If $A \subseteq B$ is a submodule such that $B = A + \mathfrak{m}B$, then $A = B$.

Proof If we consider B/A then (2) is a special case of (1). Let m be the smallest integer such that B is generated $b_1, \cdots, b_m$; as $B \neq 0$, we have $m \neq 0$. If $B = \mathfrak{m}B$, then there are $r_i \in \mathfrak{m}$ such that $b_m = \sum r_i b_i$. This yields

$$(1 - r_m)b_m = r_1 b_1 + \cdots + r_{m-1}b_{m-1}.$$

Since $1 - r_m \notin \mathfrak{m}$, it is a unit of R. Multiplying by its inverse writes b_m as a linear combination of $\{b_1, \cdots, b_{m-1}\}$, so this set also generates B. This contradicts the choice of m. $\diamond$

Remark If R is any ring, the set

$$J = \{r \in R : (\forall s \in R) \; 1 - rs \quad \text{is a unit of} \; R\}$$

is a 2-sided ideal of R, called the *Jacobson radical* of R (see [BAII, 4.2]). The above proof actually proves the following:

General Version of Nakayama's Lemma 4.3.10 Let B be a nonzero finitely generated module over R and J the Jacobson radical of R. Then $B \neq JB$.

Proposition 4.3.11 *A finitely generated projective module P over a commutative local ring R is a free module.*

Proof Choose $u_1, \cdots, u_n \in P$ whose images form a basis of the k-vector space $P/\mathfrak{m}P$. By Nakayama's lemma the u's generate P, so the map $\epsilon: R^n \to P$ sending $(r_1, \cdots, r_n)$ to $\sum r_i u_i$ is onto. As P is projective, ϵ is split, that is, $R^n \cong P \oplus \ker(\epsilon)$. As $k^n = R^n/\mathfrak{m}R^n \cong P/\mathfrak{m}P$, we have $\ker(\epsilon) \subseteq \mathfrak{m}R^n$. But then considering P as a submodule of R^n we have $R^n = P + \mathfrak{m}R^n$, so Nakayama's lemma yields $R^n = P$. $\diamond$

Third Change of Rings Theorem 4.3.12 *Let R be a commutative noetherian local ring with unique maximal ideal $\mathfrak{m}$, and let A be a finitely generated R-module. If $x \in \mathfrak{m}$ is a nonzerodivisor on both A and R, then*

$$pd_R(A) = pd_{R/x}(A/xA).$$

Proof We know $\geq$ holds by the second Change of Rings theorem 4.3.5, and we shall prove equality by induction on $n = pd_{R/x}(A/xA)$. If $n = 0$, then A/xA is projective, hence a free R/x-module because R/x is local.

Lemma 4.3.13 *If A/xA is a free R/x-module, A is a free R-module.*

Proof Pick elements $u_1, \cdots, u_n$ mapping onto a basis of A/xA; we claim they form a basis of A. Since $(u_1, \cdots, u_n)R + xA = A$, Nakayama's lemma states that $(u_1, \cdots, u_n)R = A$, that is, the u's span A. To show the u's are linearly independent, suppose $\sum r_i u_i = 0$ for $r_i \in R$. In A/xA, the images of the u's are linearly independent, so $r_i \in xR$ for all i. As x is a nonzerodivisor on R and A, we can divide to get $r_i/x \in R$ such that $\sum (r_i/x)u_i = 0$. Continuing this process, we get a sequence of elements $r_i, r_i/x, r_i/x^2, \cdots$ which generates a strictly ascending chain of ideals of R, unless $r_i = 0$. As R is noetherian, all the r_i must vanish. $\diamond$

Resuming the proof of the theorem, we establish the inductive step $n \neq 0$. Map a free R-module F onto A with kernel K. As $\operatorname{Tor}_1^R(A, R/x) = \{a \in A : xa = 0\} = 0$, tensoring with R/x yields the exact sequence

$$0 \to K/xK \to F/xF \to A/xA \to 0.$$

As F/xF is free, $pd_{R/x}(K/xK) = n - 1$ when $n \neq 0$. As R is noetherian, K is finitely generated, so by induction, $pd_R(K) = n - 1$. This implies that $pd_R(A) = n$, finishing the proof of the third Change of Rings theorem.   ◇

Remark The third Change of Rings theorem holds in the generality that R is right noetherian, and $x \in R$ is a central element lying in the Jacobson radical of R. To prove this, reread the above proof, using the generalized version 4.3.10 of Nakayama's lemma.

Corollary 4.3.14 *Let R be a commutative noetherian local ring, and let A be a finitely generated R-module with $pd_R(A) < \infty$. If $x \in \mathfrak{m}$ is a nonzerodivisor on both A and R, then*

$$pd_R(A/xA) = 1 + pd_R(A).$$

Proof Combine the first and third Change of Rings theorems.   ◇

Exercise 4.3.3 (Injective Change of Rings Theorems) Let x be a central nonzerodivisor in a ring R and let A be an R-module. Prove the following.

First Theorem. If $A \neq 0$ is an R/xR-module with $id_{R/xR}(A)$ finite, then

$$id_R(A) = 1 + id_{R/xR}(A).$$

Second Theorem. If x is a nonzerodivisor on both R and A, then either A is injective (in which case $A/xA = 0$) or else

$$id_R(A) \geq 1 + id_{R/xR}(A/xA).$$

Third Theorem. Suppose that R is a commutative noetherian local ring, A is finitely generated, and that $x \in \mathfrak{m}$ is a nonzerodivisor on both R and A. Then

$$id_R(A) = id_R(A/xA) = 1 + id_{R/xR}(A/xA).$$

4.4 Local Rings

In this section a *local ring* R will mean a commutative noetherian local ring R with a unique maximal ideal $\mathfrak{m}$. The residue field of R will be denoted $k = R/\mathfrak{m}$.

Definitions 4.4.1 The *Krull dimension* of a ring R, $\dim(R)$, is the length d of the longest chain $\mathfrak{p}_0 \subset \mathfrak{p}_1 \subset \cdots \subset \mathfrak{p}_d$ of prime ideals in R; $\dim(R) < \infty$ for every local ring R. The *embedding dimension* of a local ring R is the finite number

$$emb. \dim(R) = \dim_k(\mathfrak{m}/\mathfrak{m}^2).$$

For any local ring we have $\dim(R) \le emb. \dim(R)$; R is called a *regular local ring* if we have equality, that is, if $\dim(R) = \dim_k(\mathfrak{m}/\mathfrak{m}^2)$. Regular local rings have been long studied in algebraic geometry because the local coordinate rings of smooth algebraic varieties are regular local rings.

Examples 4.4.2 A regular local ring of dimension 0 must be a field. Every 1-dimensional regular local ring is a discrete valuation ring. The power series ring $k[[x_1, \cdots, x_n]]$ over a field k is regular local of dimension n, as is the local ring $k[x_1, \cdots, x_n]_\mathfrak{m}$, $\mathfrak{m} = (x_1, \cdots, x_n)$.

Let R be the local ring of a complex algebraic variety X at a point P. The embedding dimension of R is the smallest integer n such that some analytic neighborhood of P in X embeds in $\mathbf{C}^n$. If the variety X is smooth as a manifold, R is a regular local ring and $\dim(R) = \dim(X)$.

More Definitions 4.4.3 If A is a finitely generated R-module, a *regular sequence on A*, or *A-sequence*, is a sequence $(x_1, \cdots, x_n)$ of elements in $\mathfrak{m}$ such that x_1 is a nonzerodivisor on A (*i.e.*, if $a \ne 0$, then $x_1 a \ne 0$) and such that each x_i $(i > 1)$ is a nonzerodivisor on $A/(x_1, \cdots, x_{i-1})A$. The *grade* of A, $G(A)$, is the length of the longest regular sequence on A. For any local ring R we have $G(R) \le \dim(R)$.

R is called *Cohen-Macaulay* if $G(R) = \dim(R)$. We will see below that regular local rings are Cohen-Macaulay; in fact, any $x_1, \cdots, x_d \in \mathfrak{m}$ mapping to a basis of $\mathfrak{m}/\mathfrak{m}^2$ will be an R-sequence; by Nakayama's lemma they will also generate $\mathfrak{m}$ as an ideal. For more details, see [KapCR].

Examples 4.4.4 Every 0-dimensional local ring R is Cohen-Macaulay (since $G(R) = 0$), but cannot be a regular local ring unless R is a field. The 1-dimensional local ring $k[[x, \epsilon]]/(x\epsilon = \epsilon^2 = 0)$ is not Cohen-Macaulay; every element of $\mathfrak{m} = (x, \epsilon)R$ kills $\epsilon \in R$. Unless the maximal ideal consists entirely of zerodivisors, a 1-dimensional local ring R is always Cohen-Macaulay; R is regular only when it is a discrete valuation ring. For example, the local ring $k[[x]]$ is a discrete valuation ring, and the subring $k[[x^2, x^3]]$ is Cohen-Macaulay of dimension 1 but is not a regular local ring.

Exercise 4.4.1 If R is a regular local ring and $x_1, \cdots, x_d \in \mathfrak{m}$ map to a basis of $\mathfrak{m}/\mathfrak{m}^2$, show that each quotient ring $R/(x_1, \cdots, x_i)R$ is regular local of dimension $d - i$.

Proposition 4.4.5 *A regular local ring is an integral domain.*

Proof We use induction on $\dim(R)$. Pick $x \in \mathfrak{m} - \mathfrak{m}^2$; by the above exercise, R/xR is regular local of dimension $\dim(R) - 1$. Inductively, R/xR is a domain, so xR is a prime ideal. If there is a prime ideal Q properly contained in xR, then $Q \subset x^n R$ for all n (inductively, if $q = rx^n \in Q$, then $r \in Q \subset xR$, so $q \in x^{n+1}R$). In this case $Q \subseteq \cap x^n R = 0$, whence $Q = 0$ and R is a domain. If R were not a domain, this would imply that xR is a minimal prime ideal of R for all $x \in \mathfrak{m} - \mathfrak{m}^2$. Hence $\mathfrak{m}$ would be contained in the union of $\mathfrak{m}^2$ and the finitely many minimal prime ideals $P_1, \cdots, P_t$ of R. This would imply that $\mathfrak{m} \subseteq P_i$ for some i. But then $\dim(R) = 0$, a contradiction. $\diamond$

Corollary 4.4.6 *If R is a regular local ring, then $G(R) = \dim(R)$, and any $x_1, \cdots, x_d \in \mathfrak{m}$ mapping to a basis of $\mathfrak{m}/\mathfrak{m}^2$ is an R−sequence.*

Proof As $G(R) \leq \dim(R)$, and $x_1 \in R$ is a nonzerodivisor on R, it suffices to prove that $x_2, \cdots, x_d$ form a regular sequence on $R/x_1 R$. This follows by induction on d. $\diamond$

Exercise 4.4.2 Let R be a regular local ring and I an ideal such that R/I is also regular local. Prove that $I = (x_1, \cdots, x_i)R$, where $(x_1, \cdots, x_i)$ form a regular sequence in R.

Standard Facts 4.4.7 Part of the standard theory of associated prime ideals in commutative noetherian rings implies that if every element of $\mathfrak{m}$ is a zerodivisor on a finitely generated R-module A, then $\mathfrak{m}$ equals $\{r \in R : ra = 0\}$ for some nonzero $a \in A$ and therefore $aR \cong R/\mathfrak{m} = k$. Hence if $G(A) = 0$, then $\operatorname{Hom}_R(k, A) \neq 0$.

If $G(A) \neq 0$ and $G(R) \neq 0$, then some element of $\mathfrak{m} - \mathfrak{m}^2$ must also be a nonzerodivisor on both R and A. Again, this follows from the standard theory of associated prime ideals. Another standard fact is that if $x \in \mathfrak{m}$ is a nonzerodivisor on R, then the Krull dimension of R/xR is $\dim(R) - 1$.

Theorem 4.4.8 *If R is a local ring and $A \neq 0$ is a finitely generated R-module, then every maximal A-sequence has the same length, $G(A)$. Moreover, $G(A)$ is characterized as the smallest n such that $\operatorname{Ext}_R^n(k, A) \neq 0$.*

Proof We saw above that if $G(A) = 0$, then $\mathrm{Hom}_R(k, A) \neq 0$. Conversely, if $\mathrm{Hom}_R(k, A) \neq 0$, then some nonzero $a \in A$ has $aR \cong k$, that is, $ax = 0$ for all $x \in \mathfrak{m}$. In this case $G(A) = 0$ is clear. We now proceed by induction on the length n of a maximal regular A-sequence $x_1, \cdots, x_n$ on A. If $n \geq 1$, $x = x_1$ is a nonzerodivisor on A, so the sequence $0 \to A \xrightarrow{x} A \to A/xA \to 0$ is exact, and $x_2, \cdots, x_n$ is a maximal regular sequence on A/xA. This yields the exact sequence

$$\mathrm{Ext}^{i-1}(k, A) \xrightarrow{x} \mathrm{Ext}^{i-1}(k, A) \to \mathrm{Ext}^{i-1}(k, A/xA) \to \mathrm{Ext}^i(k, A) \xrightarrow{x} \mathrm{Ext}^i(k, A).$$

Now $xk = 0$, so $\mathrm{Ext}^i(k, A)$ is an R/xR-module. Hence the maps "x" in this sequence are zero. By induction, this proves that $\mathrm{Ext}^i(k, A) = 0$ for $0 \leq i < n$ and that $\mathrm{Ext}^n(k, A) \neq 0$. This finishes the inductive step, proving the theorem.
$\diamond$

Remark The injective dimension $id(A)$ is the largest integer n such that $\mathrm{Ext}_R^n(k, A) \neq 0$. This follows from the next result, which we cite without proof from [KapCR, section 4.5] because the proof involves more ring theory than we want to use.

Theorem 4.4.9 *If R is a local ring and A is a finitely generated R-module, then*

$$id(A) \leq d \Leftrightarrow \mathrm{Ext}_R^n(k, A) = 0 \text{ for all } n > d.$$

Corollary 4.4.10 *If R is a Gorenstein local ring (i.e., $id_R(R) < \infty$), then R is also Cohen-Macaulay. In this case $G(R) = id_R(R) = \dim(R)$ and*

$$\mathrm{Ext}_R^q(k, R) \neq 0 \Leftrightarrow q = \dim(R).$$

Proof The last two theorems imply that $G(R) \leq id(R)$. Now suppose that $G(R) = 0$ but that $id(R) \neq 0$. For each $s \in R$ and $n \geq 0$ we have an exact sequence

$$\mathrm{Ext}_R^n(R, R) \to \mathrm{Ext}_R^n(sR, R) \to \mathrm{Ext}_R^{n+1}(R/sR, R).$$

For $n = id(R) > 0$, the outside terms vanish, so $\mathrm{Ext}_R^n(sR, R) = 0$ as well. Choosing $s \in R$ so that $sR \cong k$ contradicts the previous theorem so if $G(R) = 0$ then $id(R) = 0$. If $G(R) = d > 0$, choose a nonzerodivisor $x \in \mathfrak{m}$ and set $S = R/xR$. By the third Injective Change of Rings theorem (exercise

4.3.3), $id_S(S) = id_R(R) - 1$, so S is also a Gorenstein ring. Inductively, S is Cohen-Macaulay, and $G(S) = id_S(S) = \dim(S) = \dim(R) - 1$. Hence $id_R(R) = \dim(R)$. If $x_2, \cdots, x_d$ are elements of $\mathfrak{m}$ mapping onto a maximal S-sequence in $\mathfrak{m}S$, then $x_1, x_2, \cdots, x_d$ forms a maximal R-sequence, that is, $G(R) = 1 + G(S) = \dim(R)$. $\diamond$

Proposition 4.4.11 *If R is a local ring with residue field k, then for every finitely generated R-module A and every integer d*

$$pd(A) \leq d \Leftrightarrow \mathrm{Tor}_{d+1}^R(A, k) = 0.$$

In particular, $pd(A)$ is the largest d such that $\mathrm{Tor}_d^R(A, k) \neq 0$.

Proof As $fd(A) \leq pd(A)$, the $\Rightarrow$ direction is clear. We prove the converse by induction on d. Nakayama's lemma 4.3.9 states that the finitely generated R-module A can be generated by $m = \dim_k(A/\mathfrak{m}A)$ elements. Let $\{u_1, \cdots, u_m\}$ be a minimal set of generators for A, and let K be the kernel of the surjection $\epsilon: R^m \to A$ defined by $\epsilon(r_1, \cdots, r_m) = \sum r_i u_i$. The inductive step is clear, since if $d \neq 0$, then

$$\mathrm{Tor}_{d+1}(A, k) = \mathrm{Tor}_d(K, k) \text{ and } pd(A) \leq 1 + pd(K).$$

If $d = 0$, then the assumption that $\mathrm{Tor}_1(A, k) = 0$ gives exactness of

$$
\begin{array}{ccccccccc}
0 & \longrightarrow & K \otimes k & \longrightarrow & R^m \otimes k & \longrightarrow & A \otimes k & \longrightarrow & 0 \\
& & \| & & \| & & \| & & \\
0 & \longrightarrow & K/\mathfrak{m}K & \longrightarrow & k^m & \xrightarrow{\epsilon \otimes k} & A/\mathfrak{m}A & \longrightarrow & 0.
\end{array}
$$

By construction, the map $\epsilon \otimes k$ is an isomorphism. Hence $K/\mathfrak{m}K = 0$, so the finitely generated R-module K must be zero by Nakayama's lemma. This forces $R^m \cong A$, so $pd(A) = 0$ as asserted. $\diamond$

Corollary 4.4.12 *If R is a local ring, then $gl.\dim(R) = pd_R(R/\mathfrak{m})$.*

Proof $pd(R/\mathfrak{m}) \leq gl.\dim(R) = \sup\{pd(R/I)\} \leq fd(R/\mathfrak{m}) \leq pd(R/\mathfrak{m})$. $\diamond$

Corollary 4.4.13 *If R is local and $x \in \mathfrak{m}$ is a nonzerodivisor on R, then either $gl.\dim(R/xR) = \infty$ or $gl.\dim(R) = 1 + gl.\dim(R/xR)$.*

Proof Set $S = R/xR$ and suppose that $gl.\dim(S) = d$ is finite. By the First Change of Rings Theorem, the residue field $k = R/\mathfrak{m} = S/\mathfrak{m}S$ has

$$pd_R(k) = 1 + pd_S(k) = 1 + d.$$ $\diamond$

Grade 0 Lemma 4.4.14 *If R is local and $G(R) = 0$ (i.e., every element of the maximal ideal $\mathfrak{m}$ is a zerodivisor on R), then for any finitely generated R-module A,*

$$\text{either } pd(A) = 0 \quad \text{or} \quad pd(A) = \infty.$$

Proof If $0 < pd(A) < \infty$ for some A then an appropriate syzygy M of A is finitely generated and has $pd(M) = 1$. Nakayama's lemma states that M can be generated by $m = \dim_k(M/\mathfrak{m}M)$ elements. If $u_1, \cdots, u_m$ generate M, there is a projective resolution $0 \to P \to R^m \xrightarrow{\epsilon} M \to 0$ with $\epsilon(r_1, \ldots, r_m) = \sum r_i u_i$; visibly $R^m/\mathfrak{m}R^m \cong k^m \cong M/\mathfrak{m}M$. But then $P \subseteq \mathfrak{m}R^m$, so $sP = 0$, where $s \in R$ is any element such that $\mathfrak{m} = \{r \in R : sr = 0\}$. On the other hand, P is projective, hence a free R-module (4.3.11), so $sP = 0$ implies that $s = 0$, a contradiction. $\diamond$

Theorem 4.4.15 (Auslander-Buchsbaum Equality) *Let R be a local ring, and A a finitely generated R-module. If $pd(A) < \infty$, then $G(R) = G(A) + pd(A)$.*

Proof If $G(R) = 0$ and $pd(A) < \infty$, then A is projective (hence free) by the Grade 0 lemma 4.4.14. In this case $G(R) = G(A)$, and $pd(A) = 0$. If $G(R) \neq 0$, we shall perform a double induction on $G(R)$ and on $G(A)$.

Suppose first that $G(R) \neq 0$ and $G(A) = 0$. Choose $x \in \mathfrak{m}$ and $0 \neq a \in A$ so that x is a nonzerodivisor on R and $\mathfrak{m}a = 0$. Resolve A:

$$0 \to K \to R^m \xrightarrow{\epsilon} A \to 0$$

and choose $u \in R^m$ with $\epsilon(u) = a$. Now $\mathfrak{m}u \subseteq K$ so $xu \in K$ and $\mathfrak{m}(xu) \subseteq xK$, yet $xu \notin xK$ as $u \notin K$ and x is a nonzerodivisor on R^m. Hence $G(K/xK) = 0$. Since K is a submodule of a free module, x is a nonzerodivisor on K. By the third Change of Rings theorem, and the fact that A is not free (as $G(R) \neq G(A)$),

$$pd_{R/xR}(K/xK) = pd_R(K) = pd_R(A) - 1.$$

Since $G(R/xR) = G(R) - 1$, induction gives us the required identity:

$$G(R) = 1 + G(R/xR) = 1 + G(K/xK) + pd_{R/xR}(K/xK) = pd_R(A).$$

Finally, we consider the case $G(R) \neq 0$, $G(A) \neq 0$. We can pick $x \in \mathfrak{m}$, which is a nonzerodivisor on both R and A (see the *Standard Facts* 4.4.7

cited above). Since we may begin a maximal A-sequence with x, $G(A/xA) = G(A) - 1$. Induction and the corollary 4.3.14 to the third Change of Rings theorem now give us the required identity:

$$G(R) = G(A/xA) + pd_R(A/xA)$$
$$= (G(A) - 1) + (1 + pd_R(A))$$
$$= G(A) + pd_R(A). \qquad \diamond$$

Main Theorem 4.4.16 *A local ring R is regular iff* $gl.\dim(R) < \infty$. *In this case*

$$G(R) = \dim(R) = emb.\dim(R) = gl.\dim(R) = pd_R(k).$$

Proof First, suppose R is regular. If $\dim(R) = 0$, R is a field, and the result is clear. If $d = \dim(R) > 0$, choose an R-sequence $x_1, \cdots, x_d$ generating $\mathfrak{m}$ and set $S = R/x_1 R$. Then $x_2, \cdots, x_d$ is an S-sequence generating the maximal ideal of S, so S is regular of dimension $d - 1$. By induction on d, we have

$$gl.\dim(R) = 1 + gl.\dim(S) = 1 + (d - 1) = d.$$

If $gl.\dim(R) = 0$, R must be semisimple and local (a field). If $gl.\dim(R) \neq 0, \infty$ then $\mathfrak{m}$ contains a nonzerodivisor x by the Grade 0 lemma 4.4.14; we may even find an $x = x_1$ not in $\mathfrak{m}^2$ (see the *Standard Facts* 4.4.7 cited above). To prove that R is regular, we will prove that $S = R/xR$ is regular; as $\dim(S) = \dim(R) - 1$, this will prove that the maximal ideal $\mathfrak{m}S$ of S is generated by an S-sequence $y_2, \cdots, y_d$. Lift the $y_i \in \mathfrak{m}S$ to elements $x_i \in \mathfrak{m}$ ($i = 2, \cdots, d$). By definition $x_1, \cdots, x_d$ is an R-sequence generating $\mathfrak{m}$, so this will prove that R is regular.

By the third Change of Rings theorem 4.3.12 with $A = \mathfrak{m}$,

$$pd_S(\mathfrak{m}/x\mathfrak{m}) = pd_R(\mathfrak{m}) = pd_R(k) - 1 = gl.\dim(R) - 1.$$

Now the image of $\mathfrak{m}/x\mathfrak{m}$ in $S = R/xR$ is $\mathfrak{m}/xR = \mathfrak{m}S$, so we get exact sequences

$$0 \to xR/x\mathfrak{m} \to \mathfrak{m}/x\mathfrak{m} \to \mathfrak{m}S \to 0 \quad \text{and} \quad 0 \to \mathfrak{m}S \to S \to k \to 0.$$

Moreover, $xR/x\mathfrak{m} \cong \mathrm{Tor}_1^R(R/xR, k) \cong \{a \in k : xa = 0\} = k$, and the image of x in $xR/x\mathfrak{m}$ is nonzero. We claim that $\mathfrak{m}/x\mathfrak{m} \cong \mathfrak{m}S \oplus k$ as S-modules. This will imply that

$$gl.\dim(S) = pd_S(k) \leq pd_S(\mathfrak{m}/x\mathfrak{m}) = gl.\dim(R) - 1.$$

By induction on global dimension, this will prove that S is regular.

To see the claim, set $r = emb.\dim(R)$ and find elements $x_2, \cdots, x_r$ in $\mathfrak{m}$ such that the image of $\{x_1, \cdots, x_r\}$ in $\mathfrak{m}/\mathfrak{m}^2$ forms a basis. Set $I = (x_2, \cdots, x_r)R + x\mathfrak{m}$ and observe that $I/x\mathfrak{m} \subseteq \mathfrak{m}/x\mathfrak{m}$ maps onto $\mathfrak{m}S$. As the kernel $xR/x\mathfrak{m}$ of $\mathfrak{m}/x\mathfrak{m} \to \mathfrak{m}S$ is isomorphic to k and contains $x \notin I$, it follows that $(xR/x\mathfrak{m}) \cap (I/x\mathfrak{m}) = 0$. Hence $I/x\mathfrak{m} \cong \mathfrak{m}S$ and $k \oplus \mathfrak{m}S \cong \mathfrak{m}/x\mathfrak{m}$, as claimed. $\diamond$

Corollary 4.4.17 *A regular ring is both Gorenstein and Cohen-Macaulay.*

Corollary 4.4.18 *If R is a regular local ring and $\mathfrak{p}$ is any prime ideal of R, then the localization $R_\mathfrak{p}$ is also a regular local ring.*

Proof We shall show that if S is any multiplicative set in R, then the localization $S^{-1}R$ has finite global dimension. As $R_\mathfrak{p} = S^{-1}R$ for $S = R - \mathfrak{p}$, this will suffice. Considering an $S^{-1}R$-module A as an R-module, there is a projective resolution $P \to A$ of length at most $gl.\dim(R)$. Since $S^{-1}R$ is a flat R-module and $S^{-1}A = A$, $S^{-1}P \to A$ is a projective $S^{-1}R$-module resolution of length at most $gl.\dim(R)$. $\diamond$

Remark The only non-homological proof of this result, due to Nagata, is very long and hard. This ability of homological algebra to give easy proofs of results outside the scope of homological algebra justifies its importance. Here is another result, quoted without proof from [KapCR], which uses homological algebra (projective resolutions) in the proof but not in the statement.

Theorem 4.4.19 *Every regular local ring is a Unique Factorization Domain.*

4.5 Koszul Complexes

An efficient way to perform calculations is to use Koszul complexes. If $x \in R$ is central, we let $K(x)$ denote the chain complex

$$0 \to R \xrightarrow{x} R \to 0$$

concentrated in degrees 1 and 0. It is convenient to identify the generator of the degree 1 part of $K(x)$ as the element e_x, so that $d(e_x) = x$. If $\boldsymbol{x} = (x_1, \cdots, x_n)$ is a finite sequence of central elements in R, we define the *Koszul complex* $K(\boldsymbol{x})$ to be the total tensor product complex (see 2.7.1):

$$K(x_1) \otimes_R K(x_2) \otimes_R \cdots \otimes_R K(x_n).$$

Notation 4.5.1 If A is an R-module, we define

$$H_q(x, A) = H_q(K(x) \otimes_R A);$$

$$H^q(x, A) = H^q(\text{Hom}(K(x), A)).$$

The degree p part of $K(x)$ is a free R-module generated by the symbols

$$e_{i_1} \wedge \cdots \wedge e_{i_p} = 1 \otimes \cdots \otimes 1 \otimes e_{x_{i_1}} \otimes \cdots \otimes e_{x_{i_p}} \otimes \cdots \otimes 1 \quad (i_1 < \cdots < i_p).$$

In particular, $K_p(x)$ is isomorphic to the pth exterior product $\Lambda^p R^n$ of R^n and has rank $\binom{n}{p}$, so $K(x)$ is often called the *exterior algebra complex*. The derivative $K_p(x) \to K_{p-1}(x)$ sends $e_{i_1} \wedge \cdots \wedge e_{i_p}$ to $\sum (-1)^{k+1} x_{i_k} e_{i_1} \wedge \cdots \wedge \hat{e}_{i_k} \wedge \cdots \wedge e_{i_p}$. As an example, $K(x, y)$ is the complex

$$0 \longrightarrow R \xrightarrow{(x, -y)} R^2 \xrightarrow{\binom{x}{y}} R \longrightarrow 0.$$

basis: $\{e_x \wedge e_y\}$ $\{e_y, e_x\}$ $\{1\}$

DG-Algebras 4.5.2 A *graded R-algebra* K_* is a family $\{K_p, p \geq 0\}$ of R-modules, equipped with a bilinear product $K_p \otimes_R K_q \to K_{p+q}$ and an element $1 \in K_0$ making K_0 and $\oplus K_p$ into associative R-algebras with unit. K_* is *graded-commutative* if for every $a \in K_p$, $b \in K_q$ we have $a \cdot b = (-1)^{pq} b \cdot a$. A *differential graded algebra*, or *DG-algebra*, is a graded R-algebra K_* equipped with a map $d: K_p \to K_{p-1}$, satisfying $d^2 = 0$ and satisfying the *Leibnitz rule:*

$$d(a \cdot b) = d(a) \cdot b + (-1)^p a \cdot d(b) \quad \text{for } a \in K_p.$$

Exercise 4.5.1

1. Let K be a DG-algebra. Show that the homology $H_*(K) = \{H_p(K)\}$ forms a graded R-algebra, and that $H_*(K)$ is graded-commutative whenever K_* is.

2. Show that the Koszul complex $K(x) \cong \Lambda^*(R^n)$ is a graded-commutative DG-algebra. If R is commutative, use this to obtain an external product $H_p(x, A) \otimes_R H_q(x, B) \to H_{p+q}(x, A \otimes_R B)$. Conclude that if A is a commutative R-algebra then the Koszul homology $H_*(x, A)$ is a graded-commutative R-algebra.

3. If $x_1, \cdots \in I$ and $A = R/I$, show that $H_*(x, A)$ is the exterior algebra $\Lambda^*(A^n)$.

Exercise 4.5.2 Show that $\{H_q(x, -)\}$ is a homological δ-functor, and that $\{H^q(x, -)\}$ is a cohomological δ-functor with

$$H_0(x, A) = A/(x_1, \cdots, x_n)A$$

$$H^0(x, A) = \mathrm{Hom}(R/xR, A) = \{a \in A : x_i a = 0 \text{ for all } i\}.$$

Then show that there are isomorphisms $H_p(x, A) \cong H^{n-p}(x, A)$ for all p.

Lemma 4.5.3 (Künneth formula for Koszul complexes) *If $C = C_*$ is a chain complex of R-modules and $x \in R$, there are exact sequences*

$$0 \to H_0(x, H_q(C)) \to H_q(K(x) \otimes_R C) \to H_1(x, H_{q-1}(C)) \to 0.$$

Proof Considering R as a complex concentrated in degree zero, there is a short exact sequence of complexes $0 \to R \to K(x) \to R[-1] \to 0$. Tensoring with C yields a short exact sequence of complexes whose homology long exact sequence is

$$H_{q+1}(C[-1]) \xrightarrow{\partial} H_q(C) \to H_q(K(x) \otimes C) \to H_q(C[-1]) \xrightarrow{\partial} H_q(C).$$

Identifying $H_{q+1}(C[-1])$ with $H_q(C)$, the map ∂ is multiplication by x (check this!), whence the result. ◇

Exercise 4.5.3 If x is a nonzerodivisor on R, that is, $H_1(K(x)) = 0$, use the Künneth formula for complexes 3.6.3 to give another proof of this result.

Exercise 4.5.4 Show that if one of the x_i is a unit of R, then the complex $K(x)$ is split exact. Deduce that in this case $H_*(x, A) = H^*(x, A) = 0$ for all modules A.

Corollary 4.5.4 (Acyclicity) *If x is a regular sequence on an R-module A, then $H_q(x, A) = 0$ for $q \neq 0$ and $H_0(x, A) = A/xA$, where $xA = (x_1, \cdots, x_n)A$.*

Proof Since x is a nonzerodivisor on A, the result is true for $n = 1$. Inductively, letting $x = x_n$, $y = (x_1, \cdots, x_{n-1})$, and $C = K(y) \otimes A$, $H_q(C) = 0$ for $q \neq 0$ and $K(x) \otimes H_0(C)$ is the complex

$$0 \to A/yA \xrightarrow{x} A/yA \to 0.$$

The result follows from 4.5.3, since x is a nonzerodivisor on A/yA. ◇

Corollary 4.5.5 (Koszul resolution) *If* x *is a regular sequence in* R, *then* $K(x)$ *is a free resolution of* R/I, $I = (x_1, \cdots, x_n)R$. *That is, the following sequence is exact:*

$$0 \to \Lambda^n(R^n) \to \cdots \to \Lambda^2(R^n) \to R^n \xrightarrow{x} R \to R/I \to 0.$$

In this case we have

$$\mathrm{Tor}_p^R(R/I, A) = H_p(x, A);$$

$$\mathrm{Ext}_R^p(R/I, A) = H^p(x, A).$$

Exercise 4.5.5 If x is a regular sequence in R, show that the external and internal products for Tor (2.7.8 and exercise 2.7.5(4)) agree with the external and internal products for $H_*(x, A)$ constructed in this section.

Exercise 4.5.6 Let R be a regular local ring with residue field k. Show that

$$\mathrm{Tor}_p^R(k, k) \cong \mathrm{Ext}_R^p(k, k) \cong \Lambda^p k^n \cong k^{\binom{n}{p}}, \quad \text{where } n = \dim(R).$$

Conclude that $id_R(k) = \dim(R)$ and that as rings $\mathrm{Tor}_*^R(k, k) \cong \Lambda^*(k^n)$.

Application 4.5.6 (Scheja-Storch) Here is a computational proof of Hilbert's Syzygy Theorem 4.3.8. Let F be a field, and set $R = F[x_1, \cdots, x_n]$, $S = R[y_1, \cdots, y_n]$. Let t be the sequence $(t_1, \cdots, t_n)$ of elements $t_i = y_i - x_i$ of S. Since $S = R[t_1, \cdots, t_n]$, t is a regular sequence, and $H_0(t, S) \cong R$, so the augmented Koszul complex of $K(t)$ is exact:

$$0 \to \Lambda^n S^n \to \Lambda^{n-1} S^n \to \cdots \to \Lambda^2 S^n \to S^n \xrightarrow{t} S \to R \to 0.$$

Since each $\Lambda^p S^n$ is a free R-module, this is in fact a split exact sequence of R-modules. Hence applying $\otimes_R A$ yields an exact sequence for every R-module A. That is, each $K(t) \otimes_R A$ is an S-module resolution of A. Set $R' = F[y_1, \cdots, y_n]$, a subring of S. Since $t_i = 0$ on A, we may identify the R-module structure on A with the R'-module structure on A. But $S \otimes_R A \cong R' \otimes_F A$ is a free R'-module because F is a field. Therefore each $\Lambda^p S^n \otimes_R A$ is a free R'-module, and $K(t) \otimes_R A$ is a canonical, natural resolution of A by free R'-modules. Since $K(t) \otimes_R A$ has length n, this proves that

$$pd_R(A) = pd_{R'}(A) \le n$$

for every R-module A. On the other hand, since $\mathrm{Tor}_n^R(F, F) \cong F$, we see that $pd_R(F) = n$. Hence the ring $R = F[x_1, \cdots, x_n]$ has global dimension n.

4.6 Local Cohomology

Definition 4.6.1 If I is a finitely generated ideal in a commutative ring R and A is an R-module, we define

$$H_I^0(A) = \{a \in A : (\exists i)I^i a = 0\} = \varinjlim \operatorname{Hom}(R/I^i, A).$$

Since each $\operatorname{Hom}(R/I^i, -)$ is left exact and $\varinjlim$ is exact, we see that H_I^0 is an additive left exact functor from R–**mod** to itself. We set

$$H_I^q(A) = (R^q H_I^0)(A).$$

Since the direct limit is exact, we also have

$$H_I^q(A) = \varinjlim \operatorname{Ext}_R^q(R/I^i, A).$$

Exercise 4.6.1 Show that if $J \subseteq I$ are finitely generated ideals such that $I^i \subseteq J$ for some i, then $H_J^q(A) \cong H_I^q(A)$ for all R-modules A and all q.

Exercise 4.6.2 (Mayer-Vietoris sequence) Let I and J be ideals in a noetherian ring R. Show that there is a long exact sequence for every R-module A:

$$\cdots \xrightarrow{\delta} H_{I+J}^q(A) \to H_I^q(A) \oplus H_J^q(A) \to H_{I\cap J}^q(A) \to H_{I+J}^{q+1}(A) \xrightarrow{\delta} \cdots.$$

Hint: Apply $\operatorname{Ext}^*(-, A)$ to the family of sequences

$$0 \to R/I^i \cap J^i \to R/I^i \oplus R/J^i \to R/(I^i + J^i) \to 0.$$

Then pass to the limit, observing that $(I + J)^{2i} \subseteq (I^i + J^i) \subseteq (I + J)^i$ and that, by the Artin-Rees lemma ([BA II, 7.13]), for every i there is an $N \geq i$ so that $I^N \cap J^N \subseteq (I \cap J)^i \subseteq I^i \cap J^i$.

Generalization 4.6.2 (Cohomology with supports; See [GLC]) Let Z be a closed subspace of a topological space X. If F is a sheaf on X, let $H_Z^0(X, F)$ be the kernel of $H^0(X, F) \to H^0(X - Z, F)$, that is, all global sections of F with support in Z. H_Z^0 is a left exact functor on Sheaves(X), and we write $H_Z^n(X, F)$ for its right derived functors.

If I is any ideal of R, then $H_I^n(A)$ is defined to be $H_Z^n(X, \tilde{A})$, where $X = \operatorname{Spec}(R)$ is the topological space of prime ideals of R, $Z = \{\mathfrak{p} : I \subseteq \mathfrak{p}\}$, and $\tilde{A}$ is the sheaf on $\operatorname{Spec}(R)$ associated to A. If I is a finitely generated ideal, this agrees with our earlier definition. For more details see [GLC], including the construction of the long exact sequence

$$0 \to H_Z^0(X, F) \to H^0(X, F) \to H^0(X - Z, F) \to H_Z^1(X, F) \to \cdots.$$

A standard result in algebraic geometry states that $H^n(\text{Spec}(R), \tilde{A}) = 0$ for $n \neq 0$, so for the *punctured spectrum* $U = \text{Spec}(R) - Z$ the sequence

$$0 \to H_I^0(A) \to A \to H^0(U, \tilde{A}) \to H_I^1(A) \to 0$$

is exact, and for $n \neq 0$ we can calculate the cohomology of $\tilde{A}$ on U via

$$H^n(U, \tilde{A}) \cong H_I^{n+1}(A).$$

Exercise 4.6.3 Let $\mathcal{A}$ be the full subcategory of R–**mod** consisting of the modules with $H_I^0(A) = A$.

1. Show that $\mathcal{A}$ is an abelian category, that $H_I^0: R$–**mod** $\to \mathcal{A}$ is right adjoint to the inclusion $\iota: \mathcal{A} \hookrightarrow R$–**mod**, and that ι is an exact functor.
2. Conclude that H_I^0 preserves injectives (2.3.10), and that $\mathcal{A}$ has enough injectives.
3. Conclude that each $H_I^n(A)$ belongs to the subcategory $\mathcal{A}$ of R–**mod**.

Theorem 4.6.3 *Let R be a commutative noetherian local ring with maximal ideal $\mathfrak{m}$. Then the grade $G(A)$ of any finitely generated R-module A is the smallest integer n such that $H_{\mathfrak{m}}^n(A) \neq 0$.*

Proof For each i we have the exact sequence

$$\text{Ext}^{n-1}(\mathfrak{m}^i/\mathfrak{m}^{i+1}, A) \to \text{Ext}^n(R/\mathfrak{m}^i, A) \to \text{Ext}^n(R/\mathfrak{m}^{i+1}, A) \to \text{Ext}^n(\mathfrak{m}^i/\mathfrak{m}^{i+1}, A).$$

We saw in 4.4.8 that $\text{Ext}^n(R/\mathfrak{m}, A)$ is zero if $n < G(A)$ and nonzero if $n = G(A)$; as $\mathfrak{m}^i/\mathfrak{m}^{i+1}$ is a finite direct sum of copies of $R/\mathfrak{m}$, the same is true for $\text{Ext}^n(\mathfrak{m}^i/\mathfrak{m}^{i+1}, A)$. By induction on i, this proves that $\text{Ext}^n(R/\mathfrak{m}^{i+1}, A)$ is zero if $n < G(A)$ and that it contains the nonzero module $\text{Ext}^n(R/\mathfrak{m}^i, A)$ if $n = G(A)$. Now take the direct limit as $i \to \infty$. $\diamond$

Application 4.6.4 Let R be a 2-dimensional local domain. Since $G(R) \neq 0$, $H_{\mathfrak{m}}^0(R) = 0$. From the exact sequence

$$0 \to \mathfrak{m}^i \to R \to R/\mathfrak{m}^i \to 0$$

we obtain the exact sequence

$$0 \to R \to \operatorname{Hom}_R(\mathfrak{m}^i, R) \to \operatorname{Ext}^1_R(R/\mathfrak{m}^i, R) \to 0.$$

As R is a domain, there is a natural inclusion of $\operatorname{Hom}_R(\mathfrak{m}^i, R)$ in the field F of fractions of R as the submodule

$$\mathfrak{m}^{-i} \equiv \{x \in F : x\mathfrak{m}^i \subseteq R\}.$$

Set $C = \cup \mathfrak{m}^{-i}$. (*Exercise:* Show that C is a subring of F.) Evidently

$$H^1_{\mathfrak{m}}(R) = \varinjlim \operatorname{Ext}^1(R/\mathfrak{m}^i, R) \cong C/R.$$

If R is Cohen-Macaulay, that is, $G(R) = 2$, then $H^1_{\mathfrak{m}}(R) = 0$, so $R = C$ and $\operatorname{Hom}_R(\mathfrak{m}^i, R) = R$ for all i. Otherwise $R \ne C$ and $G(R) = 1$. When the integral closure of R is finitely generated as an R-module, C is actually a Cohen-Macaulay ring—the smallest Cohen-Macaulay ring containing R [EGA, IV.5.10.17].

Here is an alternative construction of local cohomology due to Serre [EGA, III.1.1]. If $x \in R$ there is a natural map from $K(x^{i+1})$ to $K(x^i)$:

$$
\begin{array}{ccccccc}
0 & \longrightarrow & R & \xrightarrow{x^{i+1}} & R & \longrightarrow & 0 \\
 & & x\downarrow & & \| & & \\
0 & \longrightarrow & R & \xrightarrow{x^i} & R & \longrightarrow & 0.
\end{array}
$$

By tensoring these maps together, and writing x^i for $(x^i_1, \cdots, x^i_n)$, this gives a map from $K(x^{i+1})$ to $K(x^i)$, hence a tower $\{H_q(K(x^i))\}$ of R-modules. Applying $\operatorname{Hom}_R(-, A)$ and taking cohomology yields a map from $H^q(x^i, A)$ to $H^q(x^{i+1}, A)$.

Definition 4.6.5 $H^q_x(A) = \varinjlim H^q(x^i, A)$.

For our next result, recall from 3.5.6 that a tower $\{A_i\}$ satisfies the *trivial Mittag-Leffler condition* if for every i there is a $j > i$ so that $A_j \to A_i$ is zero.

Exercise 4.6.4 If $\{A_i\} \to \{B_i\} \to \{C_i\}$ is an exact sequence of towers of R-modules and both $\{A_i\}$ and $\{C_i\}$ satisfy the trivial Mittag-Leffler condition, then $\{B_i\}$ also satisfies the trivial Mittag-Leffler condition (3.5.6).

Proposition 4.6.6 *Let R be a commutative noetherian ring and A a finitely generated R-module. Then the tower $\{H_q(x^i, A)\}$ satisfies the trivial Mittag-Leffler condition for every $q \ne 0$.*

Proof We proceed by induction on the length n of x. If $n = 1$, one sees immediately that $H_1(x^i, A)$ is the submodule $A_i = \{a \in A : x^i a = 0\}$. The submodules A_i of A form an ascending chain, which must be stationary since R is noetherian and A is finitely generated. This means that there is an integer k such that $A_k = A_{k+1} = \cdots$, that is, $x^k A_i = 0$ for all i. Since the map $A_{i+j} \to A_i$ is multiplication by x^j, it is zero whenever $j \geq k$. Thus the lemma holds if $n = 1$.

Inductively, set $y = (x_1, \ldots, x_{n-1})$ and write x for x_n. Since $K(x^i) \otimes K(y^i) = K(x^i)$, the Künneth formula for Koszul complexes 4.5.3 (and its proof) yields the following exact sequences of towers:

$$\{H_q(y^i, A)\} \to \{H_q(x^i, A)\} \to \{H_{q-1}(y^i, A)\};$$

$$\{H_1(y^i, A)\} \to \{H_1(x^i, A)\} \to \{H_1(x^i, A/y^i A)\} \to 0.$$

If $q \geq 2$, the outside towers satisfy the trivial Mittag-Leffler condition by induction, so $\{H_q(x^i, A)\}$ does too. If $q = 1$ and we set $A_{ij} = \{a \in A/y^i A : x^j a = 0\} = H_1(x^j, A/y^i A)$, it is enough to show that the diagonal tower $\{A_{ii}\}$ satisfies the trivial Mittag-Leffler condition. For fixed i, we saw above that there is a k such that every map $A_{ij} \to A_{i,j+k}$ is zero. Hence the map $A_{ii} \to A_{i,i+k} \to A_{i+k,i+k}$ is zero, as desired. $\diamond$

Corollary 4.6.7 *Let R be commutative noetherian, and let E be an injective R-module. Then $H_x^q(E) = 0$ for all $q \neq 0$.*

Proof Because E is injective, $\text{Hom}_R(-, E)$ is exact. Therefore

$$H^q(x^i, E) = H^q \text{Hom}_R(K(x^i, R), E) \cong \text{Hom}_R(H_q(x^i, R), E).$$

Because the tower $\{H_q(x^i, R)\}$ satisfies the trivial Mittag-Leffler condition,

$$H_x^q(E) \cong \varinjlim \text{Hom}_R(H_q(x^i, R), E) = 0. \qquad \diamond$$

Theorem 4.6.8 *If R is commutative noetherian, $x = (x_1, \cdots, x_n)$ is any sequence of elements of R, and $I = (x_1, \cdots, x_n)R$, then for every R-module A*

$$H_I^q(A) \cong H_x^q(A).$$

Proof Both H_I^q and H_x^q are universal δ-functors, and

$$H_I^0(A) = \varinjlim \text{Hom}(R/x^i R, A) = \varinjlim H^0(x^i, A) = H_x^0(A). \qquad \diamond$$

Corollary 4.6.9 *If R is a noetherian local ring, then $H_{\mathfrak{m}}^q(A) \neq 0$ only when $G(A) \leq q \leq \dim(R)$. In particular, if R is a Cohen-Macauley local ring, then*

$$H_{\mathfrak{m}}^q(R) \neq 0 \Leftrightarrow q = \dim(R).$$

Proof Set $d = \dim(R)$. By standard commutative ring theory ([KapCR, Thm.153]), there is a sequence $x = (x_1, \cdots, x_d)$ of elements of $\mathfrak{m}$ such that $\mathfrak{m}^j \subseteq I \subseteq \mathfrak{m}$ for some j, where $I = (x_1, \cdots, x_d)R$. But then $H_{\mathfrak{m}}^q(A) = H_I^q(A) = H_x^q(A)$, and this vanishes for $q > d$ because the Koszul complexes $K(x^i)$ have length d. Now use (4.6.3). $\diamond$

Exercise 4.6.5 If I is a finitely generated ideal of R and $R \to S$ is a ring map, show that $H_I^q(A) \cong H_{IS}^q(A)$ for every S-module A. This result is rather surprising, because there isn't any nice relationship between the groups $\mathrm{Ext}_R^*(R/I^i, A)$ and $\mathrm{Ext}_S^*(S/I^i, A)$. Consequently, if $\mathrm{ann}_R(A)$ denotes $\{r \in R : rA = 0\}$, then $H_I^q(A) = 0$ for $q > \dim(R/\mathrm{ann}_R(A))$.

Application 4.6.10 (Hartshorne) Let $R = \mathbb{C}[x_1, x_2, y_1, y_2]$, $P = (x_1, x_2)R$, $Q = (y_1, y_2)R$, and $I = P \cap Q$. As P, Q, and $\mathfrak{m} = P + Q = (x_1, x_2, y_1, y_2)R$ are generated by regular sequences, the outside terms in the Mayer-Vietoris sequence (exercise 4.6.2)

$$H_P^3(R) \oplus H_Q^3(R) \to H_I^3(R) \to H_{\mathfrak{m}}^4(R) \to H_P^4(R) \oplus H_Q^4(R)$$

vanish, yielding $H_I^3(R) \cong H_{\mathfrak{m}}^4(R) \neq 0$. This implies that the union of two planes in $\mathbb{C}^4$ that meet in a point cannot be described as the solutions of only two equations $f_1 = f_2 = 0$. Indeed, if this were the case, then we would have $I^i \subseteq (f_1, f_2)R \subseteq I$ for some i, so that $H_I^3(R)$ would equal $H_f^3(R)$, which is zero.

5

Spectral Sequences

5.1 Introduction

Spectral sequences were invented by Jean Leray, as a prisoner of war during World War II, in order to compute the homology (or cohomology) of a chain complex [Leray]. They were made algebraic by Koszul in 1945.

In order to motivate their construction, consider the problem of computing the homology of the total chain complex T_* of a first quadrant double complex E_{**}. As a first step, it is convenient to forget the horizontal differentials and add a superscript zero, retaining only the vertical differentials d^v along the columns E^0_{p*}.

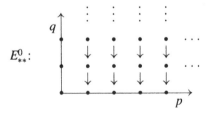

If we write E^1_{pq} for the vertical homology $H_q(E^0_{p*})$ at the (p, q) spot, we may once again arrange the data in a lattice, this time using the horizontal diffentials d^h.

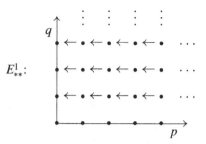

Now we write E^2_{pq} for the horizontal homology $H_p(E^1_{*q})$ at the (p, q) spot. In a sense made clearer by the following exercises, the elements of E^2_{pq} are a second-order approximation of the homology of $T_* = \text{Tot}(E_{**})$.

Exercise 5.1.1 Suppose that the double complex E consists solely of the two columns p and $p - 1$. Fix n and set $q = n - p$, so that an element of $H_n(T)$ is represented by an element $(a, b) \in E_{p-1,q+1} \times E_{pq}$. Show that we have calculated the homology of $T = \text{Tot}(E)$ up to extension in the sense that there is a short exact sequence

$$0 \to E^2_{p-1,q+1} \to H_{p+q}(T) \to E^2_{pq} \to 0.$$

Exercise 5.1.2 (Differentials at the E^2 stage)

1. Show that E^2_{pq} can be presented as the group of all pairs (a, b) in $E_{p-1,q+1} \times E_{pq}$ such that $0 = d^v b = d^v a + d^h b$, modulo the relation that these pairs are trivial: $(a, 0)$; $(d^h x, d^v x)$ for $x \in E_{p,q+1}$; and $(0, d^h c)$ for all $c \in E_{p+1,q}$ with $d^v c = 0$.
2. If $d^h(a) = 0$, show that such a pair (a, b) determines an element of $H_{p+q}(T)$.
3. Show that the formula $d(a, b) = (0, d^h(a))$ determines a well-defined map

$$d: E^2_{pq} \to E^2_{p-2,q+1}.$$

Exercise 5.1.3 (Exact sequence of low degree terms) Recall that we have assumed that E^0_{pq} vanishes unless both $p \geq 0$ and $q \geq 0$. By diagram chasing, show that $E^2_{00} = H_0(T)$ and that there is an exact sequence

$$H_2(T) \to E^2_{20} \xrightarrow{\;d\;} E^2_{01} \to H_1(T) \to E^2_{10} \to 0.$$

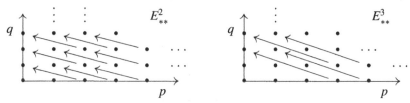

Figure 5.1.   The steps E^2 and E^3 of the spectral sequence.

There is an algorithm for computing $H_*(T)$ up to extension, called a *spectral sequence*, and we have just performed the first two steps of this algorithm. The next two steps are illustrated in Figure 5.1.

5.2 Terminology

Definition 5.2.1 A *homology spectral sequence* (*starting with E^a*) in an abelian category $\mathcal{A}$ consists of the following data:

1. A family $\{E_{pq}^r\}$ of objects of $\mathcal{A}$ defined for all integers p, q, and $r \geq a$
2. Maps $d_{pq}^r: E_{pq}^r \to E_{p-r,q+r-1}^r$ that are differentials in the sense that $d^r d^r = 0$, so that the "lines of slope $-(r+1)/r$" in the lattice E_{**}^r form chain complexes (we say the differentials go "to the left")
3. Isomorphisms between E_{pq}^{r+1} and the homology of E_{**}^r at the spot E_{pq}^r:

$$E_{pq}^{r+1} \cong \ker(d_{pq}^r)/\text{image }(d_{p+r,q-r+1}^r)$$

Note that E_{pq}^{r+1} is a subquotient of E_{pq}^r. The *total degree* of the term E_{pq}^r is $n = p + q$; the terms of total degree n lie on a line of slope -1, and each differential d_{pq}^r decreases the total degree by one.

There is a category of homology spectral sequences; a morphism $f: E' \to E$ is a family of maps $f_{pq}^r: E'^r_{pq} \to E_{pq}^r$ in $\mathcal{A}$ (for r suitably large) with $d^r f^r = f' d^r$ such that each f_{pq}^{r+1} is the map induced by f_{pq}^r on homology.

Example 5.2.2 A *first quadrant (homology) spectral sequence* is one with $E_{pq}^r = 0$ unless $p \geq 0$ and $q \geq 0$, that is, the point (p, q) belongs to the first quadrant of the plane. (If this condition holds for $r = a$, it clearly holds for all r.) If we fix p and q, then $E_{pq}^r = E_{pq}^{r+1}$ for all large r ($r > \max\{p, q+1\}$ will do), because the d^r landing in the (p, q) spot come from the fourth quadrant, while the d^r leaving E_{pq}^r land in the second quadrant. We write E_{pq}^∞ for this stable value of E_{pq}^r.

Dual Definition 5.2.3 A *cohomology spectral sequence (starting with E_a)* in $\mathcal{A}$ is a family $\{E_r^{pq}\}$ of objects ($r \geq a$), together with maps d_r^{pq} going "to the right":

$$d_r^{pq}: E_r^{pq} \to E_r^{p+r,q-r+1},$$

which are differentials in the sense that $d_r d_r = 0$, and isomorphisms between E_{r+1} and the homology of E_r. In other words, it is the same thing as a homology spectral sequence, reindexed via $E_r^{pq} = E_{-p,-q}^r$, so that d_r *increases* the total degree $p + q$ of E_{pq}^r by one.

There is a category of cohomology spectral sequences; a morphism $f: E' \to E$ is a family of maps $f_r^{pq}: E_r'^{pq} \to E_r^{pq}$ in $\mathcal{A}$ (for r suitably large) with $d_r f_r = f_r d_r$ such that each f_{r+1}^{pq} is the map induced by f_r^{pq}.

Mapping Lemma 5.2.4 *Let $f: \{E_{pq}^r\} \to \{E_{pq}'^r\}$ be a morphism of spectral sequences such that for some fixed r, $f^r : E_{pq}^r \cong E_{pq}'^r$ is an isomorphism for all p and q. The 5-lemma implies that $f^s: E_{pq}^s \cong E_{pq}'^s$ for all $s \geq r$ as well.*

Bounded Convergence 5.2.5 A homology spectral sequence is said to be *bounded* if for each n there are only finitely many nonzero terms of total degree n in E_{**}^a. If so, then for each p and q there is an r_0 such that $E_{pq}^r = E_{pq}^{r+1}$ for all $r \geq r_0$. We write E_{pq}^∞ for this stable value of E_{pq}^r.

We say that a bounded spectral sequence *converges to H_** if we are given a family of objects H_n of $\mathcal{A}$, each having a *finite* filtration

$$0 = F_s H_n \subseteq \cdots \subseteq F_{p-1} H_n \subseteq F_p H_n \subseteq F_{p+1} H_n \subseteq \cdots \subseteq F_t H_n = H_n,$$

and we are given isomorphisms $E_{pq}^\infty \cong F_p H_{p+q} / F_{p-1} H_{p+q}$. The traditional symbolic way of describing such a bounded convergence is like this:

$$E_{pq}^a \Rightarrow H_{p+q}.$$

Similarly, a cohomology spectral sequence is called *bounded* if there are only finitely many nonzero terms in each total degree in E_a^{**}. In a bounded cohomology spectral sequence, we write E_∞^{pq} for the stable value of the terms E_r^{pq} and say the (bounded) spectral sequence *converges to H^** if there is a *finite* filtration

$$0 = F^t H^n \subseteq \cdots F^{p+1} H^n \subseteq F^p H^n \cdots \subseteq F^s H^n = H^n \text{ so that}$$

$$E_\infty^{pq} \cong F^p H^{p+q} / F^{p+1} H^{p+q}.$$

Example 5.2.6 If a first quadrant homology spectral sequence converges to H_*, then each H_n has a finite filtration of length $n + 1$:

$$0 = F_{-1}H_n \subseteq F_0 H_n \subseteq \cdots \subseteq F_{n-1}H_n \subseteq F_n H_n = H_n.$$

The bottom piece $F_0 H_n = E_{0n}^\infty$ of H_n is located on the y-axis, and the top quotient $H_n / F_{n-1} H_n \cong E_{n0}^\infty$ is located on the x-axis. Note that each arrow landing on the x-axis is zero, and each arrow leaving the y-axis is zero. Therefore each E_{0n}^∞ is a subobject of E_{0n}^a, and each E_{n0}^∞ is a quotient of E_{n0}^a. The terms E_{0n}^r on the y-axis are called the *fiber* terms, and the terms E_{n0}^r on the x-axis are called the *base* terms for reasons that will become apparent in the next section. The resulting maps $E_{0n}^a \to E_{0n}^\infty \subset H_n$ and $H_n \to E_{n0}^\infty \subset E_{n0}^a$ are known as the *edge homomorphisms* of the spectral sequence for the obvious visual reason. Similarly, if a first quadrant cohomology spectral sequence converges to H^*, then H^n has a finite filtration:

$$0 = F^{n+1}H^n \subseteq F^n H^n \subseteq \cdots \subseteq F^1 H^n \subseteq F^0 H^n = H^n.$$

In this case, the bottom piece $F^n H^n \cong E_\infty^{n0}$ is located on the x-axis, and the top quotient $H^n / F^1 H^n \cong E_\infty^{0n}$ is located on the y-axis. In this case, the edge homomorphisms are the maps $E_a^{n0} \to E_\infty^{n0} \subset H^n$ and $H^n \to E_\infty^{0n} \subset E_a^{0n}$.

Definition 5.2.7 A (homology) spectral sequence *collapses at* $E^r (r \geq 2)$ if there is exactly one nonzero row or column in the lattice $\{E_{pq}^r\}$. If a collapsing spectral sequence converges to H_*, we can read the H_n off: H_n is the unique nonzero E_{pq}^r with $p + q = n$. The overwhelming majority of all applications of spectral sequences involve spectral sequences that collapse at E^1 or E^2.

Exercise 5.2.1 (2 columns) Suppose that a spectral sequence converging to H_* has $E_{pq}^2 = 0$ unless $p = 0, 1$. Show that there are exact sequences

$$0 \to E_{1,n-1}^2 \to H_n \to E_{0n}^2 \to 0.$$

Exercise 5.2.2 (2 rows) Suppose that a spectral sequence converging to H_* has $E_{pq}^2 = 0$ unless $q = 0, 1$. Show that there is a long exact sequence

$$\cdots H_{p+1} \to E_{p+1,0}^2 \xrightarrow{d} E_{p-1,1}^2 \to H_p \to E_{p0}^2 \xrightarrow{d} E_{p-2,1}^2 \to H_{p-1} \cdots.$$

If a spectral sequence is not bounded, everything is more complicated, and there is no uniform terminology in the literature. For example, a filtration in [CE] is "regular" if for each n there is an N such that $H_n(F_p C) = 0$ for $p < N$, and all filtrations are exhaustive. In [MacH] exhaustive filtrations are called "convergent above." In [EGA, $0_{III}(11.2)$] even the definition of spectral sequence is different, and "regular" spectral sequences are not only convergent but also bounded below. In what follows, we shall mostly follow the terminology of Bourbaki [BX, p.175].

E^∞ **Terms 5.2.8** Given a homology spectral sequence, we see that each E_{pq}^{r+1} is a subquotient of the previous term E_{pq}^r. By induction on r, we see that there is a nested family of subobjects of E_{pq}^a:

$$0 = B_{pq}^a \subseteq \cdots \subseteq B_{pq}^r \subseteq B_{pq}^{r+1} \subseteq \cdots \subseteq Z_{pq}^{r+1} \subseteq Z_{pq}^r \subseteq \cdots \subseteq Z_{pq}^a = E_{pq}^a$$

such that $E_{pq}^r \cong Z_{pq}^r / B_{pq}^r$. We introduce the intermediate objects

$$B_{pq}^\infty = \bigcup_{r=a}^\infty B_{pq}^r \quad \text{and} \quad Z_{pq}^\infty = \bigcap_{r=a}^\infty Z_{pq}^r$$

and define $E_{pq}^\infty = Z_{pq}^\infty / B_{pq}^\infty$. In a bounded spectral sequence both the union and intersection are finite, so $B_{pq}^\infty = B_{pq}^r$ and $Z_{pq}^\infty = Z_{pq}^r$ for large r. Thus we recover our earlier definition: $E_{pq}^\infty = E_{pq}^r$ for large r.

Warning: In an unbounded spectral sequence, we will tacitly assume that B_{pq}^∞, Z_{pq}^∞, and E_{pq}^∞ exist! The reader who is willing to only work in the category of modules may ignore this difficulty. The queasy reader should assume that the abelian category $\mathcal{A}$ satisfies axioms $(AB4)$ and $(AB4^*)$.

Exercise 5.2.3 (Mapping Lemma for E^∞) Let $f: \{E_{pq}^r\} \to \{E_{pq}^{'r}\}$ be a morphism of spectral sequences such that for some r (hence for all large r by 5.2.4) $f^r : E_{pq}^r \cong E_{pq}^{'r}$ is an isomorphism for all p and q. Show that $f^\infty : E_{pq}^\infty \cong E_{pq}^{'\infty}$ as well.

Definition 5.2.9 (Bounded below) Bounded below spectral sequences have good convergence properties. A homology spectral sequence is said to be *bounded below* if for each n there is an integer $s = s(n)$ such that the terms E_{pq}^a of total degree n vanish for all $p < s$. Bounded spectral sequences are bounded below. Right half-plane homology spectral sequences are bounded below but not bounded.

Dually, a cohomology spectral sequence is said to be *bounded below* if for each n the terms of total degree n vanish for large p. A left half-plane cohomology spectral sequence is bounded below but not bounded.

Definition 5.2.10 (Regular) Regularity is the most useful general condition for convergence used in practice; bounded below spectral sequences are also regular. We say that a spectral sequence is *regular* if for each p and q the differentials d_{pq}^r (or d_r^{pq}) leaving E_{pq}^r (or E_r^{pq}) are zero for all large r. Note that a spectral sequence is regular iff for each p and q: $Z_{pq}^\infty = Z_{pq}^r$ for all large r.

Convergence 5.2.11 We say the spectral sequence *weakly converges to* H_* if we are given objects H_n of $\mathcal{A}$, each having a filtration

$$\cdots \subseteq F_{p-1}H_n \subseteq F_p H_n \subseteq F_{p+1}H_n \subseteq \cdots \subseteq H_n,$$

together with isomorphisms $\beta_{pq}: E_{pq}^\infty \cong F_p H_{p+q}/F_{p-1}H_{p+q}$ for all p and q. Note that a weakly convergent spectral sequence cannot detect elements of $\cap F_p H_n$, nor can it detect elements in H_n that are not in $\cup F_p H_n$.

We say that the spectral sequence $\{E_{pq}^r\}$ *approaches* H_* (or *abuts to* H_*) if it weakly converges to H_* and we also have $H_n = \cup F_p H_n$ and $\cap F_p H_n = 0$ for all n. Every weakly convergent spectral sequence approaches $\cup F_p H_* / \cap F_p H_*$.

We say that the spectral sequence *converges to* H_* if it approaches H_*, it is regular, and $H_n = \lim_{\longleftarrow}(H_n/F_p H_n)$ for each n. A bounded below spectral sequence converges to H_* whenever it approaches H_*, because the inverse limit condition is always satisfied in a bounded below spectral sequence.

To show that our notion of convergence is a good one, we offer the following Comparison Theorem. If $\{E_{pq}^r\}$ and $\{E_{pq}^{'r}\}$ weakly converge to H_* and H_*', respectively, we say that a map $h: H_* \to H_*'$ is *compatible* with a morphism $f: E \to E'$ if h maps $F_p H_n$ to $F_p H_n'$ and the associated maps $F_p H_n/F_{p-1}H_n \to F_p H_n'/F_{p-1}H_n'$ correspond under β and β' to $f_{pq}^\infty: E_{pq}^\infty \to E_{pq}^{'\infty}$ $(q = n - p)$.

Comparison Theorem 5.2.12 *Let* $\{E_{pq}^r\}$ *and* $\{E_{pq}^{'r}\}$ *converge to* H_* *and* H_*', *respectively. Suppose given a map* $h: H_* \to H_*'$ *compatible with a morphism* $f: E \to E'$ *of spectral sequences. If* $f^r: E_{pq}^r \cong E_{pq}^{'r}$ *is an isomorphism for all* p *and* q *and some* r *(hence for* $r = \infty$ *by the Mapping Lemma), then* $h: H_* \to H_*'$ *is an isomorphism.*

Proof Weak convergence gives exact sequences

$$0 \longrightarrow F_{p-1}H_n/F_s H_n \longrightarrow F_p H_n/F_s H_n \longrightarrow E_{p,n-p}^\infty \longrightarrow 0$$

$$\downarrow \qquad\qquad \downarrow \qquad\qquad \downarrow \cong$$

$$0 \longrightarrow F_{p-1}H_n'/F_s H_n' \longrightarrow F_p H_n'/F_s H_n' \longrightarrow E_{p,n-p}^{'\infty} \longrightarrow 0.$$

Fixing s, induction on p shows that $F_p H_n / F_s H_n \cong F_p H_n' / F_s H_n'$ for all p. Since $H_n = \cup F_p H_n$, this yields $H_n / F_s H_n \cong H_n' / F_s H_n'$ for all s. Taking inverse limits yields the desired isomorphism $H_n \cong H_n'$. ◇

Remark The same spectral sequence may converge to two different graded groups H_*, and it can be very difficult to reconstruct a picture of H_* from this data. For example, knowing that a first quadrant spectral sequence has $E_{pq}^\infty \cong \mathbb{Z}/2$ for all p and q does not allow us to determine whether H_3 is $\mathbb{Z}/16$ or $\mathbb{Z}/2 \oplus \mathbb{Z}/8$, or even the group $(\mathbb{Z}/2)^4$. The Comparison Theorem 5.2.12 helps us reconstruct H_* without the need for convergence.

Multiplicative Structures 5.2.13 Suppose that for $r = a$ we are given a bi-graded product

$(*)$ $$ E_{p_1 q_1}^r \times E_{p_2 q_2}^r \to E_{p_1+p_2, q_1+q_2}^r $$

such that the differential d^r satisfies the Leibnitz relation

$(**)$ $$ d^r(x_1 x_2) = d^r(x_1)x_2 + (-1)^{p_1} x_1 d^r(x_2), \quad x_i \in E_{p_i q_i}^r. $$

Then the product of two cycles (boundaries) is again a cycle (boundary), and by induction we have $(*)$ and $(**)$ for every $r \geq a$. We shall call this a *multiplicative structure* on the spectral sequence. Clearly this can be a useful tool in explicit calculations.

5.3 The Leray-Serre Spectral Sequence

Before studying the algebraic aspects of spectral sequences, we shall illustrate their computational power by citing the topological applications that led to their creation by Leray. The material in this section is taken from [MacH, XI.2].

Definition 5.3.1 A sequence $F \xrightarrow{i} E \xrightarrow{\pi} B$ of based topological spaces is called a *Serre fibration* if F is the inverse image $\pi^{-1}(*_B)$ of the basepoint of B and if π has the following "homotopy lifting property": if P is any finite polyhedron and I is the unit interval $[0, 1]$, $g: P \to E$ is a map, and $H: P \times I \to B$ is a homotopy between $\pi g = H(-, 0)$ and $h_1 = H(-, 1)$,

$$
\begin{array}{ccc}
P & \xrightarrow{g} & E \\
{\scriptstyle \times 0}\downarrow & {\scriptstyle G}\nearrow & \downarrow{\scriptstyle \pi} \\
P \times I & \xrightarrow[H]{} & B
\end{array}
$$

there is a homotopy $G: P \times I \to E$ between g and a map $g_1 = G(-, 1)$ which lifts H in the sense that $\pi G = H$. The spaces F, E, and B are called the *Fiber*, *total space* (*Espace totale* for Leray), and *Base space*, respectively. The importance of Serre fibrations lies in the fact (proven in Serre's thesis) that associated to each fibration is a long exact sequence of homotopy groups

$$\cdots \pi_{n+1}(B) \xrightarrow{\partial} \pi_n(F) \to \pi_n(E) \to \pi_n(B) \xrightarrow{\partial} \cdots .$$

In order to simplify the presentation below, we shall assume that B is *simply connected*, that is, that $\pi_0(B) = \pi_1(B) = 0$. Without this assumption, we would have to introduce the action of $\pi_1(B)$ on the homology of F and talk about the homology of B with "local coefficients" in the twisted bundles $H_q(F)$.

Theorem 5.3.2 (Leray-Serre spectral sequence) *Let* $F \xrightarrow{i} E \xrightarrow{\pi} B$ *be a Serre fibration such that B is simply connected. Then there is a first quadrant homology spectral sequence starting with E^2 and converging to $H_*(E)$:*

$$E^2_{pq} = H_p(B; H_q(F)) \Rightarrow H_{p+q}(E).$$

Addendum 1 $H_0(B) = \mathbb{Z}$, so along the y-axis we have $E^2_{0q} = H_q(F)$. Because $E^2_{pq} = 0$ for $p < 0$, the groups $E^3_{0q}, \cdots, E^{n+1}_{0q} = E^\infty_{0q}$ are successive quotients of E^2_{0q}. The theorem states that $E^\infty_{0q} \cong F_0 H_q(E)$, so there is an "edge map"

$$H_q(F) = E^2_{0q} \twoheadrightarrow E^\infty_{0q} \subseteq H_q(E).$$

This edge map is the map $i_*: H_q(F) \to H_q(E)$.

Addendum 2 Suppose that $\pi_0(F) = 0$, so that $H_0(F) = \mathbb{Z}$. Along the x-axis we then have $E^2_{p0} = H_p(B)$. Because $E^2_{pq} = 0$ for $q < 0$, the groups $E^3_{p0}, \cdots, E^{n+1}_{p0} = E^\infty_{p0}$ are successive subgroups of E^2_{p0}. The theorem states that $E^\infty_{p0} \cong H_p(E)/F_{p-1}H_p(E)$, so there is an "edge map"

$$H_p(E) \twoheadrightarrow E^\infty_{p0} \hookrightarrow E^2_{p0} = H_p(B).$$

This edge map is the map $\pi_*: H_p(E) \to H_p(B)$.

Remark The Universal Coefficient Theorem 3.6.4 tells us that

$$H_p(B; H_q(F)) \cong H_p(B) \otimes H_q(F) \oplus \mathrm{Tor}_1^{\mathbb{Z}}(H_{p-1}(B), H_q(F)).$$

Therefore the terms E^2_{pq} are not hard to calculate. In particular, since $\pi_1(B) = 0$ we have $H_1(B) = H_1(B; H_q(F)) = 0$ for all q. By the Hurewicz homomorphism, $\pi_2(B) \cong H_2(B)$ and therefore $H_2(B; H_q(F)) \cong H_2(B) \otimes H_q(F)$ for all q as well.

Application 5.3.3 (Exact sequence of low degree terms) In the lower left corner of this spectral sequence we find

$$E^2_{**}: \quad \begin{array}{l} H_2(F) \leftarrow 0 \quad \bullet \\ H_1(F) \leftarrow 0 \quad \bullet \leftarrow \bullet \\ \mathbb{Z} \quad 0 \quad H_2(B) \quad H_3(B) \quad H_4(B) \end{array}$$

The kernel of the map $d^2 = d^2_{20}$ is the quotient E^∞_{20} of $H_2(E)$, because the maps d^r_{20} are zero for $r \geq 3$. Similarly, the cokernel of d^2 is the subgroup E^∞_{01} of $H_1(E)$. From this we obtain the exact homology sequence in the following diagram:

$$
\begin{array}{ccccccccccc}
\pi_3(B) & \to & \pi_2(F) & \to & \pi_2(E) & \to & \pi_2(B) & \to & \pi_1(F) & \to & \pi_1(E) & \to & 0 \\
\downarrow & & \downarrow & & \downarrow & & \downarrow{\scriptstyle\cong} & & \downarrow & & \downarrow \\
X & \hookrightarrow & H_2(F) & \to & H_2(E) & \to & H_2(B) & \xrightarrow{d^2} & H_1(F) & \to & H_1(E) & \to & 0.
\end{array}
$$

Here the group labeled X contains the image in $H_2(F)$ of $E^2_{21} \cong H_2(B) \otimes H_1(F)$ and elements related to $E^3_{30} = H_3(B)$. Thus $H_2(B) \otimes H_1(F)$ is the first obstruction involved in finding a long exact sequence for the homology of a fibration.

Application 5.3.4 (Loop spaces) Let PB denote the space of *based paths* in B, that is, maps $[0, 1] \to B$ sending 0 to $*_B$. The subspace of *based loops* in B (maps $[0, 1] \to B$ sending 0 and 1 to $*_B$) is written ΩB. There is a fibration $\Omega B \to PB \xrightarrow{\pi} B$, where π is evaluation at $1 \in [0, 1]$. The space PB is contractible, because paths may be pulled back along themselves to the basepoint, so $H_n(PB) = 0$ for $n \neq 0$. Therefore, except for $E^\infty_{00} = \mathbb{Z}$, we have a spectral sequence converging to zero. From the low degree terms (assuming that $\pi_1(B) = 0$!), we see that $H_1(\Omega B) \cong H_2(B)$ and that

$$H_4(B) \xrightarrow{d^2} H_2(B) \otimes H_2(B) \xrightarrow{d^2} H_2(\Omega B) \to H_3(B) \to 0$$

is exact. We can use induction on n to estimate the size of $H_n(\Omega B)$.

Exercise 5.3.1 Show that if $n \geq 2$ the loop space ΩS^n has

$$H_p(\Omega S^n) \cong \begin{cases} \mathbb{Z} & \text{if } (n-1) \text{ divides } p, \, p \geq 0 \\ 0 & \text{otherwise.} \end{cases}$$

Application 5.3.5 (Wang sequence) If $F \xrightarrow{i} E \xrightarrow{\pi} S^n$ is a fibration whose base space is an n-sphere $(n \neq 0, 1)$, there is a long exact sequence

$$\cdots \to H_q(F) \xrightarrow{i} H_q(E) \to H_{q-n}(F) \xrightarrow{d^n} H_{q-1}(F) \xrightarrow{i} H_{q-1}(E) \to \cdots.$$

In particular, $H_q(F) \cong H_q(E)$ if $0 \leq q \leq n - 2$.

Proof $H_p(S^n) = 0$ for $p \neq 0, n$ and $H_n(S^n) = H_0(S^n) = \mathbb{Z}$. Therefore the nonzero terms E^2_{pq} all lie on the two vertical lines $p = 0, n$ and $E^2_{pq} = H_q(F)$ for $p = 0$ or n. All the differentials d^r_{pq} must therefore vanish for $r \neq n$, so $E^2_{pq} = E^n_{pq}$ and $E^{n+1}_{pq} = E^\infty_{pq}$. The description of E^{n+1} as the homology of E^n amounts to the exactness of the sequences

$$0 \longrightarrow E^\infty_{n,q} \longrightarrow H_q(F) \xrightarrow{d^n} H_{q+n-1}(F) \longrightarrow E^\infty_{0,q+n-1} \longrightarrow 0.$$

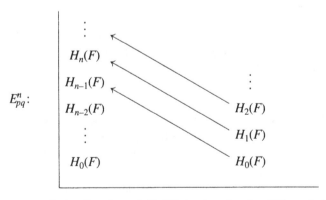

On the other hand, the filtration of $H_q(E)$ is given by the E^∞_{pq}, so it is determined by the short exact sequence

$$0 \to E^\infty_{0q} \to H_q(E) \to E^\infty_{n,q-n} \to 0.$$

The Wang sequence is now obtained by splicing together these two families of short exact sequences.   ◇

Example 5.3.6 The *special orthogonal group* $SO(3)$ is a 3-dimensional Lie group acting on $S^2 \subseteq \mathbb{R}^3$. This action gives rise to the Serre fibration

$$SO(1) \to SO(3) \to S^2.$$

Because $SO(1) = S^1$, we get $H_3(SO(3)) \cong \mathbb{Z}$ and the exact sequence

$$0 \to H_2(SO(3)) \to \mathbb{Z} \xrightarrow{d^2} \mathbb{Z} \to H_1(SO(3)) \to 0.$$

Classically, we know that $\pi_1 SO(3) = \mathbb{Z}/2$, so that $H_1(SO(3)) = \mathbb{Z}/2$. Therefore $H_2(SO(3)) \cong \mathbb{Z}$, although $H_2(SO(3)) \to H_2(S^2)$ is not an isomorphism.

Application 5.3.7 (Gysin sequence) If $S^n \to E \xrightarrow{\pi} B$ is a fibration with B simply connected and $n \neq 0$, there is an exact sequence

$$\cdots \to H_{p-n}(B) \to H_p E \xrightarrow{\pi} H_p(B) \xrightarrow{d^{n+1}} H_{p-n-1}(B) \to H_{p-1}(E) \xrightarrow{\pi} \cdots.$$

In particular, $H_p(E) \cong H_p(B)$ for $0 \leq p < n$.

Proof This is similar to the Wang sequence 5.3.5, except that now the nonzero terms E^2_{pq} all lie on the two rows $q = 0, n$. The only nontrivial differentials are d^{n+1}_{p0} from $H_p(B) = E^{n+1}_{p0}$ to $E^{n+1}_{p-n-1,n} \cong H_{p-n-1}(B)$. $\diamond$

Exercise 5.3.2 If $n \neq 0$, the complex projective n-space $\mathbb{CP}^n$ is a simply connected manifold of dimension $2n$. As such $H_p(\mathbb{CP}^n) = 0$ for $p > 2n$. Given that there is a fibration $S^1 \to S^{2n+1} \to \mathbb{CP}^n$, show that for $0 \leq p \leq 2n$

$$H_p(\mathbb{CP}^n) \cong \begin{cases} \mathbb{Z} & p \text{ even} \\ 0 & p \text{ odd} \end{cases}.$$

5.4 Spectral Sequence of a Filtration

A *filtration* F on a chain complex C is an ordered family of chain subcomplexes $\cdots \subseteq F_{p-1}C \subseteq F_p C \subseteq \cdots$ of C. In this section, we construct a spectral sequence associated to every such filtration; we will discuss convergence of the spectral sequence in the next section.

We say that a filtration is *exhaustive* if $C = \cup F_p C$. It will be clear from the construction that both $\cup F_p C$ and C give rise to the same spectral sequence. In practice, therefore, we always insist that filtrations be exhaustive.

Construction Theorem 5.4.1 *A filtration F of a chain complex C naturally determines a spectral sequence starting with $E^0_{pq} = F_pC_{p+q}/F_{p-1}C_{p+q}$ and $E^1_{pq} = H_{p+q}(E^0_{p*})$.*

Before constructing the spectral sequence, let us make some elementary remarks about the "shape" of the spectral sequence.

Definition 5.4.2 A filtration on a chain complex C is called *bounded* if for each n there are integers $s < t$ such that $F_sC_n = 0$ and $F_tC_n = C_n$. In this case, there are only finitely many nonzero terms of total degree n in E^0_{**}, so the spectral sequence is bounded. We will see in 5.5.1 that the spectral sequence always converges to $H_*(C)$.

A filtration on a chain complex C is called *bounded below* if for each n there is an integer s so that $F_sC_n = 0$, and it is called *bounded above* if for each n there is a t so that $F_tC_n = C_n$. Bounded filtrations are bounded above and below. Being bounded above is merely an easy way to ensure that a filtration is exhaustive. Bounded below filtrations give rise to bounded below spectral sequences. The Classical Convergence Theorem 5.5.1 of the next section says that the spectral sequence always converges to $H_*(C)$ when the filtration is bounded below and exhaustive.

Example 5.4.3 (First quadrant spectral sequences) We call the filtration *canonically bounded* if $F_{-1}C = 0$ and $F_nC_n = C_n$ for each n. As $E^0_{pq} = F_pC_{p+q}/F_{p-1}C_{p+q}$, every canonically bounded filtration gives rise to a first quadrant spectral sequence (converging to $H_*(C)$). For example, the Leray-Serre spectral sequence 5.3.2 arises from a canonically bounded filtration of the singular chain complex $S_*(E)$.

Here are some related notions, which we introduce now in order to give a better perspective on the construction of the spectral sequence.

Definition 5.4.4 A filtration on a chain complex C is called *Hausdorff* if $\cap F_pC = 0$. It will be clear from the construction that both C and its Hausdorff quotient $C^h = C/\cap F_pC$ give rise to the same spectral sequence.

A filtration on C is called *complete* if $C = \varprojlim C/F_pC$. Complete filtrations are Hausdorff because $\cap F_pC$ is the kernel of the map from C to its completion $\widehat{C} = \varprojlim C/F_pC$ (which is also a filtered complex: $F_n\widehat{C} = \varprojlim F_nC/F_pC$). Bounded below filtrations are complete, and hence Hausdorff, because $F_sH_n(C) = 0$ for each n. The following addendum to the Construction Theorem 5.4.1 explains why the most interesting applications of spectral sequences arise from complete filtrations. It will follow from exercise 5.4.1.

Addendum 5.4.5 The two spectral sequences arising from C and $\widehat{C}$ are the same.

The Construction 5.4.6 For legibility, we drop the bookkeeping subscript q and write η_p for the surjection $F_pC \to F_pC/F_{p-1}C = E_p^0$. Next we introduce

$$A_p^r = \{c \in F_pC : d(c) \in F_{p-r}C\},$$

the elements of F_pC that are cycles modulo $F_{p-r}C$ ("approximately cycles") and their images $Z_p^r = \eta_p(A_p^r)$ in E_p^0 and $B_{p-r}^{r+1} = \eta_{p-r}(d(A_p^r))$ in E_{p-r}^0. The indexing is chosen so that Z_p^r and $B_p^r = \eta_p(d(A_{p+r-1}^{r-1}))$ are subobjects of E_p^0. Set $Z_p^\infty = \cap_{r=1}^\infty Z_p^r$ and $B_p^\infty = \cup_{r=1}^\infty B_p^r$. Assembling the above definitions, we see that we have defined a tower of subobjects of each E_p^0:

$$0 = B_p^0 \subseteq B_p^1 \subseteq \cdots \subseteq B_p^r \subseteq \cdots \subseteq B_p^\infty \subseteq Z_p^\infty \subseteq \cdots \subseteq Z_p^r \subseteq \cdots \subseteq Z_p^1 \subseteq Z_p^0 = E_p^0.$$

Note that $A_p^r \cap F_{p-1}C = A_{p-1}^{r-1}$, so that $Z_p^r \cong A_p^r/A_{p-1}^{r-1}$. Hence

$$E_p^r = \frac{Z_p^r}{B_p^r} \cong \frac{A_p^r + F_{p-1}(C)}{d(A_{p+r-1}^{r-1}) + F_{p-1}(C)} \cong \frac{A_p^r}{d(A_{p+r-1}^{r-1}) + A_{p-1}^{r-1}}.$$

Let $d_p^r: E_p^r \to E_{p-r}^r$ be the map induced by the differential of C. To define the spectral sequence, we only need to give the isomorphism between E^{r+1} and $H_*(E^r)$.

Lemma 5.4.7 *The map d determines isomorphisms*

$$Z_p^r/Z_p^{r+1} \xrightarrow{\cong} B_{p-r}^{r+1}/B_{p-r}^r.$$

Proof This is largely an exercise in decoding notation. First, note that $d(A_p^r) \cap F_{p-r-1}C = d(A_p^{r+1})$, so that $B_{p-r}^{r+1} \cong d(A_p^r)/d(A_p^{r+1})$ and hence B_{p-r}^{r+1}/B_{p-r}^r is isomorphic to $d(A_p^r)/d(A_p^{r+1} + A_{p-1}^{r-1})$. The other term Z_p^r/Z_p^{r+1} is isomorphic to $A_p^r/(A_p^{r+1} + A_{p-1}^{r-1})$. As the kernel of $d: A_p^r \to F_{p-r}C$ is contained in A_p^{r+1}, the two sides are isomorphic. $\diamond$

Resuming the construction of the spectral sequence, the kernel of d_p^r is

$$\frac{\left\{ z \in A_p^r : d(z) \in d(A_{p-1}^{r-1}) + A_{p-r-1}^{r-1} \right\}}{d(A_{p+r-1}^{r-1}) + A_{p-1}^{r-1}} = \frac{A_{p-1}^{r-1} + A_p^{r+1}}{d(A_{p+r-1}^{r-1}) + A_{p-1}^{r-1}} \cong \frac{Z_p^{r+1}}{B_p^r}.$$

By lemma 5.4.7, the map d_p^r factors as

$$E_p^r = Z_p^r/B_p^r \to Z_p^r/Z_p^{r+1} \xrightarrow{\cong} B_{p-r}^{r+1}/B_{p-r}^r \hookrightarrow Z_{p-r}^r/B_{p-r}^r = E_{p-r}^r.$$

From this we see that the image of d_p^r is B_{p-r}^{r+1}/B_{p-r}^r; replacing p with $p+r$, the image of d_{p+r}^r is B_p^{r+1}/B_p^r. This provides the isomorphism

$$E_p^{r+1} = Z_p^{r+1}/B_p^{r+1} \cong \ker(d_p^r)/\operatorname{im}(d_{p+r}^r)$$

needed to complete the construction of the spectral sequence.   ◇

Observation Fix p and $k \geq 1$, and set $C' = C/F_{p-k}C$, $C'' = F_{p+k}C/F_{p-k}C$. The complex C' is bounded below, C'' is bounded, and there are maps $C \to C' \leftarrow C''$. For $0 \leq r \leq k$ these maps induce isomorphisms on the associated groups $A_p^r/F_{p-k}C$ and $\{d(A_{p+r-1}^{r-1}) + F_{p-k}C\}/F_{p-k}C$. (Check this!) Hence the associated groups Z_p^r, B_p^r and E_p^r are isomorphic. That is, the associated spectral sequences for C, C', and C'' agree in the (p,q) spots through the E^k terms.

Exercise 5.4.1 Recall that the completion $\widehat{C}$ is also a filtered complex. Show that $C/F_{p-k}C$ and $\widehat{C}/F_{p-k}\widehat{C}$ are naturally isomorphic.

We can now establish the addendum 5.4.5. For each p, q, and k, we have shown that the maps $C \to \widehat{C} \to C'$ induce isomorphisms between the corresponding E_{pq}^k terms. Letting k go to infinity, we see that the map $\{f_{pq}^r: E_{pq}^r(C) \to E_{pq}^r(\widehat{C})\}$ of spectral sequences is an isomorphism, because each f_{pq}^r is an isomorphism.

Exercise 5.4.2 Show that the spectral sequences for C, $\cup F_p C$, and $C/\cap F_p C$ are all isomorphic.

Multiplicative Structure 5.4.8 Suppose that C is a differential graded algebra (4.5.2) and that the filtration is multiplicative in the sense that for every s and t, $(F_s C)(F_t C) \subseteq F_{s+t}C$. Since $E_{p,n-p}^0$ is $F_p C_n/F_{p-1}C_n$, it is clear that we have a product

$$E_{p_1 q_1}^0 \times E_{p_2 q_2}^0 \to E_{p_1+p_2, q_1+q_2}^0$$

satisfying the Leibnitz relation. Hence the spectral sequence has a multiplicative structure in the sense of 5.2.13. Moreover, we saw in exercise 4.5.1 that $H_*(C)$ is an algebra and that the images $F_p H_*(C)$ of the $H_*(F_p C)$ form a multiplicative system of ideals in $H_*(C)$. Therefore whenever the spectral sequence (weakly) converges to $H_*(C)$ it follows that E^∞ is the associated graded algebra of $H_*(C)$. This convergence is the topic of the next section.

Exercise 5.4.3 (Shifting or Décalage) Given a filtration F on a chain complex C, define two new filtrations $\tilde{F}$ and $\mathrm{Dec}\,F$ on C by $\tilde{F}_p C_n = F_{p-n} C_n$ and $(\mathrm{Dec}\,F)_p C_n = \{x \in F_{p+n} C_n : dx \in F_{p+n-1} C_{n-1}\}$. Show that the spectral sequences for these three filtrations are isomorphic after reindexing: $E^r_{pq}(F) \cong E^{r+1}_{p+n,q-n}(\tilde{F})$ for $r \geq 0$, and $E^r_{pq}(F) \cong E^{r-1}_{p-n,q+n}(\mathrm{Dec}\,F)$ for $r \geq 2$.

Exercise 5.4.4 (Eilenberg-Moore) Let $f : B \to C$ be a map of filtered chain complexes. For each $r \geq 0$, define a filtration on the mapping cone $\mathrm{cone}(f)$ 1.5.1 by

$$F_p \mathrm{cone}(f)_n = F_{p-r} B_{n-1} \oplus F_p C_n.$$

Show that $E^r_p(\mathrm{cone}\,f)$ is the mapping cone of $f^r : E^r_p(B) \to E^r_p(C)$. By 1.5.2 this gives a long exact sequence

$$\cdots E^r_{p+r}(\mathrm{cone}\,f) \to E^r_p(B) \to E^r_p(C) \to E^r_p(\mathrm{cone}\,f) \cdots.$$

5.5 Convergence

A filtration on a chain complex C induces a filtration on the homology of $C : F_p H_n(C)$ is the image of the map $H_n(F_p C) \to H_n(C)$. If the filtration on C is exhaustive, then the filtration on H_n is also exhaustive ($H_n = \cup F_p H_n$), because every element of H_n is represented by an element c of some $F_p C_n$ such that $d(c) = 0$. If the filtration on C is bounded below then the filtration on each $H_n(C)$ is also bounded below, since $F_p C = 0$ implies that $F_p H_n(C) = 0$.

Exercise 5.5.1 Give an example of a complete Hausdorff filtered complex C such that the filtration on $H_0(C)$ is not Hausdorff, that is, such that $\cap F_p H_0(C) \neq 0$.

Here are the two classical criteria used to establish convergence; we will discuss convergence for complete filtrations later on.

Classical Convergence Theorem 5.5.1

1. *Suppose that the filtration on C is* bounded. *Then the spectral sequence is bounded and converges to $H_*(C)$:*

$$E^1_{pq} = H_{p+q}(F_p C / F_{p-1} C) \Rightarrow H_{p+q}(C).$$

2. *Suppose that the filtration on C is bounded below* and *exhaustive. Then the spectral sequence is bounded below and also converges to $H_*(C)$.*

   Moreover, the convergence is natural in the sense that if $f : C \to C'$ is a map of filtered complexes, then the map $f_* : H_*(C) \to H_*(C')$ is compatible with the corresponding map of spectral sequences.

Example 5.5.2 (First quadrant spectral sequences) Suppose that the filtration is *canonically bounded* ($F_{-1} C = 0$ and $F_n C_n = C_n$ for each n), so that the spectral sequence lies in the first quadrant. Then it converges to $H_*(C)$. Along the y-axis of E^1 we have $E^1_{0q} = H_q(F_0 C)$, and E^∞_{0q} is a quotient of this (see 5.2.6). Along the x-axis, E^1_{p0} is the homology $H_p(\bar{C})$ of C's top quotient chain complex $\bar{C}$, $\bar{C}_n = C_n / F_{n-1} C_n$; E^∞_{p0} is therefore a subobject of $H_p(\bar{C})$.

Corollary 5.5.3 *If the filtration is canonically bounded, then E^∞_{0q} is the image of $H_q(F_0 C)$ in $H_q(C)$ and E^∞_{p0} is the image of $H_p(C)$ in $H_p(\bar{C})$.*

Proof By definition, $E_{0q}^\infty = F_0 H_q(C)$ is the image of $H_q(F_0 C)$ in $H_q(C)$. Now consider the exact sequence of chain complexes $0 \to F_{p-1}C \to C_p \to \bar{C}_p \to 0$. From the associated homology exact sequence we see that the image of $H_p(C)$ in $H_p(\bar{C})$ is the cokernel of the map from $H_p(F_{p-1}C)$ to $H_p(C)$, which by definition is $E_{p0}^\infty = H_p(C)/F_{p-1}H_p(C)$. $\diamond$

Proof of Classical Convergence Theorem Suppose that the filtration is exhaustive and bounded below (resp. bounded). Then the filtration on H_* is exhaustive and bounded below (resp. bounded), and the spectral sequence is bounded below (resp. bounded). By Definition 5.2.11, the spectral sequence will converge to H_* whenever it weakly converges. For this, we observe that since the filtration is bounded below and p and n are fixed, the groups $A_p^r = \{c \in F_p C_n : d(c) \in F_{p-r}C_{n-1}\}$ stabilize for large r; write A_p^∞ for this stable value, and observe that since $Z_p^r = \eta_p(A_p^r)$ we have $Z_p^\infty = \eta_p(A_p^\infty)$. Now A_p^∞ is the kernel of $d: F_p C_n \to F_p C_{n-1}$, $(dC) \cap F_p C$ is the union of the $d(A_{p+r}^r)$, and A_{p-1}^∞ is the kernel of the map $\eta_p: A_p^\infty \to E_{pq}^0$. Thus

$$F_p H_n(C)/F_{p-1}H_n(C) \cong A_p^\infty/\{A_{p-1}^\infty + d(\cup A_{p+r}^r)\}$$

$$\cong \eta_p(A_p^\infty)/\,\eta_p d(\cup A_{p+r}^r)$$

$$= Z_p^\infty/B_p^\infty = E_p^\infty.$$ $\diamond$

When the filtration is not bounded below, convergence is more delicate. Of course we have to work within an abelian category such as R-mod, because we need axiom $(AB4^*)$ in order to even talk about E^∞ (see 5.2.8). But there are more basic problems. For example, the filtration on $H_*(C)$ need not be Hausdorff. This is not surprising, since by 5.4.5 the completion $\hat{C}$ has the same spectral sequence but different homology. (And see exercise 5.5.1.)

Example 5.5.4 Let C be the chain complex $0 \to \mathbb{Z} \xrightarrow{3} \mathbb{Z} \to 0$, and let $F_p C$ be $2^p C$. Then the Hausdorff quotient of $H_*(C)$ is zero, because $F_p H_*(C) = H_*(C)$ for all p, even though $H_0(C) = \mathbb{Z}/3$. Each row of E^0 is $\mathbb{Z}/2 \xleftarrow{3} \mathbb{Z}/2$ and the spectral sequence collapses to zero at E^1, so the spectral sequence is weakly converging (but not converging) to $H_*(C)$. It converges to $H_*(\hat{C}) = 0$.

Theorem 5.5.5 (Eilenberg-Moore Filtration Sequence for complete complexes) *Suppose that C is complete with respect to a filtration by subcomplexes. Associated to the tower $\{C/F_p C\}$ is the sequence of 3.5.8:*

$$0 \to \varprojlim{}^1 H_{n+1}(C/F_p C) \to H_n(C) \xrightarrow{\pi} \varprojlim H_n(C/F_p C) \to 0.$$

This sequence is associated to the filtration on $H_*(C)$ as follows. The left-hand term $\varprojlim^1 H_{n+1}(C/F_pC)$ is $\cap F_pH_n(C)$, and the right-hand term is the Hausdorff quotient of $H_*(C)$:

$$H_*(C)/\cap F_pH_n(C) \cong \varprojlim H_n(C)/F_pH_n(C) \cong \varprojlim H_n(C/F_pC).$$

Proof Taking the inverse limit of the exact sequences of towers

$$0 \to \{F_pH_*(C)\} \to H_*(C) \to \{H_*(C)/F_pH_*(C)\} \to 0;$$

$$0 \to \{H_*(C)/F_pH_*(C)\} \to \{H_*(C/F_pC)\}$$

shows that $H_*(C)/\cap F_pH_*(C)$ is a subobject of $\varprojlim H_*(C)/F_pH_*(C)$, which is in turn a subobject of $\varprojlim H_n(C/F_pC)$. Now combine this with the $\varprojlim^1$ sequence of 3.5.8. ◇

Corollary 5.5.6 *If the spectral sequence weakly converges, then $H_*(C) \cong H_*(\widehat{C})$.*

A careful reading of the proof of the Classical Convergence Theorem 5.5.1 yields the following lemma for all Hausdorff, exhaustive filtrations. To avoid confusion, we reintroduce the fixed subscripts q and $n = p + q$. Write $A_{pq}^\infty = \cap_{r=1}^\infty A_{pq}^r$, recalling that in our notation $A_{pq}^r = \{c \in F_pC_n : d(c) \in F_{p-r}C_{n-1}\}$. In $E_{pq}^0 = F_pC_n/F_{p-1}C_n$, $\eta_p(A_{pq}^\infty)$ is contained in Z_{pq}^∞ and contains $B_{pq}^\infty = \eta_p(F_pC \cap d(C))$. (Check this!) Hence $e_{pq}^\infty = \eta_p(A_{pq}^\infty)/B_{pq}^\infty$ is contained in E_{pq}^∞.

Lemma 5.5.7 *Assume that the filtration on C is Hausdorff and exhaustive. Then*

1. *A_{pq}^∞ is the kernel of $d: F_pC_n \to F_pC_{n-1}$;*
2. *$F_pH_n(C) \cong A_{pq}^\infty / \cup_{r=1}^\infty d(A_{p+r,q-r+1}^r)$;*
3. *The subgroup e_{pq}^∞ of E_{pq}^∞ is related to $H_*(C)$ by*

$$e_{pq}^\infty \cong F_pH_n(C)/F_{p-1}H_n(C).$$

Proof Recall that $F_pH_n(C)$ is the image of the map $H_n(F_pC) \to H_n(C)$. Since $\cap F_pC = 0$, the kernel of $d: F_pC_n \to F_pC_{n-1}$ is A_{pq}^∞, so $H_n(F_pC) \cong$

$A^\infty_{pq}/d(F_pC_{n+1})$. As $\cup F_pC = C$, the kernel of $A^\infty_{pq} \to H_n(C)$ is the union $\cup d(A^r_{p+r,q-r+1})$. For part (3) observe that $A^\infty_{pq} \cap F_{p-1}C_n = A^\infty_{p-1,q+1}$ by definition, so that $\eta_p A^\infty_{pq} = A^\infty_{pq}/A^\infty_{p-1,q+1}$. Hence we may calculate in E^0_{pq}

$$F_p H_n(C)/F_{p-1}H_n(C) \cong A^\infty_{pq}/A^\infty_{p-1,q+1} + \cup d(A^r_{p+r,q-r+1})$$
$$\cong \eta_p(A^\infty_{pq})/\cup \eta_p d(A^r_{p+r,q-r+1})$$
$$= \eta_p(A^\infty_{pq})/B^\infty_{pq} = e^\infty_{pq}. \qquad \diamond$$

Corollary 5.5.8 (Boardman's Criterion) *Let Q_p denote $\varprojlim^1\{A^r_{pq}\}$ for fixed p and q. The inclusions $A^{r-1}_{p-1,q+1} \subset A^r_{pq}$ induce a map $a: Q_{p-1} \to Q_p$, and there is an exact sequence*

$$0 \to e^\infty_{pq} \to E^\infty_{pq} \to Q_{p-1} \xrightarrow{a} Q_p \to \varprojlim{}^1\{Z^r_{pq}\} \to 0.$$

In particular, if the filtration is Hausdorff and exhaustive, then the spectral sequence weakly converges to $H_*(C)$ if and only if the maps $a: Q_{p-1} \to Q_p$ are all injections.

Proof The short exact sequence of towers from 5.4.6

$$0 \to \{A^{r-1}_{p-1}\} \to \{A^r_p\} \xrightarrow{\eta} \{Z^r_p\} \to 0$$

yields

$$0 \to A^\infty_{p-1} \to A^\infty_p \xrightarrow{\eta} Z^\infty_p \to Q_{p-1} \xrightarrow{a} Q_p \to \varprojlim{}^1\{Z^r_p\} \to 0.$$

Now mod out by B^∞_p, recalling that e^∞_{pq} is $\eta(A^\infty_{pq})/B^\infty_{pq}$. $\qquad \diamond$

Exercise 5.5.2 Set $R_p = \cap_r \text{image}\{H(F_rC) \to H(F_pC)\}$. Show that the spectral sequence is weakly convergent iff the maps $R_{p-1} \to R_p$ are injections for all p. *Hint:* $R_p \subset Q_p$.

Exercise 5.5.3 Suppose that the filtration on C is Hausdorff and exhaustive. If for any $p+q = n$ we have $E^r_{pq} = 0$, show that $F_p H_n(C) = F_{p-1}H_n(C)$. Conclude that $H_n(C) = \cap F_p H_n(C)$, provided that every $E^r_{p,q}$ with $p + q$ equalling n vanishes.

Proposition 5.5.9 (Boardman) *Suppose that the filtration on C_n is complete, and form the tower of groups $Q_p = \varprojlim^1\{A^r_{p,n-p}\}$ as in 5.5.8 along the maps $a: Q_{p-1} \to Q_p$. Then $\varprojlim Q_p = 0$.* $\qquad \diamond$

Proof Let I denote the poset of negative numbers $\cdots < p - 1 < p < p + 1 < \cdots < 0$. For each negative p and t, the subgroups $A(p, t) = A_p^{t-p} = \{c \in F_pC_n : d(c) \in F_tC_{n-1}\}$ of C_n form a functor $A: I \times I \to \textbf{Ab}$, that is, a "double tower" of subgroups. If we fix t and vary p, then for $p \le t$ we have $A(p, t) = F_pC_n$. Hence we have $\varprojlim_p A(p, t) = \varprojlim_p F_pC_n = 0$ and $\varprojlim_p^1 A(p, t) = \varprojlim_p^1 F_pC_n = 0$ (see 3.5.7). We assert that the derived functor $R^1 \varprojlim_{I \times I}$ from double towers to abelian groups fits into two short exact sequences:

(†)
$$0 \to \varprojlim_t^1 (\varprojlim_p A(p, t)) \to R^1 \varprojlim_{I \times I} A(p, t) \to \varprojlim_t (\varprojlim_p^1 A(p, t)) \to 0,$$

$$0 \to \varprojlim_p^1 (\varprojlim_t A(p, t)) \to R^1 \varprojlim_{I \times I} A(p, t) \to \varprojlim_p (\varprojlim_t^1 A(p, t)) \to 0.$$

We will postpone the proof of this assertion until 5.8.7 below, even though it follows from the Classical Convergence Theorem 5.5.1, as it is an easy application of the Grothendieck spectral sequence 5.8.3. The first of the sequences in (†) implies that $R^1 \varprojlim_{I \times I} A(p, t) = 0$, so from the second sequence in (†) we deduce that $\varprojlim_p (\varprojlim_t^1 A(p, t)) = 0$.

To finish, it suffices to prove that $\varprojlim_t^1 A(p, t)$ is isomorphic to Q_p for each $p < 0$. Fix p, so that there is a short exact sequence of towers in t:

(∗) $0 \to \{A(p, p + t)\} \to \{A(p, t)\} \to \{A(p, t)/A(p, p + t)\} \to 0.$

If $t' < p + t$ the map $A(p, t')/A(p, p + t') \to A(p, t)/A(p, p + t)$ is obviously zero. Therefore the third tower of (∗) satisfies the trivial Mittag-Leffler condition (3.5.6), which means that

$$\varprojlim_t A(p, t)/A(p, p + t) = \varprojlim_t^1 A(p, t)/A(p, p + t) = 0.$$

From the $\varprojlim$ exact sequence of (∗) we obtain the described isomorphism

$$Q_p = \varprojlim_t A_p^t = \varprojlim_t A(p, p + t) \cong \varprojlim_t A(p, t). \qquad \diamond$$

Complete Convergence Theorem 5.5.10 *Suppose that the filtration on C is complete and exhaustive and the spectral sequence is regular (5.2.10). Then*

1. *The spectral sequence weakly converges to $H_*(C)$.*
2. *If the spectral sequence is bounded above, it converges to $H_*(C)$.*

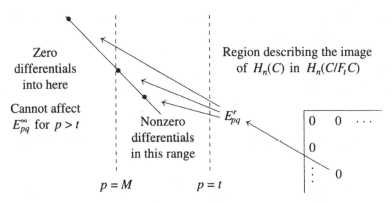

Figure 5.2.   Complete convergence for regular, bounded above spectral sequences.

Proof When the spectral sequence is regular, Z_{pq}^{∞} equals $Z_{pq}^{r} = \eta_p A_{pq}^{r}$ for large r. By Boardman's criterion 5.5.8, all the maps $Q_{p-1} \to Q_p$ are onto, and the spectral sequence weakly converges if and only if $Q_p = 0$ for all p. This is indeed the case since the group $\varprojlim Q_p$ maps onto each Q_p (3.5.3), and we have just seen in 5.5.9 that $\varprojlim Q_p = 0$. This proves (1).

To see that the spectral sequence converges to $H_*(C)$, it suffices to show that the filtration on $H_*(C)$ is Hausdorff. By the Eilenberg-Moore Filtration Sequence 5.5.5, it suffices to show that the tower $\{H_n(C/F_tC)\}$ is Mittag-Leffler for every n, since then its $\lim^1$ groups vanish by 3.5.7. Each C/F_tC has a bounded below filtration, so it has a convergent spectral sequence whose associated graded groups $E_{pq}^{\infty}(C/F_tC)$ are subquotients of $E_{pq}^0(C)$ for $p > t$. For $m < t$, the images of the maps $E_{pq}^{\infty}(C/F_mC) \to E_{pq}^{\infty}(C/F_tC)$ are the associated graded groups of the image of $H_*(C/F_mC) \to H_*(C/F_tC)$, so it suffices to show that these images are independent of m as $m \to -\infty$.

Now assume that the spectral sequence for C is regular and bounded above. Then for each n and t there is an M such that the differentials $E_{pq}^r(C) \to E_{p-r,q+1-r}^r(C)$ are zero whenever $p + q = n$, $p > t$, and $p - r \leq M$. By inspection, this implies that $E_{pq}^{\infty}(C) = E_{pq}^{\infty}(C/F_mC)$ for every $p + q = n$ with $p > t$ and every $m \leq M$. Thus the image of $E_{pq}^{\infty}(C/F_mC) \to E_{pq}^{\infty}(C/F_tC)$ is independent of $m \leq M$ for $p + q = n$ and $p > t$, as was to be shown.   ◇

Exercise 5.5.4 (Complete nonconverging spectral sequences) Let $\mathbb{Z} < x >$ denote an infinite cyclic group with generator x, and let C be the chain complex with

$$C_1 = \bigoplus_{i=1}^{\infty} \mathbb{Z} < x_i >, \quad C_0 = \prod_{i=-\infty}^{i=0} \mathbb{Z} < y_i >, \quad C_n = 0 \quad \text{for } n \neq 0, 1$$

and $d: C_1 \to C_0$ defined by $d(x_i) = y_{1-i} - y_{-i}$. For $p \leq 0$ define $F_pC_1 = 0$ and $F_pC_0 = \prod_{i \leq p} \mathbb{Z} < y_i >$; this is a complete filtration on C.

1. Show that $F_p H_0(C) = H_0(C)$ for every $p \le 0$, so that the filtration on $H_0(C)$ is not Hausdorff. (Since C_1 is countable and C_0 is not, we have $H_0(C) \ne 0$.) Hence no spectral sequence constructed with this filtration can approach $H_*(C)$, let alone converge to it; such a spectral sequence will weakly converge to $H_*(C)$ if and only if it converges to zero.

2. Here is an example of an (essentially) second quadrant spectral sequence that weakly converges but does not converge to $H_*(C)$. For $p \ge 1$ define $F_p C_1 = C_1$ and $F_p C_0 = C_0$. The resulting spectral sequence has $E_{10}^0 = C_1$, $E_{p,-p}^0 = \mathbb{Z} < y_p >$ for $p \le 0$ and $E_{pq}^0 = 0$ otherwise. Show that $d^r(x_r)$ is $[y_{1-r}]$ and $d^r(x_i) = 0$ for $i \ne r$, and conclude that $E_{pq}^\infty = 0$ for every p and q.

3. Here is a regular spectral sequence that does not converge to $H_*(C)$. For $p \ge 1$ let $F_p C_1$ be the subgroup of C_1 spanned by $x_1, \cdots, x_p$ and set $F_p C_0 = C_0$. The resulting spectral sequence has $E_{p,1-p}^0 = \mathbb{Z} < x_p >$ for $p \ge 1$, $E_{p,-p}^0 = \mathbb{Z} < y_p >$ for $p \le 0$ and $E_{pq}^0 = 0$ otherwise. Show that this spectral sequence is regular and converges to zero.

The following result generalizes the Comparison Theorem 5.2.12 to nonconvergent spectral sequences.

Eilenberg-Moore Comparison Theorem 5.5.11 *Let $f : B \to C$ be a map of filtered complexes of modules, where both B and C are complete and exhaustive. Fix $r \ge 0$. Suppose that $f^r : E_{pq}^r(B) \cong E_{pq}^r(C)$ is an isomorphism for all p and q. Then $f : H_*(B) \to H_*(C)$ is an isomorphism.*

Proof Consider the filtration on the mapping cone complex given by the formula $F_p \mathrm{cone}(f) = F_{p-r} B[-1] \oplus F_p C$. This filtration is complete and exhaustive. Since f^r is an isomorphism, the long exact sequence of Exercise 5.4.4 shows that $E_{pq}^r(\mathrm{cone}\, f) = 0$ for all p and q. By 5.5.10, this spectral sequence converges to $H_* \mathrm{cone}(f)$. Hence $\mathrm{cone}(f)$ is an exact complex, and 1.5.4 applies. $\diamond$

5.6 Spectral Sequences of a Double Complex

There are two filtrations associated to every double complex C, resulting in two spectral sequences related to the homology of $\mathrm{Tot}(C)$. Playing these spectral sequences off against each other is an easy way to calculate homology.

Definition 5.6.1 (Filtration by columns) If $C = C_{**}$ is a double complex, we may filter the (product or direct sum) total complex $\mathrm{Tot}(C)$ by the columns of C, letting $^I F_n \mathrm{Tot}(C)$ be the total complex of the double subcomplex

$$\left(^I \tau_{\le n} C \right)_{pq} = \begin{cases} C_{pq} & \text{if } p \le n \\ 0 & \text{if } p > n \end{cases}$$

$\cdots$	$*$	$*$	0	0
$\cdots$	$*$	$*$	0	0
$\cdots$	$*$	$*$	0	0
$\cdots$	$*$	$*$	0	0

of C. This gives rise to a spectral sequence $\{{}^I E^r_{pq}\}$, starting with ${}^I E^0_{pq} = C_{pq}$. The maps d^0 are just the vertical differentials d^v of C, so

$$ {}^I E^1_{pq} = H^v_q(C_{p*}). $$

The maps $d^1\colon H^v_q(C_{p*}) \to H^v_q(C_{p-1,*})$ are induced on homology from the horizontal differentials d^h of C, so we may use the suggestive notation:

$$ {}^I E^2_{pq} = H^h_p H^v_q(C). $$

If C is a first quadrant double complex, the filtration is canonically bounded, and we have the convergent spectral sequence discussed in section 5.1:

$$ {}^I E^2_{pq} = H^h_p H^v_q(C) \Rightarrow H_{p+q}(\mathrm{Tot}(C)). $$

If C is a fourth quadrant double complex (or more generally if $C_{pq} = 0$ in the second quadrant), the filtration on $\mathrm{Tot}^\Pi(C)$ is bounded below but is not exhaustive. The filtration on the direct sum total complex $\mathrm{Tot}^\oplus(C)$ is both bounded below and exhaustive, so by the Classical Convergence Theorem 5.5.1 the spectral sequence ${}^I E^r_{**}$ converges to $H_*(\mathrm{Tot}^\oplus C)$ and not to $H_*(\mathrm{Tot}^\Pi C)$.

If C is a second quadrant double complex (or more generally if $C_{pq} = 0$ in the fourth quadrant), the filtration on the product total complex $\mathrm{Tot}^\Pi(C)$ is complete and exhaustive. By the Complete Convergence Theorem 5.5.10, the spectral sequence ${}^I E^r_{**}$ weakly converges to $H_*(\mathrm{Tot}^\Pi C)$, and we have the Eilenberg-Moore filtration sequence (5.5.5)

$$ 0 \to \varprojlim{}^1 H_{n+1}(C/\tau_{\le n}C) \to H_n(\mathrm{Tot}^\Pi C) \to \varprojlim H_n(C/\tau_{\le n}C) \to 0. $$

We will encounter a spectral sequence of this type in Chapter 9, 9.6.17.

Definition 5.6.2 (Filtration by rows) If C is a double complex, we may also filter $\mathrm{Tot}(C)$ by the rows of C, letting ${}^{II} F_n \, \mathrm{Tot}(C)$ be the total complex of

$$ \left({}^{II}\tau_{\le n}C\right)_{pq} = \begin{cases} C_{pq} & \text{if } q \le n \\ 0 & \text{if } q > n \end{cases} $$

$$
\begin{array}{cccccc}
\cdots & & & & \cdots & \\
0 & 0 & 0 & 0 & 0 & 0 \\
0 & 0 & 0 & 0 & 0 & 0 \\
\hline
* & * & * & * & * & * \\
* & * & * & * & * & * \\
\cdots & & & & \cdots &
\end{array}
$$

Since $F_p \, \mathrm{Tot}(C)/F_{p-1}\,\mathrm{Tot}(C)$ is the *row* C_{*p}, ${}^{II}E^0_{pq} = C_{qp}$ and ${}^{II}E^1_{pq} = H^h_q(C_{*p})$. (Beware the interchange of p and q in the notation!) The maps d^1 are induced from the vertical differentials d^v of C, so we may use the suggestive notation

$$ {}^{II}E^2_{pq} = H^v_p H^h_q(C). $$

Of course, this should not be surprising, since interchanging the roles of p and q converts the filtration by rows into the filtration by columns, and interchanges the spectral sequences ^IE and ^{II}E.

As before, if C is a first quadrant double complex, this filtration is canonically bounded, and the spectral sequence converges to $H_* \operatorname{Tot}(C)$. If C is a second quadrant double complex (or more generally if $C_{pq} = 0$ in the fourth quadrant), the spectral sequence $^{II}E^r_{**}$ converges to $H_* \operatorname{Tot}^\oplus(C)$. If C is a fourth quadrant double complex (or if $C_{pq} = 0$ in the second quadrant), then the spectral sequence $^{II}E^r_{**}$ weakly converges to $H_* \operatorname{Tot}^\Pi(C)$. $\Diamond$

Application 5.6.3 (Balancing Tor) In Chapter 2, 2.7.1, we used a disguised spectral sequence argument to prove that $L_n(A\otimes)(B) \cong L_n(\otimes B)(A)$, that is, that $\operatorname{Tor}_*(A, B)$ could be computed by taking either a projective resolution $P \to A$ or a projective resolution $Q \to B$. In our new vocabulary, there are two spectral sequences converging to the homology of $\operatorname{Tot}(P \otimes Q)$. Since $H^v_q(P_p \otimes Q) = P_p \otimes H_q(A)$, the first has

$$^IE^2_{pq} = \left\{ \begin{matrix} H^h_p(P \otimes B) = L_p(\otimes B)(A) & \text{if } q = 0 \\ 0 & \text{otherwise} \end{matrix} \right\}.$$

This spectral sequence collapses to yield $H_p(P \otimes Q) = L_p(\otimes B)(A)$. Therefore the second spectral sequence converges to $L_p(\otimes B)(A)$. Since $H^h_q(P \otimes Q_n) = H_q(P) \otimes Q_n$,

$$^{II}E^2_{pq} = \left\{ \begin{matrix} H^v_p(A \otimes Q) = L_q(A\otimes)(B) & \text{if } q = 0 \\ 0 & \text{otherwise} \end{matrix} \right\}.$$

This spectral sequence collapses to yield $H_p(P \otimes Q) = L_p(A\otimes)(B)$, whence the result.

Theorem 5.6.4 (Künneth spectral sequence) *Let P be a bounded below complex of flat R-modules and M an R-module. Then there is a boundedly converging right half-plane spectral sequence*

$$E^2_{pq} = \operatorname{Tor}^R_p(H_q(P), M) \Rightarrow H_{p+q}(P \otimes_R M).$$

Proof Let $Q \to M$ be a projective resolution and consider the upper half-plane double complex $P \otimes Q$. Since P_p is flat, $H^v_q(P \otimes Q) = P_p \otimes H_q(Q)$, so the first spectral sequence has

$$^IE^2_{pq} = \left\{ \begin{matrix} H_p(P \otimes M) & \text{if } q = 0 \\ 0 & \text{otherwise} \end{matrix} \right\}.$$

This spectral sequence collapses to yield $H_p(P \otimes Q) = H_p(P \otimes M)$. Since Q_q is flat, $H_q(P \otimes Q_n) = H_q(P) \otimes Q_n$, so the second spectral sequence has the desired E^2 term

$$^{II}E^2_{pq} = H_p(H_q(P) \otimes Q) = \text{Tor}^R_p(H_q(P), M).\qquad\diamond$$

Künneth Formula 5.6.5 In Chapter 3, 3.6.1, we could have given the following spectral sequence argument to compute $H_*(P \otimes M)$, assuming that $d(P)$ (and hence Z) is flat. The flat dimension of $H_q(P)$ is at most 1, since

$$0 \to d(P_{q+1}) \to Z_q \to H_q(P) \to 0$$

is a flat resolution. In this case only the columns $p = 0, 1$ are nonzero, so all the differentials vanish and $E^2_{pq} = E^\infty_{pq}$. The 2-stage filtration of $H_p(P \otimes Q)$ yields the Künneth formula.

0	0	...	...	0	0
0	0	$H_q(P) \otimes M$	$\text{Tor}_1(H_q(P), M)$	0	0
0	0	$H_{q-1}(P) \otimes M$	$\text{Tor}_1(H_{q-1}(P), M)$	0	0
0	0	...	...	0	0

Exercise 5.6.1 Give a spectral sequence proof of the Universal Coefficient Theorem 3.6.5 for cohomology.

Theorem 5.6.6 (Base-change for Tor) *Let* $f: R \to S$ *be a ring map. Then there is a first quadrant homology spectral sequence*

$$E^2_{pq} = \text{Tor}^S_p(\text{Tor}^R_q(A, S), B) \Rightarrow \text{Tor}^R_{p+q}(A, B)$$

for every $A \in \textbf{mod}–R$ *and* $B \in S–\textbf{mod}$.

Proof Let $P \to A$ be an R-module projective resolution, and $Q \to B$ an S-module projective resolution. As in 2.7.1, form the first quadrant double complex $P \otimes Q$ and write $H_*(P \otimes Q)$ for $H_*(\text{Tot}(P \otimes_R Q))$. Since $P_p \otimes_R$ is an exact functor, the pth column of $P \otimes Q$ is a resolution of $P_p \otimes B$. Therefore the first spectral sequence 5.6.1 collapses at $^IE^1 = H^v_q(P \otimes Q)$ to yield $H_*(P \otimes Q) \cong H_*(P \otimes B) = \text{Tor}^R_*(A, B)$. Therefore the second spectral sequence 5.6.2 converges to $\text{Tor}^R_*(A, B)$ and has

$$^{II}E^1_{pq} = H_q(P \otimes_R Q_p) = H_q((P \otimes_R S) \otimes_S Q_p)$$

$$= H_q(P \otimes_R S) \otimes_S Q_p = \text{Tor}^R_q(A, S) \otimes_S Q_p$$

and hence the prescribed E^2_{pq} term: $H_p(^{II}E^1_{pq}) = \text{Tor}^S_p(\text{Tor}^R_q(A, S), B).\qquad\diamond$

Exercise 5.6.2 (Bourbaki) Given rings R and S, let L be a right R-module, M an R-S bimodule, and N a left S-module, so that the tensor product $L \otimes_R M \otimes_S N$ makes sense.

1. Show that there are two spectral sequences, such that

$$^{I}E^2_{pq} = \text{Tor}^R_p(L, \text{Tor}^S_q(M, N)) \qquad ^{II}E^2_{pq} = \text{Tor}^S_p(\text{Tor}^R_q(L, M), N)$$

   converging to the same graded abelian group H_*. *Hint:* Consider a double complex $P \otimes M \otimes Q$, where $P \to L$ and $Q \to N$.
2. If M is a flat S-module, show that the spectral sequence ^{II}E converges to $\text{Tor}^R_*(L, M \otimes_S N)$. If M is a flat R-module, show that the spectral sequence ^{I}E converges to $\text{Tor}^S_*(L \otimes_R M, N)$.

Exercise 5.6.3 (Base-change for Ext) Let $f: R \to S$ be a ring map. Show that there is a first quadrant cohomology spectral sequence

$$E^{pq}_2 = \text{Ext}^p_S(A, \text{Ext}^q_R(S, B)) \Rightarrow \text{Ext}^{p+q}_R(A, B)$$

for every S-module A and every R-module B.

Exercise 5.6.4 Use spectral sequences to prove the Acyclic Assembly Lemma 2.7.3.

5.7 Hyperhomology

Definition 5.7.1 Let $\mathcal{A}$ be an abelian category that has enough projectives. A (left) *Cartan-Eilenberg* resolution P_{**} of a chain complex A_* in $\mathcal{A}$ is an upper half-plane double complex ($P_{pq} = 0$ if $q < 0$), consisting of projective objects of $\mathcal{A}$, together with a chain map ("augmentation") $P_{*0} \xrightarrow{\epsilon} A_*$ such that for every p

1. If $A_p = 0$, the column P_{p*} is zero.
2. The maps on boundaries and homology

$$B_p(\epsilon): B_p(P, d^h) \to B_p(A)$$

$$H_p(\epsilon): H_p(P, d^h) \to H_p(A)$$

are projective resolutions in $\mathcal{A}$. Here $B_p(P, d^h)$ denotes the horizontal boundaries in the (p, q) spot, that is, the chain complex whose q^{th} term is $d^h(P_{p+1,q})$. The chain complexes $Z_p(P, d^h)$ and $H_p(P, d^h) = Z_p(P, d^h)/B_p(P, d^h)$ are defined similarly.

Exercise 5.7.1 In a Cartan-Eilenberg resolution show that the induced maps

$$\epsilon^p \colon P_{p*} \to A_p$$

$$Z^p(\epsilon) \colon Z_p(P, d^h) \to Z_p(A)$$

are projective resolutions in $\mathcal{A}$. Then show that the augmentation $\mathrm{Tot}^{\oplus}(P) \to A$ is a quasi-isomorphism in $\mathcal{A}$; when A isn't bounded below, you will need to assume axiom (AB4) holds.

Lemma 5.7.2 *Every chain complex A_* has a Cartan-Eilenberg resolution $P_{**} \to A$.*

Proof For each p select projective resolutions P^B_{p*} of $B_p(A)$ and P^H_{p*} of $H_p(A)$. By the Horseshoe Lemma 2.2.8 there is a projective resolution P^Z_{p*} of $Z_p(A)$ so that

$$0 \to P^B_{p*} \to P^Z_{p*} \to P^H_{p*} \to 0$$

is an exact sequence of chain complexes lying over

$$0 \to B_p(A) \to Z_p(A) \to H_p(A) \to 0.$$

Applying the Horseshoe Lemma again, we find a projective resolution P^A_{p*} of A_p fitting into an exact sequence

$$0 \to P^Z_{p*} \to P^A_{p*} \to P^B_{p-1,*} \to 0.$$

We now define P_{**} to be the double complex whose p^{th} column is P^A_{p*} except that (using the Sign Trick 1.2.5) the vertical differential is multiplied by $(-1)^p$; the horizontal differential of P_{**} is the composite

$$P^A_{p+1,*} \to P^B_{p*} \hookrightarrow P^Z_{p*} \hookrightarrow P^A_{p*}.$$

The construction guarantees that the maps $\epsilon_p \colon P_{p0} \to A_p$ assemble to give a chain map ϵ, and that each $B_p(\epsilon)$ and $H_p(\epsilon)$ give projective resolutions (check this!). $\diamond$

Exercise 5.7.2 If $f \colon A \to B$ is a chain map and $P \to A$, $Q \to B$ are Cartan-Eilenberg resolutions, show that there is a double complex map $\tilde{f} \colon P \to Q$ over f. *Hint:* Modify the proof of 2.4.6 that $L_* f$ is a homological δ-functor.

Definition 5.7.3 Let $f, g \colon D \to E$ be two maps of double complexes. A *chain homotopy* from f to g consists of maps $s^h_{pq} \colon D_{pq} \to E_{p+1,q}$ and $s^v_{pq} \colon D_{pq} \to E_{p,q+1}$ so that

$$g - f = (d^h s^h + s^h d^h) + (d^v s^v + s^v d^v)$$

$$s^v d^h + d^h s^v = s^h d^v + d^v s^h = 0.$$

This definition is set up so that $\{s^h + s^v : \mathrm{Tot}(D)_n \to \mathrm{Tot}(E)_{n+1}\}$ forms an ordinary chain homotopy between the maps $\mathrm{Tot}(f)$ and $\mathrm{Tot}(g)$ from $\mathrm{Tot}^{\oplus}(D)$ to $\mathrm{Tot}^{\oplus}(E)$.

Exercise 5.7.3

1. If $f, g : A \to B$ are homotopic maps of chain complexes, and $\tilde{f}, \tilde{g} : P \to Q$ are maps of Cartan-Eilenberg resolutions lying over them, show that $\tilde{f}$ is chain homotopic to $\tilde{g}$.
2. Show that any two Cartan-Eilenberg resolutions P, Q of A are chain homotopy equivalent. Conclude that for any additive functor F the chain complexes $\mathrm{Tot}^{\oplus}(F(P))$ and $\mathrm{Tot}^{\oplus}(F(Q))$ are chain homotopy equivalent.

Definition 5.7.4 ($\mathbb{L}_* F$) Let $F : \mathcal{A} \to \mathcal{B}$ be a right exact functor, and assume that $\mathcal{A}$ has enough projectives. If A is a chain complex in $\mathcal{A}$ and $P \to A$ is a Cartan-Eilenberg resolution, define $\mathbb{L}_i F(A)$ to be $H_i \mathrm{Tot}^{\oplus}(F(P))$. Exercise 5.7.3 shows that $\mathbb{L}_i F(A)$ is independent of the choice of P.

If $f : A \to B$ is a chain map and $\tilde{f} : P \to Q$ is a map of Cartan-Eilenberg resolutions over f, define $\mathbb{L}_i F(f)$ to be the map $H_i(\mathrm{Tot}(\tilde{f}))$ from $\mathbb{L}_i F(A)$ to $\mathbb{L}_i F(B)$. The exercise above implies that $\mathbb{L}_i F$ is a functor from $\mathbf{Ch}(\mathcal{A})$ to $\mathcal{B}$, at least when $\mathcal{B}$ is cocomplete. The $\mathbb{L}_i F$ are called the *left hyper-derived functors* of F.

Warning: If $\mathcal{B}$ is not cocomplete, $\mathrm{Tot}^{\oplus}(F(P))$ and $\mathbb{L}_i F(A)$ may not exist for all chain complexes A. In this case we restrict to the category $\mathbf{Ch}_+(\mathcal{A})$ of all chain complexes A which are bounded below in the sense that there is a p_0 such that $A_p = 0$ for $p < p_0$. Since $P_{pq} = 0$ if $p < p_0$ or $q < 0$, $\mathrm{Tot}^{\oplus}(F(P))$ exists in $\mathbf{Ch}(\mathcal{B})$ and we may consider $\mathbb{L}_i F$ to be a functor from $\mathbf{Ch}_+(\mathcal{A})$ to $\mathcal{B}$.

$\diamond$

Exercises 5.7.4

1. If A is an object of $\mathcal{A}$, considered as a chain complex concentrated in degree zero, show that $\mathbb{L}_i F(A)$ is the ordinary derived functor $L_i F(A)$.
2. Let $\mathbf{Ch}_{\geq 0}(\mathcal{A})$ be the subcategory of complexes A with $A_p = 0$ for $p < 0$. Show that the functors $\mathbb{L}_i F$ restricted to $\mathbf{Ch}_{\geq 0}(\mathcal{A})$ are the left derived functors of the right exact functor $H_0 F$.
3. (Dimension shifting) Show that $\mathbb{L}_i F(A[n]) = \mathbb{L}_{n+i} F(A)$ for all n. Here $A[n]$ is the translate of A with $A[n]_i = A_{n+i}$.

Lemma 5.7.5 *If* $0 \to A \to B \to C \to 0$ *is a short exact sequence of bounded below complexes, there is a long exact sequence*

$$\cdots \mathbb{L}_{i+1} F(C) \xrightarrow{\delta} \mathbb{L}_i F(A) \to \mathbb{L}_i F(B) \to \mathbb{L}_i F(C) \xrightarrow{\delta} \cdots.$$

Proof By dimension shifting, we may assume that A, B, and C belong to $\mathbf{Ch}_{>0}(\mathcal{A})$. The sequence in question is just the long exact sequence for the derived functors of the right exact functor $H_0 F$.   ◇

Proposition 5.7.6 *There is always a convergent spectral sequence*

$$^{II}E^2_{pq} = (L_p F)(H_q(A)) \Rightarrow \mathbb{L}_{p+q} F(A).$$

If A is bounded below, there is a convergent spectral sequence

$$^{I}E^2_{pq} = H_p(L_q F(A)) \Rightarrow \mathbb{L}_{p+q} F(A).$$

Proof We have merely written out the two spectral sequences arising from the upper half-plane double chain complex $F(P)$.   ◇

Corollary 5.7.7

  1. *If A is exact, $\mathbb{L}_i F(A) = 0$ for all i.*
  2. *Any quasi-isomorphism $f: A \to B$ induces isomorphisms*

$$\mathbb{L}_* F(A) \cong \mathbb{L}_* F(B).$$

  3. *If each A_p is F-acyclic (2.4.3), that is, $L_q F(A_p) = 0$ for $q \neq 0$, and A is bounded below, then*

$$\mathbb{L}_p F(A) = H_p(F(A)) \text{ for all } p.$$

Application 5.7.8 (Hypertor) Let R be a ring and B a left R-module. The *hypertor* groups $\mathbf{Tor}^R_i(A_*, B)$ of a chain complex A_* of right R-modules are defined to be the hyper-derived functors $\mathbb{L}_i F(A_*)$ for $F = \otimes_R B$. This extends the usual Tor to chain complexes, and if A is a bounded below complex of flat modules, then $\mathbf{Tor}^R_i(A_*, B) = H_i(A_* \otimes B)$ for all i. The hypertor spectral sequences coming from 5.7.6 are

$$^{II}E^2_{pq} = \mathrm{Tor}_p(H_q(A), B) \Rightarrow \mathbf{Tor}^R_{p+q}(A_*, B)$$

and (when A is bounded below)

$$^IE^1_{pq} = \text{Tor}_q(A_p, B), \quad ^IE^2_{pq} = H_p \text{Tor}_q(A_*, B) \Rightarrow \text{Tor}^R_{p+q}(A_*, B).$$

Even more generally, if B_* is also a chain complex, we can define the hypertor of the bifunctor $A \otimes_R B$ to be

$$\text{Tor}^R_i(A_*, B_*) = H_i \text{Tot}^{\oplus}(P \otimes_R Q),$$

where $P \to A$ and $Q \to B$ are Cartan-Eilenberg resolutions. Since $\text{Tot}(P \otimes Q)$ is unique up to chain homotopy equivalence, the hypertor is independent of the choice of P and Q. If B is a module, considered as a chain complex, this agrees with the above definition (exercise!); by symmetry the same is true for A. By definition, hypertor is a balanced functor in the sense of 2.7.7. A lengthy discussion of hypertor may be found in [EGA, III.6].

Exercise 5.7.5 Show that there is a convergent spectral sequence

$$^{II}E^2_{pq} = \bigoplus_{q'+q''=q} \text{Tor}^R_p \left(H_{q'}(A_*), H_{q''}(B_*) \right) \Rightarrow \text{Tor}^R_{p+q}(A_*, B_*).$$

If A_* and B_* are bounded below, show that there is a spectral sequence

$$^IE^2_{pq} = H_p \text{Tot}^{\oplus} \text{Tor}_q(A_*, B_*) \Rightarrow \text{Tor}^R_{p+q}(A_*, B_*).$$

Exercise 5.7.6 Let A be the mapping cone complex $0 \to A_1 \xrightarrow{f} A_0 \to 0$ with only two nonzero rows. Show that there is a long exact sequence:

$$\cdots \mathbb{L}_{i+1}F(A) \to L_i F(A_1) \xrightarrow{f} L_i F(A_0) \to \mathbb{L}_i F(A) \to L_{i-1} F(A_1) \cdots.$$

Cohomology Variant 5.7.9 Let $\mathcal{A}$ be an abelian category that has enough injectives. A (right) *Cartan-Eilenberg resolution* of a cochain complex A^* in $\mathcal{A}$ is an upper half-plane complex I^{**} of injective objects of $\mathcal{A}$, together with an augmentation $A^* \to I^{*0}$ such that the maps on coboundaries and cohomology are injective resolutions of $B^p(A)$ and $H^p(A)$. Every cochain complex has a Cartan-Eilenberg resolution $A \to I$. If $F: \mathcal{A} \to \mathcal{B}$ is a left exact functor, we define $\mathbb{R}^i F(A)$ to be $H^i \text{Tot}^{\Pi}(F(I))$, at least when $\text{Tot}^{\Pi}(F(I))$ exists in $\mathcal{B}$. By appealing to the functor $F^{op}: \mathcal{A}^{op} \to \mathcal{B}^{op}$, we see that $\mathbb{R}^i F$ is a functor from $\mathbf{Ch}^+(\mathcal{A})$ (the complexes A^* with $A^p = 0$ for $p << 0$) to $\mathcal{B}$, and even from

Ch$(\mathcal{A})$ to $\mathcal{B}$ when $\mathcal{B}$ is complete. The $\mathbb{R}^i F$ are called the *right hyper-derived functors* of F.

If A is in **Ch**$(\mathcal{A})$, the two spectral sequences arising from the upper half-plane double cochain complex $F(I)$ become

$$^{II}E_2^{pq} = (R^p F)(H^q(A)) \Rightarrow \mathbb{R}^{p+q} F(A), \text{ weakly convergent; and}$$

$$^{I}E_2^{pq} = H^p(R^q F(A)) \Rightarrow \mathbb{R}^{p+q} F(A), \text{ if } A \text{ is bounded below.}$$

Hence $\mathbb{R}^* F$ vanishes on exact complexes and sends quasi-isomorphisms of (bounded below) complexes to isomorphisms.

Application 5.7.10 (Hypercohomology) Let X be a topological space and $\mathcal{F}^*$ a cochain complex of sheaves on X. The *hypercohomology* $\mathbb{H}^i(X, \mathcal{F}^*)$ is $\mathbb{R}^i \Gamma(\mathcal{F}^*)$, where Γ is the global sections functor 2.5.4. This generalizes sheaf cohomology to complexes of sheaves, and if $\mathcal{F}^*$ is a bounded below complex of injective sheaves, then $\mathbb{H}^i(X, \mathcal{F}^*) = H^i(\Gamma(\mathcal{F}^*))$. The hypercohomology spectral sequence is $^{II}E_2^{pq} = H^p(X, H^q(\mathcal{F}^*)) \Rightarrow \mathbb{H}^{p+q}(X, \mathcal{F}^*)$.

5.8 Grothendieck Spectral Sequences

In his classic paper [Tohoku], Grothendieck introduced a spectral sequence associated to the composition of two functors. Today it is one of the organizational principles of Homological Algebra.

Cohomological Setup 5.8.1 Let $\mathcal{A}$, $\mathcal{B}$, and $\mathcal{C}$ be abelian categories such that both $\mathcal{A}$ and $\mathcal{B}$ have enough injectives. We are given left exact functors $G: \mathcal{A} \to \mathcal{B}$ and $F: \mathcal{B} \to \mathcal{C}$.

$$\begin{array}{ccc}
\mathcal{A} & \overset{G}{\longrightarrow} & \mathcal{B} \\
& {}_{FG}\searrow \quad \swarrow{}^{F} & \\
& \mathcal{C} &
\end{array}$$

Definition 5.8.2 Let $F: \mathcal{B} \to \mathcal{C}$ be a left exact functor. An object B of $\mathcal{B}$ is called *F-acyclic* if the derived functors of F vanish on B, that is, if $R^i F(B) = 0$ for $i \neq 0$. (Compare with 2.4.3.)

Grothendieck Spectral Sequence Theorem 5.8.3 *Given the above cohomological setup, suppose that G sends injective objects of $\mathcal{A}$ to F-acyclic objects* of $\mathcal{B}$. Then there exists a convergent first quadrant cohomological spectral sequence for each A in $\mathcal{A}$:

$$^IE_2^{pq} = (R^pF)(R^qG)(A) \Rightarrow R^{p+q}(FG)(A).$$

The edge maps in this spectral sequence are the natural maps

$$(R^pF)(GA) \to R^p(FG)(A) \quad \text{and} \quad R^q(FG)(A) \to F(R^qG(A)).$$

The exact sequence of low degree terms is

$$0 \to (R^1F)(GA) \to R^1(FG)A \to F(R^1G(A)) \to (R^2F)(GA) \to R^2(FG)A.$$

Proof Choose an injective resolution $A \to I$ of A in $\mathcal{A}$, and apply G to get a cochain complex $G(I)$ in $\mathcal{B}$. Using a first quadrant Cartan-Eilenberg resolution of $G(I)$, form the hyper-derived functors $\mathbb{R}^nF(G(I))$ as in 5.7.9. There are two spectral sequences converging to these hyper-derived functors. The first spectral sequence is

$$^IE_2^{pq} = H^p((R^qF)(GI)) \Rightarrow (\mathbb{R}^{p+q}F)(GI).$$

By hypothesis, each $G(I^p)$ is F-acyclic, so $(R^qF)(G(I^p)) = 0$ for $q \neq 0$. Therefore this spectral sequence collapses to yield

$$(\mathbb{R}^pF)(GI) \cong H^p(FG(I)) = R^p(FG)(A).$$

The second spectral sequences is therefore

$$^{II}E_2^{pq} = (R^pF)H^q(G(I)) \Rightarrow R^p(FG)(A).$$

Since $H^q(G(I)) = R^pG(A)$, it is Grothendieck's spectral sequence.   ◇

Corollary 5.8.4 (Homology spectral sequence) *Let $\mathcal{A}$, $\mathcal{B}$, and $\mathcal{C}$ be abelian categories such that both $\mathcal{A}$ and $\mathcal{B}$ have enough projectives. Suppose given right exact functors $G: \mathcal{A} \to \mathcal{B}$ and $F: \mathcal{B} \to \mathcal{C}$ such that G sends projective objects of $\mathcal{A}$ to F-acyclic objects of $\mathcal{B}$. Then there is a convergent first quadrant homology spectral sequence for each A in $\mathcal{A}$:*

$$E_{pq}^2 = (L_pF)(L_qG)(A) \Rightarrow L_{p+q}(FG)(A).$$

The exact sequence of low degree terms is

$$L_2(FG)A \to (L_2F)(GA) \to F(L_1G(A)) \to L_1(FG)A \to (L_1F)(GA) \to 0.$$

Proof Dualizing allows us to consider $G^{op}: \mathcal{A}^{op} \to \mathcal{B}^{op}$ and $F^{op}: \mathcal{B}^{op} \to \mathcal{C}^{op}$, and the corollary is just translation of Grothendieck's spectral sequence using the dictionary $L_p F = R^p F^{op}$, and so on.   ◇

Applications 5.8.5 The base-change spectral sequences for Tor and Ext of section 5.7 are actually special instances of the Grothendieck spectral sequence: Given a ring map $R \to S$ and an S-module B, one considers the composites

$$R\text{--mod} \xrightarrow{\otimes_S R} S\text{--mod} \xrightarrow{\otimes_S B} \mathbf{Ab}$$

and

$$R\text{--mod} \xrightarrow{\operatorname{Hom}_R(S,-)} S\text{--mod} \xrightarrow{\operatorname{Hom}_S(B,-)} \mathbf{Ab}.$$

Leray Spectral Sequence 5.8.6 Let $f: X \to Y$ be a continuous map of topological spaces. The direct image sheaf functor f_* (2.6.6) has the exact functor f^{-1} as its left adjoint (exercise 2.6.2), so f_* is left exact and preserves injectives by 2.3.10. If $\mathcal{F}$ is a sheaf of abelian groups on X, the global sections of $f_*\mathcal{F}$ is the group $(f_*\mathcal{F})(Y) = \mathcal{F}(f^{-1}Y) = \mathcal{F}(X)$. Thus we are in the situation

$$\text{Sheaves}(X) \xrightarrow{f_*} \text{Sheaves}(Y)$$
$$\Gamma \searrow \qquad \swarrow \Gamma$$
$$\mathbf{Ab}$$

The Grothendieck spectral sequence in this case is called the *Leray spectral sequence:* Since $R^p \Gamma$ is sheaf cohomology (2.5.4), it is usually written as

$$E_2^{pq} = H^p(Y; R^q f_* \mathcal{F}) \Rightarrow H^{p+q}(X; \mathcal{F}).$$

This spectral sequence is a central tool to much of modern algebraic geometry.

We will see other applications of the Grothendieck spectral sequence in 6.8.2 and 7.5.2. Here is one we needed in section 5.5.9.

Recall from Chapter 3, section 5 that a tower $\cdots A_1 \to A_0$ of abelian groups is a functor $I \to \mathbf{Ab}$, where I is the poset of whole numbers in reverse order. A *double tower* is a functor $A: I \times I \to \mathbf{Ab}$; it may be helpful to think of the groups A_{ij} as forming a lattice in the first quadrant of the plane.

Proposition 5.8.7 ($\lim^1$ of a double tower) *For each double tower* $A: I \times I \to \mathbf{Ab}$ *we have* $\varprojlim_{I \times I} A_{ij} = \varprojlim_i \varprojlim_j A_{ij}$, *a short exact sequence*

$$0 \to \varprojlim_i^1 (\varprojlim_j A_{ij}) \to \left(R^1 \varprojlim_{I \times I} \right) A_{ij} \to \varprojlim_i (\varprojlim_j^1 A_{ij}) \to 0,$$

$$\left(R^2 \varprojlim_{I \times I} \right) A_{ij} = \varprojlim_i^1 (\varprojlim_j^1 A_{ij}), \quad \text{and} \quad \left(R^n \varprojlim_{I \times I} \right) A_{ij} = 0 \quad \text{for } n \geq 3.$$

Proof We may form the inverse limit as $\lim A_{ij} = \varprojlim_i \varprojlim_j A_{ij}$, that is, as the composition of $\varprojlim_j : (\mathbf{Ab}^I)^I \to \mathbf{Ab}^I$ and $\varprojlim_i : \mathbf{Ab}^I \to \mathbf{Ab}$. From 2.3.10 and 2.6.9 we see that $\varprojlim$ preserves injectives; it is right adjoint to the "constant tower" functor. Therefore we have a Grothendieck spectral sequence

$$E_2^{pq} = \varprojlim_i^p \varprojlim_j^q A_{ij} \Rightarrow (R^{p+q} \lim) A_{ij}.$$

Since both $\mathbf{Ab}$ and $\mathbf{Ab}^I$ satisfy $(AB4^*)$, $\lim^p = \lim^q = 0$ for $p, q \neq 0, 1$. Thus the spectral sequence degenerates as described. $\diamond$

5.9 Exact Couples

An alternative construction of spectral sequences can be given via "exact couples" and is due to Massey [Massey]. It is often encountered in algebraic topology but rarely in commutative algebra.

It is convenient to forget all subscripts for a while and to work in the category of modules over some ring (or more generally in any abelian category satisfying axiom $AB5$). An *exact couple* $\mathcal{E}$ is a pair (D, E) of modules, together with three morphisms i, j, k

$$\mathcal{E}: \quad \begin{array}{ccc} D & \xrightarrow{\ i\ } & D \\ & {}_k \nwarrow \quad \swarrow {}_j & \\ & E & \end{array}$$

which form an *exact triangle* in the sense that kernel = image at each vertex.

Definition 5.9.1 (Derived couple) The composition jk from E to itself satisfies $(jk)(jk) = j(kj)k = 0$, so we may form the homology module $H(E) =$

$\ker(jk)/\mathrm{image}(jk)$. Construct the triangle

$$
\begin{array}{ccc}
i(D) & \xrightarrow{\ i'\ } & i(D) \\
 & \mathcal{E}':\ k'\nwarrow\quad\swarrow j' & \\
 & H(E) &
\end{array}
$$

where i' is the restriction of i to $i(D)$, while j' and k' are given by

$$j'(i(d)) = [j(d)], \quad k'([e]) = k(e).$$

The map j' is well defined since $i(d) = 0$ implies that for some $e \in E$ $d = k(e)$ and $j(d) = jk(e)$ is a boundary. Similarly, $k(jk(e)) = 0$ implies that the map k' is well defined. We call $\mathcal{E}'$ the *derived couple* of $\mathcal{E}$. A diagram chase (left to the reader) shows that $\mathcal{E}'$ is also an exact couple.

If we iterate the process of taking exact couples r times, the result is called the rth *derived couple* $\mathcal{E}^r$ of $\mathcal{E}$.

$$
\begin{array}{ccc}
D^r & \xrightarrow{\ i\ } & D^r \\
 & \mathcal{E}^r:\ k\nwarrow\quad\swarrow j^{(r)} & \\
 & E^r &
\end{array}
$$

Here $D^r = i^r(D)$ is a submodule of D, and $E^r = H(E^{r-1})$ is a subquotient of E. The maps i and k are induced from the i and k of $\mathcal{E}$, while $j^{(r)}$ sends $[i^r(d)]$ to $[j(d)]$.

Exercise 5.9.1 Show that $H(E) = k^{-1}(iD)/j(\ker(i))$ and more generally, that $E^r = Z^r/B^r$, with $Z^r = k^{-1}(i^r D)$ and $B^r = j(\ker(i^r))$.

With this generic background established, we now introduce subscripts (for D_{pq} and E_{pq}) in such a way that i has bidegree $(1, -1)$, k has bidegree $(-1, 0)$, and

$$\mathrm{bidegree}(j) = (-a, a).$$

Thus i and j preserve total degree $(p + q)$, while k drops the total degree by 1. Setting $D'_{pq} = i(D_{p-1,q+1}) \subseteq D_{pq}$ and letting E'_{pq} be the corresponding subquotient of E_{pq}, it is easy to see that in $\mathcal{E}'$ the maps i and k still have bidegrees $(1, -1)$ and $(-1, 0)$, while j' now has bidegree $(-1 - a, 1 + a)$. It is convenient to reindex so that $\mathcal{E} = \mathcal{E}^a$ and $\mathcal{E}^r$ denotes the $(r - a)^{th}$ derived couple of $\mathcal{E}$, so that $j^{(r)}$ has bidegree $(-r, r)$ and the E^r-differential has bidegree

$(-r, r - 1)$.

$$E^r_{pq} \xrightarrow{\ k\ } D^r_{p-1,q} \xrightarrow{\ i\ } D^r_{p,q-1} \xrightarrow{\ j^{(r)}\ } E^r_{p-r,q+r-1}.$$

In summary, we have established the following result.

Proposition 5.9.2 *An exact couple $\mathcal{E}$ in which i, k, and j have bidegrees $(1, -1)$, $(-1, 0)$, and $(-a, a)$ determines a homology spectral sequence $\{E^r_{pq}\}$ starting with E^a. A morphism of exact couples induces a morphism of the corresponding spectral sequences.*

Example 5.9.3 (Exact couple of a filtration) Let C_* be a filtered chain complex of modules, and consider the bigraded homology modules

$$D^1_{pq} = H_n(F_pC), \quad E^1_{pq} = H_n(F_pC/F_{p-1}C), \quad n = p + q.$$

Then the short exact sequences $0 \to F_{p-1} \to F_p \to F_p/F_{p-1} \to 0$ may be rolled up into an exact triangle of complexes (see Chapter 10 or 1.3.6)

$$
\begin{array}{ccc}
\oplus F_pC & \xrightarrow{\quad i \quad} & \oplus F_pC \\
 & {}^{0}\nwarrow \qquad \swarrow {}^{\oplus \eta_p} & \\
 & \oplus F_pC/F_{p-1}C &
\end{array}
$$

whose homology forms an exact couple

$$
\mathcal{E}^1: \quad
\begin{array}{ccc}
\oplus H_{p+q}(F_pC) & \xrightarrow{\quad i \quad} & \oplus H_{p+q}(F_pC) \\
 & {}_{k}\nwarrow \qquad \swarrow {}_{j} & \\
 & \oplus H_{p+q}(F_pC/F_{p-1}C) &
\end{array}
$$

Theorem 5.9.4 *Let C_* be a filtered chain complex. The spectral sequence arising from the exact couple $\mathcal{E}^1$ (which starts at E^1) is naturally isomorphic to the spectral sequence constructed in section 5.4 (which starts at E^0).*

Proof In both spectral sequences, the groups E^r_{pq} are subquotients of $E^0_p = F_pC_{p+q}/F_{p-1}C_{p+q}$; we shall show they are the same subquotients. Since the differentials in both are induced from $d: C \to C$, this will establish the result.

In the exact couple spectral sequence, we see from exercise 5.9.1 that the numerator of E^r in E^1 is $k^{-1}(i^{r-1}D^1)$ and the denominator is $j(\ker i^{r-1})$.

If $c \in F_pC_n$ represents $[c] \in H_n(F_pC/F_{p-1}C)$, then $d(c) \in F_{p-1}C$ and $k([c])$ is the class of $d(c)$. Therefore the numerator in F_p/F_{p-1} for E^r is $Z_p^r = \{c \in F_pC : d(c) = a + d(b) \text{ for some } a \in F_{p-r}C, b \in F_pC\}/F_{p-1}C$. Similarly, the kernel of $i^{r-1}: H_n(F_pC) \to H_n(F_{p+r-1}C)$ is represented by those cycles $c \in F_pC$ with $c = d(b)$ for some $b \in F_{p+r-1}C$. That is, $\ker(i^{r-1})$ is the image of A_{p+r-1}^{r-1} in $H_n(F_pC)$. Since j is induced on homology by η_p, we see that the denominator is $B_p^r = \eta_p d(A_{p+r-1}^{r-1})$. Since the spectral sequence of section 5.4 had $E_p^r = Z_p^r/B_p^r$, we have finished the proof. $\diamond$

Convergence 5.9.5 Let $\mathcal{E}$ be an exact couple in which i, j, and k have bidegrees $(-1, 1)$, $(-a, a)$ and, $(-1, 0)$, respectively. The associated spectral sequence is related to the direct limits $H_n = \varinjlim D_{p,n-p}$ of the D_{pq} along the maps $i: D_{pq} \to D_{p+1,q-1}$. Let F_pH_n denote the image of $D_{p+a,q-a}$ in H_n $(p + q = n)$; the system $\ldots F_{p-1}H_n \subseteq F_pH_n \subseteq \ldots$ forms an exhaustive filtration of H_n.

Proposition 5.9.6 *There is a natural inclusion of $F_pH_n/F_{p-1}H_n$ in $E_{p,n-p}^\infty$. The spectral sequence E_{pq}^r weakly converges to H_* if and only if:*

$$Z^\infty = \cap_r k^{-1}(i^r D) \text{ equals } k^{-1}(0) = j(D).$$

Proof Fix p, q, and $n = p + q$. The kernel $K_{p+a,q-a}$ of $D_{p+a,q-a} \to H_n$ is the union of the $\ker(i^r)$, so $j(K_{p+a,q-a}) = \cup j(\ker(i^r)) = \cup B_{pq}^r = B_{pq}^\infty$. (This is where axiom $AB5$ is used.) Applying the Snake Lemma to the diagram

$$
\begin{array}{ccccccccc}
0 & \longrightarrow & K_{p-1+a} & \longrightarrow & D_{p+a-1} & \longrightarrow & F_{p-1}H_n & \longrightarrow & 0 \\
& & \downarrow & & \downarrow{\scriptstyle i} & & \downarrow & & \\
0 & \longrightarrow & K_{p+a} & \longrightarrow & D_{p+a} & \longrightarrow & F_pH_n & \longrightarrow & 0
\end{array}
$$

yields the exact sequence

$$0 \to B_{pq}^\infty \to j(D_{p+a,q-a}) \to F_pH_n/F_{p-1}H_n \to 0.$$

But $j(D_{p+a,q-a}) = k^{-1}(0)$, so it is contained in $Z_{pq}^r = k^{-1}(i^r D_{p-r-1,q+r})$ for all r. The result now follows. $\diamond$

We say that an exact couple is *bounded below* if for each n there is an integer $f(n)$ such that $D_{p,q} = 0$ whenever $p < f(p+q)$. In this case, for each p and q there is an r such that $i^r(D_{p-r-1,q+r}) = i^r(0) = 0$, i.e., $Z_{pq}^r = k^{-1}(0)$. As an immediate corollary, we obtain the following convergence result.

Classical Convergence Theorem 5.9.7 *If an exact couple is bounded below, then the spectral sequence is bounded below and converges to $H_* = \lim_{\longrightarrow} D$.*

$$E_{pq}^a \Rightarrow H_{p+q}$$

The spectral sequence is bounded and converges to H_* if for each n there is a p such that $D_{p,n-p} \xrightarrow{\cong} H_n$.

Exercise 5.9.2 (Complete convergence) Let $\mathcal{E}$ be an exact couple that is bounded above ($D_{p,q} = 0$ whenever $p > f(p+q)$). Suppose that the spectral sequence is regular (5.2.10). Show that the spectral sequence converges to $\widehat{D}_n = \lim_{\longleftarrow} D_{p,n-p}$.

Application 5.9.8 Here is an exact couple that does not arise from a filtered chain complex. Let C_* be an exact sequence of left R-modules and M a right R-module. Let $Z_p \subset C_p$ be the kernel of $d: C_p \to C_p$; associated to the short exact sequences $0 \to Z_p \to C_p \to Z_{p-1} \to 0$ are the long exact sequences

$$\cdots \operatorname{Tor}_q(M, Z_p) \xrightarrow{j} \operatorname{Tor}_q(M, C_p) \xrightarrow{k} \operatorname{Tor}_q(M, Z_{p-1}) \xrightarrow{i} \operatorname{Tor}_{q-1}(M, Z_p) \cdots$$

which we can assemble into an exact couple $\mathcal{E} = \mathcal{E}^0$ with

$$D_{pq}^0 = \operatorname{Tor}_q(M, Z_p) \quad \text{and} \quad E_{pq}^0 = \operatorname{Tor}_q(M, C_p).$$

By inspection, the map $d = jk: \operatorname{Tor}_q(M, C_p) \to \operatorname{Tor}_q(M, C_{p-1})$ is induced via $\operatorname{Tor}_q(M, -)$ by the differential $d: C_p \to C_{p-1}$, so we may write

$$E_{pq}^1 = H_p(\operatorname{Tor}_q(M, C_*)).$$

More generally, if we replace $\operatorname{Tor}_*(M, -)$ by the derived functors L_*F of any right exact functor, the exact couple yields a spectral sequence with $E_{pq}^0 = L_qF(C_p)$ and $E_{pq}^1 = H_p(L_qF(C))$. These are essentially the hyperhomology sequences of section 5.7 related to the hyperhomology modules $\mathbb{L}_*F(C)$, which are zero. Therefore this spectral sequence converges to zero whenever C_* is bounded below.

Bockstein Spectral Sequence 5.9.9 Fix a prime ℓ and let H_* be a (graded) abelian group. Suppose that multiplication by ℓ fits into a long exact sequence

$$\cdots E_{n+1} \xrightarrow{\partial} H_n \xrightarrow{\ell} H_n \xrightarrow{j} E_n \xrightarrow{\partial} H_{n-1} \xrightarrow{\ell} \cdots.$$

If we roll this up into the exact couple

$$
\begin{array}{ccc}
H_* & \xrightarrow{\ell} & H_* \\
{\scriptstyle \partial}\nwarrow & & \swarrow{\scriptstyle j} \\
& E_* &
\end{array}
$$

then we obtain a spectral sequence with $E_*^0 = E_*$, called the *Bockstein spectral sequence* associated to H_*. This spectral sequence was first studied by W. Browder in [Br], who noted the following applications:

1. $H_* = H_*(X; \mathbb{Z})$ and $E_* = H_*(X; \mathbb{Z}/\ell)$ for a topological space X
2. $H_* = \pi_*(X)$ and $E_* = \pi_*(X; \mathbb{Z}/\ell)$ for a topological space X
3. $H_* = H_*(G; \mathbb{Z})$ and $E_* = H_*(G; \mathbb{Z}/\ell)$ for a group G
4. $H_* = H_*(C)$ for a torsionfree chain complex C, and $E_* = H_*(C/\ell C)$

We note that the differential $d = j\partial$ sends E_n^r to E_{n-1}^r, so that the bigrading subscripts we formally require for a spectral sequence are completely artificial. The next result completely describes the convergence of the Bockstein spectral sequence. To state it, it is convenient to adapt the notation that for $q \in \mathbb{Z}$

$$
{}_q H_* = \{x \in H_* : qx = 0\}.
$$

Proposition 5.9.10 *For every $r \geq 0$, there is an exact sequence*

$$
0 \to \frac{H_n}{\ell H_n + \ell^r H_n} \xrightarrow{j} E_n^r \xrightarrow{\partial} (\ell^r H_{n-1}) \cap (\ell H_{n-1}) \to 0.
$$

In particular, if T_n denotes the ℓ-primary torsion subgroup of H_n and Q_n denotes the infinitely ℓ-divisible part of ${}_\ell H_n$, then there is an exact sequence

$$
0 \to \frac{H_n}{\ell H_n + T_n} \xrightarrow{j} E_n^\infty \xrightarrow{\partial} Q_{n-1} \to 0.
$$

Proof For $r = 0$ we are given an extension

$$
0 \to H_n/\ell H_n \xrightarrow{j} E_n^0 \xrightarrow{\partial} {}_\ell H_{n-1} \to 0.
$$

Now E^r is the subquotient of E^0 with numerator $\partial^{-1}(\ell^r H)$ and denominator $j(\ell^r H)$ by the above exercise, so from the extension

$$
0 \to H/\ell H \xrightarrow{j} \partial^{-1}(\ell^r H) \xrightarrow{\partial} (\ell^r H \cap {}_\ell H) \to 0
$$

the result is immediate. $\diamond$

Corollary 5.9.11 *If each H_n is finitely generated and* $\dim(H_n \otimes \mathbb{Q}) = d_n$, *then the Bockstein spectral sequence converges to* $E_n^\infty = (\mathbb{Z}/p)^{d_n}$ *and is bounded in the sense that* $E_n^\infty = E_n^r$ *for large r.*

Actually, it turns out that the Bockstein spectral sequence can be used to completely describe H_* when each H_n is finitely generated. For example, if X is a simply connected H-space whose homology is finitely generated (such as a Lie group), Browder used the Bockstein spectral sequence in [Br] to prove that $\pi_2(X) = 0$.

For this, note that j induces a map $H_n \to E_n^r$ for each r. If $X \in E_n^r$ has $\alpha(x) = p^r y$, then $d(x) = j^{(r)}\alpha(x) = \overline{j(y)}$ in the notation of the proposition. In particular, x is a cycle if and only if $\alpha(x)$ is divisible by p^{r+1}. We can summarize these observations as follows.

Corollary 5.9.12 *In the Bockstein spectral sequence*

1. *Elements of E_n that survive to E^r but not to E^{r+1} (because they are not cycles) correspond to elements of exponent p in H_{n-1}, which are divisible by p^r but not by p^{r+1}.*
2. *An element $y \in H_n$ yields an element $j(y)$ of E^r for all r; if $j(y) \neq 0$ in E^{r-1}, but $j(y) = 0$ in E^r, then y generates a direct summand of H_n isomorphic to $\mathbb{Z}/p^r$.*

Exercise 5.9.3 Study the exact couple for $H = \mathbb{Z}/p^3$, and show directly that $E^2 \neq 0$ but $E^3 = 0$.

6

Group Homology and Cohomology

6.1 Definitions and First Properties

Let G be a group. A (left) *G-module* is an abelian group A on which G acts
by additive maps on the left; if $g \in G$ and $a \in A$, we write ga for the action of
g on a. Letting $\operatorname{Hom}_G(A, B)$ denote the G-set maps from A to B, we obtain a
category *G*–**mod** of left G-modules. The category *G*–**mod** may be identified
with the category $\mathbb{Z}G$–**mod** of left modules over the integral group ring $\mathbb{Z}G$.
It may also be identified with the functor category $\mathbf{Ab}^G$ of functors from the
category "G" (one object, G being its endomorphisms) to the category **Ab** of
abelian groups.

A *trivial G-module* is an abelian group A on which G acts "trivially," that is,
$ga = a$ for all $g \in G$ and $a \in A$. Considering an abelian group as a trivial G-
module provides an exact functor from **Ab** to *G*–**mod**. Consider the following
two functors from *G*–**mod** to **Ab**:

1. The *invariant subgroup* A^G of a G-module A,

$$A^G = \{a \in A : ga = a \text{ for all } g \in G \text{ and } a \in A\}.$$

2. The *coinvariants* A_G of a G-module A,

$$A_G = A/\text{submodule generated by } \{(ga - a) : g \in G, a \in A\}.$$

Exercise 6.1.1

1. Show that A^G is the maximal trivial submodule of A, and conclude that
   the invariant subgroup functor $-^G$ is right adjoint to the trivial module
   functor. Conclude that $-^G$ is a left exact functor.

2. Show that A_G is the largest quotient module of A that is trivial, and conclude that the coinvariants functor $-_G$ is left adjoint to the trivial module functor. Conclude that $-_G$ is a right exact functor.

Lemma 6.1.1 *Let A be any G-module, and let $\mathbb{Z}$ be the trivial G-module. Then $A_G \cong \mathbb{Z} \otimes_{\mathbb{Z}G} A$ and $A^G \cong \mathrm{Hom}_G(\mathbb{Z}, A)$.*

Proof Considering $\mathbb{Z}$ as a $\mathbb{Z}$–$\mathbb{Z}G$ bimodule, the "trivial module functor" from $\mathbb{Z}$–**mod** to $\mathbb{Z}G$–**mod** is the functor $\mathrm{Hom}_{\mathbb{Z}}(\mathbb{Z}, -)$. We saw in 2.6.3 that $\mathbb{Z} \otimes_{\mathbb{Z}G} -$ is its left adjoint; this functor must agree with its other left adjoint $(-)_G$. For the second equation, we use adjointness: $A^G \cong \mathrm{Hom}_{\mathbf{Ab}}(\mathbb{Z}, A^G) \cong \mathrm{Hom}_G(\mathbb{Z}, A)$. $\diamond$

Definition 6.1.2 Let A be a G-module. We write $H_*(G; A)$ for the left derived functors $L_*(-_G)(A)$ and call them the *homology groups of G with coefficients in A*; by the lemma above, $H_*(G; A) \cong \mathrm{Tor}_*^{\mathbb{Z}G}(\mathbb{Z}, A)$. By definition, $H_0(G; A) = A_G$. Similarly, we write $H^*(G; A)$ for the right derived functors $R^*(-^G)(A)$ and call them the *cohomology groups of G with coefficients in A*; by the lemma above, $H^*(G; A) \cong \mathrm{Ext}_{\mathbb{Z}G}^*(\mathbb{Z}, A)$. By definition, $H^0(G; A) = A^G$.

Example 6.1.3 If $G = 1$ is the trivial group, $A_G = A^G = A$. Since the higher derived functors of an exact functor vanish, $H_*(1; A) = H^*(1; A) = 0$ for $* \neq 0$.

Example 6.1.4 Let G be the infinite cyclic group T with generator t. We may identify $\mathbb{Z}T$ with the Laurent polynomial ring $\mathbb{Z}[t, t^{-1}]$. Since the sequence

$$0 \to \mathbb{Z}T \xrightarrow{t-1} \mathbb{Z}T \to \mathbb{Z} \to 0$$

is exact,

$$H_n(T; A) = H^n(T; A) = 0 \text{ for } n \neq 0, 1, \text{ and}$$

$$H_1(T; A) \cong H^0(T; A) = A^T, \, H^1(T; A) \cong H_0(T; A) = A_T.$$

In particular, $H_1(T; \mathbb{Z}) = H^1(T; \mathbb{Z}) = \mathbb{Z}$. We will see in the next section that all free groups display similar behavior, because $pd_G(\mathbb{Z}) = 1$.

Exercise 6.1.2 (kG-modules) As a variation, we can replace $\mathbb{Z}$ by any commutative ring k and consider the category kG–**mod** of k-modules on which G acts k-linearly. The functors A_G and A^G from kG–**mod** to k–**mod** are left (resp. right) exact and may be used to form the derived functors Tor_*^{kG} and Ext_{kG}^*. Prove that if A is a kG-module, then we have isomorphisms of abelian groups

$$H_*(G; A) \cong \text{Tor}_*^{kG}(k, A) \quad \text{and} \quad H^*(G; A) \cong \text{Ext}_{kG}^*(k, A).$$

This proves that $H_*(G; A)$ and $H^*(G; A)$ are k-modules whenever A is a kG-module. *Hint:* If $P \to \mathbb{Z}$ is a projective $\mathbb{Z}G$-resolution, consider $P \otimes_{\mathbb{Z}} k \to k$.

We now return our attention to H_0 and H^0.

Definition 6.1.5 The *augmentation ideal* of $\mathbb{Z}G$ is the kernel $\mathfrak{J}$ of the ring map $\mathbb{Z}G \xrightarrow{\epsilon} \mathbb{Z}$ which sends $\sum n_g g$ to $\sum n_g$. Because $\{1\} \cup \{g - 1 : g \in G, g \neq 1\}$ is a basis for $\mathbb{Z}G$ as a free $\mathbb{Z}$-module, it follows that $\mathfrak{J}$ is a free $\mathbb{Z}$-module with basis $\{g - 1 : g \in G, g \neq 1\}$.

Example 6.1.6 Since the trivial G-module $\mathbb{Z}$ is $\mathbb{Z}G/\mathfrak{J}$, $H_0(G; A) = A_G$ is isomorphic to $\mathbb{Z} \otimes_{\mathbb{Z}G} A = \mathbb{Z}G/\mathfrak{J} \otimes_{\mathbb{Z}G} A \cong A/\mathfrak{J}A$ for every G-module A. For example, $H_0(G; \mathbb{Z}) = \mathbb{Z}/\mathfrak{J}\mathbb{Z} = \mathbb{Z}$, $H_0(G; \mathbb{Z}G) = \mathbb{Z}G/\mathfrak{J} \cong \mathbb{Z}$, and $H_0(G; \mathfrak{J}) = \mathfrak{J}/\mathfrak{J}^2$.

Example 6.1.7 ($A = \mathbb{Z}G$) Because $\mathbb{Z}G$ is a projective object in $\mathbb{Z}G$–mod, $H_*(G; \mathbb{Z}G) = 0$ for $* \neq 0$ and $H_0(G; \mathbb{Z}G) = \mathbb{Z}$. When G is a finite group, Shapiro's Lemma (6.3.2 below) implies that $H^*(G; \mathbb{Z}G) = 0$ for $* \neq 0$. This fails when G is infinite; for example, we saw in 6.1.4 that $H^1(T; \mathbb{Z}T) \cong \mathbb{Z}$ for the infinite cyclic group T.

The following discussion clarifies the situation for $H^0(G; \mathbb{Z}G)$: If G is finite, then $H^0(G; \mathbb{Z}G) \cong \mathbb{Z}$, but $H^0(G; \mathbb{Z}G) = 0$ if G is infinite.

The Norm Element 6.1.8 Let G be a finite group. The *norm element* N of the group ring $\mathbb{Z}G$ is the sum $N = \sum_{g \in G} g$. The norm is a central element of $\mathbb{Z}G$ and belongs to $(\mathbb{Z}G)^G$, because for every $h \in G$ $hN = \sum_g hg = \sum_{g'} g' = N$, and $Nh = N$ similarly.

Lemma 6.1.9 *The subgroup* $H^0(G; \mathbb{Z}G) = (\mathbb{Z}G)^G$ *of* $\mathbb{Z}G$ *is the 2-sided ideal* $\mathbb{Z} \cdot N$ *of* $\mathbb{Z}G$ *(isomorphic to* $\mathbb{Z}$*) generated by* N.

Proof If $a = \sum n_g g$ is in $(\mathbb{Z}G)^G$, then $a = ga$ for all $g \in G$. Comparing coefficients of g shows that all the n_g are the same. Hence $a = nN$ for some $n \in \mathbb{Z}$. $\diamond$

Exercise 6.1.3

1. Show that if G is an infinite group, then $H^0(G; \mathbb{Z}G) = (\mathbb{Z}G)^G = 0$.
2. When G is a finite group, show that the natural map $\mathbb{Z} \cdot N = (\mathbb{Z}G)^G \to (\mathbb{Z}G)_G \cong \mathbb{Z}$ sends the norm N to the order $\#G$ of G. In particular, it is an injection.
3. Conclude that $\mathfrak{I}$ is $\ker(\mathbb{Z}G \xrightarrow{N} \mathbb{Z}G) = \{a \in \mathbb{Z}G : Na = 0\}$ when G is finite.

Proposition 6.1.10 *Let G be a finite group of order m, and N the norm. Then $e = N/m$ is a central idempotent element of $\mathbb{Q}G$ and of $\mathbb{Z}G[\frac{1}{m}]$. If A is a $\mathbb{Q}G$-module, or any G-module on which multiplication by m is an isomorphism,*

$$H_0(G; A) = H^0(G; A) = eA \quad \text{and} \quad H_*(G; A) = H^*(G; A) = 0 \text{ for } * \neq 0.$$

Proof $N^2 = (\sum g) \cdot N = m \cdot N$, so $e^2 = e$ in $R = \mathbb{Z}G[\frac{1}{m}]$. Note that $R \cong eR \times (1-e)R$ as a ring, that $eR = \mathbb{Z}[\frac{1}{m}]$, and that the projection e from R–**mod** to (eR)–**mod** $\subseteq$ **Ab** is an exact functor. Let A be an R-module; we first show that $eA = A_G = A^G$. Clearly $N \cdot A \subseteq A^G$, and if $a \in A^G$, then $N \cdot a = m \cdot a$, that is, $a = e \cdot a$. Therefore $eA = A^G$. By exercise 6.1.3 (3), $\mathfrak{I}[\frac{1}{m}] = \ker(R \xrightarrow{e} R) = (1-e)R$. Hence $(1-e)A = (1-e)R \otimes_R A$ equals $\mathfrak{I}[\frac{1}{m}] \otimes_R A = \mathfrak{I}A$; therefore $A_G = A/\mathfrak{I}A = A/(1-e)A = eA$.

Because eR is projective over R, $\operatorname{Tor}_n^R(eR, A) = \operatorname{Ext}_R^n(eR, A) = 0$ if $n \neq 0$. Since R is flat over $\mathbb{Z}G$, flat base change for Tor (3.2.29) yields

$$H_n(G; A) = \operatorname{Tor}_n^{\mathbb{Z}G}(\mathbb{Z}, A) = \operatorname{Tor}_n^R(\mathbb{Z} \otimes R, A) = \operatorname{Tor}_n^R(eR, A) = 0 \text{ if } n \neq 0.$$

For cohomology, we modify the argument used in 3.3.11 for localization of Ext. If $P \to \mathbb{Z}$ is a resolution of $\mathbb{Z}$ by projective $\mathbb{Z}G$-modules, then $P[\frac{1}{m}] \to \mathbb{Z}[\frac{1}{m}]$ is a resolution of $\mathbb{Z}[\frac{1}{m}] = eR$ by projective R-modules. Because A is an R-module, adjointness yields $\operatorname{Hom}_G(P, A) \cong \operatorname{Hom}_R(P[\frac{1}{m}], A)$. Thus for $n \neq 0$ we have

$$H^n(G; A) = H^n \operatorname{Hom}_G(P, A) \cong H^n \operatorname{Hom}_R(P[\frac{1}{m}], A) = \operatorname{Ext}_R^n(eR, A) = 0. \quad \diamond$$

We now turn our attention to the first homology group H_1.

Exercise 6.1.4

1. Define $\theta: G \to \mathfrak{I}/\mathfrak{I}^2$ by $\theta(g) = g - 1$. Show that θ is a group homomorphism and that the commutator subgroup $[G, G]$ of G maps to zero.

2. Define $\sigma: \mathfrak{J} \to G/[G, G]$ by $\sigma(g - 1) = \bar{g}$, the (left) coset of g. Show that $\sigma(\mathfrak{J}^2) = 1$, and deduce that θ and σ induce an isomorphism $\mathfrak{J}/\mathfrak{J}^2 \cong G/[G, G]$.

Theorem 6.1.11 *For any group G, $H_1(G; \mathbb{Z}) \cong \mathfrak{J}/\mathfrak{J}^2 \cong G/[G, G]$.*

Proof The sequence $0 \to \mathfrak{J} \to \mathbb{Z}G \to \mathbb{Z} \to 0$ induces an exact sequence

$$H_1(G; \mathbb{Z}G) \to H_1(G; \mathbb{Z}) \to \mathfrak{J}_G \to (\mathbb{Z}G)_G \to \mathbb{Z} \to 0.$$

Since $\mathbb{Z}G$ is projective, $H_1(G; \mathbb{Z}G) = 0$. The right-hand map is the isomorphism $(\mathbb{Z}G)_G \cong \mathbb{Z}G/\mathfrak{J} \cong \mathbb{Z}$, so evidently $H_1(G; \mathbb{Z})$ is isomorphic to $\mathfrak{J}_G = \mathfrak{J}/\mathfrak{J}^2$. By the previous exercise, this is isomorphic to $G/[G, G]$. $\diamond$

Theorem 6.1.12 *If A is any trivial G-module, $H_0(G; A) \cong A$, $H_1(G; A) \cong G/[G, G] \otimes_{\mathbb{Z}} A$, and for $n \geq 2$ there are (noncanonical) isomorphisms:*

$$H_n(G; A) \cong H_n(G; \mathbb{Z}) \otimes_{\mathbb{Z}} A \oplus \mathrm{Tor}_1^{\mathbb{Z}}(H_{n-1}(G; \mathbb{Z}), A).$$

Proof If $P \to \mathbb{Z}$ is a free right $\mathbb{Z}G$-module resolution, $H_*(G; A)$ is the homology of $P \otimes_{\mathbb{Z}G} A = (P \otimes_{\mathbb{Z}G} \mathbb{Z}) \otimes_{\mathbb{Z}} A$. Now use the Universal Coefficient Theorem. $\diamond$

Exercises 6.1.5 Let A be a trivial G-module.

1. Show that $H^1(G; A)$ is isomorphic to the group $\mathrm{Hom}_{\mathbf{Groups}}(G, A) \cong \mathrm{Hom}_{\mathbf{Ab}}(G/[G, G], A)$ of all group homomorphisms from G to A.
2. Conclude that $H^1(G; \mathbb{Z}) = 0$ for every finite group.
3. Show that in general there is a split exact sequence

$$0 \to \mathrm{Ext}^1_{\mathbb{Z}}(H_{n-1}(G; \mathbb{Z}), A) \to H^n(G; A) \to \mathrm{Hom}_{\mathbf{Ab}}(H_n(G; \mathbb{Z}), A) \to 0.$$

Exercise 6.1.6 If G is finite, show that $H^1(G; \mathbb{C}) = 0$ and that $H^2(G; \mathbb{Z})$ is isomorphic to the group $H^1(G, \mathbb{C}^*) \cong \mathrm{Hom}_{\mathbf{Groups}}(G, \mathbb{C}^*)$ of all 1-dimensional representations of G. Here G acts trivially on $\mathbb{Z}$, $\mathbb{C}$, and on the group $\mathbb{C}^*$ of complex units.

We now turn to the product $G \times H$ of two groups G and H. First note that $\mathbb{Z}[G \times H] \cong \mathbb{Z}G \otimes \mathbb{Z}H$. Indeed, the ring maps from $\mathbb{Z}G$ and $\mathbb{Z}H$ to $\mathbb{Z}[G \times H]$ induce a ring map from $\mathbb{Z}G \otimes \mathbb{Z}H$ to $\mathbb{Z}[G \times H]$. Both rings have the set $G \times H$ as a $\mathbb{Z}$-basis, so this map is an isomorphism. The Künneth formula gives the homology of $G \times H$:

Proposition 6.1.13 (Products) *For every G and H there is a split exact sequence:*

$$0 \to \bigoplus_{\substack{p+q \\ =n}} H_p(G; \mathbb{Z}) \otimes H_q(H; \mathbb{Z}) \to H_n(G \times H; \mathbb{Z})$$

$$\to \bigoplus_{\substack{p+q \\ =n-1}} \mathrm{Tor}_1^{\mathbb{Z}}(H_p(G; \mathbb{Z}), H_q(H; \mathbb{Z})) \to 0.$$

Proof Let $P \to \mathbb{Z}$ be a free $\mathbb{Z}G$-resolution and $Q \to \mathbb{Z}$ a free $\mathbb{Z}H$-resolution, and write $P \otimes_{\mathbb{Z}} Q$ for the total tensor product chain complex (2.7.1), which is a complex of $\mathbb{Z}G \otimes \mathbb{Z}H$-modules. By the Künneth formula for complexes (3.6.3), the homology of $P \otimes_{\mathbb{Z}} Q$ is zero except for $H_0(P \otimes_{\mathbb{Z}} Q) = \mathbb{Z}$. Hence $P \otimes_{\mathbb{Z}} Q \to \mathbb{Z}$ is a free $\mathbb{Z}G \otimes \mathbb{Z}H$-module resolution of $\mathbb{Z}$, and $H_*(G \times H; \mathbb{Z})$ is the homology of

$$(P \otimes_{\mathbb{Z}} Q) \otimes_{\mathbb{Z}G \otimes \mathbb{Z}H} \mathbb{Z} \cong (P \otimes_{\mathbb{Z}G} \mathbb{Z}) \otimes_{\mathbb{Z}} (Q \otimes_{\mathbb{Z}H} \mathbb{Z}).$$

Moreover, $H_*(G; \mathbb{Z}) = H_*(P \otimes_{\mathbb{Z}G} \mathbb{Z})$ and $H_*(H; \mathbb{Z}) = (Q \otimes_{\mathbb{Z}H} \mathbb{Z})$. As each $P_n \otimes_{\mathbb{Z}G} \mathbb{Z}$ is a free $\mathbb{Z}$-module, the proposition follows from the Künneth formula for complexes. $\diamond$

Exercise 6.1.7 (kG-**modules**) Let k be a field, considered as a trivial module. Modify the above proof to show that $H_n(G \times H; k) \cong \bigoplus H_p(G; k) \otimes_k H_{n-p}(H; k)$ for all n.

Cohomology Cross Product 6.1.14 Keeping the notation of the preceding proposition, there is a natural homomorphism of tensor product double complexes:

$$\mu : \mathrm{Hom}_G(P, \mathbb{Z}) \otimes \mathrm{Hom}_H(Q, \mathbb{Z}) \to \mathrm{Hom}_{G \times H}(P \otimes_{\mathbb{Z}} Q, \mathbb{Z}),$$

$$\mu(f \otimes f')(x \otimes y) = f(x)f'(y), x \in P_p, y \in Q_q.$$

The *cross product* $\times : H^p(G; \mathbb{Z}) \otimes H^q(H; \mathbb{Z}) \to H^{p+q}(G \times H; \mathbb{Z})$ is the composite obtained by taking the cohomology of the total complexes.

$$
\begin{array}{ccc}
H^p(G; \mathbb{Z}) \otimes H^q(H; \mathbb{Z}) & \longrightarrow & H^{p+q}[\mathrm{Hom}_G(P, \mathbb{Z}) \otimes \mathrm{Hom}_H(Q, \mathbb{Z})], \\
\times \downarrow & & \downarrow \mu \\
H^{p+q}(G \times H; \mathbb{Z}) & = & H^{p+q}[\mathrm{Hom}_{G \times H}(P \otimes Q, \mathbb{Z})]
\end{array}
$$

Exercise 6.1.8 Suppose that each P_p is a finitely generated $\mathbb{Z}G$-module. (For example, this can be done when G is finite; see section 6.5 below.) Show in this case that μ is an *isomorphism*. Then deduce from the Künneth formula 3.6.3 that the cross product fits into a split short exact sequence:

$$0 \to \bigoplus_{\substack{p+q \\ =n}} H^p(G;\mathbb{Z}) \otimes H^q(H;\mathbb{Z}) \xrightarrow{\times} H^n(G \times H;\mathbb{Z})$$

$$\to \bigoplus_{\substack{p+q \\ =n+1}} \operatorname{Tor}_1^{\mathbb{Z}}(H^p(G;\mathbb{Z}), H^q(H;\mathbb{Z})) \to 0.$$

Exercises 6.1.9

1. Show that the cross product is independent of the choice of P and Q.
2. If $H = 1$, show that cross product with $1 \in H^0(1;\mathbb{Z})$ is the identity map.
3. Show that the cross product is associative in the sense that the two maps

$$H^p(G;\mathbb{Z}) \otimes H^q(H;\mathbb{Z}) \otimes H^r(I;\mathbb{Z}) \to H^{p+q+r}(G \times H \times I;\mathbb{Z})$$

   given by the formulas $(x \times y) \times z$ and $x \times (y \times z)$ agree.

Exercise 6.1.10 Let k be a commutative ring.

1. Modify the above construction to obtain cross products $H^p(G;k) \otimes_k H^q(H;k) \to H^{p+q}(G \times H;k)$. Then verify that this cross product is independent of the choice of P and Q, that it is associative, and that the cross product with $1 \in H^0(1;k) = k$ is the identity.
2. If k is a field, show that $H^n(G \times H;k) \cong \bigoplus H^p(G;k) \otimes_k H^{n-p}(H;k)$ for all n.

We will return to the cross product in section 6.7, when we introduce the restriction map $H^*(G \times G) \to H^*(G)$ and show that the cross product makes $H^*(G;\mathbb{Z})$ into a ring.

Hyperhomology 6.1.15 If A_* is a chain complex of G-modules, the hyper-derived functors $\mathbb{L}_i(-_G)(A_*)$ of 5.7.4 are written as $\mathbb{H}_i(G;A_*)$ and called the *hyperhomology* groups of G. Similarly, if A^* is a cochain complex of G-modules, the *hypercohomology* groups $\mathbb{H}^i(G;A^*)$ are just the hyper-derived functors $\mathbb{R}^i(-^G)(A^*)$. The generalities of Chapter 5, section 7 become the following facts in this case. The hyperhomology spectral sequences are

$$^{II}E_{pq}^2 = H_p(G;H_q(A_*)) \Rightarrow \mathbb{H}_{p+q}(G;A_*); \quad \text{and}$$

$$^{I}E_{pq}^2 = H_p(H_q(G;A_*)) \Rightarrow \mathbb{H}_{p+q}(G;A_*) \quad \text{when } A_* \text{ is bounded below,}$$

and the hypercohomology spectral sequences are

$$^{II}E_2^{pq} = H^p(G; H^q(A^*)) \Rightarrow \mathbb{H}^{p+q}(G; A^*), \text{ weakly convergent; and}$$

$$^{I}E_2^{pq} = H^p(H^q(G; A^*)) \Rightarrow \mathbb{H}^{p+q}(G; A^*) \text{ if } A \text{ is bounded below.}$$

In particular, suppose that A is bounded below. If each A_i is a flat $\mathbb{Z}G$-module, then $\mathbb{H}_i(G; A_*) = H_i((A_*)_G)$; if each A^i is a projective $\mathbb{Z}G$-module, then $\mathbb{H}^i(G; A^*) = H^i((A^*)^G)$.

Exercise 6.1.11 Let T be the infinite cyclic group. Show that there are short exact sequences

$$0 \to H_q(A_*)_T \to \mathbb{H}_q(T; A_*) \to H_{q-1}(A_*)^T \to 0;$$

$$0 \to H^{q-1}(A^*)_T \to \mathbb{H}^q(T; A^*) \to H^q(A^*)^T \to 0.$$

Exercise 6.1.12 Let k be a commutative ring and G a group such that all the k-modules $H_*(G; k)$ are flat. (For example, this is true for $G = T$.) Use the hypertor spectral sequence (5.7.8) to show that $H_n(G \times H; k) \cong \bigoplus H_p(G; k) \otimes_k H_{n-p}(H; k)$ for all n and H.

6.2 Cyclic and Free Groups

Cyclic and free groups are two classes of groups for which explicit calculations are easy to make. We first consider cyclic groups.

Calculation 6.2.1 (Cyclic groups) Let C_m denote the cyclic group of order m on generator σ. The norm in $\mathbb{Z}C_m$ is the element $N = 1 + \sigma + \sigma^2 + \cdots + \sigma^{m-1}$, so $0 = \sigma^m - 1 = (\sigma - 1)N$ in $\mathbb{Z}C_m$. I claim that the trivial C_m-module $\mathbb{Z}$ has the periodic free resolution

$$0 \leftarrow \mathbb{Z} \xleftarrow{\epsilon} \mathbb{Z}C_m \xleftarrow{\sigma-1} \mathbb{Z}C_m \xleftarrow{N} \mathbb{Z}C_m \xleftarrow{\sigma-1} \mathbb{Z}C_m \xleftarrow{N} \cdots.$$

Indeed, since $\mathbb{Z} \cdot N = (\mathbb{Z}G)^G$ and $\mathfrak{I} = \{a \in \mathbb{Z}G : Na = 0\}$ by exercise 6.1.3, there are exact sequences

$$0 \leftarrow \mathbb{Z} \cdot N \xleftarrow{N} \mathbb{Z}G \leftarrow \mathfrak{I} \leftarrow 0 \quad \text{and} \quad 0 \leftarrow \mathfrak{I} \xleftarrow{\sigma-1} \mathbb{Z}C_m \leftarrow \mathbb{Z} \cdot N \leftarrow 0.$$

The periodic free resolution is obtained by splicing these sequences together. Applying $\otimes_{\mathbb{Z}G} A$ and $\operatorname{Hom}_G(-, A)$ and taking homology, we find the following result:

Theorem 6.2.2 *If A is a module for the cyclic group $G = C_m$, then*

$$H_n(C_m; A) = \begin{cases} A/(\sigma - 1)A & \text{if } n = 0 \\ A^G/NA & \text{if } n = 1, 3, 5, 7, \ldots \\ \{a \in A : Na = 0\}/(\sigma - 1)A & \text{if } n = 2, 4, 6, 8, \ldots \end{cases};$$

$$H^n(C_m; A) = \begin{cases} A^G & \text{if } n = 0 \\ \{a \in A : Na = 0\}/(\sigma - 1)A & \text{if } n = 1, 3, 5, 7, \ldots \\ A^G/NA & \text{if } n = 2, 4, 6, 8, \ldots \end{cases}.$$

Exercise 6.2.1 Show for $G = C_m$ that when $H^1(G; A) = 0$ there is an exact sequence

$$0 \to A^G \to A \xrightarrow{\sigma - 1} A \xrightarrow{N} A^G \to H^2(G; A) \to 0.$$

Example 6.2.3 Taking $A = \mathbb{Z}$ we find that

$$H_n(C_m; \mathbb{Z}) = \begin{cases} \mathbb{Z} & \text{if } n = 0 \\ \mathbb{Z}/m & \text{if } n = 1, 3, 5, 7, \ldots \\ 0 & \text{if } n = 2, 4, 6, 8, \ldots \end{cases};$$

$$H^n(C_m; \mathbb{Z}) = \begin{cases} \mathbb{Z} & \text{if } n = 0 \\ 0 & \text{if } n = 1, 3, 5, 7, \ldots \\ \mathbb{Z}/m & \text{if } n = 2, 4, 6, 8, \ldots \end{cases}.$$

Exercise 6.2.2 Calculate $H_*(C_m \times C_n; \mathbb{Z})$ and $H^*(C_m \times C_n; \mathbb{Z})$.

Definition 6.2.4 (Tate cohomology) Taking full advantage of this periodicity, we set

$$\widehat{H}^n(C_m; A) = \begin{cases} A^G/NA & \text{if } n \in \mathbb{Z} \text{ is even} \\ \{a \in A : NA = 0\}/(\sigma - 1)A & \text{if } n \in \mathbb{Z} \text{ is odd} \end{cases}.$$

More generally, if G is a finite group and A is a G-module, we define the *Tate cohomology groups* of G to be the groups

$$\widehat{H}^n(G; A) = \begin{cases} H^n(G; A) & \text{if } n \geq 1 \\ A^G/NA & \text{if } n = 0 \\ \{a \in A : Na = 0\}/JA & \text{if } n = -1 \\ H_{1-n}(G; A) & \text{if } n \leq -2 \end{cases}.$$

Exercise 6.2.3 If G is a finite group and $0 \to A \to B \to C \to 0$ is an exact sequence of G-modules, show that there is a long exact sequence

$$\cdots \widehat{H}^{n-1}(G; C) \to \widehat{H}^{n}(G; A) \to \widehat{H}^{n}(G; B) \to \widehat{H}^{n}(G; C) \to \widehat{H}^{n+1}(G; A) \cdots.$$

Application 6.2.5 (Dimension-shifting) Given a G-module A, choose a short exact sequence $0 \to K \to P \to A \to 0$ with P projective. Shapiro's Lemma (6.3.2 below) implies that $\widehat{H}^{*}(G, P) = 0$ for all $* \in \mathbb{Z}$. Therefore $\widehat{H}^{n}(G; A) \cong \widehat{H}^{n+1}(G; K)$. This shows that every Tate cohomology group $\widehat{H}^{n}(G; A)$ determines the entire theory.

Proposition 6.2.6 *Let G be the free group on the set X, and consider the augmentation ideal $\mathfrak{I}$ of $\mathbb{Z}G$. Then $\mathfrak{I}$ is a free $\mathbb{Z}G$-module with basis the set $X - 1 = \{x - 1 : x \in X\}$.*

Proof We have seen that $\mathfrak{I}$ is a free abelian group with $\mathbb{Z}$-basis $\{g - 1 : g \in G, g \neq 1\}$. We claim that another $\mathbb{Z}$-basis is $\{g(x - 1) : g \in G, x \in X\}$. Every $g \in G$ may be written uniquely as a reduced word in the symbols $\{x, x^{-1} : x \in X\}$; write $G(x)$ (resp. $G(x^{-1})$) for the subset of all $g \in G$ ending in the symbol x (resp. in x^{-1}) so that $G - \{1\}$ is the disjoint union (over all $x \in X$) of the sets $G(x)$ and $G(x^{-1})$. The formulas

$$(gx - 1) = g(x - 1) + (g - 1) \quad \text{if } gx \in G(x)$$
$$(gx^{-1} - 1) = -(gx^{-1})(x - 1) + (g - 1) \quad \text{if } gx^{-1} \in G(x^{-1})$$

and induction on word length allow us to uniquely rewrite the basis $\{g - 1 : g \neq 1\}$ in terms of the set $\{g(x - 1)\}$, and vice versa. Therefore $\{g(x - 1) : g \in G, x \in X\}$ is a $\mathbb{Z}$-basis of $\mathfrak{I}$, and $X - 1 = \{x - 1 : x \in X\}$ is a $\mathbb{Z}G$-basis. $\diamond$

Corollary 6.2.7 *If G is a free group on X, then $\mathbb{Z}$ has the free resolution*

$$0 \to \mathfrak{I} \to \mathbb{Z}G \to \mathbb{Z} \to 0.$$

Consequently, $pd_G(\mathbb{Z}) = 1$, that is, $H_n(G; A) = H^n(G; A) = 0$ for $n \neq 0, 1$. Moreover, $H_0(G; \mathbb{Z}) \cong H^0(G; \mathbb{Z}) \cong \mathbb{Z}$, while

$$H_1(G; \mathbb{Z}) \cong \bigoplus_{x \in X} \mathbb{Z} \quad \text{and} \quad H^1(G; \mathbb{Z}) \cong \prod_{x \in X} \mathbb{Z}.$$

Proof $H_*(G; A)$ is the homology of $0 \to \mathfrak{I} \otimes_{\mathbb{Z}G} A \to A \to 0$, and $H^*(G; A)$ is the cohomology of $0 \to A \to \operatorname{Hom}_G(\mathfrak{I}, A) \to 0$. For $A = \mathbb{Z}$, the differentials are zero. $\diamond$

Remark Conversely, Stallings [St] and Swan [SwCd1] proved that if $H^n(G, A)$ vanishes for all $n \neq 0, 1$ and all G-modules A, then G is a free group.

Exercise 6.2.4 Let G be the free group on $\{s, t\}$, and let $T \subseteq G$ be the free group on $\{t\}$. Let $\mathbb{Z}'$ denote the abelian group $\mathbb{Z}$, made into a G-module (and a T-module) by the formulas $s \cdot a = t \cdot a = -a$.

1. Show that $H_0(G, \mathbb{Z}') = H_0(T, \mathbb{Z}') = \mathbb{Z}/2$.
2. Show that $H_1(T, \mathbb{Z}') = 0$ but $H_1(G, \mathbb{Z}') \cong \mathbb{Z}$.

Free Products 6.2.8 Let $G*H$ denote the free product (or coproduct) of the groups G and H. By [BAII, 2.9], every element of $G*H$ except 1 has a unique expression as a "reduced" word, either of the form $g_1 h_1 g_2 h_2 g_3 \cdots$ or of the form $h_1 g_1 h_2 g_2 h_3 \cdots$ with all $g_i \in G$ and all $h_i \in H$ (and all $g_i, h_i \neq 1$).

Proposition 6.2.9 *Let $\mathfrak{I}_G$, $\mathfrak{I}_H$, and $\mathfrak{I}_{G*H}$ denote the augmentation ideals of $\mathbb{Z}G$, $\mathbb{Z}H$, and $\Lambda = \mathbb{Z}(G*H)$, respectively. Then*

$$\mathfrak{I}_{G*H} \cong (\mathfrak{I}_G \otimes_{\mathbb{Z}G} \Lambda) \oplus (\mathfrak{I}_H \otimes_{\mathbb{Z}H} \Lambda).$$

Proof As a left $\mathbb{Z}G$-module, $\Lambda = \mathbb{Z}(G*H)$ has a basis consisting of $\{1\}$ and the set of all reduced words beginning with an element of H. Therefore $\mathfrak{I}_G \otimes_{\mathbb{Z}G} \Lambda$ has a $\mathbb{Z}$-basis $\mathcal{B}_1$ consisting of the basis $\{g - 1 | g \in G, g \neq 1\}$ of $\mathfrak{I}_G$ and the set of all terms

$$(g - 1)(h_1 g_1 h_2 \cdots) = (g h_1 g_1 h_2 \cdots) - (h_1 g_1 h_2 \cdots).$$

Similarly, $\mathfrak{I}_H \otimes_{\mathbb{Z}H} \Lambda$ has a $\mathbb{Z}$-basis $\mathcal{B}_2$ consisting of $\{h - 1\}$ and the set of all terms

$$(h - 1)(g_1 h_1 g_2 \cdots) = (h g_1 h_1 g_2 \cdots) - (g_1 h_1 g_2 \cdots).$$

By induction on the length of a reduced word w in $G*H$, we see that $w - 1$ can be written as a sum of terms in $\mathcal{B}_1$ and $\mathcal{B}_2$. This proves that $\mathcal{B} = \mathcal{B}_1 \cup \mathcal{B}_2$ generates $\mathfrak{I}_{G*H}$. In any nontrivial sum of elements of $\mathcal{B}$, the coefficients of the longest words must be nonzero, so $\mathcal{B}$ is linearly independent. This proves that $\mathcal{B}$ forms a $\mathbb{Z}$-basis for $\mathfrak{I}_{G*H}$, and hence that $\mathfrak{I}_{G*H}$ has the decomposition we described. $\diamond$

Corollary 6.2.10 *For every left $(G*H)$-module A, and $n \geq 2$:*

$$H_n(G*H; A) \cong H_n(G; A) \oplus H_n(H; A);$$

$$H^n(G*H; A) \cong H^n(G; A) \oplus H^n(H; A).$$

Remark When $n = 0$, the conclusion fails even for $A = \mathbb{Z}$. We gave an example above of a $(T*T)$-module $\mathbb{Z}'$ for which the conclusion fails when $n = 1$.

Proof We give the proof of the homology assertion, the cohomology part being entirely analogous. Write Λ for $\mathbb{Z}(G*H)$. Because $\operatorname{Tor}_n^\Lambda(\Lambda, A) = 0$ for $n \geq 1$, we see that $\operatorname{Tor}_n^\Lambda(\mathbb{Z}, A) \cong \operatorname{Tor}_{n-1}^\Lambda(\mathfrak{I}_{G*H}, A)$ for $n \geq 2$. Hence in this range

$$H_n(G*H; A) = \operatorname{Tor}_n^\Lambda(\mathbb{Z}, A) \cong \operatorname{Tor}_{n-1}^\Lambda(\mathfrak{I}_{G*H}, A)$$

$$\cong \operatorname{Tor}_{n-1}^\Lambda(\mathfrak{I}_G \otimes_{\mathbb{Z}G} \Lambda, A) \oplus \operatorname{Tor}_{n-1}^\Lambda(\mathfrak{I}_H \otimes_{\mathbb{Z}H} \Lambda, A).$$

Since Λ is free over $\mathbb{Z}G$ and $\mathbb{Z}H$, base-change for Tor (3.2.9 or 5.6.6) implies that

$$\operatorname{Tor}_{n-1}^\Lambda(\mathfrak{I}_G \otimes_{\mathbb{Z}G} \Lambda, A) \cong \operatorname{Tor}_{n-1}^{\mathbb{Z}G}(\mathfrak{I}_G, A) \cong \operatorname{Tor}_n^{\mathbb{Z}G}(\mathbb{Z}, A) = H_n(G; A).$$

By symmetry, $\operatorname{Tor}_{n-1}^\Lambda(\mathfrak{I}_H \otimes_{\mathbb{Z}H} \Lambda, A) \cong H_n(H; A)$. ◇

Exercise 6.2.5 Show that if A is a trivial $G*H$-module, then for $n = 1$ we also have

$$H_1(G*H; A) \cong H_1(G; A) \oplus H_1(H; A);$$

$$H^1(G*H; A) \equiv H^1(G; A) \oplus H^1(H; A).$$

6.3 Shapiro's Lemma

For actually performing calculations, Shapiro's Lemma is a fundamental tool. Suppose that H is a subgroup of G and A is a left $\mathbb{Z}H$-module. We know (2.6.2) that $\mathbb{Z}G \otimes_{\mathbb{Z}H} A$ and $\operatorname{Hom}_H(\mathbb{Z}G, A)$ are left $\mathbb{Z}G$-modules. Here are their names:

Definition 6.3.1 $\mathbb{Z}G \otimes_{\mathbb{Z}H} A$ is called the *induced G-module* and is written $\operatorname{Ind}_H^G(A)$. Similarly, $\operatorname{Hom}_H(\mathbb{Z}G, A)$ is called the *coinduced G-module* and is written $\operatorname{Coind}_H^G(A)$.

Shapiro's Lemma 6.3.2 Let H be a subgroup of G and A an H-module. Then

$$H_*(G; \operatorname{Ind}_H^G(A)) \cong H_*(H; A); \text{ and } H^*(G; \operatorname{Coind}_H^G(A)) \cong H^*(H; A).$$

Proof Note that $\mathbb{Z}G$ is a free $\mathbb{Z}H$-module (any set of coset representatives will form a basis). Hence any projective right $\mathbb{Z}G$-module resolution $P \to \mathbb{Z}$ is also a projective $\mathbb{Z}H$-module resolution. Therefore the homology of the chain complex

$$P \otimes_{\mathbb{Z}G} (\mathbb{Z}G \otimes_{\mathbb{Z}H} A) \cong P \otimes_{\mathbb{Z}H} A$$

is both

$$\mathrm{Tor}_*^{\mathbb{Z}G}(\mathbb{Z}, \mathbb{Z}G \otimes_{\mathbb{Z}H} A) \cong H_*(G; \mathrm{Ind}_H^G(A))$$

and $\mathrm{Tor}_*^{\mathbb{Z}H}(\mathbb{Z}, A) \cong H_*(H; A)$. Similarly, if $P \to \mathbb{Z}$ is a projective left $\mathbb{Z}G$-module resolution, then there is an adjunction isomorphism of cochain complexes:

$$\mathrm{Hom}_G(P, \mathrm{Hom}_H(\mathbb{Z}G, A)) \cong \mathrm{Hom}_H(P, A).$$

The cohomology of this complex is both

$$\mathrm{Ext}_{\mathbb{Z}G}^*(\mathbb{Z}, \mathrm{Hom}_H(\mathbb{Z}G, A)) \cong H^*(G; \mathrm{Coind}_H^G(A))$$

$$\text{and} \quad \mathrm{Ext}_{\mathbb{Z}H}^*(\mathbb{Z}, A) \cong H^*(H; A). \qquad \diamond$$

Corollary 6.3.3 (Shapiro's Lemma for $H = 1$) *If A is an abelian group, then*

$$H_*(G; \mathbb{Z}G \otimes_{\mathbb{Z}} A) = H^*(G; \mathrm{Hom}_{\mathbf{Ab}}(\mathbb{Z}G, A)) = \left\{ \begin{array}{ll} A & \text{if } * = 0 \\ 0 & \text{if } * \neq 0 \end{array} \right\}.$$

Lemma 6.3.4 *If the index $[G : H]$ is finite, $\mathrm{Ind}_H^G(A) \cong \mathrm{Coind}_H^G(A)$.*

Proof Let X be a set of left coset representatives for G/H, so that X forms a basis for the right H-module $\mathbb{Z}G$. $\mathrm{Ind}_H^G(A)$ is the sum over X of copies $x \otimes A$ of A, with $g(x \otimes a) = y \otimes ha$ if $gx = yh$ in G. Now $X^{-1} = \{x^{-1} : x \in X\}$ is a basis of $\mathbb{Z}G$ as a left H-module, so $\mathrm{Coind}_H^G(A)$ is the product over X of copies $\pi_x A$ of A, where $\pi_x a$ represents the H-map from $\mathbb{Z}G$ to A sending x^{-1} to $a \in A$ and z^{-1} to 0 for all $z \neq x$ in X. Therefore if $gx = yh$, that is, $y^{-1}g = hx^{-1}$, the map $g(\pi_x a)$ sends y^{-1} to

$$(\pi_x a)(y^{-1}g) = (\pi_x a)(hx^{-1}) = h \cdot (\pi_x a)(x^{-1}) = ha$$

and z^{-1} to 0 if $z \neq y$ in X. That is, $g(\pi_x a) = \pi_y(ha)$. Since $X \cong [G : H]$ is finite, the map $\mathrm{Ind}_H^G(A) \to \mathrm{Coind}_H^G(A)$ sending $x \otimes a$ to $\pi_x a$ is an H-module isomorphism. $\qquad \diamond$

Corollary 6.3.5 *If G is a finite group, then $H^*(G; \mathbb{Z}G \otimes_{\mathbb{Z}} A) = 0$ for $* \neq 0$ and all A.*

Corollary 6.3.6 (Tate cohomology) *If G is finite and P is a projective G-module,*

$$\widehat{H}^*(G; P) = 0 \quad \text{for all } *.$$

Proof It is enough to prove the result for free G-modules, that is, for modules of the form $P = \mathbb{Z}G \otimes_{\mathbb{Z}} F$, where F is free abelian. Shapiro's Lemma gives vanishing for $* \neq 0, -1$. Since $P^G = (\mathbb{Z}G)^G \otimes F = N \cdot P$, we get $\widehat{H}^0(G; P) = 0$. Finally, $\widehat{H}^{-1}(G; P) = 0$ follows from the fact that $N = \#G$ on the free abelian group $P_G = P/\mathfrak{J}P \cong F$. $\diamond$

Hilbert's Theorem 90 6.3.7 (Additive version) Let $K \subset L$ be a finite Galois extension of fields, with Galois group G. Then L is a G-module, $L^G \cong L_G \cong K$, and

$$H^*(G; L) = H_*(G; L) = 0 \quad \text{for } * \neq 0.$$

Proof The Normal Basis Theorem [BAI, p. 283] asserts that there is an $x \in L$ such that the set $\{g(x) : g \in G\}$ of its conjugates forms a basis of the K-vector space L. Hence $L \cong \mathbb{Z}G \otimes_{\mathbb{Z}} K$ as a G-module. We now cite Shapiro's Lemma. $\diamond$

Example 6.3.8 (Cyclic Galois extensions) Suppose that G is cyclic of order m, generated by σ. The *trace* $tr(x)$ of an element $x \in L$ is the element $x + \sigma x + \cdots + \sigma^{m-1} x$ of K. In this case, Hilbert's Theorem 90 states that there is an exact sequence

$$0 \to K \to L \xrightarrow{\sigma - 1} L \xrightarrow{tr} K \to 0.$$

Indeed, we saw in the last section that for $* \neq 0$ every group $H_*(G; L)$ and $H^*(G; L)$ is either $K/tr(L)$ or $\ker(tr)/(\sigma - 1)K$.

As an application, suppose that $\operatorname{char}(K) = p$ and that $[L : K] = p$. Since $tr(1) = p \cdot 1 = 0$, there is an $x \in L$ such that $(\sigma - 1)x = 1$, that is, $\sigma x = x + 1$. Hence $L = K(x)$ and $x^p - x \in K$ because

$$\sigma(x^p - x) = (x + 1)^p - (x + 1) = x^p - x.$$

Remark If G is not cyclic, we will see in the next section that the vanishing of $H^1(G; L)$ is equivalent to Noether's Theorem [BAI, p. 287] that if $D: G \to L$ is a map satisfying $D(gh) = D(g) + g \cdot D(h)$, then there is an $x \in L$ such that $D(g) = g \cdot x - x$.

Application 6.3.9 (Transfer) Let H be a subgroup of finite index in G. Considering a G-module A as an H-module, we obtain a canonical map from A to $\operatorname{Hom}_H(\mathbb{Z}G, A) = \operatorname{Coind}_H^G(A) \cong \operatorname{Ind}_H^G(A)$ and from $\operatorname{Coind}_H^G(A) \cong \mathbb{Z}G \otimes_{\mathbb{Z}H} A$ to A. Applying Shapiro's Lemma, we obtain *transfer maps* $H_*(G; A) \to H_*(H; A)$ and $H^*(H; A) \to H^*(G; A)$. We will return to these maps in exercise 6.7.7 when we discuss restriction.

6.4 Crossed Homomorphisms and H^1

If A is a bimodule over any ring R, a *derivation* of R in A is an abelian group homomorphism $D: R \to A$ satisfying the *Leibnitz rule:* $D(rs) = rD(s) + D(r)s$. When $R = \mathbb{Z}G$ and A is a *left* $\mathbb{Z}G$-module, made into a bimodule by giving it a trivial right G-module structure, this definition simplifies as follows:

Definition 6.4.1 A *derivation* (or *crossed homomorphism*) of G in a left G-module A is a set map $D: G \to A$ satisfying $D(gh) = gD(h) + D(g)$. The family $\operatorname{Der}(G, A)$ of all derivations is an abelian group in an obvious way: $(D + D')(g) = D(g) + D'(g)$.

Example 6.4.2 (Principal derivations) If $a \in A$, define $D_a(g) = ga - a$; D_a is a derivation because

$$D_a(gh) = (gha - ga) + (ga - a) = gD_a(h) + D_a(g).$$

The D_a are called the *principal derivations* of G in A. Since $D_a + D_b = D_{(a+b)}$, the set $\operatorname{PDer}(G, A)$ of principal derivations forms a subgroup of $\operatorname{Der}(G, A)$.

Exercise 6.4.1 Show that $\operatorname{PDer}(G, A) \cong A/A^G$.

Example 6.4.3 If $\varphi: \mathfrak{I} \to A$ is a G-map, let $D_\varphi: G \to A$ be defined by $D_\varphi(g) = \varphi(g - 1)$. This is a derivation, since

$$D_\varphi(gh) = \varphi(gh - 1) = \varphi(gh - g) + \varphi(g - 1) = gD_\varphi(h) + D_\varphi(g).$$

Lemma 6.4.4 *The map $\varphi \mapsto D_\varphi$ is a natural isomorphism of abelian groups*

$$\text{Hom}_G(\mathfrak{I}, A) \cong \text{Der}(G, A).$$

Proof The formula defines a natural homomorphism from $\text{Hom}_G(\mathfrak{I}, A)$ to $\text{Der}(G, A)$, so it suffices to show that this map is an isomorphism. Since $\{g - 1 : g \neq 1\}$ forms a basis for the abelian group $\mathfrak{I}$, if $D_\varphi(g) = 0$ for all g, then $\varphi = 0$. Therefore the map in question is an injection. If D is a derivation, define $\varphi(g - 1) = D(g) \in A$. Since $\{g - 1 : g \neq 1\}$ forms a basis of $\mathfrak{I}$, φ extends to an abelian group map $\varphi \colon \mathfrak{I} \to A$. Since

$$\varphi(g(h - 1)) = \varphi(gh - 1) - \varphi(g - 1)$$
$$= D(gh) - D(g) = gD(h)$$
$$= g\varphi(h - 1),$$

φ is a G−map. As $D_\varphi = D$, the map in question is also a surjection.   ◇

Theorem 6.4.5 $H^1(G; A) \cong \text{Der}(G, A)/\text{PDer}(G, A)$.

Proof The sequence $0 \to \mathfrak{I} \to \mathbb{Z}G \to \mathbb{Z} \to 0$ induces an exact sequence

$$0 \;\longrightarrow\; \text{Hom}_G(\mathbb{Z}, A) \;\longrightarrow\; \text{Hom}_G(\mathbb{Z}G, A) \;\longrightarrow\; \text{Hom}_G(\mathfrak{I}, A) \;\longrightarrow\; \text{Ext}^1_{\mathbb{Z}G}(\mathbb{Z}, A) \;\longrightarrow\; 0.$$

$$\| \qquad\qquad \| \qquad\qquad \| \qquad\qquad \|$$

$$A^G \;\hookrightarrow\; A \;\longrightarrow\; \text{Der}(G, A) \;\longrightarrow\; H^1(G; A)$$

Now $A \to \text{Hom}_G(\mathfrak{I}, A)$ sends $a \in A$ to the map φ sending $(g - 1)$ to $(g - 1)a$. Under the identification of $\text{Hom}_G(\mathfrak{I}, A)$ with $\text{Der}(G, A)$, φ corresponds to the principal derivation $D_\varphi = D_a$. Hence the image of A in $\text{Der}(G, A)$ is $\text{PDer}(G, A)$, as claimed.   ◇

Corollary 6.4.6 *If A is a trivial G-module,*

$$H^1(G; A) \cong \text{Der}(G, A) \cong \text{Hom}_{\textbf{Groups}}(G, A).$$

Proof $\text{PDer}(G, A) \cong A/A^G = 0$ and a derivation is a group homomorphism.
     ◇

Hilbert's Theorem 90 6.4.7 (Multiplicative version) *Let $K \subset L$ be a finite Galois extension of fields, with Galois group G. Let L^* denote the group of units in L. Then L^* is a G-module, and $H^1(G; L^*) = 0$.*

Proof Using multiplicative notation, a derivation is a map $\theta: G \to L^*$ such that $\theta(gh)/\theta(g) = g \cdot \theta(h)$. These are "Noether's equations"; the usual Theorem 90 [BAI, p. 286] states that if θ satisfies Noether's equations then $\theta(g) = (g \cdot x)/x$ for some $x \in L^*$, that is, θ is a principal derivation. $\diamond$

Example 6.4.8 (Cyclic Galois extensions) Hilbert originally proved his Theorem 90 for cyclic field extensions in his 1897 report, *Theorie der Algebraische Zahlkörper*. Let $K \subset L$ be a cyclic Galois extension of fields, with Galois group C_m. The *norm* Nx of an element $x \in L$ is the product $\Pi g(x)$; as $H^1(C_m; L^*) = \{x : Nx = 1\}/(\sigma - 1)L^*$ (see 6.2.2), we may rephrase Hilbert's Theorem 90 as stating that whenever $Nx = 1$, there is a $y \in L$ such that $x = (\sigma y)/y$. Since $H^2(C_m; L^*) = L^{*G}/\{Nx : x \in L^*\} = K^*/NL^*$,

$$1 \to K^* \to L^* \xrightarrow{\sigma-1} L^* \xrightarrow{N} K^* \to H^2(C_m; L^*) \to 1$$

is exact. (See exercise 6.2.1.) For the cyclic extension $\mathbb{R} \subset \mathbb{C}$ it is easy to calculate that $H^2(C_2; \mathbb{C}^*) \cong \mathbb{Z}/2$, so the higher analogue of the additive version of Theorem 90 fails for $H^*(G; L^*)$.

Remark The group $H^2(G; L^*)$ is usually nonzero. We will return to this topic in 6.6.11, identifying $H^2(G; L^*)$ with the *relative Brauer group* $Br(L/K)$ of all simple algebras Λ with center K and $\dim_K \Lambda = n^2$, $n = [L : K]$, such that $\Lambda \otimes_K L$ is the matrix ring $M_n(L)$. The nonzero element of $Br(\mathbb{C}/\mathbb{R}) \cong H^2(C_2; \mathbb{C}^*) \cong \mathbb{Z}/2$ corresponds to the 4-dimensional quaternion algebra $\mathbb{H}$, which has center $\mathbb{R}$ and $\mathbb{H} \otimes_\mathbb{R} \mathbb{C} \cong M_2(\mathbb{C})$.

In order to indicate the historical origins of the terminology "crossed homomorphism," we introduce the *semidirect product* $A \rtimes G$ of a group G with a G-module A. $A \rtimes G$ is a group whose underlying set is the product $A \times G$, and whose multiplication is given by the formula

$$(a, g) \cdot (b, h) = (a + gb, gh).$$

The semidirect product contains $A = A \times 1$ as a normal subgroup. It also contains the subgroup $0 \times G$, which maps isomorphically onto the quotient $G = (A \rtimes G)/A$.

Definition 6.4.9 If σ is an automorphism of $A \rtimes G$, we say that σ *stabilizes A and G* if $\sigma(a) = a$ for $a \in A$ and the induced automorphism on $G \cong (A \rtimes G)/A$ is the identity.

Exercise 6.4.2 If D is a derivation of G in A, show that σ_D, defined by

$$\sigma_D(a, g) = (a + D(g), g),$$

is an automorphism of $A \rtimes G$ stabilizing A and G, and that $\text{Der}(G, A)$ is isomorphic to the subgroup of $\text{Aut}(A \rtimes G)$ consisting of automorphisms stabilizing A and G. Show that $\text{PDer}(G, A)$ corresponds to the inner automorphisms of $A \rtimes G$ obtained by conjugating by elements of A, with the principal derivation D_a given by $D_a(g) = a^{-1}ga$. Conclude that $H^1(G; A)$ is the group of outer automorphisms stabilizing A and G.

Example 6.4.10 (Dihedral groups) Let C_2 act on the cyclic group $\mathbb{Z}/m = C_m$ by $\sigma(a) = -a$. The semidirect product $C_m \rtimes C_2$ is the *dihedral group* D_m of symmetries of the regular m–gon. Our calculations in section 6.2 show that $H^1(C_2; C_m) \cong C_m/2C_m$. If m is even, D_m has an outer ($=$ *not* inner) automorphism with $\varphi(0, \sigma) = (1, \sigma)$. If m is odd, every automorphism of D_m is inner.

6.5 The Bar Resolution

There are two canonical resolutions B_* and B_*^u of the trivial G-module $\mathbb{Z}$ by free left $\mathbb{Z}G$-modules, called the *normalized* and *unnormalized bar resolutions*, respectively. We shall now describe these resolutions.

$$(*) \qquad 0 \leftarrow \mathbb{Z} \xleftarrow{\epsilon} B_0 \xleftarrow{d} B_1 \xleftarrow{d} B_2 \xleftarrow{d} \cdots .$$

$$(**) \qquad 0 \leftarrow \mathbb{Z} \xleftarrow{\epsilon} B_0^u \xleftarrow{d} B_1^u \xleftarrow{d} B_2^u \xleftarrow{d} \cdots .$$

B_0 and B_0^u are $\mathbb{Z}G$. Letting the symbol [ ] denote $1 \in \mathbb{Z}G$, the map $\epsilon: B_0 \to \mathbb{Z}$ sends [ ] to 1. For $n \geq 1$, B_n^u is the free $\mathbb{Z}G$-module on the set of all symbols $[g_1 \otimes \cdots \otimes g_n]$ with $g_i \in G$, while B_n is the free $\mathbb{Z}G$-module on the (smaller) set of all symbols $[g_1| \cdots |g_n]$ with the $g_i \in G - \{1\}$. We shall frequently identify B_n with the quotient of B_n^u by the submodule S_n generated by the set of all symbols $[g_1 \otimes \cdots \otimes g_n]$ with some g_i equal to 1.

Definition 6.5.1 For $n \geq 1$, define the differential $d: B_n^u \to B_{n-1}^u$ to be $d = \sum_{i=0}^{n}(-1)^i d_i$, where:

$$d_0([g_1 \otimes \cdots \otimes g_n]) = g_1[g_2 \otimes \cdots \otimes g_n];$$
$$d_i([g_1 \otimes \cdots \otimes g_n]) = [g_1 \otimes \cdots \otimes g_i g_{i+1} \otimes \cdots \otimes g_n] \quad \text{for } i = 1, \ldots, n-1;$$
$$d_n([g_1 \otimes \cdots \otimes g_n]) = [g_1 \otimes \cdots \otimes g_{n-1}].$$

The differential for B_* is given by formulas similar for those on B_*^u, except that for $i = 1, \ldots, n - 1$

$$d_i([g_1|\cdots|g_n]) = \begin{cases} [g_1|\cdots|g_i g_{i+1}|\cdots|g_n] & \text{when } g_i g_{i+1} \neq 1 \\ 0 & \text{when } g_i g_{i+1} = 1. \end{cases}$$

To avoid the clumsy case when $g_i g_{i+1} = 1$, we make the convention that $[g_1|\cdots|g_n] = 0$ if any $g_i = 1$. *Warning:* With this convention, the above formula for $d_i([g_1|\cdots])$ does not hold when g_i or $g_{i+1} = 1$; the formula for the alternating sum d does hold because the d_i and d_{i-1} terms cancel.

Examples 6.5.2

1. The image of the map $d: B_1 \to B_0$ is the augmentation ideal $\mathfrak{I}$ because $d([g]) = g[\,] - [\,] = (g - 1)[\,]$. Therefore (∗) and (∗∗) are exact at B_0.
2. $d([g|h]) = g[h] - [gh] + [g]$.
3. $d([f|g|h]) = f[g|h] - [fg|h] + [f|gh] - [f|g]$.
4. If $G = C_2$, then $B_n = \mathbb{Z}G$ for all n on $[\sigma|\cdots|\sigma]$ and (∗) is familiar from 6.2.1:

$$0 \leftarrow \mathbb{Z} \xleftarrow{\epsilon} \mathbb{Z}G \xleftarrow{\sigma-1} \mathbb{Z}G \xleftarrow{\sigma+1} \mathbb{Z}G \xleftarrow{\sigma-1} \cdots.$$

Exercises 6.5.1

1. Show that $d \circ d = 0$, so that B_*^u is a chain complex. *Hint:* If $i \leq j - 1$, show that $d_i d_j = d_{j-1} d_i$.
2. Show that $d(S_n)$ lies in S_{n-1}, so that S_* is a subcomplex of B_*^u.
3. Conclude that B_* is a quotient chain complex of B_*^u.

Theorem 6.5.3 *The sequences* (∗) *and* (∗∗) *are exact. Thus both B_* and B_*^u are resolutions of $\mathbb{Z}$ by free left $\mathbb{Z}G$-modules.*

Proof It is enough to prove that (∗) and (∗∗) are split exact as chain complexes of abelian groups. As the proofs are the same, we give the proof in the B_* case. Consider the abelian group maps s_n determined by

$$s_{-1}: \mathbb{Z} \to B_0, \qquad s_{-1}(1) = [\,];$$

$$s_n: B_n \to B_{n+1}, \qquad s_n(g_0[g_1|\cdots|g_n]) = [g_0|g_1|\cdots|g_n].$$

Visibly, $\epsilon s_{-1} = 1$ and $d s_0 + s_{-1}\epsilon$ is the identity map on B_0. If $n \geq 1$, the first term of $d s_n(g_0[g_1|\cdots|g_n])$ is $g_0[g_1|\cdots|g_n]$, and the remaining terms are exactly the terms of $s_{n-1}d(g_0[g_1|\cdots|g_n])$ with a sign change. This yields the final identity $ds_n + s_{n-1}d = 1$ needed to show that $\{s_n\}$ forms a chain contraction of $(*)$. $\diamond$

Application 6.5.4 (Homology) For every right G-module A, $H_*(G; A)$ is the homology of the chain complex $A \otimes B_*$. (If A is a left G-module, we must take the homology of $B'_* \otimes A$, where B'_* is the mirror image bar resolution.) In particular, we see that $H_1(G; \mathbb{Z})$ is the quotient of the free abelian group on the symbols $[g]$, $g \in G$, by the relations that $[1] = 0$ and $[f] + [g] = [fg]$ for all $f, g \in G$. This recovers the calculation in 6.1.11 that

$$H_1(G; \mathbb{Z}) = G/[G, G].$$

Application 6.5.5 (Cohomology) If A is a left G-module, $H^*(G; A)$ is the cohomology of either $\mathrm{Hom}_G(B^u_*, A)$ or $\mathrm{Hom}_G(B_*, A)$. An n-*cochain* is a set map φ from $G^n = G \times \cdots \times G$ to A; elements of $\mathrm{Hom}_G(B^u_n, A)$ are just n-cochains. A cochain φ is *normalized* if $\varphi(g_1, \cdots)$ vanishes whenever some $g_i = 1$; these are the elements of $\mathrm{Hom}_G(B_n, A)$. The differential $d\varphi$ of an n-cochain is the $(n + 1)$-cochain

$$(d\varphi)(g_0, \cdots, g_n) = g_0\varphi(g_1, \cdots, g_n) + \sum (-1)^i \varphi(\cdots, g_i g_{i+1}, \cdots)$$
$$+ \varphi(g_0, \cdots, g_{n-1}).$$

The n-cochains such that $d\varphi = 0$ are n-*cocycles*, and the n-cochains $d\varphi$ are called n-*coboundaries*. We write $Z^n(G; A)$ and $B^n(G; A)$ for the groups of all n-cocycles and n-coboundaries, respectively. Thus $H^n(G; A) = Z^n(G; A)/B^n(G; A)$.

Example 6.5.6 A 0-cochain is a map $1 \to A$, that is, an element of A. If $a \in A$, then da is the map $G \to A$ sending g to $ga - a$. Thus a is a 0-cocycle iff $a \in A^G$, and the set $B^1(G; A)$ of 1-coboundaries is the set $\mathrm{PDer}(G, A)$ of principal derivations.

The set $Z^1(G; A)$ of 1-cocycles is $\mathrm{Der}(G, A)$, because a 1-cocycle is a function D with $D(1) = 0$ and $gD(h) - D(gh) + D(g) = D(d[g|h]) = 0$. Therefore, the bar resolution provides a direct proof of the isomorphism $H^1(G; A) \cong \mathrm{Der}(G, A)/\mathrm{PDer}(G, A)$ of 6.4.5.

Example 6.5.7 $B^2(G; A)$ is the set of all $\psi: G \times G \to A$ such that $\psi(1, g) = \psi(g, 1)$ and

$$\psi(f, g) = \beta(d[f|g]) = f \cdot \beta(g) - \beta(fg) + \beta(f) \quad \text{for some } \beta: G \to A.$$

$Z^2(G; A)$ is the set of all 2-cochains $\psi: G \times G \to A$ such that $\psi(1, g) = \psi(g, 1)$ and

$$f \cdot \psi(g, h) - \psi(fg, h) + \psi(f, gh) - \psi(f, g) = 0 \quad \text{for every } f, g, h \in G.$$

Theorem 6.5.8 *Let G be a finite group with m elements. Then for $n \neq 0$ and every G-module A, both $H_n(G; A)$ and $H^n(G; A)$ are annihilated by m, that is, they are $\mathbb{Z}/m$-modules.*

Proof Let η denote the endomorphism of B_*, which is multiplication by $(m - N)$ on B_0 and multiplication by m on B_n, $n \neq 0$. We claim that η is null homotopic. Applying $A \otimes$ or $\text{Hom}(-, A)$, will then yield a null homotopic map, which must become zero upon taking homology, proving the theorem.
   Define $\nu_n: B_n \to B_{n+1}$ by the formula

$$\nu_n([g_1| \cdots |g_n]) = (-1)^{n+1} \sum_{g \in G} [g_1| \cdots |g_n|g].$$

Setting $\omega = [g_1| \cdots |g_n]$ and $\epsilon = (-1)^{n+1}$, we compute for $n \neq 0$

$$d\nu_n(\omega) = \epsilon \sum \{g_1[\cdots |g] + \sum (-1)^i [\cdots |g_i g_{i+1}| \cdots |g] - \epsilon[\cdots |g_{n-1}|g_n g] + \epsilon \omega\}$$

$$\nu_{n-1} d(\omega) = -\epsilon \sum \{g_1[\cdots |g] + \sum (-1)^i [\cdots |g_i g_{i+1}| \cdots |g] - \epsilon[\cdots |g_{n-1}|g]\}.$$

As the sums over all $g \in G$ of $[\cdots |g_n g]$ and $[\cdots |g]$ agree, we see that $(d\nu + \nu d)(\omega)$ is $\epsilon^2 \sum \omega = m\omega$. Now $d\nu_0([\,]) = d(-\sum[g]) = (m - N)[\,]$, where $N = \sum g$ is the norm. Thus $\{\nu_n\}$ provides the chain contraction needed to make η null homotopic.   ◇

Corollary 6.5.9 *Let G be a finite group of order m, and A a G-module. If A is a vector space over $\mathbb{Q}$, or a $\mathbb{Z}[\frac{1}{m}]$-module, then $H_n(G; A) = H^n(G; A) = 0$ for $n \neq 0$. (We had already proven this result in 6.1.10 using a more abstract approach.)*

Corollary 6.5.10 *If G is a finite group and A is a finitely generated G-module, then $H_n(G; A)$ and $H^n(G; A)$ are finite abelian groups for all $n \neq 0$.*

Proof Each $A \otimes_{\mathbb{Z}G} B_n$ and $\text{Hom}_G(B_n, A)$ is a finitely generated abelian group. Hence $H_n(G; A)$ and $H^n(G; A)$ are finitely generated $\mathbb{Z}/m$-modules when $n \neq 0$.   ◇

Shuffle Product 6.5.11 When G is an abelian group, the normalized bar complex B_* is actually a graded-commutative differential graded algebra (or DG-algebra; see 4.5.2) under a product called the *shuffle product*. If $p \geq 0$ and $q \geq 0$ are integers, a (p, q)-*shuffle* is a permutation σ of the set $\{1, \cdots, p+q\}$ of integers in such a way that $\sigma(1) < \sigma(2) < \cdots < \sigma(p)$ and $\sigma(p+1) < \cdots < \sigma(p+q)$. The name comes from the fact that the (p, q)-shuffles describe all possible ways of shuffling a deck of $p + q$ cards, after first cutting the deck between the p and $(p+1)^{st}$ cards.

If G is any group, we define the *shuffle product* $*: B_p \otimes_{\mathbb{Z}} B_q \to B_{p+q}$ by

$$a[g_1|\cdots|g_p] * b[g_{p+1}|\cdots|g_{p+q}] = \sum_\sigma (-1)^\sigma ab[g_{\sigma^{-1}1}|g_{\sigma^{-1}2}|\cdots|g_{\sigma^{-1}(p+q)}],$$

where the summation is over all (p, q)-shuffles σ. The shuffle product is clearly bilinear, and $[\ ] *[g_1|\cdots|g_q] = [g_1|\cdots|g_q]$, so B_* is a graded ring with unit $[\ ]$, and the inclusion of $\mathbb{Z}G = B_0$ in B_* is a ring map.

Examples 6.5.12 $[g] * [h] = [g|h] - [h|g]$, and

$$[f] * [g|h] = [f|g|h] - [g|f|h] + [g|h|f].$$

Exercise 6.5.2

1. Show that the shuffle product is associative. Conclude that B_* and $\mathbb{Z} \otimes_{\mathbb{Z}G} B_*$ are associative rings with unit.
2. Recall (from 4.5.2) that a graded ring R_* is called *graded-commutative* if $x * y = (-1)^{pq} y * x$ for all $x \in R_p$ and $y \in R_q$. Show that B_* is graded-commutative if G is an abelian group.

Theorem 6.5.13 *If G is an abelian group, then B_* is a differential graded algebra.*

Proof We have already seen in exercise 6.5.2 that B_* is an associative graded-commutative algebra, so all that remains is to verify that the Leibnitz identity 4.5.2 holds, that is, that

$$d(x * y) = (dx) * y + (-1)^p x * dy,$$

where x and y denote $a[g_1|\cdots|g_p]$ and $b[g_{p+1}|\cdots|g_{p+q}]$, respectively. Contained in the expansion of $x*y$, we find the expansions for $(dx)*y$ and $(-1)^p x*dy$. The remaining terms are paired for each $i \leq p < j$, and each (p, q)-shuffle σ which puts i immediately just before j, as

$$(-1)^\sigma ab[\cdots|g_i g_j|\cdots] \text{ and } (-1)^{\sigma+1} ab[\cdots|g_j g_i|\cdots].$$

(The terms with j just before i arise from the composition of σ with a transposition.) As G is abelian, these terms cancel.   ◇

Corollary 6.5.14 *For every abelian group G and commutative $\mathbb{Z}G$-algebra R, $H_*(G; R)$ is a graded-commutative ring.*

Proof $B_* \otimes_{\mathbb{Z}G} R$ is a graded-commutative DG-algebra (check this!); we saw in exercise 4.5.1 that the homology of such a DG-algebra is a graded-commutative ring.   ◇

6.6 Factor Sets and H^2

The origins of the theory of group cohomology go back—at least in nascent form—to the landmark 1904 paper [Schur]. For any field k, the *projective linear group* $PGL_n(k)$ is the quotient of the *general linear group* $GL_n(k)$ by the diagonal copy of the units k^* of k. If G is any group, a group map $\rho: G \to PGL_n(k)$ is called a *projective representation* of G. The pullback

$$E = \{(\alpha, g) \in GL_n(k) \times G : \bar{\alpha} = \rho(g)\}$$

is a group, containing $k^* \cong k^* \times 1$, and there is a diagram

$$
\begin{array}{ccccccccc}
1 & \longrightarrow & k^* & \longrightarrow & E & \longrightarrow & G & \longrightarrow & 1 \\
& & \| & & \downarrow{\rho'} & & \downarrow{\rho} & & \\
1 & \longrightarrow & k^* & \longrightarrow & GL_n(k) & \longrightarrow & PGL_n(k) & \longrightarrow & 1.
\end{array}
$$

Schur's observation was that the projective representation ρ of G may be replaced by an ordinary representation ρ' if we are willing to replace G by the larger group E, and it raises the issue of when E is a semidirect product, so that there is a representation $G \hookrightarrow E \to GL_n(k)$ lifting the projective representation. (See exercise 6.6.5.)

Definition 6.6.1 A *group extension* (of G by A) is a short exact sequence

$$0 \to A \to E \xrightarrow{\pi} G \to 1$$

of groups in which A is an abelian group; it is convenient to write the group law in A as addition, whence the term "0" on the left. The extension *splits* if $\pi: E \to G$ has a section $\sigma: G \to E$.

Given a group extension of G by A, the group G acts on A by conjugation in E; to avoid notational confusion, we shall write $^g a$ for the conjugate gag^{-1} of a in E. This induced action makes A into a G-module.

Exercise 6.6.1 Show that an extension $0 \to A \to E \to G \to 1$ splits if and only if E is isomorphic to the semidirect product $A \rtimes G$ (6.4.9).

Exercise 6.6.2 Let $G = \mathbb{Z}/2$ and $A = \mathbb{Z}/3$. Show that there are two extensions of G by A, the (split) product $\mathbb{Z}/6 = A \times G$ and the dihedral group D_3. These extensions correspond to the two possible G-module structures on A.

Exercise 6.6.3 (Semidirect product) Let A be a G-module and form the split extension

$$0 \to A \to A \rtimes G \to G \to 1.$$

Show that the induced action of G on A agrees with the G-module structure.

Extension Problem 6.6.2 Given a G-module A, we would like to determine how many extensions of G by A exist in which the induced action of G on A agrees with the given G-module structure, that is, in which $^g a = g \cdot a$.

In order to avoid duplication and set-theoretic difficulties, we say that two extensions $0 \to A \to E_i \to G \to 1$ are *equivalent* if there is an isomorphism $\varphi : E_1 \cong E_2$ so

$$
\begin{array}{ccccccccc}
0 & \longrightarrow & A & \longrightarrow & E_1 & \longrightarrow & G & \longrightarrow & 0 \\
  &   & \| &   & \varphi\downarrow &   & \| &   &   \\
0 & \longrightarrow & A & \longrightarrow & E_2 & \longrightarrow & G & \longrightarrow & 0
\end{array}
$$

commutes, and we ask for the set of equivalence classes of extensions. Here is the main result of this section:

Classification Theorem 6.6.3 *The equivalence classes of extensions are in 1–1 correspondence with the cohomology group $H^2(G; A)$.*

Here is the canonical approach to classifying extensions. Suppose given an extension $0 \to A \to E \xrightarrow{\pi} G \to 1$; choose a set map $\sigma: G \to E$ such that $\sigma(1)$ is the identity element of E and $\pi\sigma(g) = g$ for all $g \in G$. Both $\sigma(gh)$ and $\sigma(g)\sigma(h)$ are elements of E mapping to $gh \in G$, so their difference lies in A. We define

$$[g, h] = \sigma(g)\sigma(h)\sigma(gh)^{-1}.$$

Note that $[g, h]$ is an element of A that depends on our choice of E and σ.

Definition 6.6.4 The set function $[\ ]: G \times G \to A$ defined above is called the *factor set* determined by E and σ.

Lemma 6.6.5 *If two extensions* $0 \to A \to E_i \to G \to 1$ *with maps* $\sigma_i: G \to E_i$ *yield the same factor set, then the extensions are equivalent.*

Proof The maps σ_i give a concrete set-theoretic identification $E_1 \cong A \times G \cong E_2$; we claim that it is a group homomorphism. Transporting the group structure from E_1 to $A \times G$, we see that the products $(a, 1) \cdot (b, 1) = (a + b, 1)$, $(a, 1) \cdot (0, g) = (a, g)$, and $(0, g) \cdot (a, 1) = (ga, g)$ are fixed. Therefore the group structure on $A \times G$ is completely determined by the products $(1, g) \cdot (1, h)$, which by construction is $([g, h], gh)$. By symmetry, this is also the group structure induced from E_2, whence the claim. $\diamond$

Corollary 6.6.6 *If E were a semidirect product and σ were a group homomorphism, then the factor set would have* $[g, h] = 0$ *for all* $g, h \in G$. *Hence if an extension has* $[\ ] = 0$ *as a factor set, the extension must be split.*

   Recall (6.5.7) that a *(normalized) 2-cocycle* is a function $[\ ] : G \times G \to A$ such that

   1. $[g, 1] = [1, g] = 0$ for all $g \in G$.
   2. $f[g, h] - [fg, h] + [f, gh] - [f, g] = 0$ for all $f, g, h \in G$.

Theorem 6.6.7 *Let A be a G-module. A set function $[\ ] : G \times G \to A$ is a factor set iff it is a normalized 2-cocycle, that is, an element of $Z^2(G, A)$.*

Remark Equations (1) and (2) are often given as the definition of factor set.

Proof If $[\ ]$ is a factor set, formulas (1) and (2) hold because $\sigma(1) = 1$ and multiplication in E is associative (check this!).
   Conversely, suppose given a normalized 2-cocycle, that is, a function $[\ ]$ satisfying (1) and (2). Let E be the set $A \times G$ with composition defined by

$$(a, g) \cdot (b, h) = (a + (g \cdot b) + [g, h], gh).$$

This product has $(0,1)$ as identity element, and is associative by (2). Since

$$(a, g) \cdot (-g^{-1} \cdot a - g^{-1} \cdot [g, g^{-1}], g^{-1}) = (0, 1),$$

E is a group. Evidently $A \times 1$ is a subgroup isomorphic to A and $E/A \times 1$ is G. Thus $0 \to A \to E \to G \to 1$ is an extension, and the factor set arising from $G \cong 0 \times G \hookrightarrow E$ is our original function [ ]. (Check this!) $\diamond$

Change of Based Section 6.6.8 Fix an extension $0 \to A \to E \xrightarrow{\pi} G \to 1$. A *based section* of π is a map $\sigma: G \to E$ such that $\sigma(1) = 1$ and $\pi \sigma(g) = g$ for all g. Let σ' be another based section of π. Since $\sigma'(g)$ is in the same coset of A as $\sigma(g)$, there is an element $\beta(g) \in A$ so that $\sigma'(g) = \beta(g)\sigma(g)$. The factor set corresponding to σ' is

$$[g, h]' = \beta(g)\sigma(g)\beta(h)\sigma(h)\sigma(gh)^{-1}\beta(gh)^{-1}$$
$$= \beta(g) + \sigma(g)\beta(h)\sigma(g)^{-1} + \sigma(g)\sigma(h)\sigma(gh)^{-1} - \beta(gh)$$
$$= [g, h] + \beta(g) - \beta(gh) + g \cdot \beta(h).$$

The difference $[g, h]' - [g, h]$ is the coboundary $d\beta(g, h) = \beta(g) - \beta(gh) + g \cdot \beta(h)$. Therefore, although the 2-cocyle [ ] is not unique, its class in $H^2(G; A) = Z^2(G, A)/B^2(G, A)$ is independent of the choice of based section. Therefore the factor set of an extension yields a well-defined set map Ψ from the set of equivalence classes of extensions to the set $H^2(G; A)$.

Proof of Classification Theorem Analyzing the above construction, we see that the formula $\sigma'(g) = \beta(g)\sigma(g)$ gives a 1–1 correspondence between the set of all possible based sections σ' and the set of all maps $\beta: G \to A$ with $\beta(1)$. If two extensions have the same cohomology class, then an appropriate choice of based sections will yield the same factor sets, and we have seen that in this case the extensions are equivalent. Therefore Ψ is an injection. We have also seen that every 2-cocycle [ ] is a factor set; therefore Ψ is onto. $\diamond$

Exercise 6.6.4 Let $\rho: G \to H$ be a group homomorphism and A an H-module. Show that there is a natural map $Z^2\rho$ on 2-cocycles from $Z^2(H, A)$ to $Z^2(G, A)$ and that $Z^2\rho$ induces a map $\rho^*: H^2(H; A) \to H^2(G; A)$. Now let $0 \to A \to E \xrightarrow{\pi} H \to 1$ be an extension and let E' denote the pullback $E \times_H G = \{(e, g) \in E \times G : \pi(e) = \rho(g)\}$. Show that ρ^* takes the class of the extension E to the class of the extension E'.

$$
\begin{array}{ccccccccc}
0 & \longrightarrow & A & \longrightarrow & E' & \longrightarrow & G & \longrightarrow & 1 \\
 & & \| & & \downarrow & & \downarrow{\scriptstyle\rho} & & \\
0 & \longrightarrow & A & \longrightarrow & E & \longrightarrow & H & \longrightarrow & 1
\end{array}
$$

Exercise 6.6.5 (Schur) For any field k and any n, let γ denote the class in $H^2(PGL_n(k); k^*)$ corresponding to the extension $1 \to k^* \to GL_n(k) \to PGL_n(k) \to 1$. If $\rho: G \to PGL_n(k)$ is a projective representation, show that ρ lifts to a linear representation $G \to GL_n(k)$ if and only if $\rho^*(\gamma) = 0$ in $H^2(G; k^*)$.

Exercise 6.6.6 If k is an algebraically closed field, and μ_m denotes the subgroup of k^* consisting of all m^{th} roots of unity in k, show that $H^2(G; \mu_m) \cong H^2(G; k^*)$ for every finite group G of automorphisms of k order m. *Hint:* Consider the "Kummer" sequence $0 \to \mu_m \to k^* \xrightarrow{m} k^* \to 1$.

Theorem 6.6.9 (Schur-Zassenhaus) *If m and n are relatively prime, any extension $0 \to A \to E \to G \to 1$ of a group G of order m by a group A of order n is split.*

Proof If A is abelian, the extensions are classified by the groups $H^2(G; A)$, one group for every G-module structure on A. These are zero as A is a $\mathbb{Z}[\frac{1}{m}]$-module (6.1.10).

In the general case, we induct on n. It suffices to prove that E contains a subgroup of order m, as such a subgroup must be isomorphic to G under $E \to G$. Choose a prime p dividing n and let S be a p-Sylow subgroup of A, hence of E. Let Z be the center of S; $Z \neq 1$ [BAI, p. 75]. A counting argument shows that m divides the order of the normalizer N of Z in E. Hence there is an extension $0 \to (A \cap N) \to N \to G \to 1$. If $N \neq E$, this extension splits by induction, so there is a subgroup of N (hence of E) isomorphic to G. If $N = E$, then $Z \lhd E$ and the extension $0 \to A/Z \to E/Z \to G \to 1$ is split by induction. Let E' denote the set of all $x \in E$ mapping onto the subgroup G' of E/Z isomorphic to G. Then E' is a subgroup of E, and $0 \to Z \to E' \to G' \to 1$ is an extension. As Z is abelian, there is a subgroup of E', hence of E, isomorphic to G'. $\diamond$

Application 6.6.10 (Crossed product algebras) Let L/K be a finite Galois field extension with $G = Gal(L/K)$. Given a factor set $[\ ]$ of G in L^*, we can form a new associative K-algebra Λ on the left L-module $L[G]$ using the "crossed" product:

$$\left(\sum_{g \in G} a_g g\right) \times \left(\sum_{h \in G} b_h g\right) = \sum_{g,h} [g, h] a_g (g \cdot b_h)(gh), \quad (a_g, b_h \in L).$$

It is a straightforward matter to verify that the factor set condition is equivalent to the associativity of the product $\times$ on Λ. Λ is called the *crossed product algebra* of L and G over K with respect to [ ]. Note that L is a subring of Λ and that $\dim_K \Lambda = n^2$, where $n = [L : K]$. As we choose to not become sidetracked, we refer the reader to [BAII, 8.4] for the following facts:

1. Λ is a simple ring with center K and $\Lambda \otimes_K L \cong M_n(L)$. By Wedderburn's Theorem there is a division algebra Δ with center K such that $\Lambda \cong M_d(\Delta)$.
2. Every simple ring Λ with center K and $\Lambda \otimes_K L \cong M_n(L)$ is isomorphic to a crossed product algebra of L and G over K for some factor set [ ].
3. Two factor sets yield isomorphic crossed product algebras if and only if they differ by a coboundary.
4. The factor set [ ] $= 1$ yields the matrix ring $M_n(K)$, where $n = [L : K]$.
5. If Λ and Λ' correspond to factor sets [ ] and [ ]', then $\Lambda \otimes_K \Lambda' \cong M_n(\Lambda'')$, where Λ'' corresponds to the factor set [ ] $+$ [ ]'.

Definition 6.6.11 The *relative Brauer group* $Br(L/K)$ is the set of all simple algebras Λ with center K such that $\Lambda \otimes_K L \cong M_n(L)$, $n = [L : K]$. By Wedderburn's Theorem it is also the set of division algebras Δ with center K and $\Delta \otimes_K L \cong M_r(L)$, $r^2 = \dim_K \Delta$. By (1)–(3), the crossed product algebra construction induces an isomorphism

$$H^2(Gal(L/K); L^*) \xrightarrow{\cong} Br(L/K).$$

The induced group structure $[\Lambda][\Lambda'] = [\Lambda'']$ on $Br(L/K)$ is given by (4) and (5).

Crossed Modules and H^3 6.6.12 Here is an elementary interpretation of the cohomology group $H^3(G; A)$. Consider a 4-term exact sequence with A central in N

$(*)$ $\qquad\qquad 0 \to A \to N \xrightarrow{\alpha} E \xrightarrow{\pi} G \to 1,$

and choose a based section $\sigma: G \to E$ of π; as in the theory of factor sets, the map [ ]: $G \times G \to \ker(\pi)$ defined by $[g, h] = \sigma(g)\sigma(h)\sigma(gh)^{-1}$ satisfies a nonabelian cocycle condition

$$[f, g][fg, h] = {}^{\sigma(f)}[g, h] \ [f, gh],$$

where ${}^{\sigma(f)}[g, h]$ denotes the conjugate $\sigma(f)[g, h]\sigma(f)^{-1}$. Since $\ker(\pi) = \alpha(N)$, we can lift each $[f, g]$ to an element $[[f, g]]$ of N and ask if an analogue of the cocycle condition holds—for some interpretation of ${}^{\sigma(f)}[[g, h]]$. This leads to the notion of crossed module.

A *crossed module* is a group homomorphism $\alpha: N \to E$ together with an action of E on N (written $(e, n) \mapsto {}^e n$) satisfying the following two conditions:

1. For all $m, n \in N$, ${}^{\alpha(m)}n = mnm^{-1}$.
2. For all $e \in E$, $n \in N$, $\alpha({}^e n) = e\alpha(n)e^{-1}$.

For example, the canonical map $N \to \text{Aut}(N)$ is a crossed module for any group N. Crossed modules also arise naturally in topology: given a Serre fibration $F \to E \to B$, the map $\pi_1(F) \to \pi_1(E)$ is a crossed module. (This was the first application of crossed modules and was discovered in 1949 by J. H. C. Whitehead.)

Given a crossed module $N \xrightarrow{\alpha} E$, we set $A = \ker(\alpha)$ and $G = \text{coker}(\alpha)$; G is a group because $\alpha(N)$ is normal in E by (2). Note that A is in the center of N and G acts on A, so that A is a G-module, and we have a sequence $(*)$.

Returning to our original situation, but now assuming that $N \to E$ is a crossed module, the failure of $[[f, g]]$ to satisfy the cocycle condition is given by the function $c: G^3 \to A$ defined by the equation

$$c(f, g, h)[[f, g]]\, [[fg, h]] = {}^{\sigma(f)}[[g, h]]\ \ [[f, gh]].$$

The reader may check that c is a 3-cocycle, whose class in $H^3(G; A)$ is independent of the choices of σ and $[[f, g]]$. As with Yoneda extensions (3.4.6), we say that $(*)$ is *elementarily equivalent* to the crossed module

$$0 \to A \to N' \to E' \to G \to 1$$

if there is a morphism of crossed modules between them, that is, a commutative diagram compatible with the actions of E and E' on N and N'

$$
\begin{array}{ccccccccc}
0 & \longrightarrow & A & \longrightarrow & N & \xrightarrow{\alpha} & E & \longrightarrow & G & \longrightarrow & 1 \\
  &   & \| &   & \downarrow & & \downarrow & & \| & & \\
0 & \longrightarrow & A & \longrightarrow & N' & \longrightarrow & E' & \longrightarrow & G & \longrightarrow & 1.
\end{array}
$$

Since our choices of σ and $[[f, g]]$ for $(*)$ dictate choices for $N' \to E'$, these choices clearly determine the same 3-cocycle c. This proves half of the following theorem; the other half may be proven by modifying the proof of the corresponding Yoneda Ext Theorem in [BX, section 7.5].

Crossed Module Classification Theorem 6.6.13 *Two crossed modules with kernel A and cokernel G determine the same class in $H^3(G; A)$ if and only if* *they are equivalent (under the equivalence relation generated by elementary equivalence). In fact, there is a 1–1 correspondence for each G and A:*

$$\left\{ \begin{array}{c} \text{equivalence classes of crossed modules} \\ 0 \to A \to N \xrightarrow{\alpha} E \to G \to 1 \end{array} \right\} \longleftrightarrow \text{elements of } H^3(G; A).$$

6.7 Restriction, Corestriction, Inflation, and Transfer

If G is fixed, $H_*(G; A)$ and $H^*(G; A)$ are covariant functors of the G-module A. We now consider them as functors of the group G.

Definition 6.7.1 If $\rho: H \to G$ is a group map, the forgetful functor $\rho^{\#}$ from G–**mod** to H–**mod** is exact. For every G-module A, there is a natural surjection $(\rho^{\#}A)_H \to A_G$ and a natural injection $A^G \to (\rho^{\#}A)^H$. These two maps extend uniquely to the two morphisms $\rho_* = \text{cor}_H^G$ (called *corestriction*) and $\rho^* = \text{res}_H^G$ (called *restriction*) of δ-functors:

$$\text{cor}_H^G: H_*(H; \rho^{\#}A) \to H_*(G; A) \quad \text{and} \quad \text{res}_H^G: H^*(G; A) \to H^*(H; \rho^{\#}A)$$

from the category G–**mod** to **Ab** (2.1.4). This is an immediate consequence of the theorem that $H_*(G; A)$ and $H^*(G; A)$ are universal δ-functors, once we notice that $T_*(A) = H_*(H; \rho^{\#}A)$ and $T^*(A) = H^*(H; \rho^{\#}A)$ are δ-functors.

Subgroups 6.7.2 The terms *restriction* and *corestriction* are normally used only when H is a subgroup of G. In this case $\mathbb{Z}G$ is actually a free $\mathbb{Z}H$-module, a basis being given by any set of coset representatives. Therefore every projective G-module is also a projective H-module, and we may use any projective G-module resolution $P \to \mathbb{Z}$ to compute the homology and cohomology of H. If A is a G-module, we may calculate cor_H^G as the homology $H_*(\alpha)$ of the chain map $\alpha: P \otimes_H A \to P \otimes_G A$; similarly, we may calculate res_H^G as the cohomology $H^*(\beta)$ of the map $\beta: \text{Hom}_G(P, A) \to \text{Hom}_H(P, A)$.

Exercise 6.7.1 Let H be the cyclic subgroup C_m of the cyclic group C_{mn}. Show that the map $\text{cor}_H^G: H_*(C_m; \mathbb{Z}) \to H_*(C_{mn}; \mathbb{Z})$ is the natural inclusion $\mathbb{Z}/m \hookrightarrow \mathbb{Z}/mn$ for $*$ odd, while $\text{res}_H^G: H^*(C_{mn}; \mathbb{Z}) \to H^*(C_m; \mathbb{Z})$ is the natural projection $\mathbb{Z}/mn \to \mathbb{Z}/m$ for $*$ even. (See 6.2.3.)

Inflation 6.7.3 Let H be a normal subgroup of G and A a G-module. The composites

$$\text{inf}: H^*(G/H; A^H) \xrightarrow{\text{res}} H^*(G; A^H) \to H^*(G; A) \quad \text{and}$$

$$\text{coinf: } H_*(G; A) \to H_*(G; A_H) \xrightarrow{\text{cor}} H_*(G/H; A_H)$$

are called the *inflation* and *coinflation* maps, respectively. Note that on H^0 we have inf: $(A^H)^{G/H} \cong A^G$ and on H_0 we have coinf: $A_G \cong (A_H)_{G/H}$.

Example 6.7.4 If A is trivial as an H-module, inflation = restriction and coinflation = corestriction. Thus by the last exercise we see that (for $*$ odd) the map coinf: $H_*(C_m; \mathbb{Z}) \to H_*(C_{mn}; \mathbb{Z})$ is the natural inclusion $\mathbb{Z}/m \hookrightarrow \mathbb{Z}/mn$, while (for $*$ even) inf: $H^*(C_{mn}; \mathbb{Z}) \to H^*(C_m; \mathbb{Z})$ is the natural projection $\mathbb{Z}/mn \to \mathbb{Z}/m$.

Exercise 6.7.2 Show that the following compositions are zero for $i \neq 0$:

$$H^*(G/H; A^H) \xrightarrow{\text{inf}} H^*(G; A) \xrightarrow{\text{res}} H^*(H; A);$$

$$H_*(H; A) \xrightarrow{\text{cor}} H_*(G; A) \xrightarrow{\text{coinf}} H_*(G/H; A_H).$$

In general, these sequences are not exact, but rather they fit into a spectral sequence, which is the topic of the next section. (See 6.8.3.)

Functoriality of H_* and Corestriction 6.7.5 Let $\mathcal{C}$ be the category of pairs (G, A), where G is a group and A is a G-module. A morphism in $\mathcal{C}$ from (H, B) to (G, A) is a pair $(\rho: H \to G, \varphi: B \to \rho^{\#} A)$, where ρ is a group homomorphism and φ is an H-module map. Such a pair (ρ, φ) induces a map $\text{cor}_H^G \circ \varphi: H_*(H; B) \to H_*(G; A)$. It follows (and we leave the verification as an exercise for the reader) that H_* is a covariant functor from $\mathcal{C}$ to **Ab**.

We have already seen some examples of the naturality of H_*. Corestriction is H_* for $(\rho, B = \rho^{\#} A)$ and coinflation is H_* for $(G \to G/H, A \to A_H)$.

Functoriality of H^* and Restriction 6.7.6 Let $\mathcal{D}$ be the category with the same objects as $\mathcal{C}$, except that a morphism in $\mathcal{D}$ from (H, B) to (G, A) is a pair $(\rho: H \to G \; \varphi: \rho^{\#} A \to B)$. (Note the reverse direction of φ!) Such a pair (ρ, φ) induces a map $\varphi \circ \text{res}_H^G: H^*(G; A) \to H^*(H; B)$. It follows (again as an exercise) that H^* is a contravariant functor from $\mathcal{D}$ to **Ab**.

We have already seen some examples of the naturality of H^*. Restriction is H^* for $(\rho, \rho^{\#} A = B)$ and inflation is H^* for $(G \to G/H, A^H \to A)$. Conjugation provides another example:

Example 6.7.7 (Conjugation) Suppose that H is a subgroup of G, so that each $g \in G$ induces an isomorphism ρ between H and its conjugate gHg^{-1}. If A is a G-module, the abelian group map $\mu_g: A \to A$ $(a \mapsto ga)$ is actually an H-module map from A to $\rho^{\#}A$ because $\mu_g(ha) = gha = (ghg^{-1})ga = \rho(h)\mu_g a$ for all $h \in H$ and $a \in A$. In the category $\mathcal{C}$ of 6.7.5, (ρ, μ_g) is an isomorphism $(H, A) \cong (gHg^{-1}, A)$. Similarly, (ρ, μ_g^{-1}) is an isomorphism $(H, A) \cong (gHg^{-1}, A)$ in $\mathcal{D}$. Therefore we have maps $H_*(H; A) \to H_*(gHg^{-1}; A)$ and $H^*(gHg^{-1}; A) \to H^*(H; A)$.

One way to compute these maps on the chain level is to choose a projective $\mathbb{Z}G$-module resolution $P \to \mathbb{Z}$. Since the P_i are also projective as $\mathbb{Z}H$-modules and as $\mathbb{Z}[gHg^{-1}]$-modules, we may compute our homology and cohomology groups using P. The maps $\mu_g: P_i \to P_i$ ($p \mapsto gp$) form an H-module chain map from P to $\rho^{\#}P$ over the identity map on $\mathbb{Z}$. Hence the map $H_*(H; A) \to H_*(gHg^{-1}; A)$ is induced from

$$P \otimes_H A \to P \otimes_{gHg^{-1}} A, \quad x \otimes a \mapsto gx \otimes ga.$$

Similarly, the map $H^*(gHg^{-1}; A) \to H^*(H; A)$ is induced from

$$\mathrm{Hom}_H(P, A) \to \mathrm{Hom}_{gHg^{-1}}(P, A), \quad \varphi \mapsto (p \mapsto g\varphi(g^{-1}p)).$$

Theorem 6.7.8 *Conjugation by an element $g \in G$ induces the identity automorphism on $H_*(G; \mathbb{Z})$ and $H^*(G; \mathbb{Z})$.*

Proof The maps $P \otimes \mathbb{Z} \to P \otimes \mathbb{Z}$ and $\mathrm{Hom}_G(P, \mathbb{Z}) \to \mathrm{Hom}_G(P, \mathbb{Z})$ are the identity.   ◇

Corollary 6.7.9 *If H is a normal subgroup of G, then the conjugation action of G on $\mathbb{Z}$ induces an action of G/H on $H_*(G; \mathbb{Z})$ and $H^*(G; \mathbb{Z})$.*

Example 6.7.10 (Dihedral groups) The cyclic group C_m is a normal subgroup of the dihedral group D_m (6.4.10), and $D_m/C_m \cong C_2$. To determine the action of C_2 on the homology of C_m, note that there is an element g of D_m such that $g\sigma g^{-1} = \sigma^{-1}$. Let $\rho: C_m \to C_m$ be conjugation by g. If P denotes the $(\sigma - 1, N)$ complex of 6.2.1, consider the following map from P to $\rho^{\#}P$:

$$0 \longleftarrow \mathbb{Z} \longleftarrow \mathbb{Z}G \overset{1-\sigma}{\longleftarrow} \mathbb{Z}G \overset{N}{\longleftarrow} \mathbb{Z}G \overset{1-\sigma}{\longleftarrow} \mathbb{Z}G \overset{N}{\longleftarrow} \mathbb{Z}G \overset{1-\sigma}{\longleftarrow} \mathbb{Z}G \cdots$$

$$\Big\| \qquad \Big\| \qquad -\sigma\Big\downarrow \qquad -\sigma\Big\downarrow \qquad \sigma^2\Big\downarrow \qquad \sigma^2\Big\downarrow \qquad (-\sigma)^3\Big\downarrow$$

$$0 \longleftarrow \mathbb{Z} \longleftarrow \mathbb{Z}G \overset{1-\sigma^{-1}}{\longleftarrow} \mathbb{Z}G \overset{N}{\longleftarrow} \mathbb{Z}G \overset{1-\sigma^{-1}}{\longleftarrow} \mathbb{Z}G \overset{N}{\longleftarrow} \mathbb{Z}G \overset{1-\sigma^{-1}}{\longleftarrow} \mathbb{Z}G \cdots$$

An easy calculation (exercise!) shows that the map induced from conjugation by g is multiplication by $(-1)^i$ on $H_{2i-1}(C_m; \mathbb{Z})$ and $H^{2i}(C_m; \mathbb{Z})$.

6.7.1 Cup Product

As another application of the naturality of H^*, we show that $H^*(G; \mathbb{Z})$ is an associative graded-commutative ring, a fact that is familiar to topologists.

In 6.1.14 we constructed a cross product map $\times$ from $H^*(G; \mathbb{Z}) \otimes H^*(H; \mathbb{Z})$ to $H^*(G \times H; \mathbb{Z})$. When $G = H$, composition with the restriction $\Delta^* = \mathrm{res}_G^{G \times G}$ along the diagonal map $\Delta: G \to G \times G$ gives a graded bilinear product on $H^*(G; \mathbb{Z})$, called the *cup product*. If $x, y \in H^*(G; \mathbb{Z})$, the cup product $x \cup y$ is just $\Delta^*(x \times y)$.

Exercise 6.7.3 (Naturality of the cross and cup product) Show that the cross product is natural in G and H in the sense that $(\rho^* x) \times (\sigma^* y) = (\rho \times \sigma)^*(x \times y)$ in $H^{p+q}(G' \times H'; \mathbb{Z})$ for every $\rho: G' \to G$ and $\sigma: H' \to H$, $x \in H^p(G; \mathbb{Z})$, and $y \in H^q(H; \mathbb{Z})$. Conclude that the cup product is natural in G, that is, that $(\rho^* x_1) \cup (\rho^* x_2) = \rho^*(x_1 \cup x_2)$.

Theorem 6.7.11 (Cohomology ring) *The cup product makes* $H^*(G; \mathbb{Z})$ *into an associative, graded-commutative ring with unit. The ring structure is natural in the group* G.

Proof Since the composites of Δ with the maps $\Delta \times 1, 1 \times \Delta: G \times G \to G \times G \times G$ are the same, and the cross product is associative (by exercise 6.1.9),

$$x \cup (y \cup z) = x \cup \Delta^*(y \times z) = \Delta^*(x \times \Delta^*(y \times z))$$
$$= \Delta^*(1 \times \Delta)^*(x \times y \times z) = \Delta^*(\Delta \times 1)^*(x \times y \times z)$$
$$= \Delta^*(\Delta^*(x \times y) \times z) = \Delta^*(x \times y) \cup z = (x \cup y) \cup z.$$

If $\pi: G \to 1$ is the projection, the compositions $(1 \times \pi)\Delta$ and $(\pi \times 1)\Delta$ are the identity on $H^*(G; \mathbb{Z})$, and the restriction π^* sends $1 \in H^0(1; \mathbb{Z})$ to $1 \in H^0(G; \mathbb{Z}) \cong \mathbb{Z}$. Since we saw in exercise 6.1.9 that the cross product with $1 \in H^0(G; \mathbb{Z})$ is the identity map,

$$x \cup 1 = \Delta^*(x \times \pi^*(1)) = \Delta^*(1 \times \pi)^*(x \times 1) = x \times 1 = x,$$

and $1 \cup x = x$ similarly. Hence the cup product is associative with unit 1.

To see that the cup product is graded-commutative, it suffices to show that the cross product (with $G = H$) is graded-commutative, that is, that $y \times x = (-1)^{ij} x \times y$ for $x \in H^i(G; \mathbb{Z})$ and $y \in H^j(G; \mathbb{Z})$. This is a consequence of the following lemma, since if τ is the involution $\tau(g, h) = (h, g)$ on $G \times G$, we have $y \cup x = \Delta^*(y \times x) = \Delta^* \tau^*(x \times y)$. $\diamond$

Lemma 6.7.12 *Let* $\tau: G \times H \to H \times G$ *be the isomorphism* $\tau(g, h) = (h, g)$ *and write* τ^* *for the associated restriction map* $H^*(H \times G, \mathbb{Z}) \to H^*(G \times H, \mathbb{Z})$. *Then for* $x \in H^p(G; \mathbb{Z})$ *and* $y \in H^q(H; \mathbb{Z})$, *we have* $\tau^*(y \times x) = (-1)^{pq}(x \times y)$.

Proof Let $P \to \mathbb{Z}$ be a free $\mathbb{Z}G$-resolution and $Q \to \mathbb{Z}$ a free $\mathbb{Z}H$-resolution. Because of the sign trick 1.2.5 used in taking total complexes, the maps $a \otimes b \mapsto (-1)^{pq}b \otimes a$ from $P_p \otimes Q_q$ to $Q_q \otimes P_p$ assemble to give a chain map $\tau': \mathrm{Tot}(P \otimes Q) \to \mathrm{Tot}(Q \otimes P)$ over τ. (Check this!) This gives the required factor of $(-1)^{pq}$, because τ^* is obtained by applying $\mathrm{Hom}(-, \mathbb{Z})$ and taking cohomology. $\diamond$

Exercise 6.7.4 Let $\beta \in H^2(C_m; \mathbb{Z}) \cong \mathbb{Z}/m$ be a generator. Show that the ring $H^*(C_m; \mathbb{Z})$ is the polynomial ring $\mathbb{Z}[\beta]$, modulo the obvious relation that $m\beta = 0$.

Exercise 6.7.5 This exercise uses exercise 6.1.10.

1. Show that there is a cup product on $H^*(G; k)$ for any commutative ring k, making H^* into an associative, graded-commutative k-algebra, natural in G.
2. Suppose that $k = \mathbb{Z}/m$ and $G = C_m$, with m odd. Show that the graded algebra $H^*(C_m; \mathbb{Z}/m)$ is isomorphic to the ring $\mathbb{Z}/m[\sigma, \beta]/(\sigma^2 = \beta\sigma = 0)$, with $\sigma \in H^1$ and $\beta \in H^2$.

Coalgebra Structure 6.7.13 Dual to the notion of a k-algebra is the notion of a *coalgebra* over a commutative ring k. We call a k-module H a *coalgebra* if there are module homomorphisms $\Delta: H \to H \otimes_k H$ (the *coproduct*) and $\varepsilon: H \to k$ (the *counit*) such that both composites $(\varepsilon \otimes 1)\Delta$ and $(1 \otimes \varepsilon)\Delta$ (mapping $H \to H \otimes H \to H$) are the identity on H. We say that the coalgebra is *coassociative* if in addition $(\Delta \otimes 1)\Delta = (1 \otimes \Delta)\Delta$ as maps $H \to H \otimes H \to H \otimes H \otimes H$. For example, $H = kG$ is a cocommutative coalgebra; the coproduct is the diagonal map from kG to $k(G \times G) \cong kG \otimes kG$ and satisfies $\Delta(g) = g \otimes g$, while the counit is the usual augmentation $\varepsilon(g) = 1$. More examples are given below in (9.10.8).

Lemma 6.7.14 *Suppose that k is a field, or more generally that $H_*(G; k)$ is flat as a k-module. Then $H_*(G; k)$ is a cocommutative coalgebra.*

Proof Recall from exercises 6.1.7 and 6.1.12 that $H_*(G \times G; k)$ is isomorphic to $H_*(G; k) \otimes_k H_*(G; k)$, so the diagonal map $\Delta: G \to G \times G$ induces a map $\Delta_*: H_*(G; k) \to H_*(G; k) \otimes_k H_*(G; k)$. The projection $\varepsilon: G \to 1$ induces a map ε_* from $H_*(G; k)$ to $H_*(1; k) = k$. Since $(\varepsilon \times 1)\Delta = (1 \times \varepsilon)\Delta$ as maps $G \to G \times G \to G$ and $(\Delta \times 1)\Delta = (1 \times \Delta)\Delta$ as maps $G \to G \times G \to G \times G \times G$, we have the required identities $(\varepsilon_* \otimes 1)\Delta_* = (1 \otimes \varepsilon_*)\Delta_*$ and $(\Delta_* \otimes 1)\Delta_* = (1 \otimes \Delta_*)\Delta_*$. $\diamond$

Definition 6.7.15 (Hopf algebras) A *bialgebra* is an algebra H, together with algebra homomorphisms Δ and ε making H into a cocommutative coalgebra. We call H a *Hopf algebra* if in addition there is a k-module homomorphism $s: H \to H$ (called the *antipode*) such that both maps $\times(s \otimes 1)\Delta$ and $\times(1 \otimes s)\Delta$ (from $H \to H \otimes H \to H \otimes H \to H$) equal the the projection $H \xrightarrow{\varepsilon} k \hookrightarrow H$.

For example, the involution $s(g) = g^{-1}$ makes kG into a Hopf algebra, because $(s \otimes 1)\Delta(g) = g^{-1} \otimes g$ and $(1 \otimes s)\Delta(g) = g \otimes g^{-1}$. We will see another example in exercise 7.3.7.

Exercise 6.7.6 Suppose that G is an abelian group, so that the product $\mu: G \times G \to G$ is a group homomorphism and that k is a field. Show that $H_*(G; k)$ and $H^*(G; k)$ are both Hopf algebras.

Transfer Maps 6.7.16 Let H be a normal subgroup of finite index in G, and let A be a G-module. The sum $\sum ga$ over the right cosets $\{Hg\}$ of H yields a well-defined map from A to A_H. This map sends $(ga - a)$ to zero, so it induces a well-defined map $tr: A_G \to A_H$. Since $H_*(G; A)$ is a universal δ-functor, tr extends to a unique map of δ-functors, called the *transfer map*:

$$tr : H_*(G; A) \to H_*(H; A).$$

Similarly, the sum $\sum ga$ over the left cosets $\{gH\}$ of H yields a well-defined map from A^H to A. The image is G-invariant, so it induces a well-defined map $tr: A^H \to A^G$. This induces a map of δ-functors, also called *the transfer map*:

$$tr : H^*(H; A) \to H^*(G; A).$$

Lemma 6.7.17 *The composite* $\mathrm{cor}_H^G \circ tr$ *is multiplication by the index* $[G : H]$ *on* $H_*(G; A)$. *Similarly, the composite* $tr \circ \mathrm{res}_H^G$ *is multiplication by* $[G : H]$ *on* $H^*(G; A)$.

Proof In A_G and A^G, the sums over the cosets are just $\sum ga = (\sum g) \cdot a =$

$[G : H] \cdot a$. The corresponding maps between the δ-functors are determined by their behavior on A_G and A^H, so they must also be multiplication by $[G : H]$.

$\diamond$

Exercise 6.7.7 Show that the transfer map defined here agrees with the transfer map defined in 6.3.9 using Shapiro's Lemma. *Hint:* By universality, it suffices to check what happens on H_0 and H^0.

Exercise 6.7.8 Use the transfer maps to give another proof of 6.5.8, that when G is a finite group of order $m = [G : 1]$ multiplication by m is the zero map on $H_n(G; A)$ and $H^n(G; A)$ for $n \neq 0$.

6.8 The Spectral Sequence

The inflation and restriction maps fit into a filtration of $H^*(G; A)$ first studied in 1946 by Lyndon. The spectral sequence codifying this relationship was found in 1953 by Hochschild and Serre. We shall derive it as a special case of the Grothendieck spectral sequence 5.8.3, using the following lemma.

Lemma 6.8.1 *If H is a normal subgroup of G, and A is a G-module, then both A_H and A^H are G/H-modules. Moreover, the forgetful functor $\rho^\#$ from G/H–**mod** to G–**mod** has $-_H$ as left adjoint and $-^H$ as right adjoint.*

Proof A G/H-module is the same thing as a G-module on which H acts trivially. Therefore A_H and A^H are G/H-modules by construction. The universal properties of $A^H \to A$ and $A \to A_H$ translate into the natural isomorphisms

$$\mathrm{Hom}_G(A, \rho^\# B) \cong \mathrm{Hom}_{G/H}(A_H, B) \text{ and}$$

$$\mathrm{Hom}_G(\rho^\# B, A) \cong \mathrm{Hom}_{G/H}(B, A^H),$$

which provide the required adjunctions.

$\diamond$

Lyndon/Hochschild-Serre Spectral Sequence 6.8.2 *For every normal subgroup H of a group G, there are two convergent first quadrant spectral sequences:*

$$E^2_{pq} = H_p(G/H; H_q(H; A)) \Rightarrow H_{p+q}(G; A);$$

$$E_2^{pq} = H^p(G/H; H^q(H; A)) \Rightarrow H^{p+q}(G; A).$$

The edge maps $H_*(G; A) \to H_*(G/H; A_H)$ and $H_*(H; A)_{G/H} \to H_*(G; A)$ in the first spectral sequence are induced from the coinflation and corestriction maps. The edge maps $H^*(G/H; A^H) \to H^*(G; A)$ and $H^*(G; A) \to H^*(H; A)^{G/H}$ in the second spectral sequence are induced from the inflation and restriction maps.

Proof We claim that the functors $-_G$ and $-^G$ factor through G/H–**mod** as follows:

$$
\begin{array}{ccc}
G\text{-mod} & \xrightarrow{\ -^H\ } & G/H\text{-mod} \qquad\qquad G\text{-mod} & \xrightarrow{\ _-H\ } & G/H\text{-mod} \\[4pt]
{}^{-G}\searrow & \swarrow^{-G/H} & \qquad\qquad _-G\searrow & \swarrow_{-G/H} \\[4pt]
\textbf{Ab} & & \qquad\qquad\qquad \textbf{Ab}
\end{array}
$$

To see this, let A be a G-module; we saw in the last lemma that A_H and A^H are G/H-modules. The abelian group $(A_H)_{G/H}$ is obtained from A by first modding out by the relations $ha - a$ with $h \in H$, and then modding out by the relations $\bar{g}a - a$ for $\bar{g} \in G/H$. If $\bar{g}$ is the image of $g \in G$ then $\bar{g}a - a \equiv ga - a$, so we see that $(A_H)_{G/H}$ is $A/\Im A = A_G$.

Similarly, $(A^H)^{G/H}$ is obtained from A by first restricting to the subgroup of all $a \in A$ with $ha = a$, and then further restricting to the subgroup of all a with $\bar{g}a = a$ for $\bar{g} \in G/H$. If $\bar{g}$ is the image of $g \in G$, $\bar{g}a = ga$. Thus $(A^H)^{G/H} = A^G$.

Finally, we proved in Lemma 6.8.1 that $-_H$ is left adjoint to an exact functor, and that $-^H$ is right adjoint to an exact functor. We saw in 2.3.10 that this implies that $-_H$ preserves projectives and that $-^H$ preserves injectives, so that the Grothendieck spectral sequence exists. The description of the edge maps is just a translation of the description given in 5.8.3. $\diamond$

Low Degree Terms 6.8.3 The exact sequences of low degree terms in the Lyndon-Hochschild-Serre spectral sequence are

$$
H_2(G; A) \xrightarrow{\text{coinf}} H_2(G/H; A_H) \xrightarrow{d} H_1(H; A)_{G/H} \xrightarrow{\text{cor}} H_1(G; A) \xrightarrow{\text{coinf}} H_1(G/H; A_H) \to 0;
$$

$$
0 \to H^1(G/H; A^H) \xrightarrow{\text{inf}} H^1(G; A) \xrightarrow{\text{res}} H^1(H; A)^{G/H} \xrightarrow{d} H^2(G/H; A^H) \xrightarrow{\text{inf}} H^2(G; A).
$$

Example 6.8.4 If H is in the center of G, G/H acts trivially on $H_*(H; A)$ and $H^*(H; A)$, so we may compute the E^2 terms from $H_*(H; \mathbb{Z})$ and Universal Coefficient theorems. For example, let G be the cyclic group C_{2m} and $H = C_m$ for m odd. Then $H_p(C_2; H_q(C_m; \mathbb{Z}))$ vanishes unless $p = 0$ or $q = 0$.

The groups $\mathbb{Z}/2$ lie along the x-axis, and the groups $\mathbb{Z}/m$ lie along the y-axis. The spectral sequence collapses at E^2 to yield the formula for $H_*(C_{2m}; \mathbb{Z})$ that we derived in 6.2.3.

```
 0                                      0
            G = C_2m                               G = D_2m
 Z/m   0                                Z/m   0

    0   0   0                              0   0   0

 Z/m   0   0   0                           0   0   0   0

    Z  Z/2   0  Z/2   0                     Z  Z/2   0  Z/2   0
```

Example 6.8.5 (Dihedral groups) Let G be the dihedral group $D_{2m} = C_m \rtimes C_2$ and set $H = C_m$. If m is odd, then once again $H_p(C_2; H_q(C_m))$ vanishes unless $p = 0$ or $q = 0$. As before, the groups $\mathbb{Z}/2$ lie along the x-axis, but along the y-axis we now have $H_q(C_m)_{C_2}$. From our calculation 6.7.10 of the action of C_2 on $H_*(C_m)$ we see that $H_q(C_m)_{C_2}$ is zero unless $q = 0$, when it is $\mathbb{Z}$, or $q \equiv 3 \pmod 4$, when it is $\mathbb{Z}/m$. Summarizing, we have computed that

$$H_n(D_{2m}; \mathbb{Z}) = \left\{ \begin{array}{ll} \mathbb{Z} & \text{if } n = 0 \\ \mathbb{Z}/2 & \text{if } n \equiv 1 \ (\mathrm{mod}\ 4) \\ \mathbb{Z}/2m & \text{if } n \equiv 3 \ (\mathrm{mod}\ 4) \\ 0 & \text{otherwise} \end{array} \right\}.$$

Example 6.8.6 (Gysin sequence) A central element t of infinite order in G generates an infinite cyclic subgroup T. As in 5.3.7 the spectral sequence collapses to the long exact "Gysin" sequence for every trivial G-module k :

$$\cdots H_n(G; k) \xrightarrow{\mathrm{coinf}} H_n(G/T; k) \xrightarrow{S} H_{n-2}(G/T; k) \to H_{n-1}(G; k) \cdots$$

Exercise 6.8.1 The *infinite dihedral group* D_∞ is the semidirect product $T \rtimes C_2$, where $\sigma \in C_2$ acts as multiplication by -1 on the infinite cyclic group T ($\sigma t \sigma^{-1} = t^{-1}$). Show that σ acts as multiplication by -1 on $H_1(T; \mathbb{Z})$, and deduce that

$$H_n(D_\infty; \mathbb{Z}) \cong \left\{ \begin{array}{ll} \mathbb{Z} & \text{if } n = 0 \\ \mathbb{Z}/2 \oplus \mathbb{Z}/2 & \text{if } n \equiv 1, 3, 5, 7, \ldots \\ 0 & \text{if } n \equiv 2, 4, 6, 8, \ldots \end{array} \right\}.$$

Hint: By naturality, $H_*(C_2)$ is a summand of $H_*(D_\infty)$.

Presentations 6.8.7 A presentation of a group by generators and relations amounts to the same thing as a short exact sequence of groups $1 \to R \to F \to G \to 1$, where F is the free group on the generators of G and R is the normal subgroup of F generated by the relations of G. Note that R is also a free group, being a subgroup of the free group F. The spectral sequence of this extension has $E^2_{pq} = 0$ for $q \neq 0, 1$ and $H_n(F; \mathbb{Z}) = 0$ for $n \neq 0, 1$. Therefore the differentials $H_{n+2}(G; \mathbb{Z}) \to H_n(G; H_1(R))$ must be isomorphisms for $n \geq 1$, and we have the low degree sequence

$$0 \to H_2(G; \mathbb{Z}) \to \left[\frac{R}{R, R}\right]_G \to \frac{F}{[F, F]} \to \frac{G}{[G, G]} \to 0.$$

The action of G on $R/[R/R]$ is given by $g \cdot r = frf^{-1}$, where $f \in F$ lifts $g \in G$ and $r \in R$. The following calculation shows that $(R/[R/R])_G = R/[F, R]$:

$$(g - 1) \cdot r = frf^{-1} - r \equiv frf^{-1}r^{-1} = [f, r].$$

By inspection of the low degree sequence, we see that we have proven the following result, which was first established in [Hopf].

Hopf's Theorem 6.8.8 *If $G = F/R$ with F free, then $H_2(G; \mathbb{Z}) \cong \frac{R \cap [F,F]}{[F,R]}$.*

6.9 Universal Central Extensions

A *central extension* of G is an extension $0 \to A \to X \xrightarrow{\pi} G \to 1$ such that A is in the center of X. (If π and A are clear from the context, we will just say that X is a central extension of G.) A homomorphism over G from X to another central extension $0 \to B \to Y \xrightarrow{\tau} G \to 1$ of G is a map $f: X \to Y$ such that $\pi = \tau f$. X is called a *universal central extension* of G if for every central extension $0 \to B \to Y \xrightarrow{\tau} G \to 1$ of G there exists a unique homomorphism f from X to Y over G.

$$
\begin{array}{ccccccccc}
0 & \longrightarrow & A & \longrightarrow & X & \xrightarrow{\pi} & G & \longrightarrow & 1 \\
& & \downarrow & & \downarrow{\scriptstyle \exists!} & & \| & & \\
0 & \longrightarrow & B & \longrightarrow & Y & \xrightarrow{\tau} & G & \longrightarrow & 1
\end{array}
$$

Clearly, a universal central extension is unique up to isomorphism over G, provided that it exists. We will show that a necessary and sufficient condition for a universal central extension to exist is that G is perfect; recall that a group G is *perfect* if it equals its commutator subgroup $[G, G]$.

Example 6.9.1 The smallest perfect group is A_5. The universal central extension of A_5 describes A_5 as the quotient $PSL_2(\mathbb{F}_5)$ of the binary icosahedral group $X = SL_2(\mathbb{F}_5)$ by the center of order 2, $A = \pm(\begin{smallmatrix}1&0\\0&1\end{smallmatrix})$ [Suz, 2.9].

$$0 \longrightarrow \mathbb{Z}/2 \xrightarrow{\left(\begin{smallmatrix}-1&0\\0&-1\end{smallmatrix}\right)} SL_2(\mathbb{F}_5) \longrightarrow PSL_2(\mathbb{F}_5) \longrightarrow 1.$$

Lemma 6.9.2 *If G has a universal central extension X, then both G and X are perfect.*

Proof If X is perfect, then so is G. If X is not perfect, then $B = X/[X, X]$ is a nonzero abelian group, $0 \to B \to B \times G \to G \to 1$ is a central extension, and there are two homomorphisms $X \to B \times G$ over G : $(0, \pi)$ and (pr, π).

$\diamond$

Exercises 6.9.1

1. If $0 \to A \to X \to G \to 1$ is any central extension in which G and X are perfect groups, show that $H_1(X; \mathbb{Z}) = 0$ and that there is an exact sequence

$$H_2(X; \mathbb{Z}) \xrightarrow{\text{cor}} H_2(G; \mathbb{Z}) \to A \to 0.$$

2. Show that if G is perfect then central extensions $0 \to A \to X \to G \to 1$ are classified by $\mathrm{Hom}(H_2(G; \mathbb{Z}), A)$. (Use exercise 6.1.5.)

Remark The above exercises suggest that $H_2(G; \mathbb{Z})$ has something to do with universal central extensions. Indeed, we shall see that the universal central extension $0 \to A \to X \to G \to 1$ has $A \cong H_2(G; \mathbb{Z})$. The group $H_2(G; \mathbb{Z})$ is called the *Schur multiplier* of G in honor of Schur, who first investigated the notion of a universal central extension of a finite group G in [Schur].

As indicated in section 6.6, Schur was concerned with central extensions with $A = \mathbb{C}^*$, and these are classified by the group $H^2(G; \mathbb{C}^*) = \mathrm{Hom}(H_2(G; \mathbb{Z}), \mathbb{C}^*)$. Since G is finite, $H^2(G; \mathbb{C}^*)$ is the Pontrjagin dual (3.2.3) of the finite group $H_2(G; \mathbb{Z})$. Hence the groups $H^2(G; \mathbb{C}^*)$ and $H_2(G; \mathbb{Z})$ are noncanonically isomorphic.

Construction of a Universal Central Extension 6.9.3 Choose a free group F mapping onto G and let $R \subset F$ denote the kernel. Then $[R, F]$ is a normal subgroup of F, and the short exact sequence $1 \to R \to F \to G \to 1$ induces a central extension

$$0 \to R/[R, F] \to F/[R, F] \to G \to 1.$$

Now suppose that G is perfect. Since $[F, F]$ maps onto G, there is a surjection from $[F, F]/[R, F]$ to G; its kernel is the subgroup $(R \cap [F, F])/[R, F]$, which Hopf's Theorem 6.8.8 states is the Schur multiplier $H_2(G; \mathbb{Z})$. We shall prove that

$$0 \to (R \cap [F, F])/[R, F] \to [F, F]/[R, F] \to G \to 1$$

is a universal central extension of G.

Lemma 6.9.4 $[F, F]/[R, F]$ *is a perfect group.*

Proof Since $[F, F]$ and F both map onto G, any $x \in F$ may be written as $x = x'r$ with $x' \in [F, F]$ and $r \in R$. Writing $y \in F$ as $y's$ with $y' \in [F, F]$ and $s \in R$, we find that in $F/[R, F]$

$$[x, y] = (x'r)(y's)(x'r)^{-1}(y's)^{-1} \equiv [x', y'].$$

Thus every generator $[x, y]$ of $[F, F]/[R, F]$ is a commutator of elements x' and y' of $[F, F]/[R, F]$. $\diamond$

Theorem 6.9.5 *A group G has a universal central extension if and only if G is perfect. In this case, the universal central extension is*

(∗) $$0 \to H_2(G; \mathbb{Z}) \to \frac{[F, F]}{[R, F]} \xrightarrow{\pi} G \to 1.$$

Here $1 \to R \to F \to G \to 1$ is any presentation of G.

Proof If G has a universal central extension, then G must be perfect by 6.9.2. Now suppose that G is perfect; we have just seen that (∗) is a central extension and that $[F, F]/[R, F]$ is perfect. In order to show that (∗) is universal, let $0 \to B \to Y \xrightarrow{\tau} G \to 1$ be another central extension. Since F is a free group, the map $F \to G$ lifts to a map $h \colon F \to Y$. Since $\tau h(R) = 1$, $h(R)$ is in the central subgroup B of Y. This implies that $h([R, F]) = 1$. Therefore h induces a map

$$\eta \colon [F, F]/[R, F] \hookrightarrow F/[R, F] \xrightarrow{h} Y$$

such that $\tau\eta = \pi$, that is, such that η is a homomorphism over G. The following lemma shows that η is unique and finishes the proof that (∗) is universal.

◇

Lemma 6.9.6 *If* $0 \to A \to X \xrightarrow{\pi} G \to 1$ *and* $0 \to B \to Y \to G \to 1$ *are central extensions, and X is perfect, there is at most one homomorphism f from X to Y over G.*

Proof If f_1 and f_2 are two such homomorphisms, define a set map $\varphi: X \to B$ by the formula $f_1(x) = f_2(x)\varphi(x)$. Since B is central,

$$f_1(xx') = f_2(x)\varphi(x)f_2(x')\varphi(x') = f_2(xx')\varphi(x)\varphi(x').$$

Hence $\varphi(xx') = \varphi(x)\varphi(x')$, that is, φ is a group homomorphism. Since B is an abelian group, φ must factor through $X/[X, X] = 1$. Hence $\varphi(x) = 1$ for all x, that is, $f = f'$.

◇

Exercise 6.9.2 (Composition) If $0 \to B \to Y \xrightarrow{\rho} X \to 1$ and $0 \to A \to X \xrightarrow{\pi} G \to 1$ are central extensions, show that the "composition" $0 \to \ker(\pi\rho) \to Y \xrightarrow{\pi\rho} G \to 1$ is a central extension of G. If X is a universal central extension of G, conclude that every central extension $0 \to B \to Y \to X \to 1$ splits.

Recognition Criterion 6.9.7 A central extension $0 \to A \to X \xrightarrow{\pi} G \to 1$ is universal if and only if X is perfect and every central extension of X splits as a direct product of X with an abelian group.

Proof The 'only if' direction follows from the preceding exercise. Now suppose that X is perfect and that every central extension of X splits. Given a central extension $0 \to B \to Y \xrightarrow{\tau} G \to 1$ of G, we can construct a homomorphism from X to Y over G as follows. Let P be the pullback group $\{(x, y) \in X \times Y : \pi(x) = \tau(y)\}$. Then in the diagram

$$
\begin{array}{ccccccccc}
0 & \longrightarrow & B & \longrightarrow & P & \xrightarrow{\;\;\overset{\exists\sigma}{\longleftarrow}\;\;} & X & \longrightarrow & 1 \\
& & \| & & \downarrow{\scriptstyle\ulcorner} & & \downarrow{\scriptstyle\pi} & & \\
0 & \longrightarrow & B & \longrightarrow & Y & \xrightarrow{\;\tau\;} & G & \longrightarrow & 1
\end{array}
$$

the top row is a central extension of X, so it is split by a map $\sigma: X \to P$. The composite $f: X \to P \to Y$ is the homomorphism over G we wanted to construct. Since X is perfect, f is unique (6.9.6); this proves that X is a universal central extension of G.   ◇

Corollary 6.9.8 *If $0 \to A \to X \to G \to 1$ is a universal central extension, then*

$$H_1(X; \mathbb{Z}) = H_2(X; \mathbb{Z}) = 0.$$

Corollary 6.9.9 *If G is a perfect group and $H_2(G; \mathbb{Z}) = 0$, then every central extension of G is a direct product of G with an abelian group.*

$$0 \to A \to A \times G \to G \to 1$$

Proof Evidently $0 \to 0 \to G = G \to 1$ is the universal central extension of G.   ◇

Example 6.9.10 (Alternating groups) It is well known that the alternating groups A_n are perfect if $n \geq 5$. From [Suz, 3.2] we see that

$$H_2(A_n; \mathbb{Z}) \cong \begin{cases} \mathbb{Z}/6 & \text{if } n = 6, 7 \\ \mathbb{Z}/2 & \text{if } n = 4, 5 \text{ or } n \geq 8 \\ 0 & \text{if } n = 1, 2, 3 \end{cases}$$

We have already mentioned (6.9.1) the universal central extension of A_5. In general, the regular representation $A_n \to SO_{n-1}$ gives rise to a central extension

$$0 \to \mathbb{Z}/2 \to \tilde{A}_n \to A_n \to 1$$

by restricting the central extension

$$0 \to \mathbb{Z}/2 \to \mathrm{Spin}_{n-1}(\mathbb{R}) \to SO_{n-1} \to 1.$$

If $n \neq 6, 7$, $\tilde{A}_n$ must be the universal central extension of A_n.

Example 6.9.11 It is known [Suz, 1.9] that if F is a field, then the special linear group $SL_n(F)$ is perfect, with the exception of $SL_2(\mathbb{F}_2) \cong D_6$ and $SL_2(\mathbb{F}_3)$, which is a group of order 24. The center of $SL_n(F)$ is the group $\mu_n(F)$ of nth roots of unity in F (times the identity matrix I), and the quotient of $SL_n(F)$ by $\mu_n(F)$ is the *projective special linear group* $PSL_n(F)$.

When $F = \mathbb{F}_q$ is a finite field, we know that $H_2(SL_n(\mathbb{F}_q); \mathbb{Z}) = 0$ [Suz, 2.9]. It follows, again with two exceptions, that

$$0 \to \mu_n(\mathbb{F}_q) \xrightarrow{\ I\ } SL_n(\mathbb{F}_q) \to PSL_n(\mathbb{F}_q) \to 1$$

is the universal central extension of the finite group $PSL_n(\mathbb{F}_q)$.

Example 6.9.12 The *elementary matrix* e_{ij}^λ in $GL_n(R)$ is the matrix that coincides with the identity matrix except for the single nonzero entry λ in the (i, j) spot. The subgroup $E_n(R)$ of $GL_n(R)$ generated by the elementary matrices is a perfect group when $n \geq 3$ because $[e_{ij}^\lambda, e_{jk}^\mu] = e_{ik}^{\lambda\mu}$ for $i \neq k$. We now describe the universal central extension of $E_n(R)$.

Definition 6.9.13 Let R be any ring. For $n \geq 3$ the *Steinberg group* $St_n(R)$ is the group that is presented as having generators x_{ij}^λ ($\lambda \in R$, $1 \leq i, j \leq n$) and relations

1. $x_{ij}^\lambda x_{ij}^\mu = x_{ij}^{\lambda+\mu}$;
2. $[x_{ij}^\lambda, x_{jk}^\mu] = x_{ik}^{\lambda\mu}$ for $i \neq k$; and
3. $[x_{ij}^\lambda, x_{k\ell}^\mu] = 1$ for $j \neq k$ and $i \neq \ell$.

There is a homomorphism $St_n(R) \to E_n(R)$ sending x_{ij}^λ to e_{ij}^λ because these relations are also satisfied by the elementary matrices. It is known [Milnor] [Swan, p. 208] that $St_n(R)$ is actually the universal central extension of $E_n(R)$ for $n \geq 5$. The kernel of $St_n(R) \to E_n(R)$ is denoted $K_2(n, R)$ and may be identified with the Schur multiplier. The direct limit $K_2(R)$ of the groups $K_2(n, R)$ is an important part of algebraic K-theory. See [Milnor] for more details and computations.

6.10 Covering Spaces in Topology

Let G be a group that acts on a topological space X. We shall assume that each translation $X \to X$ arising from multiplication by an element $g \in G$ is a continuous map and that the action is *proper* in the sense that every point of X is contained in a small open subset U such that every translate gU is disjoint from U. Under these hypotheses, the quotient topology on the orbit space X/G is such that the projection $p: X \to X/G$ makes X into a covering space of X/G. Indeed, every small open set U is mapped homeomorphically onto its image in X/G.

Example 6.10.1 Let Y be a connected, locally simply connected space, so that its universal covering space $\tilde{Y} \to Y$ exists. The group $G = \pi_1(Y)$ acts properly on $X = \tilde{Y}$, and $\tilde{Y}/G = Y$.

Lemma 6.10.2 *If G acts properly on X, the singular complex $S_*(X)$ of X is a chain complex of free $\mathbb{Z}G$-modules, and $S_*(X)_G$ is the singular complex of X/G.*

Proof Let $\mathcal{B}_n$ denote the set of continuous maps $\sigma: \Delta_n \to X$. G acts on $\mathcal{B}_n$, with $g\sigma$ being the composition of σ with translation by $g \in G$. Since $S_n(X)$ is the free $\mathbb{Z}$-module with basis $\mathcal{B}$, $S_n(X)$ is a G-module. Since translation by g sends the faces of σ to the faces of $g\sigma$, the boundary map $d: S_n(X) \to S_{n-1}(X)$ is a G-map, so $S_n(X)$ is a G-module complex.

Let $\mathcal{B}'_n$ denote the set of continuous maps $\sigma': \Delta_n \to X/G$. The unique path lifting property of a covering space implies that any $\sigma': \Delta_n \to X/G$ may be lifted to a map $\sigma: \Delta_n \to X$ and that every other lift is $g\sigma$ for some $g \in G$. As the $g\sigma$ are distinct, this proves that $\mathcal{B} \cong G \times \mathcal{B}'$ as a G-set. Choosing one lift for each σ' gives a map $\mathcal{B}' \to \mathcal{B}$, hence a basis for $S_n(X)$ as a free $\mathbb{Z}G$-module. This proves that the natural map from $S_n(X)$ to $S_n(X/G)$ induces an isomorphism $S_n(X)_G \cong S_n(X/G)$. $\diamond$

Corollary 6.10.3 *If G acts properly on X, $H_*(X, \mathbb{Z})$ and $H^*(X, \mathbb{Z})$ are G-modules.*

Definition 6.10.4 (Classifying space) A CW complex with fundamental group G and contractible universal covering space is called a *classifying space* for G, or a *model for BG*; by abuse of notation, we will call such a space BG, and write EG for its universal covering space. From the Serre fibration $G \to EG \to BG$ we see that

$$\pi_i(BG) = \begin{cases} G & \text{if } i = 1 \\ 0 & \text{otherwise} \end{cases}.$$

It is well known that any two classifying spaces for G are homotopy equivalent. One way to find a model for BG is to find a contractible CW complex X on which G acts properly (and cellularly) and take $BG = X/G$.

Theorem 6.10.5 $H_*(BG; \mathbb{Z}) \cong H_*(G; \mathbb{Z})$ and $H^*(BG; \mathbb{Z}) \cong H^*(G; \mathbb{Z})$.

Proof Since $H_*(EG) \cong H_*(\text{point})$ is 0 for $* \neq 0$ and $\mathbb{Z}$ for $* = 0$, the chain complex $S_*(EG)$ is a free $\mathbb{Z}G$-module resolution of $\mathbb{Z}$. Hence $H_*(G; \mathbb{Z}) =$

$H_*(S_*(EG) \otimes_{\mathbb{Z}G} \mathbb{Z}) = H_*(S_*(EG)_G) = H_*(S_*(BG)) = H_*(BG; \mathbb{Z})$. Similarly, $H^*(G; \mathbb{Z})$ is the cohomology of

$$\text{Hom}_G(S_*(EG), \mathbb{Z}) = \text{Hom}_{\mathbf{Ab}}(S_*(EG)_G, \mathbb{Z}) = \text{Hom}_{\mathbf{Ab}}(S_*(BG), \mathbb{Z}),$$

the chain complex whose cohomology is $H^*(BG; \mathbb{Z})$.   ◇

Remark The relationship between the homology (resp. cohomology) of G and BG was worked out during World War II by Hopf and Freudenthal (resp. by Eilenberg and MacLane). MacLane asserts in [MacH] that this interplay "was the starting point of homological algebra." Here are some useful models of classifying spaces.

Example 6.10.6 The circle S^1 and the complex units $\mathbb{C}^*$ are two models for $B\mathbb{Z}$; the extensions $0 \to \mathbb{Z} \to \mathbb{R} \to S^1 \to 1$ and $0 \to \mathbb{Z} \xrightarrow{2\pi i} \mathbb{C} \xrightarrow{\exp} \mathbb{C}^* \to 1$ expressing $\mathbb{R}$ (resp. $\mathbb{C}$) as the universal cover of S^1 (resp. $\mathbb{C}^*$) are well known.

Example 6.10.7 The infinite sphere S^∞ is contractible, and $G = C_2$ acts properly in such a way that $S^\infty/G = \mathbb{RP}^\infty$. Hence we may take $\mathbb{RP}^\infty$ as our model for BC_2.

Example 6.10.8 Let S be a Riemann surface of genus $g \neq 0$. The fundamental group $G = \pi_1(S)$ has generators $a_1, \cdots, a_g, b_1, \cdots, b_g$ and the single defining relation $[a_1, b_1][a_2, b_2] \cdots [a_g, b_g] = 1$. One knows that the universal cover X of S is the hyperbolic plane, which is contractible. Thus S is the classifying space BG.

Example 6.10.9 Any connected Lie group L has a maximal compact subgroup K, and the homogeneous space $X = L/K$ is diffeomorphic to $\mathbb{R}^d$, where $d = \dim(L) - \dim(K)$. If Γ is a discrete torsionfree subgroup of L, then $\Gamma \cap K = \{1\}$, so Γ acts properly on X. Consequently, the double coset space $\Gamma \backslash X = \Gamma \backslash L/K$ is a model for the classifying space $B\Gamma$.

For example, the special linear group $SL_n(\mathbb{R})$ has $SO_n(\mathbb{R})$ as maximal compact, so $X = SO_n(\mathbb{R}) \backslash SL_n(\mathbb{R}) \cong \mathbb{R}^d$ where $d = \frac{n(n+2)}{2} - 1$. $SL_n(\mathbb{Z})$ is a discrete but not torsionfree subgroup of $SL_n(\mathbb{R})$. For $N \geq 3$, the *principal congruence subgroup* $\Gamma(N)$ *of level* N is the subgroup of all matrices in $SL_n(\mathbb{Z})$ congruent to the identity matrix modulo N. One knows that $\Gamma(N)$ is torsionfree, so $X/\Gamma(N)$ is a model for $B\Gamma(N)$.

Theorem 6.10.10 *Let G act properly on a space X with $\pi_0(X) = 0$. Then for every abelian group A there are spectral sequences*

$$^I E^2_{pq} = H_p(G; H_q(X, A)) \Rightarrow H_{p+q}(X/G, A);$$

$$^{II} E^{pq}_2 = H^p(G; H^q(X, A)) \Rightarrow H^{p+q}(X/G, A).$$

Proof Let us write $\mathbb{H}_*(G; -)$ for the hyperhomology functors $\mathbb{L}_*(-_G)$ defined in 6.1.15 (or 5.7.4). Since $C = S_*(X) \otimes_{\mathbb{Z}} A$ is a chain complex of G-modules, there are two spectral sequences converging to the group hyperhomology $\mathbb{H}_*(G; C)$. Shapiro's Lemma 6.3.2 tells us that $H_q(S_n(X) \otimes_{\mathbb{Z}} A)$ is 0 for $q \neq 0$ and $S_n(X/G) \otimes_{\mathbb{Z}} A$ for $q = 0$ (6.10.2). Hence the first spectral sequence collapses to yield

$$\mathbb{H}_p(G; C) = H_p(S_*(X/G) \otimes A) = H_p(X/G, A).$$

The second spectral sequence has the desired E^2 term

$$^{II} E^2_{pq} = H_p(G; H_q C) = H_p(G; H_q(X, A)).$$

Similarly, if we write $\mathbb{H}^*(G; -)$ for the group hypercohomology $\mathbb{R}^*(-^G)$ and D for $\mathrm{Hom}_{\mathbf{Ab}}(S(X), A)$, there are two spectral sequences (6.1.15) converging to $\mathbb{H}^*(G; D)$. Since

$$D_n = \mathrm{Hom}(\mathbb{Z}G \otimes S_n(X/G), A) = \mathrm{Hom}(\mathbb{Z}G, \mathrm{Hom}(S_n(X/G), A)),$$

Shapiro's Lemma tells us that the first spectral sequence collapses to yield $\mathbb{H}^*(G; D) \cong H^*(X/G, A)$, and the second spectral sequence has the desired E_2 term

$$^{II} E^{pq}_2 = H^p(G; H^q(D)) = H^p(G; H^q(A)). \qquad \diamond$$

Remark There is a map from X/G to BG such that $X \to X/G \to BG$ has the homotopy type of a Serre fibration. The spectral sequences (6.10.10) may then be viewed as special cases of the Serre spectral sequence 5.3.2.

6.11 Galois Cohomology and Profinite Groups

The notion of profinite group encodes many of the important properties of the Galois group $\mathrm{Gal}(L/K)$ of a *Galois field extension* (i.e., an algebraic extension that is separable and normal but not necessarily finite). The largest Galois extension of any field K is the *separable closure* K_s of K; K_s is the subfield of the algebraic closure $\bar{K}$ consisting of all elements separable over K, and $K_s = \bar{K}$ if char$(K) = 0$.

K_s is also the union $\cup L_i$ of the partially ordered set $\{L_i : i \in I\}$ of all finite Galois field extensions of K. If $K \subset L_i \subset L_j$, the Fundamental Theorem of finite Galois theory [BAI, 4.5] states that there is a natural surjection from Gal(L_j/K) to Gal(L_i/K) with kernel Gal(L_j/L_i). In other words, there is a contravariant functor Gal$(-/K)$ from the filtered poset I to the category of finite groups.

Krull's Theorem 6.11.1 *The Galois group* Gal(K_s/K) *of all field automorphisms of* $\bar{K}$ *fixing* K *is isomorphic to the inverse limit* $\varprojlim$ Gal(L_i/K) *of finite groups.*

Proof Since the L_i are splitting fields over K, any automorphism α of K_s over K restricts to an automorphism α_i of L_i. The resulting restriction maps Gal$(K_s/K) \to$ Gal(L_i/K) are compatible and yield a group homomorphism ϕ from Gal(K_s/K) to the set $\varprojlim$ Gal(L_i/K) of all compatible families $(\alpha_i) \in \Pi$ Gal(L_i/K). If $\alpha \ne 1$, then $\alpha(x) \ne x$ for some $x \in K_s = \cup L_i$; if $x \in L_i$, then $\alpha_i(x) = \alpha(x) \ne x$. Therefore $\phi(\alpha) \ne 1$, that is, ϕ is injective. Conversely, if we are given (α_i) in $\varprojlim$ Gal(L_i/K), define $\alpha \in$ Gal(K_s/K) as follows. If $x \in K_s$, choose L_i containing x and set $\alpha(x) = \alpha_i(x)$; compatibility of the α_i's implies that $\alpha(x)$ is independent of the choice of i. Since any $x, y \in K_s$ lie in some L_i, α is a field automorphism of K_s, that is, an element of Gal(K_s/K). By construction, $\phi(\alpha) = (\alpha_i)$. Hence ϕ is surjective and so an isomorphism.

$\diamond$

Example 6.11.2 If $\mathbb{F}_q$ is a finite field, its separable and algebraic closures coincide. The poset of finite extensions $\mathbb{F}_{q^n}$ of $\mathbb{F}_q$ is the poset of natural numbers, partially ordered by divisibility, and Gal$(\bar{\mathbb{F}}_q/\mathbb{F}_q)$ is $\varprojlim(\mathbb{Z}/n\mathbb{Z}) = \widehat{\mathbb{Z}} \cong \prod_p \widehat{\mathbb{Z}}_p$. For every prime p, let K be the union of all the $\mathbb{F}_{q^n}$ with $(p, n) = 1$; then Gal$(\bar{\mathbb{F}}_q/K)$ is $\widehat{\mathbb{Z}}_p$.

There is a topology on Gal$(K_s/K) = \varprojlim$ Gal(L_i/K) that makes it into a compact Hausdorff group: the profinite topology. To define it, recall that the *discrete topology* on a set X is the topology in which every subset of X is both open and closed. If we are given an inverse system $\{X_i\}$ of topological spaces, we give the inverse limit $\varprojlim X_i$ the topology it inherits as a subspace of the product ΠX_i. If the X_i are all finite discrete sets, the resulting topology on $X = \varprojlim X_i$ is called the *profinite topology* on X. Since each $\text{Gal}(L_i/K)$ is a finite discrete set, this defines the profinite topology on $\text{Gal}(K_s/K)$. To show that this is a compact Hausdorff group, we introduce the concepts of profinite set and profinite group.

Profinite Sets 6.11.3 A *profinite set* is a set X that is the inverse limit $\varprojlim X_i$ of some system $\{X_i\}$ of finite sets, made into a topological space using the profinite topology described above. The choice of the inverse system is not part of the data; we will see below that the profinite structure is independent of this choice.

The Cantor set is an interesting example of a profinite set; the subspace $\{0, 1, \frac{1}{2}, \ldots, \frac{1}{n}, \ldots\}$ of $\mathbb{R}$ is another. Profinite groups like $\widehat{\mathbb{Z}}_p$ and $\text{Gal}(K_s/K)$ form another important class of profinite sets.

Some elementary topological remarks are in order. Any discrete space is Hausdorff; as a subspace of ΠX_i, $\varprojlim X_i$ is Hausdorff. A discrete space is compact iff it is finite. A topological space X is called *totally disconnected* if every point of X is a connected component, and discrete spaces are totally disconnected.

Exercise 6.11.1 Suppose that $\{X_i\}$ is an inverse system of compact Hausdorff spaces. Show that $\varprojlim X_i$ is also compact Hausdorff. Then show that if each of the X_i is totally disconnected, $\varprojlim X_i$ is also totally disconnected. This proves one direction of the following theorem; the converse is proven in [Magid].

Theorem 6.11.4 *Profinite spaces are the same thing as totally disconnected, compact Hausdorff topological spaces. In particular, the profinite structure of $X \cong \varprojlim X_i$ depends only upon the topology and not upon the choice of inverse system $\{X_i\}$.*

Exercise 6.11.2 Let X be a profinite set.

1. Show that there is a canonical choice of the inverse system $\{X_i\}$ making X profinite, namely the system of its finite topological quotient spaces.
2. Show that every closed subspace of X is profinite.
3. If X is infinite, show that X has an open subspace U that is not profinite.

Definition 6.11.5 A *profinite group* is a group G that is an inverse limit of finite groups, made into a topological space using the profinite topology. Clearly

G is a profinite set that is also a compact Hausdorff topological group. In fact, the converse is true: Every totally disconnected compact Hausdorff group is a profinite group. A proof of this fact may be found in [Shatz], which we recommend as a good general reference for profinite groups and their cohomology.

Examples 6.11.6 (Profinite groups)

1. Any finite group is trivially profinite.
2. The p-adic integers $\widehat{\mathbb{Z}}_p = \lim\limits_{\leftarrow} \mathbb{Z}/p^i\mathbb{Z}$ are profinite by birthright.
3. Krull's Theorem 6.11.1 states that $\mathrm{Gal}(K_s/K)$ is a profinite group.
4. (Profinite completion) Let G be any (discrete) group. The *profinite completion* $\widehat{G}$ of G is the inverse limit of the system of all finite quotient groups G/H of G. For example, the profinite completion of $G = \mathbb{Z}$ is $\widehat{\mathbb{Z}} = \lim\limits_{\leftarrow}(\mathbb{Z}/n\mathbb{Z})$, but the profinite completion of $G = \mathbb{Q}/\mathbb{Z}$ is 0. The kernel of the natural map $G \to \widehat{G}$ is the intersection of all subgroups of finite index in G.

Exercise 6.11.3 Show that the category of profinite abelian groups is dual to the category of torsion abelian groups. *Hint:* Show that A is a torsion abelian group iff its Pontrjagin dual $\mathrm{Hom}(A, \mathbb{Q}/\mathbb{Z})$ is a profinite group.

Exercise 6.11.4 Let G be a profinite group, and let H be a subgroup of G.

1. If H is closed in G, show that H is also a profinite group.
2. If H is closed and normal, show that G/H is a profinite group.
3. If H is open in G, show that the index $[G : H]$ is finite, that H is closed in G, and therefore that H is profinite.

It is useful to have a canonical way of writing a profinite group G as the inverse limit of finite groups, and this is provided by the next result.

Lemma 6.11.7 *If G is a profinite group, let $\mathcal{U}$ be the poset of all open normal subgroups U of G. Then $\mathcal{U}$ forms a fundamental system of neighborhoods of 1, each G/U is a finite group, and $G \cong \lim\limits_{\leftarrow} G/U$.*

Proof If $G = \lim\limits_{\leftarrow} G_i$, then the $U_i = \ker(G \to G_i)$ are open normal subgroups of G and the natural map $G \to \lim\limits_{\leftarrow} G_i$ factors through $\lim\limits_{\leftarrow} G/U_i$. Since $\lim\limits_{\leftarrow}$ is left exact, this yields $G \cong \lim\limits_{\leftarrow} G/U_i$ and shows that $\{U_i\}$ (hence $\mathcal{U}$) forms a fundamental system of neighborhoods of 1. Hence every open subgroup U of G contains some U_i, and this suffices to show that $G \cong \lim\limits_{\leftarrow}\{G/U : U \in \mathcal{U}\}$. (Check this!) $\diamondsuit$

Exercise 6.11.5 (Fundamental Theorem of Galois theory) Prove that the usual correspondence of Galois theory induces a bijection between the set of topologically closed subgroups H of $G = \mathrm{Gal}(K_s/K)$ and the set of intermediate fields $K \subset L \subset K_s$. (Here $L = (K_s)^H$ and $H = \{g|gx = x$ for all $x \in L\}$.) Show that the closed normal subgroups of G correspond to the Galois extensions L of K. Conclude that if L/K is any Galois field extension, then $\mathrm{Gal}(L/K)$ is a profinite group: $\mathrm{Gal}(L/K) = G/H$.

To connect this result to more familiar Galois theory, show that the open subgroups H of $\mathrm{Gal}(K_s/K)$ correspond to the finite field extensions of K, and that the open normal subgroups of $\mathrm{Gal}(K_s/K)$ correspond to the finite Galois extensions of K.

In order to discuss the cohomology of profinite groups, we need to introduce an appropriate notion of G-module.

Definition 6.11.8 Let G be a profinite group. A *discrete G-module* is a G-module A such that, when A is given the discrete topology, the multiplication map $G \times A \to A$ is continuous. The next exercise provides a more elementary description of this.

Exercise 6.11.6

1. If A is a discrete G-module, show that for every $a \in A$ the stabilizer $U = \{g \in G : ga = a\}$ is an open subgroup of G, and $a \in A^U$, the submodule fixed by U.
2. If A is any G-module, let $\cup A^U$ denote the union of all subgroups A^U as U runs over the set of open subgroups of G. Show that A is a discrete G-module $\Longleftrightarrow \cup A^U = A$.

Examples 6.11.9 The field K_s is a discrete $\mathrm{Gal}(K_s/K)$-module for every K. If G is a finite group, every G-module is discrete, because $G \times A$ has the discrete topology.

A map of discrete G-modules is defined to be just a G-module map, so that the category $\mathbf{C}_G$ of discrete G-modules is a full additive subcategory of G–**mod**. The following exercise shows that in fact $\mathbf{C}_G$ is an abelian subcategory of G–**mod**.

Exercise 6.11.7 Let $f: A \to B$ be a map of discrete G-modules. Show that the G-modules $\ker(f) = \{a \in A : f(a) = 0\}$, $f(A)$, and $\mathrm{coker}(f) = B/f(A)$ are discrete G-modules. Conclude that $\mathbf{C}_G$ is an abelian category and that the inclusion $\mathbf{C}_G \subset G\text{-mod}$ is an exact functor. Then show that for all discrete G-modules A and all G-modules B,

$$\operatorname{Hom}_G(A, B) = \operatorname{Hom}_G(A, \cup B^U).$$

Conclude that the inclusion $\mathbf{C}_G \subset G\text{-mod}$ has the functor $\cup(\cdot)^U$ as right adjoint.

Lemma 6.11.10 *The abelian category $\mathbf{C}_G$ has enough injectives.*

Proof We may embed any discrete G-module A in an injective G-module I. By the above exercise, $A \subseteq \cup I^U \subseteq I$. Since $\cup(-)^U$ is right adjoint to the exact functor $\mathbf{C}_G \subset G\text{-mod}$, it preserves injectives (2.3.10). Consequently $\cup I^U$ is an injective object in $\mathbf{C}_G$. $\diamond$

Remark $\mathbf{C}_G$ does *not* have enough projectives.

Profinite Cohomology 6.11.11 The cohomology groups $H^*(G; A)$ of a profinite group G with coefficients in a discrete G-module A are defined to be the right derived functors of the functor $\mathbf{C}_G \to \mathbf{Ab}$ sending A to A^G, applied to A.

From this definition, we see that $H^0(G; A) = A^G$ and that when G is a finite group, $H^*(G; A)$ agrees with the usual group cohomology.

In fact, many of the results for the cohomology of finite groups carry over to profinite groups. For example, there is a category of profinite groups, a morphism being a continuous group homomorphism, and $H^*(G; A)$ is contravariant in G via the restriction maps. Indeed, the entire discussion of the functoriality of H^* in sections 6.3 and 6.7 carries through verbatim to our context. Of course, the inflation maps $\operatorname{inf}: H^*(G/H; A^H) \to H^*(G; A)$ are only defined when H is a closed normal subgroup of G, because the map $G \to G/H$ is only continuous when H is a closed normal subgroup of G. Similarly, whenever H is a closed normal subgroup of G, we can construct a Lyndon/Hochschild-Serre spectral sequence (6.8.2):

$$E_2^{pq} = H^p(G/H; H^q(H; A)) \Rightarrow H^{p+q}(G; A).$$

Since $\mathbf{C}_G$ doesn't have enough projectives, we need to modify the discussion in section 6.5 about the bar construction in order to talk about cocycles.

Cochains and cocycles 6.11.12 If A is a discrete G-module, let $C^n(G, A)$ denote the set of continuous maps from G^n to A. (When $n = 0$, $C^0(G, A) = A$

because $G^0 = \{1\}$.) Under pointwise addition, $C^n(G, A)$ becomes an abelian group, a subgroup of the group of n-cochains $\mathrm{Hom}_G(B^u_n, A)$ described in 6.5.4. The explicit formula for d shows that $C^*(G, A)$ is a subcomplex of the cochain complex $\mathrm{Hom}_G(B^u_*, A)$.

Exercise 6.11.8 Show that a map $\varphi\colon G^n \to A$ is continuous iff φ is locally constant, that is, iff each point of G^n has a neighborhood on which φ is constant.

Exercise 6.11.9 Show that $C^n(G, -)$ is an exact functor from $\mathbf{C}_G$ to $\mathbf{Ab}$. *Hint:* If $g\colon B \to C$ is onto, use the fact that every continuous $\varphi\colon G^n \to C$ is locally constant to lift φ to $C^n(G, B)$.

Exercise 6.11.10 Show that $C^n(G, A) = \varinjlim C^n(G/U, A^U)$, where U runs through all open normal subgroups of G.

Theorem 6.11.13 *Let G be a profinite group and A a discrete G-module. Then*

$$H^*(G; A) \cong H^*(C^*(G, A))$$

$$\cong \varinjlim H^*(G/U; A^U),$$

where U runs through all open normal subgroups of G.

Proof For simplicity, set $T^n(A) = H^n(C^*(G, A))$. We first calculate that

$$T^0(A) = \ker(A \xrightarrow{\ d\ } C^1(G, A))$$
$$= \{a \in A : (\forall g \in G) \quad 0 = (da)(g) = ga - a\}$$
$$= A^G.$$

Since $C^*(G, A) = \varinjlim C^*(G/U; A^U)$, and $\varinjlim$ commutes with cohomology (2.6.15), we see that $T^n(A) = \varinjlim H^n(C^*(G/U, A^U)) = \varinjlim H^n(G/U; A^U)$.

It now suffices to show that the $\{T^n\}$ form a universal cohomological δ-functor in the sense of 2.1.4, for this will imply that $T^n(A) \cong H^n(G; A)$. To see that they form a δ-functor, let $0 \to A \to B \to C \to 0$ be a short exact sequence of discrete G-modules. By exercise 6.11.9, each sequence

$$0 \to C^n(G, A) \to C^n(G, B) \to C^n(G, C) \to 0$$

is naturally exact, so we get a short exact sequence of cochain complexes. The associated long exact cohomology sequence with its natural coboundary $\delta^n \colon T^n(C) \to T^{n+1}(A)$ makes $\{T^n\}$ into a cohomological δ-functor.

To see that $\{T^n\}$ is universal, it suffices to show that each T^n (except T^0) vanishes on injective objects, for then T^n will be effaceable in the sense of exercise 2.4.5. If I is an injective object in $\mathbf{C}_G$ and U is an open normal subgroup of G, then I^U is an injective object in $\mathbf{C}_{G/U} = G/U-\mathbf{mod}$ because (as in 6.8.1) $-^U$ is right adjoint to the forgetful functor. Hence if $n \neq 0$, then

$$T^n(I) = \varinjlim H^n(G/U; I^U) = 0. \qquad \diamond$$

Corollary 6.11.14 *For $n \geq 1$, the $H^n(G; A)$ are torsion abelian groups.*

Proof Each G/U is a finite group, so $H^n(G/U, A^U)$ is a torsion group. $\diamond$

Exercise 6.11.11 Let G be the profinite group $\widehat{\mathbb{Z}}_p$. Show that

$$H^i(G; \mathbb{Z}) = \begin{Bmatrix} \widehat{\mathbb{Z}}_p & i \text{ even} \\ 0 & i \text{ odd} \end{Bmatrix}.$$

Low Dimensions 6.11.15 We have already seen that $H^0(G; A) = A^G$. A calculation using the complex $C^*(G, A)$ shows that $H^1(G; A)$ is the group of continuous derivations of G in A, modulo the (ctn.) principal derivations, and that $H^1(G; \mathbb{Z})$ is the group of continuous maps from G to $\mathbb{Z}$. Similarly, $H^2(G; A)$ is the group of classes of continuous factor sets of G in A. If A is *finite*, $H^2(G; A)$ classifies the profinite extensions of G by A. (The discrete group A is only profinite when it is finite.)

Hilbert's Theorem 90 6.11.16 *Let K be a field and set $G = \mathrm{Gal}(K_s/K)$. Then K_s and its units K_s^* are discrete G-modules with $(K_s)^G = K$ and $(K_s^*)^G = K^*$. Moreover*

1. *$H^n(G; K_s) = 0$ for all $n \neq 0$.*
2. *$H^1(G; K_s^*) = 0$.*

Proof Let U be an open normal subgroup of G and $L = K_s^U$ the corresponding Galois extension of K, so that $G/U = \mathrm{Gal}(L/K)$ and $(K_s^*)^U = L^*$. By Hilbert's Theorem 90 for L/K (6.3.7, 6.4.7), we see that

$$H^n(G/U; L) = 0 \text{ for } n \neq 0,$$

$$H^1(G/U; L^*) = 0.$$

Now take the limit over all U to get the result.   ◇

Brauer group 6.11.17 The classical *Brauer group* of K is the set of all equivalence classes of central simple K-algebras Λ (with equivalence relation $M_i(\Lambda) \approx M_j(\Lambda')$). It is also isomorphic to the set of all finite-dimensional division K-algebras Δ with center K. The relative Brauer groups $Br(L/K)$ of 6.6.11 were constructed so that $Br(K)$ is the union of the relative groups $Br(L/K)$. On the other hand, since $Br(L/K) = H^2(\mathrm{Gal}(L/K), L^*)$ by 6.6.11, $H^2(G; K_s^*)$ is also the direct limit of the $Br(L/K)$, because if U is an open normal subgroup and $L = (K_s)^U$, then $G/U = \mathrm{Gal}(L/K)$ and $(K_s^*)^U = L^*$. Therefore $Br(K)$ is naturally isomorphic to the profinite cohomology group $H^2(G; K_s^*)$. The following result provides a cohomological proof of the fact that each $Br(L/K)$ is a subgroup of $Br(K)$.

Proposition 6.11.18 *If $K \subset L$ is a Galois field extension with Galois group $G = \mathrm{Gal}(L/K)$, there is an exact sequence*

$$0 \to Br(L/K) \xrightarrow{\mathrm{inf}} Br(K) \xrightarrow{\mathrm{res}} Br(L)^G \to H^3(G; L^*) \to H^3(K, K_s^*).$$

In particular, $Br(L/K)$ is the kernel of $Br(K) \to Br(L)$.

Proof Let $H \subset \mathrm{Gal}(K_s/K)$ be the closed normal subgroup corresponding to L, so that $G = \mathrm{Gal}(K_s/K)/H$. The Hochschild-Serre spectral sequence 6.11.11 is

$$E_2^{pq} = H^p(G; H^q(H; K_s^*)) \Rightarrow H^*(\mathrm{Gal}(K_s/K); K_s^*).$$

Along the x-axis we find $H^p(G; L^*)$. By Hilbert's Theorem 90 for L, the row $q = 1$ vanishes. The exact sequence of low degree terms is the sequence in question.   ◇

Exercise 6.11.12 Let $\mathbb{F}_q$ be a finite field. Show that $Br(L/\mathbb{F}_q) = 0$ for every finite extension L of $\mathbb{F}_q$ and conclude that $Br(\mathbb{F}_q) = 0$. *Hint:* $\mathrm{Gal}(L/\mathbb{F}_q)$ is cyclic of order $n = [L : \mathbb{F}_q]$ and the norm map $N: L^* \to K^*$ is onto (6.4.8).

Vista 6.11.19 Many deep results about the Brauer group can be established more easily using cohomological machinery. We list a few here, referring the reader to [Shatz] for more details.

- If $\mathrm{char}(K) = p \neq 0$, $Br(K)$ is divisible by p.
- (Tsen's Theorem) If K is a function field in one variable over an algebraically closed field, then $Br(K) = 0$.

- $Br(\mathbb{R}) = \mathbb{Z}/2$, the quaternion algebra $\mathbb{H}$ being nontrivial. (See 6.4.8.)
- (Hasse) If K is a local field, that is, the p-adic rationals $\widehat{\mathbb{Q}}_p$, or a finite extension of $\widehat{\mathbb{Q}}_p$, then there is a canonical isomorphism $Br(K) \cong \mathbb{Q}/\mathbb{Z}$. The element of $\mathbb{Q}/\mathbb{Z}$ corresponding to a central simple K-algebra Λ is called the *Hasse invariant* of Λ.
- The Brauer group of $\mathbb{Q}$ injects into $Br(\mathbb{R}) \cong \mathbb{Z}/2$ plus the direct sum over all primes p of the groups $Br(\widehat{\mathbb{Q}}_p) \cong \mathbb{Q}/\mathbb{Z}$, with cokernel $\mathbb{Q}/\mathbb{Z}$. Thus the Hasse invariants uniquely determine $Br(\mathbb{Q})$, and the sum of the Hasse invariants is zero.

7

Lie Algebra Homology and Cohomology

Lie algebras were introduced by Sophus Lie in connection with his studies of Lie groups; Lie groups are not only groups but also smooth manifolds, the group operations being smooth. If G is a Lie group, the tangent space $\mathfrak{g}$ of G at the identity $e \in G$ is a Lie algebra over $\mathbb{R}$. The vector space of left invariant vector fields on G is canonically isomorphic to $\mathfrak{g}$, and the Lie bracket $[X, Y]$ of vector fields X and Y may be defined as a vector field:

$$[X, Y]f = X(Yf) - Y(Xf), \quad f \text{ a smooth function on } G.$$

This rich interplay with Differential Geometry forms the original motivation for the abstract study of Lie algebras. More history is given in 7.8.14 below.

7.1 Lie Algebras

Let k be a fixed commutative ring. A *nonassociative algebra* A is a k-module equipped with a bilinear product $A \otimes_k A \to A$. Note that we do not assume the existence of a unit, so that 0 is the smallest possible nonassociative algebra. A *Lie algebra* $\mathfrak{g}$ is a nonassociative algebra whose product, written as $[xy]$ or $[x, y]$ and called the *Lie bracket*, satisfies (for $x, y, z \in \mathfrak{g}$):

Skew-symmetry: $[x, x] = 0$ (and hence $[x, y] = -[y, x]$);
Jacobi's Identity: $[x, [y, z]] + [y, [z, x]] + [z, [x, y]] = 0$.

An *ideal* of $\mathfrak{g}$ is a k-submodule $\mathfrak{h}$ such that $[\mathfrak{g}, \mathfrak{h}] \subseteq \mathfrak{h}$, that is, for all $g \in \mathfrak{g}$ and $h \in \mathfrak{h}$ we have $[g, h] \in \mathfrak{h}$. Note that an ideal is a Lie algebra in its own right, and that the quotient $\mathfrak{g}/\mathfrak{h}$ inherits the structure of a Lie algebra as well. There is a category whose objects are $(k\text{-})$Lie algebras; a morphism $\varphi: \mathfrak{g} \to \mathfrak{h}$ is a product-preserving k-module homomorphism. Thus every ideal $\mathfrak{h} \subset \mathfrak{g}$ yields a short exact sequence (!) of Lie algebras:

$$0 \to \mathfrak{h} \to \mathfrak{g} \to \mathfrak{g}/\mathfrak{h} \to 0.$$

Example 7.1.1 An *abelian* Lie algebra is one in which all the Lie brackets $[x, y] = 0$. Every k-module has the structure of an abelian Lie algebra.

If $\mathfrak{g}$ is any Lie algebra, define $[\mathfrak{g}, \mathfrak{g}]$ to be the k-submodule of $\mathfrak{g}$ generated by all Lie brackets $[x, y]$ with $x, y \in \mathfrak{g}$. Then $[\mathfrak{g}, \mathfrak{g}]$ is an ideal of $\mathfrak{g}$, and the quotient $\mathfrak{g}^{ab} = \mathfrak{g}/[\mathfrak{g}, \mathfrak{g}]$ is an abelian Lie algebra. Obviously, $\mathfrak{g}^{ab}$ is the largest quotient Lie algebra of $\mathfrak{g}$ that is abelian.

Example 7.1.2 The primordial Lie algebra is the Lie algebra $\mathfrak{a} = \text{Lie}(A)$ of an associative k-algebra A (even if A is an algebra without a unit). This is the underlying k-module A, given the commutator product $[x, y] = xy - yx$. We leave it to the reader (exercise!) to verify Jacobi's identify, that is, that $\mathfrak{a}$ is a Lie algebra, and to check that this defines a functor "Lie" from the category of (associative, possibly nonunital) k-algebras to the category of Lie algebras.

Examples 7.1.3 If A is an associative k-algebra, so is $M_m(A)$, the $m \times m$ matrices with coefficients in A. We write $\mathfrak{gl}_m(A)$ for the Lie algebra $\text{Lie}(M_m(A))$. If $A = k$, we write $\mathfrak{gl}_m$ for $\mathfrak{gl}_m(k)$.

Here are some famous Lie subalgebras of $\mathfrak{gl}_m(A)$; if $A = k$, it is traditional to drop the reference to A, writing merely, $\mathfrak{o}_m$, $\mathfrak{sl}_m$, $\mathfrak{t}_m$, and $\mathfrak{n}_m$ instead of $\mathfrak{o}_m(k)$, $\mathfrak{sl}_m(k)$, and so on.

1. The *orthogonal algebra* $\mathfrak{o}_m(A)$ of all skew-symmetric matrices: $\{g : g_{ij} = -g_{ji}\}$.
2. The *special linear algebra* $\mathfrak{sl}_m(A)$. If A is commutative, this is the algebra of all matrices of trace 0. If A is not commutative, then we must consider the trace as taking values in $A/[A, A]$, because a matrix change of basis changes the trace $\sum g_{ii}$ by an element of $[A, A]$. Thus $\mathfrak{sl}_m(A)$ is the kernel of the trace map, yielding the short exact sequence of Lie algebras:

$$0 \to \mathfrak{sl}_m(A) \to \mathfrak{gl}_m(A) \xrightarrow{\text{trace}} A/[A, A] \to 0.$$

3. The *upper triangular matrices* $\mathfrak{t}_m(A) : \{g : g_{ij} = 0 \text{ if } i < j\}$.
4. The *strictly upper triangular matrices* $\mathfrak{n}_m(A) : \{g : g_{ij} = 0 \text{ if } i \le j\}$.

Example 7.1.4 (Derivation algebras) Let A be a nonassociative (= not necessarily associative) k-algebra. A *derivation* D of A (into itself) is a k-module endomorphism of A such that the Leibnitz formula holds

$$D(ab) = (Da)b + a(Db) \quad (a, b \in A).$$

The set $\mathrm{Der}(A)$ of derivations of A is clearly a k-submodule of $\mathrm{End}_k(A)$. Moreover, the commutator $[D_1, D_2]$ of two derivations is a derivation, since

$$
\begin{aligned}
[D_1 D_2]ab &= D_1(D_2(ab)) - D_2(D_1(ab)) \\
&= D_1((D_2a)b) + D_1(a(D_2b)) - D_2((D_1a)b) - D_2(a(D_1b)) \\
&= (D_1 D_2 a)b + a(D_1 D_2 b) - (D_2 D_1 a)b - a(D_2 D_1 b) \\
&= ([D_1 D_2]a)b + a([D_1 D_2]b).
\end{aligned}
$$

Hence $\mathrm{Der}(A)$ is a Lie algebra; it is called the *derivation algebra* of A.

Example 7.1.5 Given a k-module M, the *free Lie algebra on M* is a Lie algebra $\mathfrak{f}(M)$, containing M as a submodule, which satisfies the usual universal property: Every k-module map $M \to \mathfrak{g}$ into a Lie algebra extends uniquely to a Lie algebra map $\mathfrak{f}(M) \to \mathfrak{g}$. In other words, as a functor $\mathfrak{f}$ is left adjoint to the forgetful functor from Lie algebras to modules

$$\mathrm{Hom}_{k-\mathbf{mod}}(M, \mathfrak{g}) \cong \mathrm{Hom}_{\mathrm{Lie}}(\mathfrak{f}(M), \mathfrak{g}).$$

The existence of $\mathfrak{f}(M)$ follows from general considerations of category theory (the Adjoint Functor Theorem); a concrete construction will be given in section 7.3. Clearly $\mathfrak{f}(M)$ is unique up to isomorphism.

If X is a set, the *free Lie algebra on X* is $\mathfrak{f}(M)$, where M is the free k-module on the set X. Clearly

$$\mathrm{Hom}_{\mathbf{Sets}}(X, \mathfrak{g}) \cong \mathrm{Hom}_{\mathrm{Lie}}(\mathfrak{f}(X), \mathfrak{g}),$$

so there is a corresponding universal property for $\mathfrak{f}(X)$.

Exercise 7.1.1 Show that the free Lie algebra $\mathfrak{f}(\{x\}) = \mathfrak{f}(k)$ on the set $\{x\}$ is the 1-dimensional abelian Lie algebra k. Then show that $\mathfrak{f}(\{x, y\})$ is a graded, free k-module having an infinite basis of monomials

$$x, y, [xy], [x[xy]], [y[xy]], [x[x[xy]]], [x[y[xy]]], [y[y[xy]]], \cdots.$$

(There are 6 monomials of degree 5. In general, there are $\frac{1}{d} \sum_{i|d} \mu(i) 2^{d/i}$ monomials of degree d, where μ denotes the Möbius function [Bour, ch. 2, sec. 3.3, thm. 2].)

Exercise 7.1.2 (Product Lie algebra) If $\mathfrak{g}$ and $\mathfrak{h}$ are Lie algebras, we can make the k-module $\mathfrak{g} \times \mathfrak{h}$ into a Lie algebra by a slotwise product: $[(g_1, h_1), (g_2, h_2)] = ([g_1, g_2], [h_1, h_2])$. Show that $\mathfrak{g} \times \mathfrak{h}$ is the product in the category of Lie algebras.

Nilpotent Lie Algebras 7.1.6 In analogy with group theory, we define the *lower central series* of a Lie algebra $\mathfrak{g}$ to be the following descending sequence of ideals:

$$\mathfrak{g} \supseteq \mathfrak{g}^2 = [\mathfrak{g}, \mathfrak{g}] \supseteq \mathfrak{g}^3 = [\mathfrak{g}^2, \mathfrak{g}] \supseteq \cdots \supseteq \mathfrak{g}^n = [\mathfrak{g}^{n-1}, \mathfrak{g}] \supseteq \cdots .$$

We say that $\mathfrak{g}$ is a *nilpotent* Lie algebra if $\mathfrak{g}^n = 0$ for some n. For example, the strictly upper triangular Lie algebra $\mathfrak{n}_m(A)$ is nilpotent for every k-algebra A; $\mathfrak{n}_m(A)^n$ is the ideal of matrices (g_{ij}) with $g_{ij} = 0$ unless $i \geq j + n$. Abelian Lie algebras are another obvious class of nilpotent Lie algebras.

Solvable Lie Algebras 7.1.7 Again following group theory, we define the *derived series* of $\mathfrak{g}$ to be the descending sequence of ideals

$$\mathfrak{g} \supseteq \mathfrak{g}' = [\mathfrak{g}, \mathfrak{g}] \supseteq \mathfrak{g}'' = (\mathfrak{g}')' \supseteq \cdots \supseteq \mathfrak{g}^{(n)} = [\mathfrak{g}^{(n-1)}, \mathfrak{g}^{(n-1)}] \supseteq \cdots .$$

We say that $\mathfrak{g}$ is a *solvable* Lie algebra if $\mathfrak{g}^{(n)} = 0$ for some n.

Lemma 7.1.8 *Every nilpotent Lie algebra is solvable.*

Proof It suffices to show that $[\mathfrak{g}^i, \mathfrak{g}^j] \subseteq \mathfrak{g}^{i+j}$, for then by induction we see that $\mathfrak{g}^{(n)} \subseteq \mathfrak{g}^n$. To see this we proceed by induction on j, the case $j = 1$ being the definition $\mathfrak{g}^{i+1} = [\mathfrak{g}^i, \mathfrak{g}]$. Inductively, we compute

$$[\mathfrak{g}^i, \mathfrak{g}^{j+1}] = [\mathfrak{g}^i, [\mathfrak{g}^j, \mathfrak{g}]] \subseteq [[\mathfrak{g}^i, \mathfrak{g}], \mathfrak{g}^j] + [[\mathfrak{g}^i, \mathfrak{g}^j], \mathfrak{g}]$$
$$\subseteq [\mathfrak{g}^{i+1}, \mathfrak{g}^j] + [\mathfrak{g}^{i+j}, \mathfrak{g}] = \mathfrak{g}^{i+j+1}. \qquad \diamond$$

Example 7.1.9 The upper triangular Lie algebra $\mathfrak{t}_m(A)$ of a commutative k-algebra A is solvable but not nilpotent.

7.2 g-Modules

Let $\mathfrak{g}$ be a Lie algebra over k. A (left) $\mathfrak{g}$-*module* M is a k-module equipped with a k-bilinear product $\mathfrak{g} \otimes_k M \to M$ (written $x \otimes m \mapsto xm$) such that

$$[x, y]m = x(ym) - y(xm) \text{ for all } x, y \in \mathfrak{g} \text{ and } m \in M.$$

Examples 7.2.1

1. If A is an associative algebra and $\mathfrak{g} = \text{Lie}(A)$, any left A-module may be thought of as a left $\mathfrak{g}$-module in an obvious way.
2. The Lie bracket makes $\mathfrak{g}$ itself into a left $\mathfrak{g}$-module (by Jacobi's identity). This module is usually called the *adjoint representation* of $\mathfrak{g}$.
3. A *trivial $\mathfrak{g}$-module* is a k-module M on which $\mathfrak{g}$ acts as zero: $xm = 0$ for all $x \in \mathfrak{g}, m \in M$.

A *$\mathfrak{g}$-module homomorphism* $f: M \to N$ is a k-module map that is product-preserving, that is, $f(xm) = xf(m)$. We write $\text{Hom}_{\mathfrak{g}}(M, N)$ for the set of all such $\mathfrak{g}$-module homomorphisms. If $\alpha \in k$, then αf is also a $\mathfrak{g}$-module map, so therefore $\text{Hom}_{\mathfrak{g}}(M, N)$ is a k-submodule of $\text{Hom}_k(M, N)$.

The left $\mathfrak{g}$-modules and $\mathfrak{g}$-module homomorphisms form a category called $\mathfrak{g}$–**mod**. By the above remarks, it is an additive category. The following exercise shows that it is in fact an abelian category.

Exercise 7.2.1

1. Let $f: M \to N$ be a $\mathfrak{g}$-module homomorphism. Show that the k-modules $\ker(f)$, $\text{im}(f)$, and $\text{coker}(f)$ are the kernel, image, and cokernel of f in $\mathfrak{g}$–**mod**.
2. Show that a monic (resp., epi) in $\mathfrak{g}$–**mod** is also a monic (resp., epi) in k–**mod**. By (1), this proves that $\mathfrak{g}$–**mod** is an abelian category.

Exercise 7.2.2 Let $E = \text{End}_k(M)$ be the associative algebra of k-module endomorphisms of a k-module M. Show that maps $\mathfrak{g} \otimes M \to M$ making M into a $\mathfrak{g}$-module are in 1–1 correspondence with Lie algebra homomorphisms $\mathfrak{g} \to \text{Lie}(E)$. Conclude that a $\mathfrak{g}$-module may also be described as a k-module M together with a Lie algebra homomorphism $\mathfrak{g} \to \text{Lie}(\text{End}_k(M))$.

Exercise 7.2.3 There is also a category **mod**–$\mathfrak{g}$ of right $\mathfrak{g}$-modules, whose definition should be obvious. If M is a right $\mathfrak{g}$-module, show that the product $xm = -mx$ $(x \in \mathfrak{g}, m \in M)$ makes M into a left $\mathfrak{g}$-module, and that this induces a natural isomorphism of categories: $\mathfrak{g}$–**mod** $\cong$ **mod**–$\mathfrak{g}$.

Many of the notions we introduced for G-modules in Chapter 6 have analogues for $\mathfrak{g}$-modules. For example, there is a *trivial $\mathfrak{g}$-module functor* from k–**mod** to $\mathfrak{g}$–**mod**; it is the exact functor obtained by considering a k-module as a trivial $\mathfrak{g}$-module. Consider the following two functors from $\mathfrak{g}$–**mod** to k–**mod**:

1. The *invariant submodule* $M^{\mathfrak{g}}$ of a $\mathfrak{g}$-module M,

$$M^{\mathfrak{g}} = \{m \in M : xm = 0 \quad \text{for all } x \in \mathfrak{g}\}.$$

Considering k as a trivial $\mathfrak{g}$-module, we have $M^{\mathfrak{g}} \cong \operatorname{Hom}_{\mathfrak{g}}(k, M)$.
2. The *coinvariants* $M_{\mathfrak{g}}$ of a $\mathfrak{g}$-module M, $M_{\mathfrak{g}} = M/\mathfrak{g}M$.

Exercise 7.2.4 Let M be a $\mathfrak{g}$-module.

1. Show that $M^{\mathfrak{g}}$ is the maximal trivial $\mathfrak{g}$-submodule of M, and conclude that $-^{\mathfrak{g}}$ is right adjoint to the trivial $\mathfrak{g}$-module functor. Conclude that $-^{\mathfrak{g}}$ is a left exact functor.
2. Show that $M_{\mathfrak{g}}$ is the largest quotient module of M that is trivial, and conclude that $-_{\mathfrak{g}}$ is left adjoint to the trivial $\mathfrak{g}$-module functor. Conclude that $-_{\mathfrak{g}}$ is a right exact functor.

We will see in the next section that the category $\mathfrak{g}$–**mod** has "enough" projectives and injectives in the sense of Chapter 2. Therefore we can form the derived functors of $-^{\mathfrak{g}}$ and $-_{\mathfrak{g}}$.

Definition 7.2.2 Let M be a $\mathfrak{g}$-module. We write $H_*(\mathfrak{g}, M)$ or $H_*^{\mathrm{Lie}}(\mathfrak{g}, M)$ for the left derived functors $L_*(-_{\mathfrak{g}})(M)$ of $-_{\mathfrak{g}}$ and call them the *homology groups* of $\mathfrak{g}$ *with coefficients in* M. By definition, $H_0(\mathfrak{g}, M) = M_{\mathfrak{g}}$.

Similarly, we write $H^*(\mathfrak{g}, M)$ or $H^*_{\mathrm{Lie}}(\mathfrak{g}, M)$ for the right derived functors $R^*(-^{\mathfrak{g}})(M)$ of $-^{\mathfrak{g}}$ and call them the *cohomology groups* of $\mathfrak{g}$ *with coefficients in* M. By definition, $H^0(\mathfrak{g}, M) = M^{\mathfrak{g}}$.

Examples 7.2.3

0. If $\mathfrak{g} = 0$, $M_{\mathfrak{g}} = M^{\mathfrak{g}} = M$. Since the higher derived functors of an exact functor vanish, $H_*^{\mathrm{Lie}}(0, M) = H^*_{\mathrm{Lie}}(0, M) = 0$ for $* \neq 0$.
1. Let $\mathfrak{g}$ be the free k-module on basis $\{e_1, \cdots, e_n\}$, made into an (abelian) Lie algebra with zero Lie bracket. Since a $\mathfrak{g}$-module is just a k-module with n commuting endomorphisms $e_1, \cdots, e_n$, it follows that $\mathfrak{g}$–**mod** is isomorphic to the category R–**mod** of left modules over the polynomial ring $R = k[e_1, \cdots, e_n]$. If k is the trivial $\mathfrak{g}$-module, considered as an R-module on which the e_i act as zero, then $M_{\mathfrak{g}} = k \otimes_R M$ and $M^{\mathfrak{g}} = \operatorname{Hom}_R(k, M)$. Therefore we have

$$H_*^{\mathrm{Lie}}(\mathfrak{g}, M) = \operatorname{Tor}_*^R(k, M) \text{and } H^*_{\mathrm{Lie}}(\mathfrak{g}, M) = \operatorname{Ext}_R^*(k, M).$$

These functors were discussed in Chapter 3.

2. Let $\mathfrak{f}$ be the free Lie algebra on a set X. In this case an $\mathfrak{f}$-module is just a k-module M with an arbitrary set $\{e_x : x \in X\}$ of endomorphisms. That is, the category $\mathfrak{f}$–**mod** is isomorphic to the category R–**mod** of left modules over the free ring $R = k\{X\}$ on the set X. If k denotes the trivial $\mathfrak{f}$-module, then $M_{\mathfrak{f}} = k \otimes_R M$ and $M^{\mathfrak{f}} = \mathrm{Hom}_R(k, M)$. Therefore

$$H_*^{\mathrm{Lie}}(\mathfrak{f}, M) = \mathrm{Tor}_*^R(k, M) \quad \text{and} \quad H_{\mathrm{Lie}}^*(\mathfrak{f}, M) = \mathrm{Ext}_R^*(k, M).$$

We end this section with a calculation of the H^* and H_* groups for $\mathfrak{f}$.

Proposition 7.2.4 *The ideal $\mathfrak{I} = Xk\{X\}$ of the free ring $k\{X\}$ is free as a right $k\{X\}$-module with basis the set X. Hence*

$$0 \to \mathfrak{I} \to k\{X\} \to k \to 0$$

is a free resolution of k as a right $k\{X\}$-module.

Proof As a free k-module, $k\{X\}$ has for basis the set $\mathcal{W}$ of words in the elements of the set X, and $\mathfrak{I}$ is a free k-module on basis $\mathcal{W} - \{1\}$. Every element of $\mathcal{W} - \{1\}$ has a unique expression of the form xw with $x \in X$ and $w \in \mathcal{W}$, so $\{xw : x \in X, w \in \mathcal{W}\}$ is another basis for $\mathfrak{I}$ as a k-module. For each $x \in X$ the k-span $xk\{X\}$ of the set $\{xw : w \in \mathcal{W}\}$ is isomorphic to $k\{X\}$, and $\mathfrak{I}$ is the direct sum of the $xk\{X\}$, both as k-modules and as right $k\{X\}$-modules. That is, $\mathfrak{I}$ is a free right $k\{X\}$-module with basis X, as claimed. $\diamond$

Corollary 7.2.5 *If $\mathfrak{f}$ is the free Lie algebra on a set X, then $H_n^{\mathrm{Lie}}(\mathfrak{f}, M) = H_{\mathrm{Lie}}^n(\mathfrak{f}, M) = 0$ for all $n \geq 2$ and all $\mathfrak{f}$-modules M. Moreover $H_0^{\mathrm{Lie}}(\mathfrak{f}, k) = H_{\mathrm{Lie}}^0(\mathfrak{f}, k) = k$, while*

$$H_1^{\mathrm{Lie}}(\mathfrak{f}, k) \cong \bigoplus_{x \in X} k \quad \text{and} \quad H_{\mathrm{Lie}}^1(\mathfrak{f}, k) = \prod_{x \in X} k.$$

Proof Using the given free resolution of k, $H_*^{\mathrm{Lie}}(\mathfrak{f}, M)$ is the homology of the complex $0 \to \mathfrak{I} \otimes_R M \to M \to 0$, and $H_{\mathrm{Lie}}^*(\mathfrak{f}, M)$ is the homology of the complex $0 \to M \to \mathrm{Hom}_{\mathfrak{f}}(\mathfrak{I}, M) \to 0$. For $M = k$, the differentials are zero. $\diamond$

Exercise 7.2.5 Let $\mathfrak{r}$ be an ideal of a free Lie algebra $\mathfrak{f}$ on a set X. Show that if $\mathfrak{r} \neq 0$, then $[\mathfrak{f}, \mathfrak{r}] \neq \mathfrak{r}$.

7.3 Universal Enveloping Algebras

The universal enveloping algebra $U\mathfrak{g}$ of a Lie algebra $\mathfrak{g}$ plays the same formal role as the group ring $\mathbb{Z}G$ of a group G does. In particular, $\mathfrak{g}$–**mod** is naturally isomorphic to the category $U\mathfrak{g}$–**mod** of left $U\mathfrak{g}$-modules. This isomorphism provides an easy proof that $\mathfrak{g}$–**mod** has enough projectives and injectives in the sense of Chapter 2, so that the derived functor definitions of $H_*(\mathfrak{g}, M)$ and $H^*(\mathfrak{g}, M)$ make sense.

In this section we will develop some of the ring-theoretic properties of $U\mathfrak{g}$. Since $U\mathfrak{g}$ will be a quotient ring of the tensor algebra $T(\mathfrak{g})$, we first describe the tensor algebra $T(M)$ of a k-module M.

Definition 7.3.1 If M is any k-module, the *tensor algebra* $T(M)$ is the following graded associative algebra with unit generated by M:

$$T(M) = k \oplus M \oplus (M \otimes M) \oplus (M \otimes M \otimes M) \oplus \cdots \oplus M^{\otimes n} \oplus \cdots.$$

Here $M^{\otimes n}$ denotes $M \otimes \cdots \otimes M$, the tensor product (over k) of n copies of M, whose elements are finite sums of terms $x_1 \otimes \cdots \otimes x_n$ ($x_i \in M$). The product $\otimes$ in $T(M)$ amounts to concatenation of terms. Writing $i\colon M \to T(M)$ for the evident inclusion, this means that $T(M)$ is generated by $i(M)$ as a k-algebra. Clearly T is a functor from k–**mod** to the category of (associative, unital) k-algebras.

Here is a presentation of $T(M)$ as an algebra. $T(M)$ is the free algebra on generators $i(x)$, $x \in M$, subject only to the k-module relations on $i(M)$:

$$\alpha i(x) = i(\alpha x) \quad \text{and} \quad i(x) + i(y) = i(x + y) \quad (\alpha \in k; x, y \in M).$$

If M is a free module with basis $\{x_1, \ldots\}$, then $T(M)$ is the free k-algebra $k\{x_1, \ldots\}$. In particular, $T(k)$ is isomorphic to the polynomial ring $k[x]$. In general $T(M)$ is not a commutative algebra except when $M = k$ or $M \cong k/I$ for some ideal I of k.

Exercise 7.3.1 Show that T is the left adjoint of the forgetful functor from k–**alg** to k–**mod**, and that $i\colon M \to T(M)$ is the unit of this adjunction. That is, show that for every associative k-algebra A,

$$\mathrm{Hom}_{k-\mathbf{mod}}(M, A) \cong \mathrm{Hom}_{k-\mathbf{alg}}(T(M), A).$$

Exercise 7.3.2 (Free Lie algebras) Given a k-module M, consider the Lie algebra $\mathrm{Lie}(T(M))$ underlying the tensor algebra $T(M)$. Let $\mathfrak{f}$ denote the Lie subalgebra generated by M. That is, elements of $\mathfrak{f}$ are sums of iterated brackets $[x_1, [x_2[\cdots, x_n]]]$ of elements $x_i \in M$. Show that $\mathfrak{f}$ satisfies the universal property of a free Lie algebra of M (see 7.1.5). This provides a constructive proof of the existence of free Lie algebras.

Definition 7.3.2 If $\mathfrak{g}$ is a Lie algebra over k, the *universal enveloping algebra* $U(\mathfrak{g})$ is the quotient of $T(\mathfrak{g})$ by the 2-sided ideal generated by the relations

$$(*) \qquad i([x, y]) = i(x)i(y) - i(y)i(x) \quad (x, y \in \mathfrak{g}).$$

Alternatively, $U\mathfrak{g}$ is the free algebra on generators $i(x)$, $x \in \mathfrak{g}$, subject to the k-module relations on $\mathfrak{g}$ as well as the relation $(*)$. The relation $(*)$ guarantees that i preserves the Lie bracket, that is, that $i: \mathfrak{g} \to \mathrm{Lie}(U\mathfrak{g})$ is a Lie algebra homomorphism and that $U\mathfrak{g}$ is a left $\mathfrak{g}$-module. Since the construction is natural in $\mathfrak{g}$, U is a functor from Lie algebras to associative k-algebras. See [BAII, section 3.9] [JLA, ch. V].

Exercise 7.3.3 Show that U is the left adjoint of the "Lie" algebra functor described in 7.1.2 and that i is the unit of the adjunction. That is, for every associative k-algebra A, there is a natural isomorphism

$$\mathrm{Hom}_{\mathrm{Lie}}(\mathfrak{g}, \mathrm{Lie}(A)) \cong \mathrm{Hom}_{k-\mathbf{alg}}(U\mathfrak{g}, A).$$

This isomorphism explains the term "universal"; any Lie algebra map $\mathfrak{g} \to \mathrm{Lie}(A)$ extends to a unique k-algebra map $U\mathfrak{g} \to A$.

Theorem 7.3.3 *If $\mathfrak{g}$ is a Lie algebra, then every left $\mathfrak{g}$-module is naturally a left $U\mathfrak{g}$-module, and conversely. The category $\mathfrak{g}$–$\mathbf{mod}$ is naturally isomorphic to the category $U\mathfrak{g}$–$\mathbf{mod}$ of left $U\mathfrak{g}$-modules.*

Proof Let M be a k-module and write $E = \mathrm{End}_k(M)$ for the k-algebra of all k-module endomorphisms of M. By adjointness,

$$\mathrm{Hom}_{\mathrm{Lie}}(\mathfrak{g}, \mathrm{Lie}(E)) \cong \mathrm{Hom}_{k-\mathbf{alg}}(U\mathfrak{g}, \mathrm{End}_k(M)).$$

A $\mathfrak{g}$-module is a k-module M together with a Lie algebra map $\mathfrak{g} \to \mathrm{Lie}(E)$ (see exercise 7.2.2). But a $U\mathfrak{g}$-module is a k-module M together with an associative algebra map $U\mathfrak{g} \to \mathrm{End}_k(M)$, so the theorem follows. $\diamond$

Corollary 7.3.4 *The category $\mathfrak{g}$–$\mathbf{mod}$ has enough projectives and enough injectives in the sense of Chapter 2. In particular, $U\mathfrak{g}$ is a projective object in $\mathfrak{g}$–$\mathbf{mod}$.*

Here is a more concrete description of the correspondence between $\mathfrak{g}$-modules and $U\mathfrak{g}$-modules. Given a $\mathfrak{g}$-module M and a monomial $x_1 \cdots x_n$ in $U\mathfrak{g}$ $(x_i \in \mathfrak{g})$, the formula

$$(x_1 \cdots x_n)m = x_1(x_2(\cdots(x_n m))), \, m \in M,$$

makes M into a $U\mathfrak{g}$-module. Conversely, if M is a $U\mathfrak{g}$-module and $x \in \mathfrak{g}$, the formula $xm = i(x)m$ $(m \in M)$ makes M into a $\mathfrak{g}$-module because of the relation $(*)$ of 7.3.2.

Example 7.3.5 (Augmentation ideal) There is a unique k-algebra homomorphism $\varepsilon: U\mathfrak{g} \to k$, sending $i(\mathfrak{g})$ to zero, called the *augmentation*. This is clear from the presentation of $U\mathfrak{g}$, and ε corresponds to the zero Lie algebra map $\mathfrak{g} \to \mathrm{Lie}(k)$ under the adjunction. It is the analogue for Lie algebras of the augmentation map $\varepsilon: \mathbb{Z}G \to \mathbb{Z}$ of a group ring. Following that analogy, we define the *augmentation ideal* $\mathfrak{I}$ to be the kernel of ε; $\mathfrak{I}$ is evidently the (2-sided) ideal of $U\mathfrak{g}$ generated (as a left ideal) by $i(\mathfrak{g})$. Therefore $\mathfrak{I}$ is a $U\mathfrak{g}$-module and $k \cong U\mathfrak{g}/\mathfrak{I} = (U\mathfrak{g})_\mathfrak{g}$.

Corollary 7.3.6 *Let M be a $\mathfrak{g}$-module. Then*

$$H_*(\mathfrak{g}, M) \cong \mathrm{Tor}_*^{U\mathfrak{g}}(k, M) \text{ and}$$

$$H^*(\mathfrak{g}, M) \cong \mathrm{Ext}_{U\mathfrak{g}}^*(k, M).$$

Proof To show any two derived functors are isomorphic, we only need show the underlying functors are isomorphic. Therefore we need only observe

$$k \otimes_{U\mathfrak{g}} M = (U\mathfrak{g}/\mathfrak{I}) \otimes_{U\mathfrak{g}} M \cong M/\mathfrak{I} M = M/\mathfrak{g}M = M_\mathfrak{g};$$

$$\mathrm{Hom}_{U\mathfrak{g}}(k, M) = \mathrm{Hom}_\mathfrak{g}(k, M) = M^\mathfrak{g}. \qquad \diamond$$

We conclude this section by stating the Poincaré-Birkhoff-Witt Theorem, which gives the structure of $U\mathfrak{g}$ when k is a field (or more generally when $\mathfrak{g}$ is a free k-module). A proof may be found in [JLA, V.2] or [CE, XIII.3]. Let $\{e_\alpha\}$ be a fixed ordered k-basis of $\mathfrak{g}$. If $I = (\alpha_1, \cdots, \alpha_p)$ is a sequence of indices, we shall use the notation e_I for the product $e_{\alpha_1} \cdots e_{\alpha_p}$ in $U\mathfrak{g}$. The sequence I is called *increasing* if $\alpha_1 \leq \cdots \leq \alpha_p$. By convention, we regard the empty sequence ϕ as increasing, and set $e_\phi = 1$. If $I = (\alpha)$ is a single index, note that $e_\alpha \in \mathfrak{g}$, but $e_{(\alpha)} = i(e_\alpha)$ is in $U\mathfrak{g}$.

Poincaré-Birkhoff-Witt Theorem 7.3.7 *If $\mathfrak{g}$ is a free k-module, then $U\mathfrak{g}$ is also a free k-module. If $\{e_\alpha\}$ is an ordered basis of $\mathfrak{g}$, then the elements e_I with I an increasing sequence form a basis of $U\mathfrak{g}$.*

Corollary 7.3.8 *The map $i: \mathfrak{g} \to U\mathfrak{g}$ is an injection, so we may identify $\mathfrak{g}$ with $i(\mathfrak{g})$.*

Corollary 7.3.9 *If $\mathfrak{h} \subset \mathfrak{g}$ is a Lie subalgebra, and k is a field, then $U\mathfrak{g}$ is a free $U\mathfrak{h}$-module.*

Proof First pick an ordered basis for $\mathfrak{h}$, and then complete it to an ordered basis of $\mathfrak{g}$. The e_I with increasing $I = (\alpha_1, \cdots, \alpha_p)$ such that no e_{α_i} is in $\mathfrak{h}$ will form a basis of $U\mathfrak{g}$ over $U\mathfrak{h}$. $\diamond$

Exercise 7.3.4 (Hom as a $\mathfrak{g}$-module) Let M and N be left $\mathfrak{g}$-modules. Then $\mathrm{Hom}_k(M, N)$ is a $\mathfrak{g}$-module by $(xf)(m) = xf(m) - f(xm)$, $x \in \mathfrak{g}$, $m \in M$. Show that there is a natural isomorphism $\mathrm{Hom}_\mathfrak{g}(M, N) \cong \mathrm{Hom}_k(M, N)^\mathfrak{g}$.

Exercise 7.3.5 (Cohomological dimension) Extend the natural isomorphism $\mathrm{Hom}_\mathfrak{g}(M, N) \cong \mathrm{Hom}_k(M, N)^\mathfrak{g}$ of exercise 7.3.4 to a natural isomorphism of δ-functors:

$$\mathrm{Ext}^*_{U\mathfrak{g}}(M, N) \cong H^*_{\mathrm{Lie}}(\mathfrak{g}, \mathrm{Hom}_k(M, N))$$

By the Global Dimension Theorem (4.1.2), this proves that the global dimension of $U\mathfrak{g}$ equals the Lie algebra cohomological dimension of $\mathfrak{g}$ (see 7.7.4).

Exercise 7.3.6 (Associated graded algebra) For any Lie algebra $\mathfrak{g}$, let $F_p = F_p U\mathfrak{g}$ be the k-submodule of $U\mathfrak{g}$ generated by all products $x_1 \cdots x_i$ of elements of $\mathfrak{g}$ with $i \le p$. By convention, $F_0 U\mathfrak{g} = k$, and clearly $F_1 U\mathfrak{g} = k + \mathfrak{g}$. Show that

$$k = F_0 U\mathfrak{g} \subseteq F_1 U\mathfrak{g} \subseteq F_2 U\mathfrak{g} \subseteq \cdots$$

is an increasing filtration in the sense that $F_p \cdot F_q \subseteq F_{p+q}$. Then show that $A = k \oplus (F_1/F_0) \oplus (F_2/F_1) \oplus \cdots \oplus (F_p/F_{p-1}) \oplus \cdots$ is a commutative, associative graded k-algebra. Finally, if $\mathfrak{g}$ is a free k-module on basis $\{e_\alpha\}$, show that $F_1/F_0 = \mathfrak{g}$ and that A is a polynomial ring on the indeterminates e_α:

$$A = k[e_1, e_2, \cdots].$$

Exercise 7.3.7 (Hopf algebra) In this exercise we show that $U\mathfrak{g}$ is a Hopf algebra (see 6.7.15).

1. Use the universal property of $U\mathfrak{g}$ to show that $U(\mathfrak{g} \times \mathfrak{h}) \cong U\mathfrak{g} \otimes_k U\mathfrak{h}$. In particular, $U(\mathfrak{g} \times \mathfrak{g}) \cong U\mathfrak{g} \otimes_k U\mathfrak{g}$.
2. Show that the diagonal map $\Delta: \mathfrak{g} \to \mathfrak{g} \times \mathfrak{g}$ induces a ring homomorphism $\Delta: U\mathfrak{g} \to U\mathfrak{g} \otimes_k U\mathfrak{g}$ with $\Delta(x) = x \otimes 1 + 1 \otimes x$ for $x \in \mathfrak{g}$.
3. Show that there is an isomorphism $s : U\mathfrak{g} \cong (U\mathfrak{g})^{op}$, called the *antipode*, and that the resulting isomorphism between left and right $\mathfrak{g}$-modules

$$\textbf{mod–}\mathfrak{g} = \textbf{mod–}U\mathfrak{g} \cong (U\mathfrak{g})^{op}\textbf{–mod} \cong U\mathfrak{g}\textbf{–mod} = \mathfrak{g}\textbf{–mod}$$

is the correspondence $xm = -mx$ of 7.2.3.
4. Show that the maps Δ and s make $U\mathfrak{g}$ into a Hopf algebra.

Exercise 7.3.8 (Products) Let $\mathfrak{g}$ and $\mathfrak{h}$ be Lie algebras. Use the Künneth formula (3.6.3) as in 6.1.13 to construct split exact sequences

$$0 \to \bigoplus_{\substack{p+q \\ =n}} H_p(\mathfrak{g}, k) \otimes H_q(\mathfrak{h}, k) \to H_n(\mathfrak{g} \times \mathfrak{h}, k) \to \bigoplus_{\substack{p+q \\ =n-1}} \mathrm{Tor}_1^k(H_p(\mathfrak{g}), H_q(\mathfrak{h})) \to 0$$

$$0 \to \bigoplus_{\substack{p+q \\ =n}} H^p(\mathfrak{g}, k) \otimes H^q(\mathfrak{h}, k) \xrightarrow{\times} H^n(\mathfrak{g} \times \mathfrak{h}, k) \to \bigoplus_{\substack{p+q \\ =n-1}} \mathrm{Tor}_1^k(H^p(\mathfrak{g}), H^q(\mathfrak{h})) \to 0.$$

The map $\times$ is called the *cross product*. Composition with $\Delta^*: H^n(\mathfrak{g} \times \mathfrak{g}) \to H^n(\mathfrak{g})$ gives a graded bilinear product on $H^*(\mathfrak{g}, k)$, called the *cup product*. Show that the cup product makes $H^*(\mathfrak{g}, k)$ into an associative graded-commutative k-algebra (see 6.7.11). Dually, when k is a field, show that $H_*(\mathfrak{g}, k)$ is a coalgebra (6.7.13).

Exercise 7.3.9 (Restricted Lie algebras) Let k be a field of characteristic $p \neq 0$. A *restricted Lie algebra* over k is a Lie algebra $\mathfrak{g}$, together with a set map $x \mapsto x^{[p]}$ of $\mathfrak{g}$ such that $[x^{[p]}, y]$ equals the p-fold product $[x[x[\cdots [xy]]]]$; $(\alpha\, x)^{[p]} = \alpha^p x^{[p]}$ for all $\alpha \in k$; $(x + y)^{[p]} = x^{[p]} + y^{[p]} + \sum_{i=1}^{p-1} s_i(x, y)$, where $i \cdot s_i(x, y)$ is the coefficient of λ^{i-1} in the formal $(p - 1)$-fold product $[\lambda x + y[\cdots [\lambda x + y, x]]]$. See [JLA, V.7].

1. If A is an associative k-algebra, show that $\mathrm{Lie}(A)$ is a restricted Lie algebra with $a^{[p]} = a^p$. In particular, this makes the abelian Lie algebra k into a restricted Lie algebra.
2. Let $u(\mathfrak{g})$ denote the quotient of $U\mathfrak{g}$ by the ideal generated by all elements $x^p - x^{[p]}$; $u(\mathfrak{g})$ is called the *restricted universal enveloping algebra* of $\mathfrak{g}$. If $\mathfrak{g}$ is n-dimensional over k, show that $u(\mathfrak{g})$ is n^p-dimensional as a vector space.
3. A *restricted* $\mathfrak{g}$-module M is a $\mathfrak{g}$-module in which the p-fold product $(x(x(\cdots(xm))))$ equals $x^{[p]}m$ for all $m \in M$ and $x \in \mathfrak{g}$. Show that the

category of restricted $\mathfrak{g}$-modules is equivalent to the category of $u(\mathfrak{g})$-modules.

4. Define the restricted cohomology groups $H^*_{\mathrm{res}}(\mathfrak{g}, M)$ to be the right derived functors of $M^{\mathfrak{g}}$ on the category of restricted $\mathfrak{g}$-modules. Show that $H^*_{\mathrm{res}}(\mathfrak{g}, M) \cong \mathrm{Ext}^*_{u(\mathfrak{g})}(k, M)$.

5. Show that there is a canonical map from $H^*_{\mathrm{res}}(\mathfrak{g}, M)$ to the ordinary cohomology $H^*(\mathfrak{g}, M)$.

7.4 H^1 and H_1

The results in Chapter 6 for $H_1(G)$ and $H^1(G)$ have analogues for $H_1(\mathfrak{g})$ and $H^1(\mathfrak{g})$. As there, we begin with the exact sequence of $\mathfrak{g}$-modules:

$$0 \to \mathfrak{I} \to U\mathfrak{g} \to k \to 0.$$

If M is a $\mathfrak{g}$-module, applying $\mathrm{Tor}^{U\mathfrak{g}}_*(-, M)$ yields

$$H_n(\mathfrak{g}, M) = \mathrm{Tor}^{U\mathfrak{g}}_n(k, M) \cong \mathrm{Tor}^{U\mathfrak{g}}_{n-1}(\mathfrak{I}, M), \quad n \geq 2$$

and the exact sequence

(†) $$0 \to H_1(\mathfrak{g}, M) \to \mathfrak{I} \otimes_{U\mathfrak{g}} M \to M \to M_{\mathfrak{g}} \to 0.$$

Exercise 7.4.1 (Compare with exercise 6.1.4.)

1. Show that $i: \mathfrak{g} \to U\mathfrak{g}$ maps $[\mathfrak{g}, \mathfrak{g}]$ to $\mathfrak{I}^2$. Conclude that it induces a map $i: \mathfrak{g}^{ab} \to \mathfrak{I}/\mathfrak{I}^2$, where $\mathfrak{g}^{ab} = \mathfrak{g}/[\mathfrak{g}, \mathfrak{g}]$.

2. Show that there is a k-module map $\sigma: U\mathfrak{g} \to \mathfrak{g}^{ab}$ sending $\mathfrak{I}^2$ to zero and $i(x)$ to $\bar{x}$. *Hint:* First define a map from the tensor algebra $T(\mathfrak{g})$ to $\mathfrak{g}^{ab}$ sending $\mathfrak{g} \otimes_k \mathfrak{g}$ to zero and then pass to the quotient $U\mathfrak{g}$.

3. Deduce from (1) and (2) that $\mathfrak{I}/\mathfrak{I}^2 \cong \mathfrak{g}^{ab}$.

Theorem 7.4.1 *For any Lie algebra* $\mathfrak{g}$, $H_1(\mathfrak{g}, k) \cong \mathfrak{g}^{ab}$.

Proof Taking $M = k$ in (†) yields the exact sequence

$$0 \to H_1(\mathfrak{g}, k) \to \mathfrak{I} \otimes_{U\mathfrak{g}} k \to k \xrightarrow{\cong} k_{\mathfrak{g}} \to 0.$$

But for the right $\mathfrak{g}$-module $\mathfrak{I}$ the exercise 7.4.1 above yields

$$\mathfrak{I} \otimes_{U\mathfrak{g}} k = \mathfrak{I} \otimes_{U\mathfrak{g}} (U\mathfrak{g}/\mathfrak{I}) \cong \mathfrak{I}/\mathfrak{I}^2 \cong \mathfrak{g}^{ab}. \qquad \diamond$$

Corollary 7.4.2 *If M is any trivial $\mathfrak{g}$-module, $H_1(\mathfrak{g}, M) \cong \mathfrak{g}^{ab} \otimes_k M$.*

Proof Since $M = M_{\mathfrak{g}}$, (†) yields $H_1(\mathfrak{g}, M) \cong \mathfrak{I} \otimes_{U\mathfrak{g}} M \cong (\mathfrak{I} \otimes_{U\mathfrak{g}} k) \otimes_k M \cong \mathfrak{g}^{ab} \otimes_k M$. $\diamond$

Exercise 7.4.2 Let $\mathfrak{g}$ be a free k-module on basis $\{e_1, \cdots, e_n\}$, made into an abelian Lie algebra. Show that $H_p(\mathfrak{g}, k) \cong \Lambda^p \mathfrak{g} \cong k^{\binom{n}{p}}$, the pth exterior power of the k-module $\mathfrak{g}$. *Hint:* $U\mathfrak{g} \cong k[e_1, \cdots, e_n]$.

Exercise 7.4.3 Consider the Lie algebra $\mathfrak{gl}_m(A)$ of $n \times n$ matrices over an associative k-algebra A.

1. Write e_{ij}^a for the matrix whose (i, j)-entry is a, all the other entries being 0. If i, j and k are distinct, show that

$$[e_{ij}^a, e_{jk}^b] = e_{ik}^{ab} \quad \text{and} \quad [e_{ij}^a, e_{ji}^b] = e_{ii}^{ab} - e_{jj}^{ba}.$$

2. Recall from 7.1.3 that the special linear Lie algebra $\mathfrak{sl}_n(A)$ is the kernel of the trace map from $\mathfrak{gl}_n(A)$ to $A/[A, A]$. Show that for $n \geq 3$

$$H_1(\mathfrak{sl}_n(A), k) = 0 \quad \text{and} \quad H_1(\mathfrak{gl}_n(A), k) \cong A/[A, A].$$

We now turn our attention to cohomology. Applying $\text{Ext}^*_{U\mathfrak{g}}(-, M)$ to the sequence $0 \to \mathfrak{I} \to U\mathfrak{g} \to k \to 0$ yields

$$H^n(\mathfrak{g}, M) \cong \text{Ext}^{n-1}_{U\mathfrak{g}}(\mathfrak{I}, M), n \geq 2$$

and the exact sequence

$$0 \to M^{\mathfrak{g}} \to M \to \text{Hom}_{\mathfrak{g}}(\mathfrak{I}, M) \to H^1(\mathfrak{g}, M) \to 0.$$

To describe $H^1(\mathfrak{g}, M)$, it remains to interpret $\text{Hom}_{\mathfrak{g}}(\mathfrak{I}, M)$ as derivations and interpret the image of M as inner derivations.

Definition 7.4.3 If M is a $\mathfrak{g}$-module, a *derivation* from $\mathfrak{g}$ into M is a k-linear map $D: \mathfrak{g} \to M$ such that the Leibnitz formula holds

$$D([x, y]) = x(Dy) - y(Dx).$$

The set of all such derivations is denoted $\text{Der}(\mathfrak{g}, M)$; it is a k-submodule of $\text{Hom}_k(\mathfrak{g}, M)$. Note that if $\mathfrak{g} = M$, then $\text{Der}(\mathfrak{g}, \mathfrak{g})$ is the derivation algebra $\text{Der}(\mathfrak{g})$ of 7.1.4. If M is a trivial $\mathfrak{g}$-module, then $\text{Der}(\mathfrak{g}, M) = \text{Hom}_k(\mathfrak{g}^{ab}, M)$.

Example 7.4.4 (Inner derivations) If $m \in M$, define $D_m(x) = xm$. D_m is a derivation:

$$D_m([x, y]) = [x, y]m = x(ym) - y(xm).$$

The D_m are called the *inner derivations* of $\mathfrak{g}$ into M, and they form a k-submodule $\text{Der}_{\text{Inn}}(\mathfrak{g}, M)$ of $\text{Der}(\mathfrak{g}, M)$.

Example 7.4.5 If $\varphi: \mathfrak{I} \to M$ is a $\mathfrak{g}$-map, let $D_\varphi: \mathfrak{g} \to M$ be defined by $D_\varphi(x) = \varphi(i(x))$. This too is a derivation:

$$D_\varphi([x, y]) = \varphi(i(x)i(y) - i(y)i(x)) = x\varphi(i(y)) - y\varphi(i(x)).$$

As in the analogous discussion for group cohomology (6.4.4), the next step is to show that every derivation is of the form D_φ.

Lemma 7.4.6 *The map* $\varphi \mapsto D_\varphi$ *is a natural isomorphism of k-modules:*

$$\text{Hom}_\mathfrak{g}(\mathfrak{I}, M) \cong \text{Der}(\mathfrak{g}, M).$$

Proof The formula $\varphi \mapsto D_\varphi$ defines a natural homomorphism, so it suffices to show that it is an isomorphism. For this we use the fact (7.3.5) that the product map $U\mathfrak{g} \otimes_k \mathfrak{g} \to (U\mathfrak{g})\mathfrak{g} = \mathfrak{I}$ is onto, and that its kernel is the k-module generated by the terms $(u \otimes [xy] - ux \otimes y + uy \otimes x)$ with $u \in U\mathfrak{g}$ and $x, y \in \mathfrak{g}$.

Given a derivation $D: \mathfrak{g} \to M$, consider the map

$$f: U\mathfrak{g} \otimes_k \mathfrak{g} \to M, \quad f(u \otimes x) = u(Dx).$$

Since D is a derivation, $f(u \otimes [xy] - ux \otimes y + uy \otimes x) = 0$ for all u, x, and y. Therefore f induces a map $\varphi: \mathfrak{I} \to M$, which is evidently a left $\mathfrak{g}$-module map. Since $D_\varphi(x) = \varphi(i(x)) = f(1 \otimes x) = Dx$, we have lifted D to an element of $\text{Hom}_\mathfrak{g}(\mathfrak{I}, M)$. On the other hand, given $D = D_h$ for some $h \in \text{Hom}_\mathfrak{g}(\mathfrak{I}, M)$, we have $\varphi(ux) = u(Dx) = uh(x) = h(ux)$ for all $u \in U\mathfrak{g}, x \in \mathfrak{g}$. Hence $\varphi = h$ as maps from $\mathfrak{I} = (U\mathfrak{g})\mathfrak{g}$ to M. $\diamond$

Theorem 7.4.7 $H^1(\mathfrak{g}, M) \cong \text{Der}(\mathfrak{g}, M) / \text{Der}_{\text{Inn}}(\mathfrak{g}, M)$.

Proof If $\varphi: \mathfrak{I} \to M$ extends to a $\mathfrak{g}$-map $U\mathfrak{g} \to M$ sending 1 to $m \in M$, then

$$D_\varphi(x) = \varphi(x \cdot 1) = xm = D_m(x).$$

Hence D_φ is an inner derivation. This shows that the image of

$$M \to \operatorname{Hom}_\mathfrak{g}(\mathfrak{I}, M) = \operatorname{Der}(\mathfrak{g}, M)$$

is the submodule of inner derivations, as desired. $\diamond$

Corollary 7.4.8 *If M is a trivial $\mathfrak{g}$-module*

$$H^1(\mathfrak{g}, M) \cong \operatorname{Der}(\mathfrak{g}, M) \cong \operatorname{Hom}_{\operatorname{Lie}}(\mathfrak{g}, M) \cong \operatorname{Hom}_k(\mathfrak{g}^{ab}, M).$$

Semidirect Products 7.4.9 Given a Lie algebra $\mathfrak{g}$ and a (left) $\mathfrak{g}$-module M, we can form the *semidirect product* Lie algebra $M \rtimes \mathfrak{g}$, much as we did in group theory. The k-module underlying $M \rtimes \mathfrak{g}$ is the product $M \times \mathfrak{g}$, and the product is given by the formula

$$[(m, g), (n, h)] = (gn - hm, [gh]).$$

As in group theory, $M \rtimes \mathfrak{g}$ is a Lie algebra and both $M \times 0$ and $0 \times \mathfrak{g}$ are Lie subalgebras.

We will study other Lie algebra extensions of $\mathfrak{g}$ by M in section 7.6 below. But first, here is an interpretation of $H^1(\mathfrak{g}, M)$ in terms of automorphisms of $M \rtimes \mathfrak{g}$; it is the analogue of a result for semidirect products of groups (exercise 6.4.2). We say that a Lie algebra automorphism σ of $M \rtimes \mathfrak{g}$ *stabilizes M* and $\mathfrak{g}$ if $\sigma(m) = m$ for all m in $M = M \times 0$ and if the induced automorphism on the quotient $\mathfrak{g} \cong (M \rtimes \mathfrak{g})/M$ is the identity, that is, if there is a commutative diagram of Lie algebras:

$$
\begin{array}{ccccccccc}
0 & \longrightarrow & M & \longrightarrow & M \rtimes \mathfrak{g} & \longrightarrow & \mathfrak{g} & \longrightarrow & 0 \\
 & & \| & & \sigma\downarrow & & \| & & \\
0 & \longrightarrow & M & \longrightarrow & M \rtimes \mathfrak{g} & \longrightarrow & \mathfrak{g} & \longrightarrow & 0.
\end{array}
$$

Exercise 7.4.4 If D is a derivation of $\mathfrak{g}$ into M, show that σ_D, defined by

$$\sigma_D(m, g) = (m + D(g), g),$$

is a Lie algebra automorphism of $M \rtimes g$ that stabilizes M and $\mathfrak{g}$. Then show that $\operatorname{Der}(\mathfrak{g}, M)$ is isomorphic to the subgroup of $\operatorname{Aut}(M \rtimes \mathfrak{g})$ of all automorphisms stabilizing M and $\mathfrak{g}$. Evidently the inner derivations correspond to the subgroup of all "inner" automorphisms of the form

$$\sigma(m, g) = (m + ga, g), a \in M.$$

In this way we can identify $H^1(\mathfrak{g}, M)$ with a subquotient of $\text{Aut}(M \rtimes \mathfrak{g})$.

Exercise 7.4.5 (Extensions of $\mathfrak{g}$-modules) Use the natural isomorphism $\text{Ext}^1_{U\mathfrak{g}}(M, N) \cong H^1(\mathfrak{g}, \text{Hom}_k(M, N))$ of exercise 7.3.5 to interpret H^1 in terms of extensions of $\mathfrak{g}$-modules. In particular, show that $H^1(\mathfrak{g}, N)$ classifies extensions of $\mathfrak{g}$-modules of the form

$$0 \to N \to M \to k \to 0.$$

Exercise 7.4.6 Let $\mathfrak{g}$ be a restricted Lie algebra over a field of characteristic $p \neq 0$, and let N be a restricted $\mathfrak{g}$-module (exercise 7.3.9). Show that $H^1_{\text{res}}(\mathfrak{g}, N)$ classifies extensions of restricted $\mathfrak{g}$-modules of the form

$$0 \to N \to M \to k \to 0.$$

Conclude that the natural map $H^1_{\text{res}}(\mathfrak{g}, N) \to H^1(\mathfrak{g}, M)$ is an injection.

7.5 The Hochschild-Serre Spectral Sequence

In this section we develop the Hochschild-Serre spectral sequence, which is the analogue of the Lyndon/Hochschild-Serre spectral sequence for groups. The analogue of a normal subgroup of a group is an ideal of a Lie algebra. If $\mathfrak{h}$ is an ideal of $\mathfrak{g}$, then $\mathfrak{g}/\mathfrak{h}$ inherits a natural Lie algebra structure from $\mathfrak{g}$, and there is an exact sequence of Lie algebra homomorphisms

$$0 \to \mathfrak{h} \to \mathfrak{g} \to \mathfrak{g}/\mathfrak{h} \to 0.$$

The proof of the following lemma is exactly the same as the proof of the corresponding result 6.8.4 for groups, and we omit it here.

Lemma 7.5.1 *If $\mathfrak{h}$ is an ideal of a Lie algebra $\mathfrak{g}$ and M is a $\mathfrak{g}$-module, then both $M_{\mathfrak{h}}$ and $M^{\mathfrak{h}}$ are $\mathfrak{g}/\mathfrak{h}$-modules. Moreover, the forgetful functor from $\mathfrak{g}/\mathfrak{h}$–mod to $\mathfrak{g}$–mod has $-_{\mathfrak{h}}$ as left adjoint and $-^{\mathfrak{h}}$ as right adjoint.*

Hochschild-Serre Spectral Sequence 7.5.2 For every ideal $\mathfrak{h}$ of a Lie algebra $\mathfrak{g}$, there are two convergent first quadrant spectral sequences:

$$E^2_{pq} = H_p(\mathfrak{g}/\mathfrak{h}, H_q(\mathfrak{h}, M)) \Rightarrow H_{p+q}(\mathfrak{g}, M)$$

$$E_2^{pq} = H^p(\mathfrak{g}/\mathfrak{h}, H^q(\mathfrak{h}, M)) \Rightarrow H^{p+q}(\mathfrak{g}, M).$$

Proof We claim that the functors $-_{\mathfrak{g}}$ and $-^{\mathfrak{g}}$ factor as follows.

$$
\mathfrak{g}\text{-mod} \xrightarrow{\;-_{\mathfrak{h}}\;} \mathfrak{g}/\mathfrak{h}\text{-mod} \qquad\qquad \mathfrak{g}\text{-mod} \xrightarrow{\;-^{\mathfrak{h}}\;} \mathfrak{g}/\mathfrak{h}\text{-mod}
$$

$$
{}_{-_{\mathfrak{g}}}\searrow \qquad \nearrow_{-_{\mathfrak{g}/\mathfrak{h}}} \qquad\qquad {}_{-^{\mathfrak{g}}}\searrow \qquad \nearrow_{-^{\mathfrak{g}/\mathfrak{h}}}
$$

$$
k\text{-mod} \qquad\qquad\qquad k\text{-mod}
$$

The proof of this claim is the same as the proof of the corresponding claim for groups, and we leave the translation to the reader. To apply the Grothendieck spectral sequence (5.8.3), we need only see that $-_{\mathfrak{h}}$ preserves projectives and that $-^{\mathfrak{h}}$ preserves injectives. This follows from the preceding lemma (see 2.3.10): $-_{\mathfrak{h}}$ is left adjoint and $-^{\mathfrak{h}}$ is right adjoint to the forgetful functor, which is an exact functor. $\diamond$

Low Degree Terms 7.5.3 The exact sequences of low degree terms in the Hochschild-Serre spectral sequence are

$$
H_2(\mathfrak{g}, M) \to H_2(\mathfrak{g}/\mathfrak{h}, M_{\mathfrak{h}}) \xrightarrow{d} H_1(\mathfrak{h}, M)_{\mathfrak{g}/\mathfrak{h}} \to H_1(\mathfrak{g}, M) \to H_1(\mathfrak{g}/\mathfrak{h}, M_{\mathfrak{h}}) \to 0;
$$

$$
0 \to H^1(\mathfrak{g}/\mathfrak{h}, M^{\mathfrak{h}}) \to H^1(\mathfrak{g}, M) \to H^1(\mathfrak{h}, M)^{\mathfrak{g}/\mathfrak{h}} \xrightarrow{d} H^2(\mathfrak{g}/\mathfrak{h}, M^{\mathfrak{h}}) \to H^2(\mathfrak{g}, M).
$$

Exercise 7.5.1

1. Show that there is an exact sequence

    $$
    H_2(\mathfrak{g}/\mathfrak{h}, k) \oplus [\mathfrak{g}, \mathfrak{h}] \to \mathfrak{h}^{ab} \to \mathfrak{g}^{ab} \to (\mathfrak{g}/\mathfrak{h})^{ab} \to 0.
    $$

2. If M is a $\mathfrak{g}/\mathfrak{h}$-module, show that there is an exact sequence

    $$
    0 \to \mathrm{Der}(\mathfrak{g}/\mathfrak{h}, M) \to \mathrm{Der}(\mathfrak{g}, M) \to \mathrm{Hom}_{\mathfrak{g}}(\mathfrak{h}^{ab}, M) \to H^2(\mathfrak{g}/\mathfrak{h}, M) \to H^2(\mathfrak{g}, M).
    $$

3. Let $\mathfrak{n}_3$ be the nilpotent Lie algebra of strictly upper triangular 3×3 matrices over k (7.1.3). Using the extension

    $$
    0 \to k e_{13} \to \mathfrak{n}_3(k) \to k e_{12} \oplus k e_{23} \to 0,
    $$

    calculate $H^*(\mathfrak{n}_3, k)$ and $H_*(\mathfrak{n}_3, k)$.

4. Let $\mathfrak{g}$ be the Lie subalgebra of $\mathfrak{gl}_3$ generated by e_{11}, e_{12}, e_{13}, and e_{23}. Use the extension $0 \to \mathfrak{n}_3 \to \mathfrak{g} \to k \to 0$ to compute $H^1(\mathfrak{g}, k)$ and $H^2(\mathfrak{g}, k)$.

Exercise 7.5.2 Suppose that $\mathfrak{f}$ is a free Lie algebra on a set of generators of a Lie algebra $\mathfrak{g}$ and that $\mathfrak{r}$ is the kernel of the natural surjection $\mathfrak{f} \to \mathfrak{g}$. Using the low degree sequence 7.5.3, show that the analogue of Hopf's theorem 6.8.8 holds, that is, that

$$H_2(\mathfrak{g}, k) \cong \frac{\mathfrak{r} \cap [\mathfrak{f}, \mathfrak{f}]}{[\mathfrak{f}, \mathfrak{r}]}.$$

Exercise 7.5.3 (Inflation and restriction) The forgetful map $\mathfrak{g}$–**mod**$\to \mathfrak{h}$–**mod** is exact for every Lie algebra homomorphism $\mathfrak{h} \to \mathfrak{g}$. Show that the natural injection $M^{\mathfrak{g}} \to M^{\mathfrak{h}}$ extends to a morphism $\mathrm{res}^{\mathfrak{g}}_{\mathfrak{h}} \colon H^*(\mathfrak{g}, M) \to H^*(\mathfrak{h}, M)$ of δ-functors, called the *restriction map*. If $\mathfrak{h}$ is an ideal of $\mathfrak{g}$, the *inflation map* is the composite

$$\mathrm{inf} \colon H^*(\mathfrak{g}/\mathfrak{h}, M) \xrightarrow{\mathrm{res}} H^*(\mathfrak{g}, m^{\mathfrak{h}}) \to H^*(\mathfrak{g}, M).$$

Show that the edge maps of the Hochschild-Serre spectral sequence for $H^*(\mathfrak{g}, M)$ are the inflation and restriction maps. (Cf. 6.7.1, 6.8.2.)

7.6 H^2 and Extensions

In Chapter 6 we showed that $H^2(G; A)$ classified extensions of groups. There is an analogous result for $H^2_{\mathrm{Lie}}(\mathfrak{g}, M)$, which we shall establish in this section.

Definition 7.6.1 An *extension of Lie algebras* (of $\mathfrak{g}$ by M) is a short exact sequence of Lie algebras

$$0 \to M \to \mathfrak{e} \xrightarrow{\pi} \mathfrak{g} \to 0$$

in which M is an abelian Lie algebra. Such an extension makes M into a $\mathfrak{g}$-module in a well-defined way: If $g \in \mathfrak{g}$ and $m \in M$, define gm to be the product $[\tilde{g}, m]$ in $\mathfrak{e}$, where $\pi(\tilde{g}) = g$. Since M is abelian, gm is independent of the choice of $\tilde{g}$.

Exercise 7.6.1 Let M be a $\mathfrak{g}$-module, and form the semidirect product

$$0 \to M \to M \rtimes \mathfrak{g} \to \mathfrak{g} \to 0.$$

1. Show that the induced $\mathfrak{g}$-module structure on M agrees with the original $\mathfrak{g}$-module structure.
2. We say an extension *splits* if π has a Lie algebra section $\sigma \colon \mathfrak{g} \to \mathfrak{e}$. Show that an extension splits if and only if $\mathfrak{e}$ is isomorphic to the semidirect product Lie algebra $M \rtimes \mathfrak{g}$ constructed in 7.4.9, and that under this isomorphism π corresponds to the projection $M \rtimes \mathfrak{g} \to \mathfrak{g}$.

3. Let $\mathfrak{e} = \mathfrak{n}_3(k)$ to be the Lie algebra of strictly upper triangular matrices. Show that $[\mathfrak{e}, \mathfrak{e}]$ is the 1-dimensional subalgebra ke_{13} of matrices supported in the (1,3) spot, and that $\mathfrak{g} = \mathfrak{e}^{ab}$ is a 2-dimensional abelian Lie algebra. Finally, show that the following extension does not split:

$$0 \to ke_{13} \to \mathfrak{n}_3(k) \to \mathfrak{g} \to 0.$$

Extension Problem 7.6.2 Given a $\mathfrak{g}$-module M, we would like to determine how many extensions of $\mathfrak{g}$ by M exist in which the induced action of $\mathfrak{g}$ on M recovers the given $\mathfrak{g}$-module structure of M. As with groups (6.6.2), we say that two extensions $0 \to M \to \mathfrak{e}_i \to \mathfrak{g} \to 0$ are *equivalent* if there is an isomorphism $\varphi \colon \mathfrak{e}_1 \cong \mathfrak{e}_2$ so that

$$
\begin{array}{ccccccccc}
0 & \longrightarrow & M & \longrightarrow & \mathfrak{e}_1 & \longrightarrow & \mathfrak{g} & \longrightarrow & 0 \\
 & & \| & & {\scriptstyle\varphi}\downarrow & & \| & & \\
0 & \longrightarrow & M & \longrightarrow & \mathfrak{e}_2 & \longrightarrow & \mathfrak{g} & \longrightarrow & 0
\end{array}
$$

commutes, and we ask for a description of the set $\mathrm{Ext}(\mathfrak{g}, M)$ of equivalence classes of extensions.

Classification Theorem 7.6.3 *Let M be a $\mathfrak{g}$-module. The set $\mathrm{Ext}(\mathfrak{g}, M)$ of equivalence classes of extensions of $\mathfrak{g}$ by M is in 1–1 correspondence with $H^2_{\mathrm{Lie}}(\mathfrak{g}, M)$.*

The canonical approach to classifying extensions of groups (Chapter 6, section 6) has an analogue only for extensions in which $\mathfrak{g}$ is a free k-module (e.g., if k is a field). Rather than pursue that method, which calls for a canonical $\mathfrak{g}$-module resolution of k and a notion of 2-cocycle (see exercise 7.7.5), we shall resort to a more functorial method.

Suppose first that $0 \to M \to \mathfrak{e} \to \mathfrak{g} \to 0$ is an extension of $\mathfrak{g}$ by an abelian Lie algebra M. The low degree terms sequence of 7.5.3 with $\mathfrak{h} = M$ is

$$0 \to H^1(\mathfrak{g}, M) \to H^1(\mathfrak{e}, M) \to \mathrm{Hom}_\mathfrak{g}(M, M) \xrightarrow{d^2} H^2(\mathfrak{g}, M) \to H^2(\mathfrak{e}, M).$$

This sequence is natural with respect to extensions, so $d^2 \colon \mathrm{Hom}_\mathfrak{g}(M, M) \to H^2(\mathfrak{g}, M)$ depends only on the equivalence class of the extension in $\mathrm{Ext}(\mathfrak{g}, M)$. Therefore assigning $d^2(\mathrm{id}_M)$ to the extension gives a well-defined set map from $\mathrm{Ext}(\mathfrak{g}, M)$ to $H^2(\mathfrak{g}, M)$, called the *classifying map*.

Before showing that the classifying map is a bijection, we consider a universal case. Choose a presentation of $\mathfrak{g}$: $0 \to \mathfrak{r} \to \mathfrak{f} \to \mathfrak{g} \to 0$ with $\mathfrak{f}$ free on some set. Modding out by the ideal $[\mathfrak{r}, \mathfrak{r}]$ of $\mathfrak{f}$ gives an extension of $\mathfrak{g}$ by $\mathfrak{r}^{ab} = \mathfrak{r}/[\mathfrak{r}, \mathfrak{r}]$:

$$0 \to \mathfrak{r}^{ab} \to \mathfrak{f}/[\mathfrak{r}, \mathfrak{r}] \to \mathfrak{g} \to 0.$$

Let $u \in H^2(\mathfrak{g}, \mathfrak{r}^{ab})$ be the image of this extension under the classifying map.

If $\mathfrak{e} \to \mathfrak{g}$ is a central extension of $\mathfrak{g}$ by M, we can lift $\mathfrak{f} \to \mathfrak{g}$ to a map $\mathfrak{f} \to \mathfrak{e}$. This yields maps of Lie algebra extensions:

$$
\begin{array}{ccccccccc}
0 & \longrightarrow & \mathfrak{r} & \longrightarrow & \mathfrak{f} & \longrightarrow & \mathfrak{g} & \longrightarrow & 0 \\
 & & \downarrow & & \downarrow & & \| & & \\
0 & \longrightarrow & \mathfrak{r}^{ab} & \longrightarrow & \mathfrak{f}/[\mathfrak{r}, \mathfrak{r}] & \longrightarrow & \mathfrak{g} & \longrightarrow & 0 \\
 & & \varphi\downarrow & & \downarrow & & \| & & \\
0 & \longrightarrow & M & \longrightarrow & \mathfrak{e} & \longrightarrow & \mathfrak{g} & \longrightarrow & 0
\end{array}
$$

Comparing low degree term sequences (for the Hochschild-Serre spectral sequences 7.5.2) and using 7.2.5 yields a diagram

$$
\begin{array}{ccc}
\mathrm{Hom}_\mathfrak{g}(M, M) & \xrightarrow{\ d^2\ } & H^2(\mathfrak{g}, M) \\
\varphi^*\downarrow & & \| \\
\mathrm{Hom}_\mathfrak{g}(\mathfrak{r}^{ab}, M) & \xrightarrow{\ d^2\ } & H^2(\mathfrak{g}, M) \\
\downarrow \cong & & \| \\
\end{array}
$$

(*)

$$H^1(\mathfrak{f}, M) \longrightarrow H^1(\mathfrak{r}, M)^\mathfrak{g} \xrightarrow{\ d^2\ } H^2(\mathfrak{g}, M) \longrightarrow 0.$$

Exercise 7.6.2 In this exercise we show that $u \in H^2(\mathfrak{g}, \mathfrak{r}^{ab})$ is universal in the sense that the class of any extension of $\mathfrak{g}$ by M is $\varphi_*(u)$ for some $\varphi \in \mathrm{Hom}_\mathfrak{g}(\mathfrak{r}^{ab}, M)$. To do this, let $\varphi: \mathfrak{r}^{ab} \to M$ be the map induced from $\mathfrak{f} \to \mathfrak{e}$. Considered as an element of $\mathrm{Hom}_\mathfrak{g}(\mathfrak{r}^{ab}, M)$ we see from (*) that $d^2(\varphi) = d^2(\mathrm{id}_M)$ in $H^2(\mathfrak{g}, M)$. Show that the corresponding map $\varphi_*: H^2(\mathfrak{g}, \mathfrak{r}^{ab}) \to H^2(\mathfrak{g}, M)$ sends u to $d^2(\mathrm{id}_M)$.

Lemma 7.6.4 *Every element of $H^2(\mathfrak{g}, M)$ arises as the class of an extension.*

Proof Since $\mathfrak{f}$ is free, $H^2(\mathfrak{f}, M) = 0$. By (*), every element of $H^2(\mathfrak{g}, M)$ is $d^2(\varphi)$ for some element φ of

$$H^1(\mathfrak{r}, M)^{\mathfrak{g}} \cong \mathrm{Hom}_k(\mathfrak{r}^{ab}, M)^{\mathfrak{g}} = \mathrm{Hom}_{\mathfrak{g}}(\mathfrak{r}^{ab}, M).$$

Regarding M as an $\mathfrak{f}$-module via $\mathfrak{f} \to \mathfrak{g}$, form the semidirect product $M \rtimes \mathfrak{f}$. The set $\mathfrak{h} = \{(\varphi(-r), r) : r \in \mathfrak{r}\}$ is an ideal of $M \rtimes \mathfrak{f}$; set $\mathfrak{e} = (M \rtimes \mathfrak{f})/\mathfrak{h}$. Evidently $\mathfrak{h} \cap M = 0$ so we have an extension

$$0 \to M \to \mathfrak{e} \to \mathfrak{g} \to 0$$

together with a map $\mathfrak{f} \to \mathfrak{e}$ over $\mathfrak{g}$. The resulting map from $\mathrm{Hom}_{\mathfrak{g}}(M, M)$ to $\mathrm{Hom}(\mathfrak{r}^{ab}, M)$ sends id_M to φ. By diagram $(*)$, the class of this extension is $d^2(\mathrm{id}_M) = d^2(\varphi)$ as desired. $\diamond$

We are now ready to prove the classification theorem. The above lemma shows that the classifying map $\mathrm{Ext}(\mathfrak{g}, M) \to H^2(\mathfrak{g}, M)$ is onto; it suffices to show that this map is an injection. Suppose that $0 \to M \to \mathfrak{e}_i \to \mathfrak{g} \to 0$ ($i = 1, 2$) are two extensions of $\mathfrak{g}$ by M that both map to $\theta \in H^2(\mathfrak{g}, M)$.

Choosing lifts $\tau_i : \mathfrak{f} \to \mathfrak{e}_i$, the above argument yields $\varphi_i \in \mathrm{Hom}_{\mathfrak{g}}(\mathfrak{r}^{ab}, M)$ with $d^2(\varphi_i) = \theta$ in diagram $(*)$. By making $\mathfrak{f}$ larger if necessary, we may assume that $\mathfrak{f}$ maps onto both $\mathfrak{e}_1$ and $\mathfrak{e}_2$. (For this it suffices to add M to the set of generators of $\mathfrak{f}$.) Since $d^2(\varphi_2 - \varphi_1) = 0$, we see from $(*)$ that there is a derivation $D : \mathfrak{f} \to M$ such that the class of D in $H^1(\mathfrak{f}, M)$ maps to $\varphi_2 - \varphi_1$ in $\mathrm{Hom}_{\mathfrak{g}}(\mathfrak{r}^{ab}, M)$. Define a map $\tau : \mathfrak{f} \to \mathfrak{e}_1$ by sending $x \in \mathfrak{f}$ to $\tau_1(x) + D(x)$. This is a Lie algebra homomorphism, since

$$
\begin{aligned}
\tau([xy]) &= \tau_1([x, y]) + D([xy]) \\
&= [\tau_1(x), \tau_1(y)] + x(Dy) - y(Dx) \\
&= [\tau_1(x) + D(x), \tau_1(y) + D(y)] \\
&= [\tau(x), \tau(y)].
\end{aligned}
$$

There is no harm in replacing τ_1 by τ, except that we replace φ_1 by $\varphi_1 + D = \varphi_2$ in $\mathrm{Hom}_{\mathfrak{g}}(\mathfrak{r}^{ab}, M)$. We are now in the situation

$$
\begin{array}{ccccccccc}
0 & \longrightarrow & \mathfrak{r} & \longrightarrow & \mathfrak{f} & \longrightarrow & \mathfrak{g} & \longrightarrow & 0 \\
& & \varphi \downarrow & & \tau_i \downarrow & & \| & & \\
0 & \longrightarrow & M & \longrightarrow & \mathfrak{e}_i & \longrightarrow & \mathfrak{g} & \longrightarrow & 0.
\end{array}
$$

As $\mathfrak{f}$ maps onto $\mathfrak{e}_i$, we see that $\ker(\varphi)$ is an ideal of $\mathfrak{f}$ and that $\mathfrak{e}_1 \cong \mathfrak{f}/\ker(\varphi) \cong \mathfrak{e}_2$. As this isomorphism is a homomorphism over $\mathfrak{g}$, $\mathfrak{e}_1$, and $\mathfrak{e}_2$ define the same element of $\mathrm{Ext}(\mathfrak{g}, M)$. $\diamond$

Exercise 7.6.3 We saw in Corollary 7.2.5 that if $\mathfrak{f}$ is a free Lie algebra on a set, then $H^n(\mathfrak{f}, M) = 0$ for $n \geq 2$ and all $\mathfrak{f}$-modules M. Give a direct proof that $H^2(\mathfrak{f}, M) = 0$ by showing that all extensions $0 \to \mathfrak{h} \to \mathfrak{e} \to \mathfrak{f} \to 0$ split. Show that conversely if $\mathfrak{g}$ is free as a k-module and $H^2(\mathfrak{g}, M) = 0$ for all $\mathfrak{g}$-modules M, then $\mathfrak{g}$ is a free Lie algebra. *Hint:* Writing $\mathfrak{g} = \mathfrak{f}/\mathfrak{r}$ for some free Lie algebra $\mathfrak{f}$, with $\mathfrak{r} \subseteq [\mathfrak{f}, \mathfrak{f}]$, it suffices to show that $\mathfrak{r} = [\mathfrak{f}, \mathfrak{r}]$ (exercise 7.2.5). But $H^2(\mathfrak{g}, \mathfrak{r}/[\mathfrak{f}, \mathfrak{r}]) = 0$.

Exercise 7.6.4 (Restricted extensions) Let k be a field of characteristic $p \neq 0$. Let $\mathfrak{g}$ be a restricted Lie algebra and M a restricted $\mathfrak{g}$-module such that $M^{[p]} = 0$. (See exercise 7.3.9.) A *restricted extension* $\mathfrak{e}$ of $\mathfrak{g}$ by M is a restricted Lie algebra $\mathfrak{e}$ containing M as a restricted ideal, together with a restricted homomorphism $\mathfrak{e} \to \mathfrak{g}$ whose kernel is M. Let $\mathrm{Ext}_{\mathrm{res}}(\mathfrak{g}, M)$ denote the equivalence classes of restricted extensions of $\mathfrak{g}$ by M, $\mathfrak{e}$ and $\mathfrak{e}'$ being equivalent if there is a restricted homomorphism $\mathfrak{e} \to \mathfrak{e}'$ over $\mathfrak{g}$. Show that there is a natural isomorphism $\mathrm{Ext}_{\mathrm{res}}(\mathfrak{g}, M) \cong H^2_{\mathrm{res}}(\mathfrak{g}, M)$ compatible with the isomorphism $\mathrm{Ext}(\mathfrak{g}, M) \cong H^2(\mathfrak{g}, M)$ of the Classification Theorem 7.6.3.

7.7 The Chevalley-Eilenberg Complex

Throughout this section $\mathfrak{g}$ will denote a Lie algebra over k that is free as a k-module. We shall construct the $U\mathfrak{g}$-module chain complex $V_*(\mathfrak{g})$ originally used by C. Chevalley and S. Eilenberg [ChE] in 1948 to define $H^*_{\mathrm{Lie}}(\mathfrak{g}, M)$.

Let $\Lambda^p\mathfrak{g}$ denote the pth-exterior product of the k-module $\mathfrak{g}$, which is generated by monomials $x_1 \wedge \cdots \wedge x_p$ with $x_i \in \mathfrak{g}$; see 4.5.1 above. Our chain complex has $V_p(\mathfrak{g}) = U\mathfrak{g} \otimes_k \Lambda^p\mathfrak{g}$; since $\Lambda^p\mathfrak{g}$ is a free k-module, $V_p(\mathfrak{g})$ is free as a left $U\mathfrak{g}$-module. By convention, $\Lambda^0\mathfrak{g} = k$ and $\Lambda^1\mathfrak{g} = \mathfrak{g}$, so $V_0 = U\mathfrak{g}$ and $V_1 = U\mathfrak{g} \otimes_k \mathfrak{g}$. We define $\varepsilon : V_0(\mathfrak{g}) = U\mathfrak{g} \to k$ to be the augmentation 7.3.5 and $d \colon V_1(\mathfrak{g}) \to V_0(\mathfrak{g})$ to be the product map $d(u \otimes x) = ux$ from $U\mathfrak{g} \otimes \mathfrak{g}$ to $U\mathfrak{g}$ whose image is the augmentation ideal $\mathfrak{J}$. By 7.3.5, we have an exact sequence

$$V_1(\mathfrak{g}) \xrightarrow{d} V_0(\mathfrak{g}) \xrightarrow{\varepsilon} k \to 0.$$

Definition 7.7.1 For $p \geq 2$, let $d \colon V_p(\mathfrak{g}) \to V_{p-1}(\mathfrak{g})$ be given by the formula $d(u \otimes x_1 \wedge \cdots \wedge x_p) = \theta_1 + \theta_2$, where (for $u \in U\mathfrak{g}$ and $x_i \in \mathfrak{g}$):

$$\theta_1 = \sum_{i=1}^{p}(-1)^{i+1}ux_i \otimes x_1 \wedge \cdots \wedge \hat{x}_i \wedge \cdots \wedge x_p;$$

$$\theta_2 = \sum_{i<j}(-1)^{i+j}u \otimes [x_i x_j] \wedge x_1 \wedge \cdots \wedge \hat{x}_i \wedge \cdots \wedge \hat{x}_j \wedge \cdots \wedge x_p.$$

(The notation $\hat{x}_i$ indicates an omitted term.) For example, if $p = 2$, then

$$d(u \otimes x \wedge y) = ux \otimes y - uy \otimes x - u \otimes [xy].$$

$V_*(\mathfrak{g})$ with this differential is called the *Chevalley-Eilenberg complex*. It is sometimes also called the *standard complex*.

Exercise 7.7.1 Verify that $d^2 = 0$, so that V_* is indeed a chain complex of $U\mathfrak{g}$-modules. *Hint:* Writing $d(\theta_i) = \theta_{i1} + \theta_{i2}$, show that $-\theta_{11}$ is the $i = 1$ part of θ_{21} and that $\theta_{22} = 0$. Then show that $-\theta_{12}$ is the $i > 1$ part of θ_{21}.

Theorem 7.7.2 $V_*(\mathfrak{g}) \xrightarrow{\varepsilon} k$ *is a projective resolution of the $\mathfrak{g}$-module k.*

Proof (Koszul) It suffices to show that $H_n(V_*(\mathfrak{g})) = 0$ for $n \neq 0$.

Choose an ordered basis $\{e_\alpha\}$ of $\mathfrak{g}$ as a k-module. By the Poincaré-Birkhoff-Witt Theorem (7.3.7), $V_n(\mathfrak{g})$ is a free k-module with a basis consisting of terms

$$(*) \quad e_I \otimes (e_{\alpha_1} \wedge \cdots \wedge e_{\alpha_n}), \quad \alpha_1 < \cdots < \alpha_n \text{ and } I = (\beta_1, \cdots, \beta_m) \text{ increasing.}$$

We filter $V_*(\mathfrak{g})$ by k-submodules, letting $F_p V_n$ be the submodule generated by terms $(*)$ with $m + n \leq p$. Since $[e_i e_j]$ is a linear combination of the e_α in $\mathfrak{g}$, this is actually a filtration by chain subcomplexes

$$0 \subset F_0 V_* \subseteq F_1 V_* \subseteq \cdots \subseteq V_*(\mathfrak{g}) = \cup F_p V_*.$$

This filtration is bounded below and exhaustive (see 5.4.2), so by 5.5.1 there is a convergent spectral sequence

$$E^0_{pq} = F_p V_{p+q} / F_{p-1} V_{p+q} \Rightarrow H_{p+q}(V_*(\mathfrak{g})).$$

This spectral sequence is concentrated in the octant $p \geq 0$, $q \leq 0$, $p + q \geq 0$. The first column is $F_0 V_*$, which is zero except in the $(0,0)$ spot, where E^0_{00} is $F_0 V_0 = k$.

We claim that each column E^0_{p*} is exact for $p \neq 0$. This will prove that the spectral sequence collapses at E^1, with $E^1_{pq} = 0$ for $(p, q) \neq (0, 0)$, yielding the desired computation: $H_n(V_*) = 0$ for $n \neq 0$.

Let A_q be the free k-submodule of $U\mathfrak{g}$ on basis

$$\{e_I : I = (\beta_1, \cdots, \beta_q) \quad \text{is an increasing sequence}\}.$$

Then $A_q \cong F_q V_0 / F_{q-1} V_0$ and $E^0_{pq} = A_{-q} \otimes_k \Lambda^{p+q} \mathfrak{g}$. Moreover, the formula for the differential in V_* shows that the differential $d^0 \colon E^0_{pq} \to E^0_{p,q-1}$ is given by

$$d^0(a \otimes e_{\alpha_1} \wedge \cdots \wedge e_{\alpha n}) = \theta_1 = \sum_{i=1}^{n} (-1)^{i+1} a e_{\alpha_i} \otimes e_{\alpha_1} \wedge \cdots \wedge \hat{e}_{\alpha_i} \wedge \cdots \wedge e_{\alpha n}.$$

We saw in exercise 7.3.6 that $A = A_0 \oplus A_1 \oplus \cdots$ is a polynomial ring on the indeterminates $e_\alpha \colon A \cong k[e_1, e_2, \cdots]$. Comparing formulas for d, we see that the direct sum $\bigoplus E^0_{p*}$ of the chain complexes E^0_{p*} is identical to the Koszul complex

$$A \otimes_k \Lambda^* \mathfrak{g} = \Lambda^* (\oplus A e_\alpha) = K(x)$$

of 4.5.1 corresponding to the sequence $x = (e_1, e_2, \cdots)$. Since x is a regular sequence, we know from *loc. cit.* that

$$H_n(x, A) = H_n(A \otimes \Lambda^* \mathfrak{g}) = \bigoplus_{p=0}^{\infty} H_{n-p}(E^0_{p*}) = \bigoplus_{p=0}^{\infty} E^1_{p,n-p}$$

is zero for $n \neq 0$ and $A/xA = k$ for $n = 0$. Since $E^1_{00} = k$, it follows that $E^1_{pq} = 0$ for $(p, q) \neq (0, 0)$, as claimed. $\diamond$

Corollary 7.7.3 (Chevalley-Eilenberg) *If M is a right $\mathfrak{g}$-module, then the homology modules $H_*(\mathfrak{g}, M)$ are the homology of the chain complex*

$$M \otimes_{U\mathfrak{g}} V_*(\mathfrak{g}) = M \otimes_{U\mathfrak{g}} U\mathfrak{g} \otimes_k \Lambda^* \mathfrak{g} = M \otimes_k \Lambda^* \mathfrak{g}.$$

If M is a left $\mathfrak{g}$-module, then the cohomology modules $H^*(\mathfrak{g}, M)$ are the cohomology of the cochain complex

$$\operatorname{Hom}_\mathfrak{g}(V(\mathfrak{g}), M) = \operatorname{Hom}_\mathfrak{g}(U\mathfrak{g} \otimes_k \Lambda^* \mathfrak{g}, M) \cong \operatorname{Hom}_k(\Lambda^* \mathfrak{g}, M).$$

In this complex, an n-cochain $f \colon \Lambda^n \mathfrak{g} \to M$ is just an alternating k-multilinear function $f(x_1, \cdots, x_n)$ of n variables in $\mathfrak{g}$, taking values in M. The coboundary δf of such an n-cochain is the $(n + 1)$-cochain

$$\delta f(x_1, \cdots, x_{n+1}) = \sum (-1)^{i+1} x_i f(x_1, \cdots, \hat{x}_i, \cdots)$$
$$+ \sum (-1)^{i+j} f([x_i x_j], x_1, \cdots, \hat{x}_i, \cdots, \hat{x}_j, \cdots).$$

Application 7.7.4 (Cohomological dimension) If $\mathfrak{g}$ is n-dimensional as a vector space over a field k, then $H^i(\mathfrak{g}, M) = H_i(\mathfrak{g}, M) = 0$ for all $i > n$. Indeed, $\Lambda^i \mathfrak{g} = 0$ in this range. The following exercise shows that $H^n(\mathfrak{g}, M) \neq 0$ for some $\mathfrak{g}$-module M, so that $\mathfrak{g}$ has cohomological dimension $n = \dim_k(\mathfrak{g})$.

Exercise 7.7.2 If k is a field and $\mathfrak{g}$ is n-dimensional as a vector space, show that $U\mathfrak{g}$ has global dimension n (4.1.2). To do this, we proceed in several steps. First note that $pd_{U\mathfrak{g}}(k) \leq n$ because $V_*(\mathfrak{g})$ is a projective resolution of k.

1. Let $\Lambda^n \mathfrak{g} \cong k$ be given the $\mathfrak{g}$-module structure

$$[y, x_1 \wedge \cdots \wedge x_n] = \sum_{i=1}^{n} x_1 \wedge \cdots \wedge [yx_i] \wedge \cdots \wedge x_n.$$

   Show that $H^n(\mathfrak{g}, \Lambda^n \mathfrak{g}) \cong k$. This proves that $pd_{U\mathfrak{g}}(\Lambda^n \mathfrak{g}) = n$ and hence that $gl.\dim(U\mathfrak{g}) \geq n$.

2. Use the natural isomorphism $\operatorname{Ext}^*_{U\mathfrak{g}}(M, N) \cong H^*_{\mathrm{Lie}}(\mathfrak{g}, \operatorname{Hom}_k(M, N))$ (exercise 7.3.5) and the Global Dimension theorem 4.1.2 to show that $gl.\dim(U\mathfrak{g}) \leq n$, and hence that $gl.\dim(U\mathfrak{g}) = n$.

Exercise 7.7.3 Use the Chevalley-Eilenberg complex to show that

$$H_3(\mathfrak{sl}_2, k) \cong H^3(\mathfrak{sl}_2, k) \cong k.$$

Exercise 7.7.4 (1-cocycles and module extensions) Let M be a left $\mathfrak{g}$-module. If $0 \to M \to N \xrightarrow{\pi} k \to 0$ is a short exact sequence of $\mathfrak{g}$-modules, and $n \in N$ is such that $\pi(n) = 1$, define $f: \mathfrak{g} \to M$ by $f(x) = xn$. Show that f is a 1-cocycle in the Chevalley-Eilenberg complex $\operatorname{Hom}_k(\Lambda^* \mathfrak{g}, M)$ and that its class $[f] \in H^1(\mathfrak{g}, M)$ is independent of the choice of n. Then show that $H^1(\mathfrak{g}, M)$ is in 1–1 correspondence with the equivalence classes of $\mathfrak{g}$-module extensions of k by M. (Compare to exercise 7.4.5.)

Exercise 7.7.5 (2-cocycles and algebra extensions) Let M be a left $\mathfrak{g}$-module, with $\mathfrak{g}$ free as a k-module.

1. If $0 \to M \to \mathfrak{e} \xrightarrow{\pi} \mathfrak{g} \to 0$ is an extension of Lie algebras, and $\sigma: \mathfrak{g} \to \mathfrak{e}$ is a k-module splitting of π, show that the Lie algebra structure on $\mathfrak{e} \cong M \times \mathfrak{g}$ may be described by an alternating k-bilinear function $f: \mathfrak{g} \times \mathfrak{g} \to M$ defined by

$$[\sigma(x), \sigma(y)] = \sigma([xy]) + f(x, y), \quad x, y \in \mathfrak{g}.$$

Show that f is a 2-cocycle for the Chevalley-Eilenberg cochain complex $\mathrm{Hom}_k(\Lambda^* \mathfrak{g}, M)$. Also, show that if σ' is any other splitting of π, then the resulting 2-cocycle f' is cohomologous to f. This shows that such an extension determines a well-defined element $[f] \in H^2_{\mathrm{Lie}}(\mathfrak{g}, M)$.

2. Using part (1), show directly that $H^2_{\mathrm{Lie}}(\mathfrak{g}, M)$ is in 1–1 correspondence with equivalence classes of Lie algebra extensions of $\mathfrak{g}$ by M. This is the same correspondence as we gave in section 7.6 by a more abstract approach.

Exercise 7.7.6 If M is a right $\mathfrak{g}$-module and $g \in \mathfrak{g}$, show that the formula

$$(m \otimes x_1 \wedge \cdots \wedge x_p)g = [mg] \otimes x_1 \wedge \cdots \wedge x_p$$
$$+ \sum m \otimes x_1 \wedge \cdots \wedge [x_i g] \wedge \cdots \wedge x_p$$

makes $M \otimes V_*(g)$ into a chain complex of right $\mathfrak{g}$-modules. Then show that the induced $\mathfrak{g}$-module structure on $H_*(\mathfrak{g}; M)$ is trivial.

7.8 Semisimple Lie Algebras

We now restrict our attention to finite-dimensional Lie algebras over a field k of characteristic 0. We will give cohomological proofs of several main theorems involving solvable and semisimple Lie algebras. First, however, we need to summarize the main notions of the classical theory of semisimple Lie algebras.

Definitions 7.8.1 An ideal of $\mathfrak{g}$ is called *solvable* if it is solvable as a Lie algebra (see 7.1.7). It is not hard to show that the family of all solvable ideals of $\mathfrak{g}$ forms a lattice, because the sum and intersection of solvable ideals is a solvable ideal [JLA, I.7]. If $\mathfrak{g}$ is finite-dimensional, there is a largest solvable ideal of $\mathfrak{g}$, called the *solvable radical* rad $\mathfrak{g}$ of $\mathfrak{g}$. Every ideal $\mathfrak{h}$ of $\mathfrak{g}$ contained in rad $\mathfrak{g}$ is a solvable ideal.

A Lie algebra $\mathfrak{g}$ is called *simple* if it has no ideals except itself and 0, and if $[\mathfrak{g}, \mathfrak{g}] \neq 0$ (i.e., $\mathfrak{g} = [\mathfrak{g}, \mathfrak{g}]$). For example, $\mathfrak{sl}_n(k)$ is a simple Lie algebra for $n \geq 2$ (as $\mathrm{char}(k) \neq 2$).

A Lie algebra $\mathfrak{g}$ is called *semisimple* if rad $\mathfrak{g} = 0$, that is, if $\mathfrak{g}$ has no nonzero solvable ideals. In fact, $\mathfrak{g}$ is semisimple iff $\mathfrak{g}$ has no nonzero abelian ideals; to see this, note that the last nonzero term $(\mathrm{rad}\ \mathfrak{g})^{(n-1)}$ in the derived series for rad $\mathfrak{g}$ is an abelian ideal of $\mathfrak{g}$. Clearly simple Lie algebras are semisimple.

Lemma 7.8.2 *If $\mathfrak{g}$ is a finite-dimensional Lie algebra, then $\mathfrak{g}/(\mathrm{rad}\ \mathfrak{g})$ is a semisimple Lie algebra.*

Proof If not, $\mathfrak{g}/\operatorname{rad}\mathfrak{g}$ contains a nonzero abelian ideal $\mathfrak{a} = \mathfrak{h}/\operatorname{rad}\mathfrak{g}$. But $[\mathfrak{a}, \mathfrak{a}] = 0$, so $\mathfrak{h}' = [\mathfrak{h}, \mathfrak{h}]$ must lie inside $\operatorname{rad}\mathfrak{g}$. Hence $\mathfrak{h}'$ is solvable, and therefore so is $\mathfrak{h}$. This contradicts the maximality of $\operatorname{rad}\mathfrak{g}$. $\diamond$

Definition 7.8.3 (Killing form) If $\mathfrak{g}$ is a Lie subalgebra of $\mathfrak{gl}_n$ we can use matrix multiplication to define the symmetric bilinear form $\beta(x, y) = \operatorname{trace}(xy)$ on $\mathfrak{g}$. This symmetric form is "$\mathfrak{g}$-invariant" in the sense that for $x, y, z \in \mathfrak{g}$ we have $\beta([xy], z) = \beta(x, [yz])$, or equivalently $\beta([xy], z) + \beta(x, [zy]) = 0$. (Exercise!)

Now suppose that $\mathfrak{g}$ is an n-dimensional Lie algebra over k. Left multiplication by elements of $\mathfrak{g}$ gives a Lie algebra homomorphism

$$ad: \mathfrak{g} \to \operatorname{Lie}(\operatorname{End}_k(\mathfrak{g})) = \mathfrak{gl}_n,$$

called the *adjoint representation* of $\mathfrak{g}$. The symmetric bilinear form on $\mathfrak{g}$ obtained by pulling back β is called the *Killing form* of $\mathfrak{g}$, that is, the Killing form is $\kappa(x, y) = \operatorname{trace}(ad(x)ad(y))$. The importance of the Killing form is summed up in the following result, which we cite from [JLA, III.4]:

Cartan's Criterion for Semisimplicity 7.8.4 *Let $\mathfrak{g}$ be a finite-dimensional Lie algebra over a field of characteristic 0.*

1. *$\mathfrak{g}$ is semisimple if and only if the Killing form is a nondegenerate symmetric bilinear form on the vector space $\mathfrak{g}$.*
2. *If $\mathfrak{g} \subseteq \mathfrak{gl}_n$ and $\mathfrak{g}$ is semisimple, then the bilinear form $\beta(x, y) = \operatorname{trace}(xy)$ is also nondegenerate on $\mathfrak{g}$.*

Structure Theorem of Semisimple Lie Algebras 7.8.5 *Let $\mathfrak{g}$ be a finite-dimensional Lie algebra over a field of characteristic 0. Then $\mathfrak{g}$ is semisimple iff $\mathfrak{g} = \mathfrak{g}_1 \times \mathfrak{g}_2 \times \cdots \times \mathfrak{g}_r$ is the finite product of simple Lie algebras $\mathfrak{g}_i$. In particular, every ideal of a semisimple Lie algebra is semisimple.*

Proof If the $\mathfrak{g}_i$ are simple, every ideal of $\mathfrak{g} = \mathfrak{g}_1 \times \cdots \times \mathfrak{g}_r$ is a product of $\mathfrak{g}_i$'s and cannot be abelian, so $\mathfrak{g}$ is semisimple.

For the converse, it suffices to show that every minimal ideal $\mathfrak{a}$ of a semisimple Lie algebra $\mathfrak{g}$ is a direct factor: $\mathfrak{g} = \mathfrak{a} \times \mathfrak{b}$. Define $\mathfrak{b}$ to be the orthogonal complement of $\mathfrak{a}$ with respect to the Killing form. To see that $\mathfrak{b}$ is an ideal of $\mathfrak{g}$, we use the $\mathfrak{g}$-invariance of κ: for $a \in \mathfrak{a}, b \in \mathfrak{b}$, and $x \in \mathfrak{g}$,

$$\kappa(a, [x, b]) = \kappa([ax], b) = 0$$

because $[ax] \in \mathfrak{a}$. Hence $[xb] \in \mathfrak{b}$ and $\mathfrak{b}$ is an ideal of $\mathfrak{g}$.

To conclude, it suffices to show that $\mathfrak{a} \cap \mathfrak{b} = 0$, since this implies $\mathfrak{g} = \mathfrak{a} \times \mathfrak{b}$. Now $\mathfrak{a} \cap \mathfrak{b}$ is an ideal of $\mathfrak{g}$; since $\mathfrak{a}$ is minimal either $\mathfrak{a} \cap \mathfrak{b} = \mathfrak{a}$ or $\mathfrak{a} \cap \mathfrak{b} = 0$. If $\mathfrak{a} \cap \mathfrak{b} = \mathfrak{a}$, then $\kappa([a_1 a_2], x) = \kappa(a_1, [a_2 x]) = 0$ for every $a_1, a_2 \in \mathfrak{a}$ and $x \in \mathfrak{g}$. Since κ is nondegenerate, this implies that $[a_1 a_2] = 0$. Thus $\mathfrak{a}$ is abelian, contradicting the semisimplicity of $\mathfrak{g}$. Hence $\mathfrak{a} \cap \mathfrak{b} = 0$, and we are done. $\diamond$

Corollary 7.8.6 *If $\mathfrak{g}$ is finite-dimensional and semisimple (and char$(k) = 0$), then $\mathfrak{g} = [\mathfrak{g}, \mathfrak{g}]$. Consequently,*

$$H_1(\mathfrak{g}, k) = H^1(\mathfrak{g}, k) = 0.$$

Proof If $\mathfrak{g} = [\mathfrak{g}, \mathfrak{g}]$, then $\mathfrak{g}^{ab} = 0$. On the other hand, we saw in 7.4.1 and 7.4.8 that $H_1(\mathfrak{g}, k) \cong \mathfrak{g}^{ab}$ and $H^1(\mathfrak{g}, k) \cong \mathrm{Hom}_k(\mathfrak{g}^{ab}, k)$. $\diamond$

Corollary 7.8.7 *If $\mathfrak{g} \subseteq \mathfrak{gl}_n$ is semisimple, then $\mathfrak{g} \subseteq \mathfrak{sl}_n = [\mathfrak{gl}_n, \mathfrak{gl}_n]$.*

Exercise 7.8.1 Suppose that k is an algebraically closed field of characteristic 0 and that $\mathfrak{g}$ is a finite-dimensional simple Lie algebra over k.

1. Use Schur's Lemma to see that $\mathrm{Hom}_{\mathfrak{g}}(\mathfrak{g}, \mathfrak{g}) \cong k$.
2. Show that $\mathfrak{g} \cong \mathrm{Hom}_k(\mathfrak{g}, k)$ as $\mathfrak{g}$-modules.
3. If $f: \mathfrak{g} \otimes \mathfrak{g} \to k$ is any $\mathfrak{g}$-invariant symmetric bilinear form, show that f is a multiple of the Killing form κ, that is, $f = \alpha \kappa$ for some $\alpha \in k$.
4. If V is any k-vector space and $f: \mathfrak{g} \otimes \mathfrak{g} \to V$ is any $\mathfrak{g}$-invariant symmetric bilinear map, show that there is a $v \in V$ such that $f(x, y) = \kappa(x, y)v$.

Exercise 7.8.2 (Counterexample to structure theorem in char. $p \neq 0$) Let k be a field of characteristic $p \neq 0$, and consider the Lie algebra $\mathfrak{gl}_n$, $n \geq 3$. Show that the only ideals of $\mathfrak{gl}_n$ are $\mathfrak{sl}_n = [\mathfrak{gl}_n, \mathfrak{gl}_n]$ and the center $k \cdot 1$. If $p | n$, show that the center is contained inside $\mathfrak{sl}_n$. This shows that $\mathfrak{pgl}_n = \mathfrak{gl}_n / k \cdot 1$ has only one ideal, namely $\mathfrak{psl}_n = \mathfrak{sl}_n / k \cdot 1$, and that $\mathfrak{psl}_n$ is simple. Conclude that $\mathfrak{pgl}_n$ is semisimple but not a direct product of simple ideals and show that $H_1(\mathfrak{pgl}_n, k) \cong H^1(\mathfrak{pgl}_n, k) \cong k$.

The Casimir Operator 7.8.8 Let $\mathfrak{g}$ be semisimple and let M be an m-dimensional $\mathfrak{g}$-module. If $\mathfrak{h}$ is the image of the structure map

$$\rho: \mathfrak{g} \to \mathrm{Lie}(\mathrm{End}_k(M)) \cong \mathfrak{gl}_m(k),$$

then $\mathfrak{g} \cong \mathfrak{h} \times \ker(\rho)$, $\mathfrak{h} \subseteq \mathfrak{gl}_m$, and the bilinear form β on $\mathfrak{h}$ is nondegenerate by Cartan's criterion 7.8.4. Choose a basis $\{e_1, \cdots, e_r\}$ of $\mathfrak{h}$; by linear algebra there is a dual basis $\{e^1, \cdots, e^r\}$ of $\mathfrak{h}$ such that $\beta(e_i, e^j) = \delta_{ij}$. The element $c_M = \sum e_i e^i \in U\mathfrak{g}$ is called the *Casimir operator* for M; it is independent of the choice of basis for $\mathfrak{h}$. The following facts are easy to prove and are left as exercises:

1. If $x \in \mathfrak{g}$ and $[e_i, x] = \sum c_{ij} e_j$, then $[x, e^j] = \sum c_{ij} e^i$.
2. c_M is in the center of $U\mathfrak{g}$ and $c_M \in \mathfrak{I}$. *Hint:* Use (1).
3. The image of c_M in the matrix ring $\operatorname{End}_k(M)$ is r/m times the identity matrix. In particular, if M is nontrivial as a $\mathfrak{g}$-module, then $r \neq 0$ and c_M acts on M as an automorphism. *Hint:* By (2) it is a scalar matrix, so it suffices to show that the trace is $r = \dim(\mathfrak{h})$.

Exercise 7.8.3 Let $\mathfrak{g} = \mathfrak{sl}_2$ with basis $x = \left(\begin{smallmatrix} 0 & 1 \\ 0 & 0 \end{smallmatrix}\right)$, $y = \left(\begin{smallmatrix} 0 & 0 \\ 1 & 0 \end{smallmatrix}\right)$, $h = \left(\begin{smallmatrix} 1 & 0 \\ 0 & -1 \end{smallmatrix}\right)$. If M is the canonical 2-dimensional $\mathfrak{g}$-module, show that $c_M = 2xy - h + h^2/2$, while its image in $\operatorname{End}(M)$ is the matrix $\left(\begin{smallmatrix} 3/2 & 0 \\ 0 & 3/2 \end{smallmatrix}\right)$.

Theorem 7.8.9 *Let $\mathfrak{g}$ be a semisimple Lie algebra over a field of characteristic 0. If M is a simple $\mathfrak{g}$-module, $M \neq k$, then*

$$H^i_{\mathrm{Lie}}(\mathfrak{g}, M) = H_i^{\mathrm{Lie}}(\mathfrak{g}, M) = 0 \quad \text{for all } i.$$

Proof Let C be the center of $U\mathfrak{g}$. We saw in 3.2.11 and 3.3.6 that $H_*(\mathfrak{g}, M) = \operatorname{Tor}_*^{U\mathfrak{g}}(k, M)$ and $H^*(\mathfrak{g}, M) = \operatorname{Ext}^*_{U\mathfrak{g}}(k, M)$ are naturally C-modules; moreover, multiplication by $c \in C$ is induced by $c: k \to k$ as well as $c: M \to M$. Since the Casimir element c_M acts by 0 on k (as $c_M \in \mathfrak{I}$) and by the invertible scalar r/m on M, we must have $0 = r/m$ on $H_*(\mathfrak{g}, M)$ and $H^*(\mathfrak{g}, M)$. This can only happen if these C-modules are zero. $\diamond$

Corollary 7.8.10 (Whitehead's first lemma) *Let $\mathfrak{g}$ be a semisimple Lie algebra over a field of characteristic 0. If M is any finite-dimensional $\mathfrak{g}$-module, then $H^1_{\mathrm{Lie}}(\mathfrak{g}, M) = 0$. That is, every derivation from $\mathfrak{g}$ into M is an inner derivation.*

Proof We proceed by induction on $\dim(M)$. If M is simple, then either $M = k$ and $H^1(\mathfrak{g}, k) = \mathfrak{g}/[\mathfrak{g}, \mathfrak{g}] = 0$ or else $M \neq k$ and $H^*(\mathfrak{g}, M) = 0$ by the theorem. Otherwise, M contains a proper submodule L. By induction, $H^1(\mathfrak{g}, L) = H^1(\mathfrak{g}, M/L) = 0$, so we are done via the cohomology exact sequence

$$\cdots H^1(\mathfrak{g}, L) \to H^1(\mathfrak{g}, M) \to H^1(\mathfrak{g}, M/L) \cdots.$$ $\diamond$

Weyl's Theorem 7.8.11 *Let $\mathfrak{g}$ be a semisimple Lie algebra over a field of characteristic 0. Then every finite-dimensional $\mathfrak{g}$-module M is completely reducible, that is, is a direct sum of simple $\mathfrak{g}$-modules.*

Proof Suppose that M is not a direct sum of simple modules. Since $\dim(M)$ is finite, M contains a submodule M_1 minimal with respect to this property. Clearly M_1 is not simple, so it contains a proper $\mathfrak{g}$-submodule M_0. By induction, both M_0 and $M_2 = M_1/M_0$ are direct sums of simple $\mathfrak{g}$-modules yet M_1 is not, so the extension M_1 of M_2 by M_0 must be represented (3.4.3) by a nonzero element of

$$\text{Ext}^1_{U\mathfrak{g}}(M_2, M_0) \cong H^1_{\text{Lie}}(\mathfrak{g}, \text{Hom}_k(M_2, M_0))$$

(see exercise 7.3.5), and this contradicts Whitehead's first lemma.　　　　◇

Corollary 7.8.12 (Whitehead's second lemma) *Let $\mathfrak{g}$ be a semisimple Lie algebra over a field of characteristic 0. If M is any finite-dimensional $\mathfrak{g}$-module, then $H^2_{\text{Lie}}(\mathfrak{g}, M) = 0$.*

Proof Since H^* commutes with direct sums, and M is a direct sum of simple $\mathfrak{g}$-modules, we may assume that M is simple. If $M \neq k$ we already know the result by 7.8.9, so it suffices to show that $H^2(\mathfrak{g}, k) = 0$, that is, that every Lie algebra extension

$$0 \to k \to \mathfrak{e} \xrightarrow{\pi} \mathfrak{g} \to 0$$

splits. We claim that $\mathfrak{e}$ can be made into a $\mathfrak{g}$-module in such a way that π is a $\mathfrak{g}$-map. To see this, let $\tilde{x}$ be any lift of $x \in \mathfrak{g}$ to $\mathfrak{e}$ and define $x \circ y$ to be $[\tilde{x}, y]$ for $y \in \mathfrak{e}$. This is independent of the choice of $\tilde{x}$ because k is in the center of $\mathfrak{e}$. The $\mathfrak{g}$-module axioms are readily defined (exercise!), and by construction $\pi(x \circ y) = [x, \pi(y)]$. This establishes the claim.

By Weyl's Theorem $\mathfrak{e}$ and $\mathfrak{g}$ split as $\mathfrak{g}$-modules, and there is a $\mathfrak{g}$-module homomorphism $\sigma: \mathfrak{g} \to \mathfrak{e}$ splitting π such that $\mathfrak{e} \cong k \times \mathfrak{g}$ as a $\mathfrak{g}$-module. If we choose $\tilde{x} = \sigma(x)$, then it is clear that σ is a Lie algebra homomorphism and that $\mathfrak{e} \cong k \times \mathfrak{g}$ as a Lie algebra. This proves that $H^2(\mathfrak{g}, k) = 0$, as desired.　◇

Remark $H^3(\mathfrak{sl}_2, k) \cong k$ (exercise 7.7.3), so there can be no "third Whitehead lemma."

Levi's Theorem 7.8.13 *If $\mathfrak{g}$ is a finite-dimensional Lie algebra over a field of characteristic zero, then there is a semisimple Lie subalgebra $\mathcal{L}$ of $\mathfrak{g}$ (called a*

Levi factor *of* $\mathfrak{g}$) *such that* $\mathfrak{g}$ *is isomorphic to the semidirect product*

$$\mathfrak{g} \cong (\text{rad } \mathfrak{g}) \rtimes \mathcal{L}.$$

Proof We know that $\mathfrak{g}/(\text{rad } \mathfrak{g})$ is semisimple, so it suffices to show that the following Lie algebra extension splits.

$$0 \to \text{rad } \mathfrak{g} \to \mathfrak{g} \to \mathfrak{g}/\text{rad } \mathfrak{g} \to 0$$

If rad $\mathfrak{g}$ is abelian then these extensions are classified by $H^2(\mathfrak{g}/(\text{rad } \mathfrak{g}), \text{rad } \mathfrak{g})$, which vanishes by Whitehead's second lemma, so every extension splits.

If rad $\mathfrak{g}$ is not abelian, we proceed by induction on the derived length of rad $\mathfrak{g}$. Let $\mathfrak{r}$ denote the ideal [rad $\mathfrak{g}$, rad $\mathfrak{g}$] of $\mathfrak{g}$. Since $\text{rad}(\mathfrak{g}/\mathfrak{r}) = (\text{rad } \mathfrak{g})/\mathfrak{r}$ is abelian, the extension

$$0 \to (\text{rad } \mathfrak{g})/\mathfrak{r} \to \mathfrak{g}/\mathfrak{r} \to \mathfrak{g}/(\text{rad } \mathfrak{g}) \to 0$$

splits. Hence there is an ideal $\mathfrak{h}$ of $\mathfrak{g}$ containing $\mathfrak{r}$ such that $\mathfrak{g}/\mathfrak{r} \cong (\text{rad } \mathfrak{g})/\mathfrak{r} \times \mathfrak{h}/\mathfrak{r}$ and $\mathfrak{h}/\mathfrak{r} \cong \mathfrak{g}/(\text{rad } \mathfrak{g})$. Now

$$\text{rad}(\mathfrak{h}) = \text{rad}(\mathfrak{g}) \cap \mathfrak{h} = \mathfrak{r},$$

and $\mathfrak{r}$ has a smaller derived length than rad $\mathfrak{g}$. By induction there is a Lie subalgebra $\mathcal{L}$ of $\mathfrak{h}$ such that $\mathfrak{h} \cong \mathfrak{r} \rtimes \mathcal{L}$ and $\mathcal{L} \cong \mathfrak{h}/\mathfrak{r} \cong \mathfrak{g}/\text{rad } \mathfrak{g}$. But then $\mathcal{L}$ is our desired Levi factor of $\mathfrak{g}$. $\diamond$

Remark Levi factors are not unique, but they are clearly all isomorphic to $\mathfrak{g}/(\text{rad } \mathfrak{g})$ and hence to each other. Malcev proved (in 1942) that the Levi factors are all conjugate by nice automorphisms of $\mathfrak{g}$.

Historical Remark 7.8.14 (see [Bour]) Sophus Lie developed the theory of Lie groups and their Lie algebras from 1874 to 1893. Semisimple Lie algebras over $\mathbb{C}$ are in 1–1 correspondence with compact, simply connected Lie groups. In the period 1888–1894 much of the structure of Lie algebras over $\mathbb{C}$ was developed, including W. Killing's discovery of the solvable radical and semisimple Lie algebras, and the introduction of the "Killing form" in E. Cartan's thesis. The existence of Levi factors was announced by Cartan but only proven (publicly) by E. E. Levi in 1905. Weyl's Theorem (1925) was originally proven using integration on compact Lie groups. An algebraic proof of Weyl's theorem was found in 1935 by Casimir and van der Waerden. This and J. H. C. Whitehead's two lemmas (1936–1937) provided the first clues that enabled Chevalley and Eilenberg (1948 [ChE]) to construct the cohomology $H^*(\mathfrak{g}, M)$. The cohomological proofs in this section are close parallels of the treatment by Chevalley and Eilenberg.

Exercise 7.8.4 If $\mathfrak{g}$ is a finite-dimensional Lie algebra over a field of characteristic 0, show that $\mathfrak{g}$ is semisimple iff $H^1(\mathfrak{g}, M) = 0$ for all finite-dimensional $\mathfrak{g}$-modules M.

Exercise 7.8.5 (Reductive Lie algebras) A Lie algebra $\mathfrak{g}$ is called *reductive* if $\mathfrak{g}$ is a completely reducible $\mathfrak{g}$-module (via the adjoint representation). That is, $\mathfrak{g}$ is reductive if $\mathfrak{g}$ is a direct sum of simple $\mathfrak{g}$-modules. Now assume that $\mathfrak{g}$ is finite-dimensional over a field of characteristic 0, so that $\mathfrak{g} \cong (\text{rad } \mathfrak{g}) \rtimes \mathcal{L}$ for some semisimple Lie algebra $\mathcal{L}$ by Levi's theorem. Show that the following are equivalent:

1. $\mathfrak{g}$ is reductive
2. $[\mathfrak{g}, \mathfrak{g}] = \mathcal{L}$
3. $\text{rad}(\mathfrak{g})$ is abelian and equals the center of $\mathfrak{g}$
4. $\mathfrak{g} \cong \mathfrak{a} \times \mathcal{L}$ where $\mathfrak{a}$ is abelian and $\mathcal{L}$ is semisimple

Then show that $\mathfrak{gl}_m$ is a reductive Lie algebra, and in fact that $\mathfrak{gl}_m \cong k \times \mathfrak{sl}_m$.

7.9 Universal Central Extensions

A *central extension* $\mathfrak{e}$ of a Lie algebra $\mathfrak{g}$ is an extension $0 \to M \to \mathfrak{e} \xrightarrow{\pi} \mathfrak{g} \to 0$ of Lie algebras such that M is in the center of $\mathfrak{e}$ (i.e., it is just an extension of Lie algebras of $\mathfrak{g}$ by a trivial $\mathfrak{g}$-module M in the sense of 7.6.1). A homomorphism over $\mathfrak{g}$ from $\mathfrak{e}$ to another central extension $0 \to M' \to \mathfrak{e}' \xrightarrow{\pi'} \mathfrak{g} \to 0$ is a map $f: \mathfrak{e} \to \mathfrak{e}'$ such that $\pi = \pi' f$. $\mathfrak{e}$ is called a *universal central extension* of $\mathfrak{g}$ if for every central extension $\mathfrak{e}'$ of $\mathfrak{g}$ there is a unique homomorphism $f: \mathfrak{e} \to \mathfrak{e}'$ over $\mathfrak{g}$. Clearly, a universal central extension of $\mathfrak{g}$ is unique up to isomorphism over $\mathfrak{g}$, provided it exists. As with groups (6.9.2), if $\mathfrak{g}$ has a universal central extension, then $\mathfrak{g}$ must be perfect, that is, $\mathfrak{g} = [\mathfrak{g}, \mathfrak{g}]$.

Construction of a Universal Central Extension 7.9.1 We may copy the construction 6.9.3 for groups. Choose a free Lie algebra $\mathfrak{f}$ mapping onto $\mathfrak{g}$ and let $\mathfrak{r} \subset \mathfrak{f}$ denote the kernel, so that $\mathfrak{g} \cong \mathfrak{f}/\mathfrak{r}$. This yields a central extension

$$0 \to \mathfrak{r}/[\mathfrak{r}, \mathfrak{f}] \to \mathfrak{f}/[\mathfrak{r}, \mathfrak{f}] \to \mathfrak{g} \to 0.$$

If $\mathfrak{g}$ is perfect, $[\mathfrak{f}, \mathfrak{f}]$ maps onto $\mathfrak{g}$, and we claim that

$$0 \to (\mathfrak{r} \cap [\mathfrak{f}, \mathfrak{f}])/[\mathfrak{r}, \mathfrak{f}] \to [\mathfrak{f}, \mathfrak{f}]/[\mathfrak{r}, \mathfrak{f}] \to \mathfrak{g} \to 0$$

is a universal central extension of $\mathfrak{g}$. Note that $H_2(\mathfrak{g}, k) = (\mathfrak{r} \cap [\mathfrak{f}, \mathfrak{f}])/[\mathfrak{r}, \mathfrak{f}]$ by exercise 7.5.2.

Theorem 7.9.2 *A Lie algebra $\mathfrak{g}$ has a universal central extension iff $\mathfrak{g}$ is perfect. In this case, the universal central extension is*

$(*)$ $\qquad\qquad 0 \to H_2(\mathfrak{g}, k) \to [\mathfrak{f}, \mathfrak{f}]/[\mathfrak{r}, \mathfrak{f}] \to \mathfrak{g} \to 0.$

Proof We have seen that $(*)$ is a central extension. Set $\mathfrak{e} = [\mathfrak{f}, \mathfrak{f}]/[\mathfrak{r}, \mathfrak{f}]$. Since $[\mathfrak{f}, \mathfrak{f}]$ maps onto $\mathfrak{g}$, any $x, y \in \mathfrak{f}$ may be written as $x = x' + r$, $y = y' + s$ with $x', y' \in [\mathfrak{f}, \mathfrak{f}]$ and $r, s \in \mathfrak{r}$. Thus in $\mathfrak{f}/[\mathfrak{r}, \mathfrak{f}]$

$$[x, y] = [x', y'] + [x', s] + [r, y'] + [r, s] = [x', y'].$$

This shows that $\mathfrak{e}$ is also a perfect Lie algebra. If $0 \to M \to \mathfrak{e}' \xrightarrow{\pi} \mathfrak{g} \to 0$ is another central extension, lift $\mathfrak{f} \to \mathfrak{g}$ to a map $\phi \colon \mathfrak{f} \to \mathfrak{e}'$. Since $\pi\phi(\mathfrak{r}) = 0$, $\phi(\mathfrak{r}) \subseteq M$. This implies that $\phi([\mathfrak{r}, \mathfrak{f}]) = 1$. As in 6.9.5, ϕ induces a map $f \colon \mathfrak{e} \to \mathfrak{e}'$ over $\mathfrak{g}$. If f_1 is another such map, the difference $\delta = f_1 - f \colon \mathfrak{e} \to M$ is zero because $\mathfrak{e} = [\mathfrak{e}, \mathfrak{e}]$ and

$$f_1([xy]) = [f(x) + \delta(x), f(y) + \delta(y)] = [f(x), f(y)] = f([x, y]).$$

Hence $f_1 = f$, that is, f is unique. $\qquad\qquad\qquad\qquad\qquad\qquad\quad \diamond$

By copying the proofs of 6.9.6 and 6.9.7, we also have the following two results.

Lemma 7.9.3 *If $0 \to M \to \mathfrak{e} \to \mathfrak{g} \to 0$ and $0 \to M' \to \mathfrak{e}' \to \mathfrak{g} \to 0$ are central extensions, and $\mathfrak{e}$ is perfect, there is at most one homomorphism from $\mathfrak{e}$ to $\mathfrak{e}'$ over $\mathfrak{g}$.*

Recognition Criterion 7.9.4 *Call a Lie algebra $\mathfrak{g}$ simply connected if every central extension $0 \to M \to \mathfrak{e} \to \mathfrak{g} \to 0$ splits in a unique way as a product Lie algebra $\mathfrak{e} = \mathfrak{g} \times M$. A central extension $0 \to M \to \mathfrak{e} \to \mathfrak{g} \to 0$ is universal iff $\mathfrak{e}$ is perfect and simply connected. Moreover, $H_1(\mathfrak{e}, k) = H_2(\mathfrak{e}, k) = 0$. In particular, if $\mathfrak{g}$ is perfect and $H_2(\mathfrak{e}, k) = 0$, then $\mathfrak{g}$ is simply connected.*

Corollary 7.9.5 *Let $\mathfrak{g}$ be a finite-dimensional semisimple Lie algebra over a field of characteristic 0. Then $H_2(\mathfrak{e}, k) = 0$ and $\mathfrak{g}$ is simply connected.*

Proof $M = H_2(\mathfrak{e}, k)$ is a finite-dimensional $\mathfrak{g}$-module because it is a subquotient of $\Lambda^2 \mathfrak{g}$ in the Chevalley-Eilenberg complex. By Whitehead's second lemma 7.8.12, $H^2(\mathfrak{g}, M) = 0$, so the universal central extension is $\mathfrak{e} = M \times \mathfrak{g}$. By universality, we must have $M = 0$. $\qquad\qquad\qquad\qquad\qquad\qquad\quad\diamond$

Exercise 7.9.1 Show that simply connected Lie algebras are perfect.

Exercise 7.9.2 If $0 \to M_i \to \mathfrak{e}_i \to \mathfrak{g}_i \to 0$ are universal central extensions, show that $0 \to M_1 \times M_2 \to \mathfrak{e}_1 \times \mathfrak{e}_2 \to \mathfrak{g}_1 \times \mathfrak{g}_2 \to 0$ is also a universal central extension.

In the rest of this section, we shall use the above ideas in the construction of Affine Lie algebras $\hat{\mathfrak{g}}$ corresponding to simple Lie algebras.

Let $\mathfrak{g}$ be a fixed finite-dimensional simple Lie algebra over a field k of characteristic 0. Write $\mathfrak{g}[t, t^{-1}]$ for the Lie algebra $\mathfrak{g} \otimes_k k[t, t^{-1}]$ over $k[t, t^{-1}]$. Elements of $\mathfrak{g}[t, t^{-1}]$ are Laurent polynomials $\sum x_i t^i$ with $x_i \in \mathfrak{g}$ and $i \in \mathbb{Z}$. Since the Chevalley-Eilenberg complex $V_*(\mathfrak{g}[t, t^{-1}])$ is $V_*(\mathfrak{g}) \otimes_k k[t, t^{-1}]$, we have

$$H_*(\mathfrak{g}[t, t^{-1}], k[t, t^{-1}]) = H_*(\mathfrak{g}, k) \otimes_k k[t, t^{-1}].$$

In particular, $H_1 = H_2 = 0$ (7.8.6, 7.8.12) so $\mathfrak{g}[t, t^{-1}]$ is perfect and simply connected as a Lie algebra over the ground ring $k[t, t^{-1}]$.

Now we wish to consider $\mathfrak{g}[t, t^{-1}]$ as an infinite-dimensional Lie algebra over k. Since $\mathfrak{g}[t, t^{-1}]$ is perfect, we still have $H_1(\mathfrak{g}[t, t^{-1}], k) = 0$, but we will no longer have $H_2(\mathfrak{g}[t, t^{-1}], k) = 0$. We now construct an example of a nontrivial central extension of $\mathfrak{g}[t, t^{-1}]$ over k.

Affine Lie Algebras 7.9.6 If $\kappa\colon \mathfrak{g} \otimes \mathfrak{g} \to k$ is the Killing form (7.8.3), set

$$\beta\left(\sum x_i t^i, \sum y_j t^j\right) = \sum i\kappa(x_i, y_{-i}).$$

Since β is alternating bilinear, it is a 2-cochain (7.7.3). Because k is a trivial $\mathfrak{g}[t, t^{-1}]$-module, β is a 2-cocycle: if $x = \sum x_i t^i$, $y = \sum y_j t^j$, and $z = \sum z_k t^k$, then the $\mathfrak{g}$-invariance of the Killing form gives

$$\delta\beta(x, y, z) = -\beta([xy], z) + \beta([xz], y) - \beta([yz], x)$$

$$= \sum_{i,j,k} -\beta([x_i y_j]t^{i+j}, z_k t^k) + \beta([x_i z_k]t^{i+k}, y_j t^j) - \beta([y_j z_k]t^{j+k}, x_i t^i)$$

$$= \sum_{i+j+k=0} -(i+j)\kappa([x_i y_j], z_k) + (i+k)\kappa([x_i z_k], y_j) - (j+k)\kappa([y_j z_k], x_i)$$

$$= \sum_{i+j+k=0} \{-(i+j) - (i+k) - (j+k)\}\kappa(x_i, [y_j, z_k])$$

$$= 0.$$

The class $[\beta] \in H^2(\mathfrak{g}[t, t^{-1}], k)$ corresponds to a central extension of Lie algebras over k:

$$0 \to k \to \hat{\mathfrak{g}} \to \mathfrak{g}[t, t^{-1}] \to 0.$$

The Lie algebra $\hat{\mathfrak{g}}$ is called the *Affine Lie algebra* corresponding to $\mathfrak{g}$. It is a special type of Kac-Moody Lie algebra. We are going to prove that $\hat{\mathfrak{g}}$ is the universal central extension of $\mathfrak{g}[t, t^{-1}]$ following the proof in [Wil].

Lemma 7.9.7 $\hat{\mathfrak{g}}$ *is perfect.*

Proof Let $\rho: \mathfrak{g}[t, t^{-1}] \to \hat{\mathfrak{g}}$ be the vector space splitting corresponding to the 2-cocycle β. If $x, y \in \mathfrak{g}$ then $[\rho(xt^i), \rho(yt^{-i})] = \rho([xy]) + i\,\kappa(x, y)$ for $i = 0, 1$ so $k \subseteq [\hat{\mathfrak{g}}, \hat{\mathfrak{g}}]$. Since $[\hat{\mathfrak{g}}, \hat{\mathfrak{g}}]$ maps onto the perfect $\mathfrak{g}[t, t^{-1}]$, we must have $\hat{\mathfrak{g}} = [\hat{\mathfrak{g}}, \hat{\mathfrak{g}}]$. $\diamond$

Now fix an arbitrary central extension $0 \to M \to \mathfrak{e} \xrightarrow{\pi} \mathfrak{g}[t, t^{-1}] \to 0$. If $\sigma: \mathfrak{g}[t, t^{-1}] \to \mathfrak{e}$ is a vector space splitting of π, recall (exercise 7.7.5) that the corresponding 2-cocycle $f_\sigma: \Lambda^2(\mathfrak{g}[t, t^{-1}]) \to M$ is defined by

$$[\sigma(x), \sigma(y)] = \sigma([xy]) + f_\sigma(x, y),$$

and that conversely every 2-cocycle f determines a σ such that $f = f_\sigma$. Let S denote the set of all splittings σ of π such that

$$f_\sigma(\sum x_i t^i, y) = 0 \quad \text{for all } x_i, y \in \mathfrak{g} \text{ and } i \in \mathbb{Z}.$$

Lemma 7.9.8 S *is nonempty for every central extension of* $\mathfrak{g}[t, t^{-1}]$.

Proof Given any splitting σ, write $f_\sigma^i(x, y)$ for $f_\sigma(xt^i, y)$. Each $f_\sigma^i(-, y)$ is an element of $\mathrm{Hom}_k(\mathfrak{g}, M)$, so we may think of f_σ^i as a 1-cochain, that is, a map from $\mathfrak{g}$ to $\mathrm{Hom}_k(\mathfrak{g}, M)$. In fact, f_σ^i is a cocycle (exercise!). But $\mathrm{Hom}_k(\mathfrak{g}, M)$ is finite-dimensional, so by Whitehead's first lemma (7.8.10) there exists $\varphi^i \in \mathrm{Hom}_k(\mathfrak{g}, M)$ such that $f_\sigma^i(x, y) = \varphi^i([xy])$. Assembling the φ^i into a k-linear map $\varphi\colon \mathfrak{g}[t, t^{-1}] \to M$ by the rule $\varphi(\sum x_i t^i) = \sum \varphi^i(x_i)$, we see that the 2-cocycle $\delta\varphi\colon \Lambda^2 \mathfrak{g}[t, t^{-1}] \to M$ satisfies

$$(\delta\varphi)(\sum x_i t^i, y) = -\sum \varphi^i([x_i y]) = -\sum f_\sigma^i(x_i, y) = -f(\sum x_i t^i, y).$$

Hence the splitting τ corresponding to the 2-cocycle $f + \delta\varphi$ is an element of S. $\diamond$

Exercise 7.9.3 Show that S contains exactly one element.

Lemma 7.9.9 *If $k = \mathbb{C}$ and $\sigma \in S$, then there exist $c_{ij} \in M$ such that*

$$f_\sigma(\sum x_i t^i, \sum y_j t^j) = \sum \kappa(x_i, y_j) c_{ij},$$

where κ is the Killing form on $\mathfrak{g}$.

Proof Because $\sigma \in S$, we have

$$0 = \delta f(x_i t^i, y_j t^j, z) = f_\sigma([x_i z] t^i, y_j t^j) - f_\sigma([x_i t^i, [z, y_j] t^j).$$

Therefore each $f_\sigma^{ij}(x, y) = f_\sigma(xt^i, yt^j)$ is a $\mathfrak{g}$-invariant bilinear form on $\mathfrak{g}$:

$$f_\sigma^{ij}([xz], y) = f_\sigma^{ij}(x, [zy]).$$

On the other hand the Killing form is a nondegenerate $\mathfrak{g}$-invariant bilinear form on $\mathfrak{g}$. Since $k = \mathbb{C}$, any $\mathfrak{g}$-invariant symmetric bilinear form must therefore be a multiple of κ (exercise 7.8.1). Thus $f_\sigma^{ij} = \kappa c_{ij}$ for some $c_{ij} \in M$. $\diamond$

Corollary 7.9.10 *If $k = \mathbb{C}$ and $\sigma \in S$, then there is a $c \in M$ such that for $x = \sum x_i t^i, y = \sum y_j t^j$ in $\mathfrak{g}[t, t^{-1}]$ we have*

$$f_\sigma(x, y) = \beta(x, y) c = \sum i\kappa(x_i, y_{-i}) c.$$

Proof Setting $c = c_{1,-1}$, it suffices to prove that $c_{i,-i} = ic$ and that $c_{ij} = 0$ if $i \neq -j$. As $\sigma \in S$, $c_{i0} = 0$ for all i; since f_σ is skew-symmetric, we have $c_{ij} = -c_{ji}$. Since κ is $\mathfrak{g}$-invariant and symmetric,

$$0 = \delta f_\sigma(xt^i, yt^j, zt^k) = -\kappa(x, [yz])(c_{i+j,k} + c_{i+k,j} + c_{j+k,i})$$

which yields $0 = c_{i+j,k} + c_{i+k,j} + c_{j+k,i}$. Taking $i + j = 1$ and $k = -1$, so that $j + k = -i$, we get

$$c_{i,-i} = -c_{-i,i} = c + c_{i-1,1-i}.$$

By induction on $|i| \geq 0$, this yields $c_{i,-i} = ic$ for all $i \in \mathbb{Z}$. Taking $i + j + k = s$ and $k = 1$, we get

$$c_{s-1,1} = c_{i,j+1} - c_{i+1,j}.$$

Summing from $i = 0$ to $s - 1$ if $s > 0$ (or from $i = s$ to -1 if $s < 0$) yields $sc_{s-1,1} = 0$, so $c_{t,1} = 0$ unless $t = -1$. This yields $c_{i,j+1} = c_{i+1,j}$ unless $i + j = -1$. Fixing $s \neq 0$, induction on $|i|$ shows that $c_{i,s-i} = 0$ for all $i \in \mathbb{Z}$. $\diamond$

Theorem 7.9.11 (H. Garland) *Let $\mathfrak{g}$ be a finite-dimensional simple Lie algebra over $k = \mathbb{C}$. Then the corresponding Affine Lie algebra $\hat{\mathfrak{g}}$ (7.9.6) is the universal central extension of $\mathfrak{g}[t, t^{-1}]$.*

Proof Let $0 \to M \to \mathfrak{e} \xrightarrow{\pi} \mathfrak{g}[t, t^{-1}] \to 0$ be a central extension. Choose a splitting σ in S (7.9.8), and let $c_{ij} \in M$ be the elements constructed in lemma 7.9.9. Recall that there is a vector space splitting $\iota: \mathfrak{g}[t, t^{-1}] \to \hat{\mathfrak{g}}$ corresponding to the 2-cocycle β, which yields a vector space decomposition $\hat{\mathfrak{g}} \cong k \times \mathfrak{g}[t, t^{-1}]$. Define $F: k \to M$ by $F(\alpha) = \alpha c_{1,-1}$ and extend this to a vector space map from $\hat{\mathfrak{g}}$ to $\mathfrak{e}$ by setting $F(\iota(x)) = \sigma(x)$ for $x \in \mathfrak{g}[t, t^{-1}]$. Since

$$F([\iota(x), \iota(y)]) = F(\iota[x, y]) + F(\beta(x, y))$$

$$= \sigma([x, y]) + \sum i\kappa(x_i, y_{-i})c_{1,-1}$$

$$= \sigma([x, y]) + f_\sigma(x, y)$$

$$= [F(\iota(x)), F(\iota(y))],$$

and k is in the center of $\hat{\mathfrak{g}}$, F is a Lie algebra homomorphism $\hat{\mathfrak{g}} \to \mathfrak{e}$ over $\mathfrak{g}[t, t^{-1}]$. Since $\hat{\mathfrak{g}}$ is perfect, there is at most one such map, so F is unique. $\diamond$

Remark 7.9.12 If $\mathfrak{g}$ is semisimple over $\mathbb{C}$, then $\mathfrak{g} = \mathfrak{g}_1 \times \cdots \times \mathfrak{g}_r$ for simple Lie algebras $\mathfrak{g}_i$. Consequently the universal central extension of $\mathfrak{g}[t, t^{-1}]$ is the product

$$0 \to k^r \to \hat{\mathfrak{g}}_1 \times \cdots \times \hat{\mathfrak{g}}_r \to \mathfrak{g}[t, t^{-1}] \to 0.$$

If k is a subfield of $\mathbb{C}$ and $\mathfrak{g}$ is simple over k, $\mathfrak{g} \otimes \mathbb{C}$ is semisimple over $\mathbb{C}$. If $\mathfrak{g} \otimes \mathbb{C}$ is simple then since $H_2(\mathfrak{g}, k) \otimes_k \mathbb{C} = H_2(\mathfrak{g} \otimes_k \mathbb{C}, \mathbb{C}) = \mathbb{C}$ it follows that $\hat{\mathfrak{g}}$ is still the universal central extension of $\mathfrak{g}[t, t^{-1}]$. However, this fails if $\mathfrak{g} \otimes \mathbb{C} = \mathfrak{g}_1 \times \cdots \times \mathfrak{g}_r$ because then $H_2(\mathfrak{g}, k) = k^r$.

8

Simplicial Methods in Homological Algebra

By now, the reader has seen several examples of chain complexes in which the boundary maps $C_n \to C_{n-1}$ are alternating sums $d_0 - d_1 + \cdots \pm d_n$. The primordial example is the singular chain complex of a topological space X; elements of $C_n(X)$ are formal sums of maps f from the n-simplex Δ_n into X, and $d_i(f)$ is the composition of f with the inclusion $\Delta_{n-1} \subset \Delta_n$ of the ith face of the simplex (1.1.4). Other examples of this phenomenon include Koszul complexes (4.5.1), the bar resolution of a group (6.5.1), and the Chevalley-Eilenberg complex of a Lie algebra (7.7.1). Complexes of this form arise from simplicial modules, which are the subject of this chapter.

8.1 Simplicial Objects

Let Δ be the category whose objects are the finite ordered sets $[n] = \{0 < 1 < \cdots < n\}$ for integers $n \geq 0$, and whose morphisms are nondecreasing monotone functions. If $\mathcal{A}$ is any category, a *simplicial object* A in $\mathcal{A}$ is a contravariant functor from Δ to $\mathcal{A}$, that is, $A: \Delta^{\mathrm{op}} \to \mathcal{A}$. For simplicity, we write A_n for $A([n])$. Similarly, a *cosimplicial object* C in $\mathcal{A}$ is a covariant functor $C: \Delta \to \mathcal{A}$, and we write A^n for $A([n])$. A morphism of simplicial objects is a natural transformation, and the category $\mathcal{SA}$ of all simplicial objects in $\mathcal{A}$ is just the functor category $\mathcal{A}^{\Delta^{\mathrm{op}}}$.

Example 8.1.1 (Constant simplicial objects) Let A be a fixed object of $\mathcal{A}$. The constant functor $\Delta \to \mathcal{A}$ sending every object to A is called the *constant simplicial object* in $\mathcal{A}$ at A. We have $A_n = A$ for all n, and $\alpha^* =$ identity morphism for every α in Δ.

We want to give a more combinational description of simplicial (and cosimplicial) objects, and for this we need to study the simplicial category Δ directly. The reader interested in more details about simplicial sets may want to read [May].

It is easy to see that for each n there are $n + 1$ maps $[0] \to [n]$ but only one map $[n] \to [0]$. There are $\binom{n+2}{2}$ maps $[1] \to [n]$ and more generally $\binom{n+i+1}{i+1}$ maps $[i] \to [n]$ in Δ. In order to make sense out of this chaos, it is useful to introduce the *face maps* ε_i and *degeneracy maps* η_i. For each n and $i = 0, \cdots, n$ the map $\varepsilon_i : [n-1] \to [n]$ is the unique injective map in Δ whose image misses i and the map $\eta_i : [n+1] \to [n]$ is the unique surjective map in Δ with two elements mapping to i. Combinationally, this means that

$$\varepsilon_i(j) = \begin{cases} j & \text{if } j < i \\ j+1 & \text{if } j \geq i \end{cases}, \quad \eta_i(j) = \begin{cases} j & \text{if } j \leq i \\ j-1 & \text{if } j > i \end{cases}.$$

Exercise 8.1.1 Verify the following identities in Δ:

$$\varepsilon_j \varepsilon_i = \varepsilon_i \varepsilon_{j-1} \quad \text{if } i < j$$

$$\eta_j \eta_i = \eta_i \eta_{j+1} \quad \text{if } i \leq j$$

$$\eta_j \varepsilon_i = \begin{cases} \varepsilon_i \eta_{j-1} & \text{if } i < j \\ \text{identity} & \text{if } i = j \text{ or } i = j+1 \\ \varepsilon_{i-1} \eta_j & \text{if } i > j+1. \end{cases}$$

Lemma 8.1.2 *Every morphism* $\alpha : [n] \to [m]$ *in* Δ *has a unique epi-monic factorization* $\alpha = \varepsilon \eta$, *where the monic* ε *is uniquely a composition of face maps*

$$\varepsilon = \varepsilon_{i_1} \cdots \varepsilon_{i_s} \quad \text{with} \quad 0 \leq i_s \leq \cdots \leq i_1 \leq m$$

and the epi η *is uniquely a composition of degeneracy maps*

$$\eta = \eta_{j_1} \cdots \eta_{j_t} \quad \text{with} \quad 0 \leq j_1 < \cdots < j_t < n.$$

Proof Let $i_s < \cdots < i_1$ be the elements of $[m]$ not in the image of α and $j_1 < \cdots < j_t$ be the elements of $[n]$ such that $\alpha(j) = \alpha(j+1)$. Then if $p = n - t = m - s$, the map α factors as

$$[n] \xrightarrow{\eta} [p] \xhookrightarrow{\varepsilon} [m].$$

The rest of the proof is straightforward. (Check this!) $\diamond$

Proposition 8.1.3 *To give a simplicial object A in $\mathcal{A}$, it is necessary and sufficient to give a sequence of objects A_0, A_1, $\cdots$ together with* face operators $\partial_i \colon A_n \to A_{n-1}$ and *degeneracy operators* $\sigma_i \colon A_n \to A_{n+1}$ $(i = 0, 1, \cdots, n)$, *which satisfy the following "simplicial" identities*

$$\partial_i \partial_j = \partial_{j-1} \partial_i \quad \text{if } i < j$$

$$\sigma_i \sigma_j = \sigma_{j+1} \sigma_i \quad \text{if } i \leq j$$

$$\partial_i \sigma_j = \begin{cases} \sigma_{j-1} \partial_i & \text{if } i < j \\ \text{identify} & \text{if } i = j \text{ or } i = j+1 \\ \sigma_j \partial_{i-1} & \text{if } i > j+1. \end{cases}$$

Under this correspondence $\partial_i = A(\varepsilon_i)$ and $\sigma_i = A(\eta_i)$.

Proof If A is simplicial, we obtain the above data by setting $A_n = A([n])$ and considering only faces and degeneracies. Conversely, given the data and a map in Δ written in the standard form $\alpha = \varepsilon_{i_1} \cdots \eta_{j_t}$ of the lemma, we set $A(\alpha) = \sigma_{j_t} \cdots \partial_{i_1}$. Since the simplicial identities control composition in Δ, this makes A into a contravariant functor, that is, a simplicial object. ◇

If we dualize the above discussion, we get cosimplicial objects. Recall that a cosimplicial object is a covariant functor $A \colon \Delta \to \mathcal{A}$.

Corollary 8.1.4 *To give a cosimplicial object A in $\mathcal{A}$, it is necessary and sufficient to give a sequence of objects A^0, A^1, $\cdots$ together with* coface operators $\partial^i \colon A^{n-1} \to A^n$ and *codegeneracy operators* $\sigma^i \colon A^{n+1} \to A^n$ $(i = 0, \cdots, n)$ *which satisfy the "cosimplicial" identities*

$$\partial^j \partial^i = \partial^i \partial^{j-i} \quad \text{if } i < j$$

$$\sigma^j \sigma^i = \sigma^i \sigma^{j+1} \quad \text{if } i \leq j$$

$$\sigma^j \partial^i = \begin{cases} \partial^i \sigma^{j-1} & \text{if } i < j \\ \text{identity} & \text{if } i = j \text{ or } i = j+1 \\ \partial^{i-1} \sigma^j & \text{if } i > j+1. \end{cases}$$

Example 8.1.5 (Simplices) The geometric n-simplex Δ^n is the subspace of $\mathbb{R}^{n+1}$

$$\Delta^n = \{(t_0, \cdots, t_n) \colon 0 \leq t_i \leq 1, \sum t_i = 1\}.$$

If we identify the elements of $[n]$ with the vertices $v_0 = (1, 0, \cdots, 0), \cdots,$ $v_n = (0, \cdots, 0, 1)$ of Δ^n, then a map $\alpha \colon [n] \to [p]$ in Δ sends the vertices of

Δ^n to the vertices of Δ^p by the rule $\alpha(v_i) = v_{\alpha(i)}$. Extending linearly gives a map $\alpha_*: \Delta^n \to \Delta^p$ and makes the sequence $\Delta^0, \Delta^1, \cdots, \Delta^n, \cdots$ into a cosimplicial topological space. Geometrically, the face map ε_i induces the inclusion of Δ^{n-1} into Δ^n as the ith face (the face opposite the vertex v_i), and the degeneracy map η_i induces the projection $\Delta^{n+1} \to \Delta^n$ onto the ith face that identifies v_i and v_{i+1}. This geometric interpretation provided the historical origins of the terms face and degeneracy operators.

Geometric Realization 8.1.6 If X is a simplicial set, its *geometric realization* $|X|$ is a topological space constructed as follows. For each $n \geq 0$, topologize the product $X_n \times \Delta^n$ as the disjoint union of copies of the n-simplex Δ^n indexed by the elements x of X_n. On the disjoint union $\coprod X_n \times \Delta^n$, define the equivalence relation $\sim$ by declaring that $(x, s) \in X_m \times \Delta^m$ and $(y, t) \in X_n \times \Delta^n$ are equivalent if there is a map $\alpha: [m] \to [n]$ in Δ such that $\alpha^*(y) = x$ and $\alpha_*(s) = t$. That is,

$$(\alpha^*(y), s) \sim (y, \alpha_*(t)).$$

The identification space $\coprod (X_n \times \Delta^n) / \sim$ is the geometric realization $|X|$. It is easy to see that in forming $|X|$ we can ignore every n-simplex of the form $\sigma_i(y) \times \Delta^n$, so we say that the elements $\sigma_i(y)$ are *degenerate*. An element $x \in X_n$ is called *non-degenerate* if it is not of the form $\sigma_i(y)$ for some $i < n$ and $y \in X_{n-1}$; the nondegenerate elements of X_n index the n-cells of $|X|$, which implies that $|X|$ is a "CW complex." A more detailed discussion of the geometric realization may be found in [May].

Example 8.1.7 (Classifying space) Let G be a group and consider the simplicial set BG defined by $BG_0 = \{1\}$, $BG_1 = G, \cdots, BG_n = G^n, \cdots$. The face and degeneracy maps are defined by insertion, deletion, and multiplication:

$$\sigma_i(g_1, \cdots, g_n) = (g_1, \cdots, g_i, 1, g_{i+1}, \cdots, g_n)$$

$$\partial_i(g_1, \cdots, g_n) = \begin{cases} (g_2, \ldots, g_n) & \text{if } i = 0 \\ (g_1, \ldots, g_i g_{i+1}, \ldots, g_n) & \text{if } 0 < i < n \\ (g_1, \ldots, g_{n-1}) & \text{if } i = n. \end{cases}$$

The geometric realization $|BG|$ of the simplicial set BG is called the *classifying space* of G. The name comes from the theory of fiber bundles; if X is a finite cell complex then the set $[X, |BG|]$ of homotopy classes of maps $X \to |BG|$ gives a complete classification of fiber bundles over X with structure group G. We will see in 8.2.3 and 8.3.3 that $|BG|$ is an Eilenberg-MacLane space whose homology is the same as the group homology $H_*(G)$ of Chapter 6. Thus we recover definition 6.10.4 as well as 6.10.5.

Example 8.1.8 (Simplicial complexes) A (combinational) *simplicial complex* is a collection K of nonempty finite subsets of some vertex set V such that if $\emptyset \neq \tau \subset \sigma \subset V$ and $\sigma \in K$ then $\tau \in K$. If the vertex set is ordered, we call K an *ordered* simplicial complex. To every such ordered simplicial complex we associate a simplicial set $SS(K)$ as follows. Let $SS_n(K)$ consist of all ordered $(n + 1)$-tuples $(v_0, \cdots, v_n)$ of vertices, possibly including repetition, such that the underlying set $\{v_0, \cdots, v_n\}$ is in K. If $\alpha: [n] \to [p]$ is a map in Δ, define $\alpha_*: SS_p(K) \to SS_n(K)$ by $\alpha_*(v_0, \cdots, v_p) = (v_{\alpha(0)}, \cdots, v_{\alpha(n)})$. Note that $v_0 \leq \cdots \leq v_n$ and that

$$\partial_i(v_0, \cdots, v_n) = (v_0, \cdots, v_{i-1}, v_{i+1}, \cdots, v_n)$$

$$\sigma^i(v_0, \cdots, v_n) = (v_0, \cdots, v_i, v_i, \cdots, v_n).$$

The following exercises explain how combinatorial simplicial complexes correspond to triangulated polyhedra. Clearly a triangulated polyhedron P gives rise to a combinatorial simplicial simplex K whose elements correspond to the faces of P, the vertices of P forming the vertex set V of K (see 1.1.3).

Exercise 8.1.2 Show that if K is an ordered combinatorial simplicial complex, then $SS(K)$ determines K, because there is a bijection between K and the subset of $SS(K)$ consisting of non-degenerate elements.

Exercise 8.1.3 Let K be the collection of all nonempty subsets of a vertex set V having $n + 1$ elements. (K is the combinational simplicial complex arising from the polyhedron Δ^n.) Show that the geometric realization $|SS(K)|$ is homeomorphic to the geometric n-simplex Δ^n.

Exercise 8.1.4 (Geometric simplicial complexes) If K is a combinatorial simplicial complex (8.1.8), let $|K|$ denote the geometric realization $|SS(K)|$ of the simplicial set $SS(K)$ associated to some ordering of K. Show that $|K|$ is a triangulated polyhedron with one face e_σ for each $\sigma \in K$. (If σ has $n + 1$ elements, then e_σ is homeomorphic to an n-simplex.) Therefore K is the combinational simplicial complex arising from $|K|$. The polyhedron $|K|$ is sometimes called the *geometric simplicial complex* associated to K.

Definition 8.1.9 (Semisimplicial objects) Let Δ_s denote the subcategory of Δ whose morphisms are the *injections* $\varepsilon: [i] \hookrightarrow [n]$. A *semi-simplicial* object K in a category $\mathcal{A}$ is a contravariant functor from Δ_s to $\mathcal{A}$.

For example, an ordered combinational simplicial complex K yields a semi-simplicial set with $K_n = \{\tau \in K : \tau$ has $n + 1$ elements$\}$. Every simplicial set becomes a semi-simplicial set by forgetting the degeneracies, but the degeneracies provide a richer combinatorial structure.

The forgetful functor from simplicial objects to semi-simplicial objects has a left adjoint L when $\mathcal{A}$ has finite coproducts; $(LK)_n$ is the coproduct $\coprod_{p \le n} \coprod_{\eta} K_p[\eta]$, where for each $p \le n$ the index η runs over all the surjections $[n] \to [p]$ in Δ and $K_p[\eta]$ denotes a copy of K_p. The maps defining the simplicial structure on LK are given in the following tedious exercise 8.1.5; LK is called the *left Kan extension* of K along $\Delta_s \subset \Delta$ in [MacCW, X.3]. When $\mathcal{A}$ is abelian we will give an alternate description of LK in exercise 8.4.3.

Exercise 8.1.5 (Left Kan extension) If $\alpha: [m] \to [n]$ is any morphism in Δ, define $LK(\alpha): LK_n \to LK_m$ by defining its restrictions to $K_p[\eta]$ for each surjection η as follows. Find the epi-monic factorization $\varepsilon\eta'$ of $\eta\alpha$ with $\eta': [m] \to [q]$ and $\varepsilon: [q] \to [n]$; the restriction of $LK(\alpha)$ to $K_p[\eta]$ is defined to be the map $K(\varepsilon)$ from K_p to the factor $K_q[\eta']$ of the coproduct $(LK)_m$. Show that these maps make LK into a simplicial object of $\mathcal{A}$.

Exercise 8.1.6 Show that a semi-simplicial object K is the same thing as a sequence of objects $K_0, K_1, \cdots$ together with *face operators* $\partial_i: K_n \to K_{n-1}$ ($i = 0, \cdots, n$) such that if $i < j$ then $\partial_i \partial_j = \partial_{j-1} \partial_i$.

$$K_0 \underset{\partial_1}{\overset{\partial_0}{\Longleftarrow}} K_1 \underset{\partial_2}{\overset{\partial_0}{\Lleftarrow}} K_2 \Lleftarrow K_3 \cdots .$$

Historical Remark 8.1.10 Simplicial sets first arose in Eilenberg and Zilber's 1950 study [EZ] under the name "complete semi-simplicial sets" (c.s.s.). For them, semi-simplicial sets (defined as above) were more natural, and the adjective "complete" reflected the addition of degeneracies. By 1954, this adjective was often dropped, and "semi-simplicial set" was a common term for a c.s.s. By the late 1960s even the prefix "semi" was deleted, influenced by the book [May], and "simplicial set" is now universally used for c.s.s. In view of modern usage, we have decided to retain the original use of "semi-simplicial" in definition 8.1.9.

8.2 Operations on Simplicial Objects

Definition 8.2.1 Let A be a simplicial (or semi-simplicial) object in an abelian category $\mathcal{A}$. The *associated*, or *unnormalized*, *chain complex* $C =$

$C(A)$ has $C_n = A_n$, and its boundary morphism $d: C_n \to C_{n-1}$ is the alternating sum of the face operators $\partial_i: C_n \to C_{n-1}$:

$$d = \partial_0 - \partial_1 + \cdots + (-1)^n \partial_n.$$

The (semi-) simplicial identities for $\partial_i \partial_j$ imply that $d^2 = 0$. (Check this!)

Example 8.2.2 (Koszul complexes) Let $x = (x_1, \cdots, x_m)$ be a sequence of central elements in a ring R. Then the sequence $R^m, \Lambda^2 R^m, \cdots, \Lambda^{n+1} R^m, \cdots$ of exterior products of R^m forms a semi-simplicial R-module with

$$\partial_i(e_{\alpha_0} \wedge \cdots \wedge e_{\alpha_n}) = x_{\alpha_i} e_{\alpha_0} \wedge \cdots \wedge \hat{e}_{\alpha_i} \wedge \cdots \wedge e_{\alpha_n}.$$

The Koszul complex $K(x)$ of 4.5.1 is obtained by augmenting the chain complex associated to the semi-simplicial module $\{\Lambda^{n+1} R^m\}$. If R is a k-algebra, this defines an action of the abelian Lie algebra $\mathfrak{g} = k^m$ on R, and $K(x)$ coincides with the Chevalley-Eilenberg complex 7.7.1 used to compute $H_*(\mathfrak{g}, R)$.

An extremely useful observation is that if we apply a functor $F: \mathcal{A} \to \mathcal{B}$ to a simplicial object A in $\mathcal{A}$, we obtain a simplicial object in $\mathcal{B}$. Similar remarks apply to semisimplicial and cosimplicial objects.

Example 8.2.3 (Simplicial homology) If R is a ring, the free module $R[X]$ on a set X is a functor **Sets** $\to$ R–**mod**. Whenever $X = \{X_n\}$ is a (semi-) simplicial set, $R[X] = \{R[X_n]\}$ is a (semi-) simplicial R-module. The chain complex associated to $R[X]$ is the chain complex used to form the simplicial homology of the cellular complex $|X|$ with coefficients in R. (See 1.1.3.)

Motivated by this example, we define the *simplicial homology* $H_*(X; R)$ of any simplicial set X to be the homology of the chain complex associated to the simplicial module $R[X]$. Thus $H_*(X; R) = H_*(|X|; R)$.

For example, consider the classifying space $|BG|$ of a group G (8.1.7). The chain complex associated to $R[BG]$ is the canonical chain complex used in 6.5.4 to compute the group homology $H_*(G; R)$ of G with coefficients in the trivial G-module R. This yields the formula

$$H_*(G; R) \cong H_*(BG; R) = H_*(|BG|; R).$$

Example 8.2.4 (Singular chain complex) Let X be a topological space. Applying the contravariant functor $\mathrm{Hom}_{\mathbf{Top}}(-, X)$ to the cosimplicial space $\{\Delta^n\}$ gives a simplicial set $S(X)$ with $S_n(X) = \mathrm{Hom}_{\mathbf{Top}}(\Delta^n, X)$, called the *singular simplicial set* of X. The singular chain complex of X used to compute the singular homology of X with coefficients in R (1.1.4) is exactly the chain complex associated to the simplicial R-module $R[S(X)]$.

Remark There is a natural continuous map $|S(X)| \to X$, which is a homotopy equivalence if (and only if) X has the homotopy type of a CW complex. It is induced from the maps $S_n(X) \times \Delta^n \to X$ sending (f, t) to $f(t)$. In fact, S is the right adjoint to geometric realization: for every simplicial set K, $\mathrm{Hom}_{\mathbf{Top}}(|K|, X) \cong \mathrm{Hom}_{\mathcal{S}\mathbf{Sets}}(K, S(X))$. These assertions are proven in [May, section 16].

Example 8.2.5 For each $n \geq 0$ a simplicial set $\Delta[n]$ is given by the functor $\mathrm{Hom}_\Delta(-, [n])$. These are universal in the following sense. For each simplicial set A, the Yoneda Embedding 1.6.10 gives a 1–1 correspondence between elements $a \in A_n$ and simplicial morphisms $f: \Delta[n] \to A$; f determines the element $a_f = f(\mathrm{id}_{[n]})$ and conversely f_a is defined on $\lambda \in \mathrm{Hom}_\Delta([m], [n])$ by $f_a(\lambda) = \lambda^*(a) \in A_m$.

Exercise 8.2.1 Show that $\Delta[n]$ is the simplicial set $SS(\Delta^n)$ associated (8.1.8) to the combinatorial simplicial complex underlying the geometric n-simplex Δ^n.

Cartesian Products 8.2.6 The cartesian product $A \times B$ of two simplicial objects A and B is defined as $(A \times B)_n = A_n \times B_n$ with face and degeneracy operators defined diagonally:

$$\partial_i(a, b) = (\partial_i a, \partial_i b) \quad \text{and} \quad \sigma_i(a, b) = (\sigma_i a, \sigma_i b).$$

If B is a simplicial set and A is a simplicial object in a category $\mathcal{A}$ having products, then we can also make sense out of $A \times B$ by defining $A_n \times B_n$ to be the product of B_n copies of A_n. This construction is most interesting when each B_n is finite, in which case $\mathcal{A}$ need only have finite products.

Exercise 8.2.2 If K and L are combinorial simplicial complexes (8.1.8), there is a combinational simplicial complex P with $|P| = |K| \times |L|$ as polyhedra, defined by $SS(P) = SS(K) \times SS(L)$; see [May, 14.3] or [EZ]. Verify this assertion by finding combinational simplicial complexes underlying the square $\Delta^1 \times \Delta^1$ and the prism $\Delta^2 \times \Delta^1$ whose associated simplicial sets are $\Delta[1] \times \Delta[1]$ and $\Delta[2] \times \Delta[1]$.

Fibrant Simplicial Sets 8.2.7 From the standpoint of homotopy theory, it is technically useful to restrict one's attention to those simplicial sets X that satisfy the following *Kan condition*:

For every n and k with $0 \le k \le n+1$, if $x_0, \cdots, x_{k-1}, x_{k+1}, \cdots, x_{n+1} \in X_n$ are such that $\partial_i x_j = \partial_{j-1} x_i$ for all $i < j$ (i and j not equal to k), then there exists a $y \in X_{n+1}$ such that $\partial_i(y) = x_i$ for all $i \ne k$.

We call such simplicial sets *fibrant*; they are sometimes called *Kan complexes* after D. Kan, who first isolated this condition in 1955 and observed that the singular simplicial set $S(X)$ of a topological space X (8.2.4) is always fibrant. The class of fibrant simplicial sets includes all simplicial groups and all simplicial abelian groups by the following calculation.

Lemma 8.2.8 *If G is a simplicial group (a simplicial object in the category of groups), then the underlying simplicial set is fibrant. A fortiori every simplicial abelian group, and every simplicial R-module, is fibrant when considered as a simplicial set.*

Proof Suppose given $x_i \in G_n$ ($i \ne k$) such that $\partial_i x_j = \partial_{j-1} x_i$ for $i < j$. We use induction on r to find $g_r \in G_{n+1}$ such that $\partial_i(g_r) = x_i$ for all $i \le r$, $i \ne k$. We begin the induction by setting $g_{-1} = 1 \in G_{n+1}$ and suppose inductively that $g = g_{r-1}$ is given. If $r = k$, we set $g_r = g$. If $r \ne k$, we consider $u = x_r^{-1}(\partial_r g)$. If $i < r$ and $i \ne k$, then $\partial_i(u) = 1$ and hence $\partial_i(\sigma_r u) = 1$. Hence $g_r = g(\sigma_k u)^{-1}$ satisfies the inductive hypothesis. The element $y = g_n$ therefore has $\partial_i(y) = x_i$ for all $i \ne k$, so the Kan condition is satisfied. ◇

Exercise 8.2.3 Show that $\Delta[n]$ is not fibrant if $n \ne 0$. Then show that any fibrant simplicial set X is either constant (8.1.1) or has a non-degenerate "n-cell" $x \in X_n$ for every n (8.1.6).

Exercise 8.2.4 Show that BG is fibrant for every group G but that BG is a simplicial group if and only if G is abelian.

Fibrations 8.2.9 A map $\pi: E \to B$ of simplicial sets is called a (Kan) *fibration* if

for every n, $b \in B_{n+1}$ and $k \le n+1$, if $x_0, \cdots, x_{k-1}, x_{k+1}, \cdots, x_{n+1} \in E_n$ are such that $\partial_i b = \pi(x_i)$ and $\partial_i x_j = \partial_{j-1} x_i$ for all $i < j$ ($i, j \ne k$), then there exists a $y \in E_{n+1}$ such that $\pi(y) = b$ and $\partial_i(y)$ for all $i \ne k$.

This notion generalizes that of a fibrant simplicial set X, which is after all just a simplicial set such that $X \to *$ is a fibration. The following two exercises give some important examples of fibrations.

Exercise 8.2.5 Show that every surjection $E \to B$ of simplicial groups is a fibration.

Exercise 8.2.6 (Principal G-fibrations) We say that a group G acts on a simplicial set X (or X is a *simplicial G-set*) if G acts on each X_n, and the action commutes with the face and degeneracy operators. The orbit spaces X_n/G fit together to form a simplicial set X/G; if G acts freely on X ($gx \neq x$ for every $g \neq 1$ and every x) we say that $X \to X/G$ is a *principal G-fibration*. Show that every principal G-fibration is a fibration.

Front-to-Back Duality 8.2.10 Simplicial constructions (e.g., homotopy in 8.3.11) always have a "front-to-back" dual formulation. Consider the involution ˇ on Δ, which fixes every object $[n]$; it is defined on the morphisms in Δ by

$$\check{\partial}_i = \partial_{n-i}\colon [n-1] \to [n] \quad \text{and} \quad \check{\sigma}_i = \sigma_{n-i}\colon [n+1] \to [n].$$

We may think of it as reversing the ordering of $[n] = (0 < 1 < \cdots < n)$ to get the ordering $(n < \cdots < 1 < 0)$. That is, if $\alpha\colon [m] \to [n]$ then $\check{\alpha}(i) = n - \alpha(m - i)$. If A is a simplicial object in $\mathcal{A}$, then its *front-to-back dual* $\check{A}$ is the composition of A with this involution.

8.3 Simplicial Homotopy Groups

Given a fibrant simplicial set X (8.2.7) and a basepoint $* \in X_0$, we define $\pi_n(X)$ as follows. By abuse of notation, we write $*$ for the element $\sigma_0^n(*)$ of X_n and set $Z_n = \{x \in X_n : \partial_i(x) = * \text{ for all } i = 0, \cdots, n\}$. We say that two elements x and x' of Z_n are *homotopic*, and write $x \sim x'$, if there is a $y \in X_{n+1}$ (called a *homotopy* from x to x') such that

$$\partial_i(y) = \begin{cases} * & \text{if } i < n \\ x & \text{if } i = n \\ x' & \text{if } i = n+1. \end{cases}$$

Lemma/Definition 8.3.1 *If X is a fibrant simplicial set, then $\sim$ is an equivalence relation, and we set $\pi_n(X) = Z_n/\!\sim$.*

Proof The relation is reflexive since $y = (\sigma_n x)$ is a homotopy from x to itself. To see that $\sim$ is symmetric and transitive, suppose given homotopies y' and y'' from x to x' and from x to x''. The Kan condition 8.2.7 applied to the elements $*, \cdots, *, y', y''$ of X_{n+1} with $k = n + 2$ yields an element $z \in X_{n+2}$ with $\partial_n z = y'$, $\partial_{n+1}z = y''$ and $\partial_i z = *$ for $i < n$. The element $y = \partial_{n+2}z$ is a homotopy from x' to x''. (Check this!) Therefore $x' \sim x''$.   ◇

Remark If X is a fibrant simplicial set, $\pi_n(X)$ agrees with the topological homotopy group $\pi_n(|X|)$; see [May, 16.1]. Since $\pi_n(|X|) \cong \pi_n(|S(X)|)$, one usually defines $\pi_n(X)$ as $\pi_n S(X)$ when X is not fibrant. Thus $\pi_1(X)$ is a group, and $\pi_n(X)$ is an abelian group for $n \geq 2$.

Example 8.3.2 $\pi_0(X) = X_0/\sim$, where for each $y \in X_1$ we declare $\partial_0(y) \sim \partial_1(y)$.

Example 8.3.3 (Classifying space) Consider the classifying space BG of a group G. By inspection $Z_n = \{1\}$ for $n \neq 1$ and $Z_1 = G$. From this we deduce that

$$\pi_n(|BG|) = \pi_n(BG) = \begin{cases} G & \text{if } n = 1 \\ 1 & \text{if } n \neq 1. \end{cases}$$

Definition 8.3.4 If G is a group, then an *Eilenberg-MacLane space* of type $K(G, n)$ is a fibrant simplicial set K such that $\pi_n K = G$ and $\pi_i K = 0$ for $i \neq n$. Note that G must be abelian if $n \geq 2$. The previous example shows that BG is an Eilenberg-MacLane space of type $K(G, 1)$. In the next section (exercise 8.4.4), we will construct Eilenberg-MacLane spaces of type $K(G, n)$ for $n \geq 2$ as an application of the Dold-Kan correspondence 8.4.1. The term "space," rather than "simplicial set," is used for historical reasons as well as to avoid a nine-syllable name.

Exercise 8.3.1 If G is a simplicial group (or simplicial module), considered as a fibrant simplicial set, show that any two choices of basepoint lead to naturally isomorphic $\pi_n(G)$. *Hint:* G_0 acts on G.

If G is a simplicial group (or simplicial module), considered (by 8.2.8) as a fibrant simplicial set with basepoint $* = 1$, it is helpful to consider the subgroups

$$N_n(G) = \{x \in G_n : \partial_i x = 1 \quad \text{for all} \quad i \neq n\}.$$

Then $Z_n = \ker(\partial_n \colon N_n \to N_{n-1})$ and the image of the homomorphism $\partial_{n+1} \colon N_{n+1} \to N_n$ is $B_n = \{x : x \sim 1\}$. Hence $\pi_n(G)$ is the homology group Z_n/B_n of the (not necessarily abelian) chain complex N_*

$$1 \leftarrow N_0 \xleftarrow{\partial_1} N_1 \xleftarrow{\partial_2} N_2 \leftarrow \cdots.$$

Exercise 8.3.2 Show that B_n is a normal subgroup of Z_n, so that $\pi_n(G)$ is a group for all $n \geq 0$. Then show that $\pi_n(G)$ is abelian for $n \geq 1$. *Hint:* Consider $(\sigma_{n-1}x)(\sigma_n y)$ and $(\sigma_n x)(\sigma_{n-1}y)$ for $x, y \in G_n$.

Exercise 8.3.3 If $G \to G''$ is a surjection of simplicial groups with kernel G', show that there is a short exact sequence of (not necessarily abelian) chain complexes $1 \to NG' \to NG \to NG'' \to 1$. By modifying the discussion in Chapter 1, section 3 show that there is a natural connecting homomorphism $\partial \colon \pi_n G'' \to \pi_{n-1} G'$ fitting into a long exact sequence

$$\cdots \pi_{n+1} G'' \xrightarrow{\partial} \pi_n G' \to \pi_n G \to \pi_n G'' \xrightarrow{\partial} \pi_{n-1} G' \cdots.$$

Remark 8.3.5 More generally, suppose that $\pi \colon E \to B$ is a fibration with E and B fibrant. Suppose given basepoints $*_E \in E_0$ and $*_B = \pi(*_E) \in B_0$; the fibers $F_n = \pi^{-1}(\sigma_0^n(*))$ form a fibrant simplicial subset F of E. Given $b \in B_n$ with $\partial_i(b) = *$ for all i, the fibration condition yields $e \in E_n$ with $\pi(e) = b$ and $\partial_i(e) = *$ for all $i < n$. The equivalence class of $\partial_n(e)$ in $\pi_{n-1}(F)$ is independent of the choices of e and induces a map $\partial_n \colon \pi_n(B) \to \pi_{n-1}(F)$ fitting into a long "exact" sequence of homotopy "groups":

$$\cdots \pi_{n+1}(B) \xrightarrow{\partial} \pi_n(F) \to \pi_n(E) \xrightarrow{\pi} \pi_n(B) \xrightarrow{\partial} \pi_{n-1}(F) \cdots.$$

For more details, see [May].

This remark and exercise 8.3.3 show that the homotopy groups π_* form a (nonabelian) homological δ-functor. This observation forms the basis for the subject of nonabelian homological algebra. We shall not pursue this subject much, referring the reader to [DP] and [Swan 1]. Instead we use it as a model to generalize the definition of homology to any abelian category $\mathcal{A}$, even if the objects of $\mathcal{A}$ have no underlying set structure.

Definition 8.3.6 (Homotopy groups) Suppose that A is a simplicial object in an abelian category $\mathcal{A}$. The *normalized*, or *Moore*, chain complex $N(A)$ is the chain complex with

$$N_n(A) = \bigcap_{i=0}^{n-1} \ker(\partial_i \colon A_n \to A_{n-1})$$

and differential $d = (-1)^n \partial_n$. By construction, $N(A)$ is a chain subcomplex of the unnormalized complex $C(A)$ and we define

$$\pi_n(A) = H_n(N(A)).$$

If $\mathcal{A}$ is the category of abelian groups or R-modules, this recovers the definition 8.3.1 of $\pi_n(A)$ obtained by regarding A as a fibrant simplicial set and taking homotopy.

Exercise 8.3.4 Show that $N(A)$ is naturally isomorphic to its front-to-back dual $\check{N}(A) = N(\check{A})$, which has $\check{N}_n(A) = \{x \in A_n : \partial_i x = 0 \text{ for all } i \neq 0\}$ and differential ∂_0. (See 8.2.10.)

Now let $D(A)$ denote the "degenerate" chain subcomplex of $C(A)$ generated by the images of the degeneracies σ_i, so that $D_n(A) = \sum \sigma_i(C_{n-1}A)$.

Lemma 8.3.7 $C(A) = N(A) \oplus D(A)$. Hence $N(A) \cong C(A)/D(A)$.

Proof We will use an element-theoretic proof, which is valid by the Freyd-Mitchell Embedding Theorem 1.6.1. An element of $D_n(A)$ is a sum $y = \sum \sigma_j(x_j)$ with $x_i \in C_{n-1}(A)$. If $y \in N_n(A)$ and i is the smallest integer such that $\sigma_i(x_i) \neq 0$, then $\partial_i(y) = x_i$, which is a contradiction. Hence $D_n \cap N_n = 0$. To see that $D_n + N_n = C_n$, we pick $y \in C_n$ and use downward induction on the smallest integer j such that $\partial_j(y) \neq 0$. The element y is congruent modulo D_n to $y' = y - \sigma_j \partial_j(y)$, and for $i < j$ the simplicial identities yield

$$\partial_i(y') = \partial_i(y) - \sigma_{j-1}\partial_{j-1}\partial_i(y) = 0.$$

Since $\partial_j(y') = 0$ as well, y' is congruent modulo D_n to an element of N_n by induction, and hence $D_n + N_n = C_n$.   ◇

Theorem 8.3.8 *In any abelian category $\mathcal{A}$, the homotopy $\pi_*(A)$ of a simplicial object A is naturally isomorphic to the homology $H_*(C)$ of the unnormalized chain complex $C = C(A)$:*

$$\pi_*(A) = H_*(N(A)) \cong H_*(C(A)).$$

Proof It suffices to prove that $D(A)$ is acyclic. Filter $D(A)$ by setting $F_0 D_n = 0$, $F_p D_n = D_n$ if $n \leq p$ and $F_p D_n = \sigma_0(C_{n-1}) + \cdots + \sigma_p(C_{n-1})$ otherwise. The simplicial identities show that each $F_p D$ is a subcomplex. (Check this!) Since this filtration is canonically bounded, we have a convergent first quadrant spectral sequence

$$E_{pq}^1 = H_{p+q}(F_p D / F_{p-1} D) \Rightarrow H_{p+q}(D).$$

Therefore it suffices to show that each complex $F_p D / F_{p-1} D$ is acyclic. Note that $(F_p D / F_{p-1} D)_n$ is a quotient of $\sigma_p(C_{n-1})$ and is zero for $n < p$. In element-theoretic language, if $x \in C_{n-1}(A)$, the simplicial identities yield in $F_p D / F_{p-1} D$:

$$d\sigma_p(x) = \sum_{i=p+1}^{n} (-1)^i \sigma_p \partial_{i-1}(x),$$

$$d\sigma_p^2(x) - \sigma_{p-1} d\sigma_p(x) = \sum_{i=p+2}^{n+1} (-1)^i \sigma_p \partial_{i-1} \sigma_p(x) - \sum_{i=p+2}^{n} (-1)^i \sigma_p^2 \partial_{i-1}(x)$$

$$= (-1)^{p+1} \sigma_p(x).$$

Hence $\{s_n = (-1)^{p+1} \sigma_p\}$ forms a chain contraction of the identity map of $F_p D / F_{p-1} D$, which is therefore null homotopic and hence acyclic (1.4.5).
$\diamond$

Application 8.3.9 (Hurewicz homomorphism) Let X be a fibrant simplicial set, and $\mathbb{Z}[X]$ the simplicial abelian group that in degree n is the free abelian group with basis the set X_n (8.2.3). The simplicial set map $h: X \to \mathbb{Z}[X]$ sending X to the basis elements of $\mathbb{Z}[X]$ is called the *Hurewicz homomorphism*, since on homotopy groups it is the map

$$\pi_*(X) \to \pi_*(\mathbb{Z}[X]) \cong H_* C(\mathbb{Z}[X]) = H_*(X; \mathbb{Z})$$

corresponding via 8.2.4 and 8.3.1 to the topological Hurewicz homomorphism $\pi_*(|X|) \to H_*(|X|; \mathbb{Z})$. (To see this, represent an element φ of $\pi_n(|X|)$ by a map $f: \Delta^n \to |X|$ and consider f as an element of $S_n(|X|)$. The class of $h(f)$ in $H_n \mathbb{Z}[S(|X|)] = H_n(|X|); \mathbb{Z})$ is the topological Hurewicz element $h(\varphi)$.)

Proposition 8.3.10 *Let A be a simplicial abelian group. Then the Hurewicz map $h_*: \pi_*(A) \to H_*(A; \mathbb{Z}) = H_*(|A|; \mathbb{Z})$ is a split monomorphism.*

Proof There is a natural surjection from the free abelian group $\mathbb{Z}[G]$ onto G for every abelian group G, defined on the basis elements as the identity. Thus there is a natural surjection of simplicial abelian groups $j: \mathbb{Z}[A] \to A$. The composite simplicial set map $jh: A \to \mathbb{Z}[A] \to A$ is the identity, so on homotopy groups $j_* h_*: \pi_*(A) \to \pi_*(\mathbb{Z}[A]) \to \pi_*(A)$ is the identity homomorphism.
$\diamond$

Remark The above proposition is the key result used to prove that every simplicial abelian group has the homotopy type of a product of Eilenberg-MacLane spaces of type $K(\pi_n A, n)$; see [May, 24.5].

8.3.1 Simplicial Homotopies

8.3.11 Let A and B be simplicial objects in a category $\mathcal{A}$. Two simplicial maps $f, g: A \to B$ are said to be *(simplicially) homotopic* if there are morphisms $h_i: A_n \to B_{n+1}$ in $\mathcal{A}$ $(i = 0, \cdots, n)$ such that $\partial_0 h_0 = f$ and $\partial_{n+1} h_n = g$, while

$$\partial_i h_j = \left\{ \begin{array}{ll} h_{j-1}\partial_i & \text{if } i < j \\ \partial_i h_{i-1} & \text{if } i = j \neq 0 \\ h_j \partial_{i-1} & \text{if } i > j+1 \end{array} \right\},$$

$$\sigma_i h_j = \left\{ \begin{array}{ll} h_{j+1}\sigma_i & \text{if } i \leq j \\ h_j \sigma_{i-1} & \text{if } i > j \end{array} \right\}.$$

We call $\{h_j\}$ a *simplicial homotopy* from f to g and write $f \simeq g$.

If $\mathcal{A}$ is an abelian category, or the category of sets, the next theorem gives a cleaner definition of simplicial homotopy using the Cartesian product $A \times \Delta[1]$ of 8.2.6 and the two maps $\varepsilon_0, \varepsilon_1: A = A \times \Delta[0] \to A \times \Delta[1]$ induced by the maps $\varepsilon_0, \varepsilon_1: [0] \to [1]$ in Δ.

Theorem 8.3.12 *Suppose that $\mathcal{A}$ is either an abelian category or the category of sets. Let A, B be simplicial objects and $f, g: A \to B$ two simplicial maps. There is a one-to-one correspondence between simplicial homotopies from f to g and simplicial maps $h: A \times \Delta[1] \to B$ such that the following diagram commutes.*

$$A \xrightarrow{\varepsilon_0} A \times \Delta[1] \xleftarrow{\varepsilon_1} A$$

$$f \searrow \quad \downarrow h \quad \swarrow g$$

$$B$$

Proof We give the proof when $\mathcal{A}$ is an abelian category. The set $\Delta[1]_n$ consists of the maps $\alpha_i: [n] \to [1]$ $(i = -1, \cdots, n)$, where α_i is characterized by $\alpha_i^{-1}(0) = \{0, 1, \cdots, i-1\}$. Thus $(A \times \Delta[1])_n$ is the direct sum of $n+2$ copies of A_n indexed by the α_i. A map $h^{(n)}: (A \times \Delta[1])_n \to B_n$ is therefore equivalent to a family of maps $h_i^{(n)}: A_n \to B_n$ $(i = -1, \cdots, n)$. Given a simplicial homotopy $\{h_j\}$ we define $h^{(n)}_{-1} = g$, $h^{(n)}_n = f$ and $h^{(n)}_i = \partial_{i+1}h_i$ for $0 \leq i < n$. It is easily verified that $\partial_i h^{(n)} = h^{(n-1)}\partial_i$ and $\sigma_i h^{(n)} = h^{(n+1)}\sigma_i$, so that the $h^{(n)}$ form a simplicial map h such that $h\varepsilon_0 = f$ and $h\varepsilon_1 = g$. (Exercise!) Conversely, given h the maps $h_i = h^{(n+1)}_i \sigma_i : A_n \to B_{n+1}$ define a simplicial homotopy from f to g. $\diamond$

Exercise 8.3.5 (Swan) Show that the above theorem fails when $\mathcal{A}$ is the category of groups, but that the theorem will hold if $A \times \Delta[1]$ is replaced by the simplicial group $A * \Delta[1]$, which in degree n is the free product of $n + 2$ copies of A_n indexed by the set $\Delta[1]_n$.

Exercise 8.3.6 In this exercise we show that simplicial homotopy is an additive equivalence relation when $\mathcal{A}$ is any abelian category. Let f, f', g, g' be simplicial maps A $\to$ B, and show that:

1. $f \simeq f$.
2. if $f \simeq g$ and $f' \simeq g'$, then $(f + f') \simeq (g + g')$.
3. if $f \simeq g$, then $(-f) \simeq (-g)$, $(f - g) \simeq 0$ and $g \simeq f$.
4. if $f \simeq g$ and $g \simeq h$, then $f \simeq h$.

Lemma 8.3.13 *Let $\mathcal{A}$ be an abelian category and $f, g : A \to B$ two simplicially homotopic maps. Then $f_*, g_* : N(A) \to N(B)$ are chain homotopic maps between the corresponding normalized chain complexes.*

Proof By exercise 8.3.6 above we may assume that $f = 0$ (replace g by $g - f$). Define $s_n = \sum(-1)^j h_j$ as a map from A_n to B_{n+1}, where $\{h_j\}$ is a simplicial homotopy from 0 to g. The restriction of s_n to $Z_n(A)$ lands in $Z_n(B)$, and we have

$$\partial_{n+1}s_n - s_{n-1}\partial_n = (-1)^n g.$$

(Check this!) Therefore $\{(-1)^n s_n\}$ is a chain homotopy from 0_* to g_*. $\diamond$

Path Spaces 8.3.14 There is a functor $P : \Delta \to \Delta$ with $P[n] = [n + 1]$ such that the natural map $\varepsilon_0 : [n] \to [n + 1] = P[n]$ is a natural transformation $id_\Delta \Rightarrow P$. It is obtained by formally adding an initial element $0'$ to each $[n]$ and then identifying $(0' < 0 < \cdots < n)$ with $[n + 1]$. Thus $P(\varepsilon_i) = \varepsilon_{i+1}$ and $P(\eta_i) = \eta_{i+1}$. If A is a simplicial object in $\mathcal{A}$, the *path space* PA is the simplicial object obtained by composing A with P. Thus $(PA)_n = A_{n+1}$, the ith face operator on PA is the ∂_{i+1} of A, and the ith degeneracy operator on PA is the σ_{i+1} of A. Moreover, the maps $\partial_0 : A_{n+1} \to A_n$ form a simplicial map

$PA \to A$. The path space will play a key role in the proof of the Dold-Kan correspondence.

Exercise 8.3.7 ($PA \simeq A_0$) Let A be a simplicial object, and write A_0 for the constant simplicial object at A_0. The natural maps $\sigma_0^{n+1} \colon A_0 \to A_{n+1}$ form a simplicial map $\iota \colon A_0 \to PA$, and the maps $A_{n+1} \to A_0$ induced by the canonical inclusion of $[0] = \{0\}$ in $[n+1] = (0 < 1 < \cdots < n+1)$ form a simplicial map $\rho \colon PA \to A_0$ such that $\rho\iota$ is the identity on A_0. Use σ_0 to construct a homotopy from $\iota\rho$ to the identity on PA. This shows that PA is homotopy equivalent to the constant object A_0.

Exercise 8.3.8 If G is a group one usually writes EG for the simplicial set $P(BG)$. By the previous exercise 8.3.7, $EG \simeq \{1\}$. Show that the surjection $\partial_0 \colon EG \to BG$ is a principal G-fibration (exercise 8.2.6). Then use the long exact homotopy sequence of a fibration (exercise 8.3.3) to recalculate $\pi_*(BG)$.

Exercise 8.3.9 (J. Moore) Let A be a simplicial object in an abelian category $\mathcal{A}$. Let ΛA denote the simplicial object of $\mathcal{A}$ which is the kernel of $\partial_0 \colon PA \to A$; ΛA is a kind of brutal "loop space" of A. To see this, let $A_0[1]$ denote the chain complex that is A_0 concentrated in degree -1, and let $\mathrm{cone}(NA)$ be the mapping cone of the identity map of NA (1.5.1). Show that $N_n(\Lambda A) \cong N_{n+1}(A)$ for all $n \geq 0$ and that there are exact sequences:

$$0 \to A_0[1] \to NA[1] \to N(\Lambda A) \to 0,$$

$$0 \to A_0[1] \to \mathrm{cone}(NA)[1] \to N(PA) \to 0.$$

That is, $N(\Lambda A)$ is the brutal truncation $\sigma_{\geq 0}NA[1]$ of $NA[1]$ and $N(PA)$ is the brutal truncation of $\mathrm{cone}(NA)[1]$, in the sense of 1.2.7 and 1.2.8.

8.4 The Dold-Kan Correspondence

Let $\mathcal{A}$ be an abelian category. The normalized chain complex $N(A)$ of a simplicial object A of $\mathcal{A}$ (8.3.6) depends naturally on A and forms a functor N from the category of simplicial objects in $\mathcal{A}$ to the category of chain complexes in $\mathcal{A}$. The following theorem, discovered independently by Dold and Kan in 1957, is called the *Dold-Kan correspondence*. (See [Dold].)

Dold-Kan Theorem 8.4.1 *For any abelian category $\mathcal{A}$, the normalized chain complex functor N is an equivalence of categories between $S\mathcal{A}$ and $\mathbf{Ch}_{\geq 0}(\mathcal{A})$.*

$$SA = \left\{ \begin{array}{c} \text{simplicial} \\ \text{objects in } \mathcal{A} \end{array} \right\} \xrightarrow{N} \mathbf{Ch}_{\geq 0}(\mathcal{A}) = \left\{ \begin{array}{c} \text{chain complexes } C \text{ in } \mathcal{A} \\ \text{with } C_n = 0 \text{ for } n < 0 \end{array} \right\}.$$

Under this correspondence, simplicial homotopy corresponds to homology (i.e., $\pi_*(A) \cong H_*(NA)$) and simplicially homotopic morphisms correspond to chain homotopic maps.

Corollary 8.4.2 (See 2.4.7) *The simplicial homotopy groups $\pi_* A$ of a simplicial object A of $\mathcal{A}$ form a universal δ-functor (the left derived functors of the functor π_0).*

Corollary/Definition 8.4.3 (Dual Dold-Kan correspondence) *For any abelian category $\mathcal{A}$, there is an equivalence*

$$\left\{ \begin{array}{c} \text{cosimplicial} \\ \text{objects in } \mathcal{A} \end{array} \right\} \xrightarrow{N^*} \mathbf{Ch}^{\geq 0}(\mathcal{A}) = \left\{ \begin{array}{c} \text{cochain complexes } C \text{ in } \mathcal{A} \\ \text{with } C^n = 0 \text{ for } n < 0 \end{array} \right\}.$$

N^*A is a summand of the unnormalized cochain CA of A. We define the cohomotopy of a cosimplicial object A to be the cohomology of N^*A, that is, as $\pi^i A = H^i(N^*A)$. Then $\pi^i A \cong H^i(CA)$. Finally, if $\mathcal{A}$ has enough injectives, the cohomotopy groups $\pi^* A$ are the right derived functors of the functor π^0.

8.4.4 The equivalence in the Dold-Kan Theorem is concretely realized by an inverse functor K:

$$\mathbf{Ch}_{\geq 0}(\mathcal{A}) \xrightarrow{K} SA = \left\{ \begin{array}{c} \text{simplicial} \\ \text{objects in } \mathcal{A} \end{array} \right\}$$

which is constructed as follows. Given a chain complex C we define $K_n(C)$ to be the finite direct sum $\bigoplus_{p \leq n} \bigoplus_\eta C_p[\eta]$, where for each $p \leq n$ the index η ranges over all surjections $[n] \to [p]$ in Δ and $C_p[\eta]$ denotes a copy of C_p.

If $\alpha: [m] \to [n]$ is any morphism in Δ, we shall define $K(\alpha): K_n(C) \to K_m(C)$ by defining its restrictions $K(\alpha, \eta): C_p[\eta] \to K_m(C)$. For each surjection $\eta: [n] \to [p]$, find the epi-monic factorization $\varepsilon \eta'$ of $\eta\alpha$ (8.1.2):

$$\begin{array}{ccc} [m] & \xrightarrow{\alpha} & [n] \\ \eta' \downarrow & & \downarrow \eta \\ [q] & \xhookrightarrow{\varepsilon} & [p]. \end{array}$$

If $p = q$ (in which case $\eta\alpha = \eta'$) we take $K(\alpha, \eta)$ to be the natural identification of $C_p[\eta]$ with the summand $C_p[\eta']$ of $K_m(C)$. If $p = q + 1$ and $\varepsilon = \varepsilon_p$

(in which case the image of $\eta\alpha$ is the subset $\{0, \cdots, p-1\}$ of $[p]$), we take $K(\alpha, \eta)$ to be the map

$$C_p \xrightarrow{d} C_{p-1} = C_q[\eta'] \subseteq K_m(C).$$

Otherwise we define $K(\alpha, \eta)$ to be zero. Here is a picture of $K(C)$:

$$C_0 \Leftleftarrows C_0 \oplus C_1 \Lleftarrow C_0 \oplus C_1 \oplus C_1 \oplus C_2 \Lleftarrow C_0 \oplus (C_1)^3 \oplus (C_2)^3 \oplus C_3 \cdots.$$

Exercise 8.4.1 Show that $K(C)$ is a simplicial object of $\mathcal{A}$. Since it is clearly natural in C, this shows that K is a functor.

It is easy to see that $NK(C) \cong C$. Indeed, if $\eta: [n] \to [p]$ and $n \neq p$, then $\eta = \eta_{i_1} \cdots \eta_{i_t}$ and $C_p[\eta] = (\sigma_{i_t} \cdots \sigma_{i_1} C_p)[\mathrm{id}_p]$ lies in the degenerate subcomplex $D(K(C))$. If η is the identity map of $[n]$, then ∂_i restricted to $C_n[\mathrm{id}_n]$ is $K(\varepsilon_i, \mathrm{id}_n)$, which is 0 if $i \neq n$ and d if $i = n$. Hence $N_n(KC) = C_n[\mathrm{id}_n]$ and the differential is d. Therefore in order to prove the Dold-Kan Theorem we must show that $KN(A)$ is naturally isomorphic to A for every simplicial object A in $\mathcal{A}$.

We first construct a natural simplicial map $\psi_A: KN(A) \to A$. If $\eta: [n] \to [p]$ is a surjection, the corresponding summand of $KN_n(A)$ is $N_p(A)$, and we define the restriction of ψ_A to this summand to be $N_p(A) \subset A_p \xrightarrow{\eta} A_n$. Given $\alpha: [m] \to [n]$ in Δ, and the epi-monic factorization $\varepsilon\eta'$ of $\eta\alpha$ in Δ (8.1.2) with $\eta': [m] \to [q]$, the diagram

$$
\begin{array}{ccccccc}
KN_n(A) & \supset & N_p(A) & \subset & A_p & \xrightarrow{\eta} & A_n \\
\alpha\downarrow & & \downarrow\varepsilon & & \downarrow\varepsilon & & \downarrow\alpha \\
KN_m(A) & \supset & N_q(A) & \subset & A_q & \xrightarrow{\eta'} & A_m
\end{array}
$$

commutes because $\varepsilon: N_p(A) \to N_q(A)$ is zero unless $\varepsilon = \varepsilon_p$. (Check this!) Hence ψ_A is a simplicial morphism from $KN(A)$ to A and is natural in A. We have to show that ψ_A is an isomorphism for all A. From the definition of ψ_A it follows that $N\psi_A: NKN(A) \to N(A)$ is the above isomorphism $NK(NA) \cong NA$. The following lemma therefore implies that ψ_A is an isomorphism, proving that N and K are inverse equivalences.

Lemma 8.4.5 *If $f: B \to A$ is a simplicial morphism such that $Nf: N(B) \to N(A)$ is an isomorphism, then f is an isomorphism.*

Proof We prove that each $f_n: B_n \to A_n$ is an isomorphism by induction on n, the case $n = 0$ being the isomorphism $B_0 = N_0 B \cong N_0 A = A$. Recall from exercise 8.3.9 that the brutal loop space ΛA is the kernel of $\partial_0: PA \to A$, $(PA)_n = A_{n+1}$, and that $N(\Lambda A)$ is the translate $((NA)/A_0)[1]$. Therefore $N\Lambda f: N(\Lambda B) \to N(\Lambda A)$ is an isomorphism. By induction both f_n and $(\Lambda f)_n$ are isomorphisms. From the 5-lemma applied to the following diagram, we deduce that f_{n+1} is an isomorphism.   ◇

$$
\begin{array}{ccccccccc}
0 & \longrightarrow & (\Lambda B)_n & \longrightarrow & B_{n+1} & \overset{\partial_0}{\longrightarrow} & B_n & \longrightarrow & 0 \\
 & & {\scriptstyle \Lambda f_n}\downarrow {\scriptstyle \cong} & & \downarrow {\scriptstyle f_{n+1}} & & {\scriptstyle f_n}\downarrow {\scriptstyle \cong} & & \\
0 & \longrightarrow & (\Lambda A)_n & \longrightarrow & A_{n+1} & \overset{\partial_0}{\longrightarrow} & A_n & \longrightarrow & 0.
\end{array}
$$

Exercise 8.4.2 Show that N and K are adjoint functors. That is, if A is a simplicial object and C is a chain complex, show that ψ induces a natural isomorphism:

$$\mathrm{Hom}_{\mathcal{SA}}(K(C), A) \cong \mathrm{Hom}_{\mathbf{Ch}}(C, NA).$$

Exercise 8.4.3 Given a semi-simplicial object B in $\mathcal{A}$, $KC(B)$ is a simplicial object. Show that KC is left adjoint to the forgetful functor from simplicial objects to semi-simplicial objects. (Cf. exercise 8.1.5.) *Hint:* Show that if A is a simplicial object, then there is a natural split surjection $KC(A) \to A$.

To conclude the proof of the Dold-Kan Theorem 8.4.1, we have to show that simplicially homotopic maps correspond to chain homotopic maps. We saw in 8.3.13 that if $f \simeq g$ then Nf and Ng were chain homotopic. Conversely suppose given a chain homotopy $\{s_n\}$ from f to g for two chain maps $f, g: C \to C'$. Define $h_i: K(C)_n \to K(C')_{n+1}$ as follows. On the summand C_n of $K(C)_n$ corresponding to $\eta = id$, set

$$
h_i \text{ on } C_n =
\begin{cases}
\sigma_i f & \text{if } i < n - 1 \\
\sigma_{n-1} f - \sigma_n s_{n-1} d & \text{if } i = n - 1 \\
\sigma_n (f - s_{n-1} d) - s_n & \text{if } i = n.
\end{cases}
$$

On the summand $C_p[\eta]$ of $K(C)_n$ corresponding to $\eta: [n] \to [p]$, $n \neq p$, we define h_i by induction on $n - p$. Let j be the largest element of $[n]$ such that $\eta(j) = \eta(j + 1)$ and write $\eta = \eta' \eta_j$. Then σ_j maps $C_p[\eta']$ isomorphically onto $C_p[\eta]$, and we have already defined the maps h_i on $C_p[\eta']$. Writing h_i' for the composite of $C_p[\eta] \cong C_p[\eta']$ with h_i restricted to $C_p[\eta']$, we define

$$h_i \text{ on } C_p[\eta] = \begin{cases} \sigma_j h'_{i-1} & \text{if } j < i \\ \sigma_{j+1} h'_i & \text{if } j \geq i. \end{cases}$$

A straightforward calculation (exercise!) shows that $\{h_i\}$ form a simplicial homotopy from $K(f)$ to $K(g)$. $\diamond$

Exercise 8.4.4 (Eilenberg-MacLane spaces) Let G be an abelian group, and write $G[-n]$ for the chain complex that is G concentrated in degree n (1.2.8).

1. Show that the simplicial abelian group $K(G[-n])$ is an Eilenberg-MacLane space of type $K(G, n)$ in the sense of 8.3.4 and that the loop space of exercise 8.3.9 satisfies $\Lambda K(G[-n-1]) \cong K(G[-n])$ for $n \geq 0$.

2. Suppose that a simplicial abelian group A is an Eilenberg-MacLane space of type $K(G, n)$. Use the truncation $\tau_{\geq n} NA$ (1.2.7) to show that there are simplicial maps $A \leftarrow B \rightarrow K(G[-n])$ that induce isomorphisms on homotopy groups. Hence A has the same simplicial homotopy type as $K(G[-n])$. A similar result holds for all Eilenberg-MacLane spaces, and is given in [May, section 23].

Exercise 8.4.5 Suppose that $\mathcal{A}$ has enough projectives, so that the category of $\mathcal{SA}$ of simplicial objects in $\mathcal{A}$ has enough projectives (exercise 2.2.2). Show that a simplicial object P is projective in $\mathcal{SA}$ if and only if (1) each P_n is projective in $\mathcal{A}$, and (2) the identity map on P is simplicially homotopic to the zero map.

Augmented Objects 8.4.6 An *augmented simplicial object* in a category $\mathcal{A}$ is a simplicial object A_* together with a morphism $\varepsilon \colon A_0 \rightarrow A_{-1}$ to a fixed object A_{-1} such that $\varepsilon \partial_0 = \varepsilon \partial_1$. If $\mathcal{A}$ is an abelian category, this allows us to augment the associated chain complexes $C(A)$ and $N(A)$ by adding A_{-1} in degree -1.

$$0 \leftarrow A_{-1} \xleftarrow{\varepsilon} A_0 \xleftarrow{\partial_0 - \partial_1} A_1 \xleftarrow{d} A_2 \xleftarrow{d} \cdots .$$

An augmented simplicial object $A_* \rightarrow A_{-1}$ is called *aspherical* if $\pi_n(A_*) = 0$ for $n \neq 0$ and $\varepsilon \colon \pi_0(A_*) \cong A_{-1}$. In an abelian category, this is equivalent to the assertion that the associated augmented chain complexes are exact, that is, that $C(A)$ and $N(A)$ are resolutions for A_{-1} in $\mathcal{A}$. For this reason, A_* is sometimes called a *simplicial resolution* of A_{-1}. We will use aspherical simplicial objects to construct canonical resolutions in 8.6.8. The following exercise gives a useful criterion for $A_* \xrightarrow{\varepsilon} A_{-1}$ to be aspherical.

An augmented simplicial object $A_* \xrightarrow{\varepsilon} A_{-1}$ is called (*right*) *contractible* if there are morphisms $f_n: A_n \to A_{n+1}$ for all n (including $f_{-1}: A_{-1} \to A_0$) such that $\varepsilon f_{-1} = \mathrm{id}$, $\partial_{n+1} f_n = \mathrm{id}$ for $n \geq 0$, $\partial_0 f_0 = f_{-1}\varepsilon$, and $\partial_i f_n = f_{n-1}\partial_i$ for all $0 \leq i \leq n$. (It is called *left contractible* if its dual $\check{A} \xrightarrow{\varepsilon} A_{-1}$ (8.2.10) is right contractible, that is, if $\varepsilon f_{-1} = \mathrm{id}$, $\partial_0 f_n = \mathrm{id}$, $\partial_{-1} f_0 = f_{-1}\varepsilon$, and $\partial_i f_n = f_{n-1}\partial_{i-1}$.)

Exercise 8.4.6 (Gersten)

1. If $\mathcal{A}$ is an abelian category, prove that every contractible augmented simplicial object is aspherical, and that the associated augmented chain complexes are split exact.

2. Now suppose that $\mathcal{A}$ is the category of sets. Let X be a fibrant simplicial set with basepoint $*$ and $\varepsilon: X \to X_{-1}$ an augmentation. Prove that if $X \to X_{-1}$ is (left or right) contractible and $f_n(*) = *$ for all n, then X is aspherical. *Hint:* Set $y = f_n(x)$ in 8.3.1.

8.5 The Eilenberg-Zilber Theorem

A *bisimplicial object* in a category $\mathcal{A}$ is a contravariant functor A from $\Delta \times \Delta$ to $\mathcal{A}$. Alternatively, it is a bigraded sequence of objects A_{pq} ($p, q \geq 0$), together with horizontal face and degeneracy operators $\partial_i^h: A_{pq} \to A_{p-1,q}$ and $\sigma_i^h: A_{pq} \to A_{p+1,q}$ as well as vertical face and degeneracy operators $\partial_i^v: A_{pq} \to A_{p,q-1}$ and $\sigma_i^v: A_{pq} \to A_{p,q+1}$. These operators must satisfy the simplicial identities (horizontally and vertically), and in addition every horizontal operator must commute with every vertical operator.

There is an (unnormalized) first quadrant double complex $CA = \{A_{pq}\}$ associated to any bisimplicial object A. The horizontal maps d^h are $\sum(-1)^i \partial_i^h$ and we use the sign trick (1.2.5) to define the vertical maps $d^v: A_{pq} \to A_{p,q-1}$ to be $(-1)^p \sum(-1)^i \partial_i^v$.

Clearly we may regard a bisimplicial object as a simplicial object in the catagory $\mathcal{SA}$ of simplicial objects in $\mathcal{A}$. The Dold-Kan correspondence implies that the category of bisimplicial objects is equivalent to the category of first quadrant double chain complexes, the normalized double complex corresponding to A being quasi-isomorphic to CA.

The *diagonal* diag(A) of a bisimplicial object A is the simplicial object obtained by composing the diagonal functor $\Delta \to \Delta \times \Delta$ with the functor A.

Thus $\mathrm{diag}(A)_n = A_{nn}$, the face operators are $\partial_i = \partial_i^h \partial_i^v$, and the degeneracy operators are $\sigma_i = \sigma_i^h \sigma_i^v$.

Eilenberg-Zilber Theorem 8.5.1 *Let A be a bisimplicial object in an abelian category $\mathcal{A}$. Then there is a natural isomorphism*

$$\pi_* \mathrm{diag}(A) \cong H_* \mathrm{Tot}(CA).$$

Moreover there is a convergent first quadrant spectral sequence

$$E_{pq}^1 = \pi_q^v(A_{p*}), \quad E_{pq}^2 = \pi_p^h \pi_q^v(A) \Rightarrow \pi_{p+q} \mathrm{diag}(A).$$

Proof We first observe that $\pi_0 \cong H_0$. By inspection, we have decompositions $A_{10} = \sigma_0^h(A_{00}) \oplus N_{10}$, $A_{01} = \sigma_0^v(A_{00}) \oplus N_{01}$, and $A_{11} = \sigma_0^v \sigma_0^h(A_{00}) \oplus \sigma_0^v(N_{10}) \oplus \sigma_0^h(N_{01}) \oplus N_{11}$. Now $H_0 \mathrm{Tot}(CA) = A_{00}/(\partial_1^h(N_{10}) + \partial_1^v(N_{01}))$ and $\pi_0 \mathrm{diag}(A)$ is the quotient of A_{00} by

$$\partial_1^h \partial_1^v(\sigma_0^v N_{10} \oplus \sigma_0^h N_{01} \oplus N_{11}) = \partial_1^h(N_{10}) + \partial_1^v(N_{01}) + 0.$$

Hence there is a natural isomorphism $\pi_0 \mathrm{diag}(A) \cong H_0 \mathrm{Tot}(A)$.

Now the functors $\mathrm{diag}(A)$ and $\mathrm{Tot}(CA)$ are exact, while π_* and H_* are δ-functors, so both $\pi_* \mathrm{diag}(A)$ and $H_* \mathrm{Tot}(CA)$ are homological δ-functors on the category of bisimplicial objects in $\mathcal{A}$. We will show that they are both universal δ-functors, which will imply that they are naturally isomorphic. (The isomorphisms are given explicitly in 8.5.4.) This will finish the proof, since canonical first quadrant spectral sequence associated to the double complex CA has $E_{pq}^1 = H_q^v(C_{p*}) = \pi_q^v(A_{p*})$ and $E_{pq}^2 = H_p^h(C(\pi_q^v(A_{p*}))) = \pi_p^h \pi_q^v(A)$ and converges to $H_{p+q} \mathrm{Tot}(CA) \cong \pi_{p+q} \mathrm{diag}(A)$.

To see that π_* diag and H_* Tot C are universal δ-functors, we may assume (using the Freyd-Mitchell Embedding Theorem 1.6.1 if necessary) that $\mathcal{A}$ has enough projectives. (Why?) We saw in exercise 2.2.2 that this implies that the category of double complexes—and hence the category of bisimplicial objects by the Dold-Kan correspondence—has enough projectives. By the next lemma, diag and Tot C preserve projectives. Therefore we have the desired result:

$$\pi_* \mathrm{diag} = (L_* \pi_0)\mathrm{diag} = L_*(\pi_* \mathrm{diag}),$$
$$H_* \mathrm{Tot}\ C = (L_* H_0) \mathrm{Tot}\ C = L_*(H_0 \mathrm{Tot}\ C). \qquad \diamond$$

Lemma 8.5.2 *The functors diag and Tot C preserve projectives.*

Proof Fix a projective bisimplicial object P. We see from exercise 8.4.5 that any bisimplicial object A is projective if and only if each A_{pq} is projective in $\mathcal{A}$, each row and column is simplicially null-homotopic, and the vertical homotopies h_i^v are simplicial maps. Therefore diag(P) is a projective simplicial object, since each diag(P)$_n = P_{nn}$ is projective and the maps $h_i = h_i^h h_i^v$ form a simplicial homotopy (8.3.11) from the identity of diag(P) to zero. Now Tot(CP) is a non-negative chain complex of projective objects, so it is projective in $\mathbf{Ch}_{\geq 0}(\mathcal{A})$ if and only if it is split exact if and only if it is exact. But every column of Tot(CP) is acyclic, since $H_*(CP_{p*}) = \pi_*(P_{p*}) = 0$, so Tot($CP$) is exact by the Acyclic Assembly lemma 2.7.3 (or a spectral sequence argument).

$\diamondsuit$

Application 8.5.3 (Künneth formula) Let A and B be simplicial right and left R-modules, respectively. Their tensor product $(A \otimes_R B) = A_p \otimes_R B_q$ is a bisimplicial abelian group, and the associated double complex $C(A \otimes B)$ is the total tensor product Tot $C(A) \otimes_R C(B)$ of 2.7.1. The Eilenberg-Zilber Theorem 8.5.1 states that

$$\pi_* \text{diag}(A \otimes_R B) \cong H_*(\text{Tot } C(A) \otimes_R C(B)).$$

This is the form in which Eilenberg and Zilber originally stated their theorem in 1953. Now suppose that X and Y are simplicial sets and set $A = R[X]$, $B = R[Y]$ 8.2.3. Then diag($A \otimes B$) $\cong R[X \times Y]$, and the computation of the homology of the product $X \times Y$ (8.2.6) with coefficients in R is

$$H_n(X \times Y; R) = \pi_n \text{diag}(A \otimes B) \cong H_n(\text{Tot } C(X) \otimes C(Y)).$$

The Künneth formula 3.6.3 yields $H_n(X \times Y) \cong \bigoplus_{p+q=n} H_p(X) \otimes H_q(Y)$ when R is a field. If $R = \mathbb{Z}$ there is an extra Tor term, as described in 3.6.4.

The Alexander-Whitney Map 8.5.4 For many applications it is useful to have an explicit formula for the isomorphisms in the Eilenberg-Zilber Theorem 8.5.1. If $p + q = n$, we define $f_{pq}: A_{nn} \to A_{pq}$ to be the map

$$\partial_{p+1}^h \cdots \partial_n^h \partial_0^v \cdots \partial_0^v.$$

The sum over p and q yields a map $f_n: A_{nn} \to \text{Tot}_n(CA)$, and the f_n assemble to yield a chain complex map f from $C(\text{diag}(A))$ to Tot(CA). (Exercise!) The map f is called the *Alexander-Whitney map*, since these two mathematicians discovered it independently while constructing the cup product in topology. Since f is defined by face operators, it is natural and induces a morphism of universal δ-functors $f_*\colon \pi_* \text{diag} A \to H_* \text{Tot}(CA)$. Moreover, $f_0\colon A_{00} = A_{00}$, so f_* induces the natural isomorphism $\pi_0 \text{diag} A \cong H_0 \text{Tot}(CA)$. Therefore the Alexander-Whitney map is the unique chain map (up to equivalence) inducing the isomorphism f_* of the Eilenberg-Zilber Theorem.

The inverse map $\nabla\colon \text{Tot}(CA) \to C(\text{diag} A)$ is related to the shuffle product on the bar complex (6.5.11). The component $\nabla_{pq}\colon A_{pq} \to A_{nn}$ $(n = p + q)$ is the sum

$$\sum_\mu (-1)^\mu \sigma^h_{\mu(n)} \cdots \sigma^h_{\mu(p+1)} \sigma^v_{\mu(p)} \cdots \sigma^v_{\mu(1)}$$

over all (p, q)-shuffles μ. The proof that ∇ is a chain map is a tedious but straightforward exercise. Clearly, ∇ is natural, and it is easy to see that ∇_* induces the natural isomorphism $H_0 \text{Tot}(CA) \cong \pi_0 \text{diag} A$. Therefore ∇_* is the unique isomorphism of universal δ-functors given by the Eilenberg-Zilber Theorem. In particular, ∇_* is the inverse of the Alexander-Whitney map f_*.

Remark The analogue of the Eilenberg-Zilber Theorem for semi-simplicial simplicial objects is false; the degeneracies are necessary. For example, if A_{pq} is zero for $p \neq 1$, then $\pi_1 \text{diag}(A) = A_{11}$ need not equal $H_1 \text{Tot}(CA) = \pi_1(A_{1*})$.

8.6 Canonical Resolutions

To motivate the machinery of this section, we begin with a simplicial description of the (unnormalized) bar resolution of a group G. By inspecting the construction in 6.5.1 we see that the bar resolution

$$0 \leftarrow \mathbb{Z} \xleftarrow{\varepsilon} B^u_0 \xleftarrow{d} B^u_1 \xleftarrow{d} B^u_2 \xleftarrow{d} \cdots$$

is exactly the augmented chain complex associated to the augmented simplicial G-module

$$\mathbb{Z} \xleftarrow{\varepsilon} B^u_0 \xLeftarrow{} B^u_1 \xLeftarrow{} B^u_2 \xLeftarrow{} B^u_3 \cdots,$$

in which B^u_n is the free $\mathbb{Z}G$-module on the set G^n. In fact, we can construct the simplicial module B^u_* directly from the trivial G-module $\mathbb{Z}$ using only the functor $F = \mathbb{Z}G \otimes_{\mathbb{Z}}\colon G\text{–mod} \to G\text{–mod}$; B^u_n is $F^{n+1}\mathbb{Z} = \mathbb{Z}G \otimes_{\mathbb{Z}} \cdots \otimes_{\mathbb{Z}} \mathbb{Z}G$, the face operators are formed from the natural map $\varepsilon\colon \mathbb{Z}G \otimes_{\mathbb{Z}} M \to M$, and the degeneracy operators are formed from the natural map $\eta\colon M = \mathbb{Z} \otimes_{\mathbb{Z}} M \to \mathbb{Z}G \otimes_{\mathbb{Z}} M$.

In this section we formalize the above process (see 8.6.11) so that it yields augmented simplicial objects whose associated chain complexes provide canonical resolutions in a wide variety of contexts. To begin the formalization, we introduce the dual concepts of triple and cotriple. (The names "triple" and "cotriple" are unfortunate because nothing occurs three times. Nonetheless it is the traditional terminology. Some authors use "monad" and "comonad", which is not much better.)

Definition 8.6.1 A *triple* (T, η, μ) on a category $\mathcal{C}$ is a functor $\mathsf{T} : \mathcal{C} \to \mathcal{C}$, together with natural transformations $\eta : \mathrm{id}_{\mathcal{C}} \Rightarrow \mathsf{T}$ and $\mu : \mathsf{TT} \Rightarrow \mathsf{T}$, such that the following diagrams commute for every object C.

$$
\begin{array}{ccc}
\mathsf{TT}(\mathsf{T}C) = \mathsf{T}(\mathsf{TT}C) \xrightarrow{\ \mathsf{T}\mu\ } \mathsf{T}(\mathsf{T}C) \\
\Big\downarrow{\mu_{\mathsf{T}C}} \qquad\qquad\qquad \Big\downarrow{\mu_C} \\
\mathsf{T}(\mathsf{T}C) \xrightarrow[\ \ \mu C\ \ ]{} \mathsf{T}C
\end{array}
\qquad
\begin{array}{ccc}
\mathsf{T}C \xrightarrow{\ \mathsf{T}\eta_C\ } \mathsf{T}(\mathsf{T}C) \xleftarrow{\ \eta_{\mathsf{T}C}\ } \mathsf{T}C \\
{}_{=}\searrow \quad \Big\downarrow{\mu} \quad \swarrow_{=} \\
\mathsf{T}C
\end{array}
$$

Symbolically, we may write these as $\mu(\mathsf{T}\mu) = \mu(\mu\mathsf{T})$ and $\mu(\mathsf{T}\eta) = \mathrm{id} = \mu(\eta\mathsf{T})$.

Dually, a *cotriple* $(\bot, \varepsilon, \delta)$ in a category $\mathcal{A}$ is a functor $\bot : \mathcal{A} \to \mathcal{A}$, together with natural transformations $\varepsilon : \bot \Rightarrow \mathrm{id}_{\mathcal{A}}$ and $\delta : \bot \Rightarrow \bot\bot$, such that the following diagrams commute for every object A.

$$
\begin{array}{ccc}
\bot A \xrightarrow{\ \ \delta_A\ \ } \bot(\bot A) \\
\Big\downarrow{\delta_A} \qquad\qquad \Big\downarrow{\delta_{\bot A}} \\
\bot(\bot A) \xrightarrow{\ \bot\delta\ } \bot(\bot\bot A) = \bot\bot(\bot A)
\end{array}
\qquad
\begin{array}{ccc}
& \bot A & \\
{}_{=}\swarrow & \Big\downarrow{\delta} & \searrow_{=} \\
\bot A \xleftarrow[\ \bot\varepsilon_A\ ]{} \bot(\bot A) \xrightarrow[\ \varepsilon_{\bot A}\ ]{} \bot A
\end{array}
$$

Symbolically, we may write these as $(\bot\,\delta)\delta = (\delta\,\bot)\delta$ and $(\bot\,\varepsilon)\delta = \mathrm{id} = (\varepsilon\,\bot)\delta$. Note the duality: a cotriple in $\mathcal{A}$ is the same as a triple in $\mathcal{A}^{\mathrm{op}}$.

Exercise 8.6.1 Provided that they exist, show that any product $\Pi\,\mathsf{T}_\alpha$ of triples T_α is a triple and that any coproduct $\amalg\,\bot_\alpha$ of cotriples $\bot_\alpha$ is again a cotriple.

Exercise 8.6.2 Show that the natural transformation ε of a cotriple satisfies the identity $\varepsilon(\varepsilon\,\bot) = \varepsilon(\bot\,\varepsilon)$. That is, for every A the following diagram commutes:

$$\perp (\perp A) \xrightarrow{\ \perp \varepsilon_A\ } \perp A$$

$$\varepsilon_{\perp A} \Big\downarrow \qquad\qquad \Big\downarrow \varepsilon_A$$

$$\perp A \xrightarrow{\ \varepsilon\ } A.$$

Main Application 8.6.2 (Adjoint functors) Suppose we are given an adjoint pair of functors (F, U) with F left adjoint to U.

$$\mathcal{B} \underset{U}{\overset{F}{\leftrightarrows}} \mathcal{C}$$

That is, $\operatorname{Hom}_{\mathcal{B}}(FC, B) \cong \operatorname{Hom}_{\mathcal{C}}(C, UB)$ for every C in $\mathcal{C}$ and B in $\mathcal{B}$. We claim that $\top = UF \colon \mathcal{C} \to \mathcal{C}$ is part of a triple $(\top, \eta, \mu)$ and that $\perp = FU \colon \mathcal{B} \to \mathcal{B}$ is part of a cotriple $(\perp, \varepsilon, \delta)$.

Recall from A.6.1 of the Appendix that such an adjoint pair determines two natural transformations: the *unit* of the adjunction $\eta \colon \mathrm{id}_{\mathcal{B}} \to UF$ and the *counit* of the adjunction $\varepsilon \colon FU \Rightarrow \mathrm{id}_{\mathcal{C}}$. We define δ and μ by

$$\delta_B = F(\eta_{UB}) \colon F(UB) \to F(UF(UB)), \quad \mu_C = U(\varepsilon_{FC}) \colon U(FU(FC)) \to U(FC).$$

In the Appendix, A.6.2 and exercise A.6.3, we see that $(\varepsilon F) \circ (F\eta) \colon FC \to FC$ and $(U\varepsilon) \circ (\eta U) \colon UB \to UB$ are the identity maps and that $\varepsilon \circ (FU\varepsilon) = \varepsilon \circ (\varepsilon FU) \colon FU(FU(B)) \to B$. From these we deduce the triple axioms for $(\top, \eta, \mu)$:

$$\mu(\top \eta) = U((\varepsilon F) \circ (F\eta)) = \mathrm{id}, \quad \mu(\eta \top) = ((U\varepsilon) \circ (\eta U))F = \mathrm{id},$$

$$\mu(\top \mu) = (U\varepsilon F) \circ (UFU\varepsilon F) = U(\varepsilon \circ UF\varepsilon)F = U(\varepsilon \circ \varepsilon UF)F = \mu(\mu \top).$$

By duality applied to the adjoint pair $(U^{\mathrm{op}}, F^{\mathrm{op}})$, $(\perp, \varepsilon, \delta)$ is a cotriple on $\mathcal{B}$.

Example 8.6.3 The forgetful functor $U \colon G\text{–}\mathbf{mod} \to \mathbf{Ab}$ has for its left adjoint the functor $F(C) = \mathbb{Z}G \otimes_{\mathbb{Z}} C$. The resulting cotriple on $G\text{–}\mathbf{mod}$ has $\perp = FU$, and $\perp (\mathbb{Z}) \cong \mathbb{Z}G$. The following construction of a simplicial object out of the cotriple $\perp$ on the trivial G-module $\mathbb{Z}$ will yield the simplicial G-module used to form the unnormalized bar resolution described at the beginning of this section; see 8.6.11.

Simplicial Object of a Cotriple 8.6.4 Given a cotriple $\perp$ on $\mathcal{A}$ and an object A, set $\perp_n A = \perp^{n+1} A$ and define face and degeneracy operators

$$\partial_i = \perp^i \varepsilon \perp^{n-i} \colon \quad \perp^{n+1} A \to \perp^n A,$$

$$\sigma_i = \perp^i \delta \perp^{n-i} \colon \quad \perp^{n+1} A \to \perp^{n+2} A.$$

We claim that $\perp_* A$ is a simplicial object in $\mathcal{A}$. To see this, note that

$$\partial_i \sigma_i = \perp^i (\varepsilon \perp)\delta \perp^{n-i} = \perp^i (1) \perp^{n-i} = \text{ identity, and}$$

$$\partial_{i+1}\sigma_i = \perp^i (\perp \varepsilon)\delta \perp^{n-i} = \perp^i (1) \perp^{n-i} = \text{identity.}$$

Similarly, we have

$$\partial_i \partial_{i+1} = \perp^i (\varepsilon(\perp \varepsilon)) \perp^{n-i} = \perp^i (\varepsilon(\varepsilon \perp)) \perp^{n-i} = \partial_i \partial_i,$$

$$\sigma_{i+1}\sigma_i = \perp^i ((\perp \delta)\delta) \perp^{n-i} = \perp^i ((\delta \perp)\delta) \perp^{n-i} = \sigma_i \sigma_i.$$

The rest of the simplicial identities are formally valid. The map $\varepsilon_A \colon \perp A \to A$ satisfies $\varepsilon\partial_0 = \varepsilon\partial_1$ (because $\varepsilon(\varepsilon \perp) = \varepsilon(\perp \varepsilon)$), so in fact $\perp_* A \to A$ is an augmented simplicial object.

Dually, given a triple $\top$ on $\mathcal{C}$, we define $L^n = \top^{n+1}C$ and $\partial^i = \top^i\eta\top^{n-i}$, $\sigma^i = \top^i\mu\top^{n-i}$. Since a triple $\top$ on $\mathcal{C}$ is the same as a cotriple $\top^{op}$ on $\mathcal{C}^{op}$, $L^* = \top^{*+1}C$ is a cosimplicial object in $\mathcal{C}$ for every object C of $\mathcal{C}$.

Definition 8.6.5 *Let* $\perp$ *be a cotriple in a category* $\mathcal{A}$. *An object* A *is called* $\perp$-projective *if* $\varepsilon_A \colon \perp A \to A$ *has a section* $f \colon A \to \perp A$ *(i.e., if* $\varepsilon_A f = \mathrm{id}_A$). *For example, if* $\perp = FU$ *for an adjoint pair* (F, U), *then every object* FC *is* $\perp$-projective *because* $F\eta \colon FC \to F(UFC) = \perp (FC)$ *is such a section.*

Paradigm 8.6.6 (Projective R-modules) If R is a ring, the forgetful functor $U \colon R\text{–mod} \to \textbf{Sets}$ has the free R-module functor F as its left adjoint; we call FU the *free module cotriple*. Since $FU(P)$ is a free module, an R-module P is FU-projective if and only if P is a projective R-module. This paradigm explains the usage of the suggestive term "$\perp$-projective." It also shows that a cotriple on R–**mod** need not be an additive functor.

$\perp$-Projective Lifting Property 8.6.7 *Let* $U \colon \mathcal{A} \to \mathcal{C}$ *have a left adjoint* F, *and set* $\perp = FU$. *An object* P *is* $\perp$-projective *if and only if it satisfies the following lifting property: given a map* $g \colon A_1 \to A_2$ *in* $\mathcal{A}$ *such that* $UA_1 \to UA_2$ *is a split surjection and a map* $\gamma \colon P \to A_2$, *there is a map* $\beta \colon P \to A_1$ *such that* $\gamma = g\beta$.

Proof The lifting property applied to $FU(P) \to P$ shows that P is $\perp$-projective. For the converse we may replace P by $FU(P)$ and observe that since $\mathrm{Hom}_{\mathcal{A}}(FU(P), A) \cong \mathrm{Hom}_{\mathcal{C}}(UP, UA)$, the map $\mathrm{Hom}_{\mathcal{A}}(FU(P), A_1) \to \mathrm{Hom}_{\mathcal{A}}(FU(P), A_2)$ is a split surjection. $\diamond$

Exercise 8.6.3 Show that an object P is $\perp$-projective if and only if there is an A in $\mathcal{A}$ such that P is a *retract* of $\perp A$. (That is, there are maps $i: P \to \perp A$ and $r: \perp A \to P$ so that $ri = \mathrm{id}_P$.)

Proposition 8.6.8 (Canonical resolution) *Let $\perp$ be a cotriple in an abelian category $\mathcal{A}$. If A is any $\perp$-projective object, then the augmented simplicial object $\perp_* A \xrightarrow{\varepsilon} A$ is aspherical, and the associated augmented chain complex is exact.*

$$0 \leftarrow A \xleftarrow{\varepsilon} \perp A \xleftarrow{\partial_0 - \partial_1} \perp^2 A \xleftarrow{d} \perp^3 A \xleftarrow{d} \cdots .$$

Proof For $n \geq 0$, set $f_n = \perp^{n+1} f: \perp^{n+1} A \to \perp^{n+2} A$, and set $f_{-1} = f$. By definition, $\partial_{n+1} f_n = \perp^{n+1} (\varepsilon f) =$ identity and $\partial_0 f_0 = (\varepsilon \perp)(\perp f) = f\varepsilon$. If $n \geq 1$ and $0 \leq i < n + 1$, then (setting $j = n - i$ and $B = \perp^j A$) naturality of ε with respect to $g = \perp^j f$ yields

$$\partial_i f_n = (\perp^i \varepsilon_{\perp B})(\perp^i \perp g) = (\perp^i g)(\perp^i \varepsilon_B) = f_{n-1} \partial_i .$$

We saw (in 8.4.6 and exercise 8.4.6) that such a family of morphisms $\{f_n\}$ makes $\perp_* A \to A$ "contractible," hence aspherical. $\diamond$

Corollary 8.6.9 *If $\mathcal{A}$ is abelian and $U: \mathcal{A} \to \mathcal{C}$ is a functor having a left adjoint $F: \mathcal{C} \to \mathcal{A}$, then for every C in $\mathcal{C}$ the augmented simplicial object $\perp_*(FC) \to FC$ is contractible, hence aspherical in $\mathcal{A}$.*

Proposition 8.6.10 *Suppose that $U: \mathcal{A} \to \mathcal{C}$ has a left adjoint $F: \mathcal{C} \to \mathcal{A}$. Then for every A in $\mathcal{A}$ the augmented simplicial object $U(\perp_* A) \xrightarrow{U\varepsilon} UA$ is left contractible in $\mathcal{C}$ and hence aspherical.*

Proof Set $f_{-1} = \eta U: UA \to UFUA = U(\perp A)$ and $f_n = \eta U \perp^n$. Then the $\{f_n\}$ make $U(\perp_* A)$ left contractible in the sense of 8.4.6. (Check this!) $\diamond$

8.6.1 Applications

Group Homology 8.6.11 If G is a group, the forgetful functor $U: G\text{–}\mathbf{mod} \to \mathbf{Ab}$ has a left adjoint $F(C) = \mathbb{Z}G \otimes_{\mathbb{Z}} C$. For every G-module M, the resulting simplicial G-module $\perp_* M \to M$ is aspherical because its underlying simplicial abelian group $U(\perp_* M) \to UM$ is aspherical by 8.6.10. Moreover by Shapiro's Lemma 6.3.2 the G-modules $\perp^{n+1} M = F(C)$ are acyclic for $H_*(G; -)$ in the sense of 2.4.3. Therefore the associated chain complex $C(\perp_* M)$ is a resolution by $H_*(G; -)$-acyclic G-modules. It follows from 2.4.3 that we can compute the homology of the G-module M according to the formula

$$H_*(G; M) = H_*(C(\perp_* M)_G) = \pi_*((\perp_* M)_G),$$

using the homotopy groups of the simplicial abelian group $(\perp_* M)_G$.

If we take $M = \mathbb{Z}$, $C(\perp_* \mathbb{Z})$ is exactly the unnormalized bar resolution of 6.5.1. The proof given in 6.5.3 that the bar resolution is exact amounts to a paraphrasing of the proof of proposition 8.6.10.

The Bar Resolution 8.6.12 Let $k \to R$ be a ring homomorphism. The forgetful functor $U: R\text{--mod} \to k\text{--mod}$ has $F(M) = R \otimes_k M$ as its left adjoint, so we obtain a cotriple $\perp = FU$ on $R\text{--mod}$. Since the homotopy groups of the simplicial R-module $\perp_* M$ may be computed using the underlying simplicial k-module $U(\perp_* M)$, it follows that $\perp_* M \to M$ is aspherical 8.4.6 ($\perp_* M$ is a simplicial resolution of M). The associated augmented chain complexes are not only exact in $R\text{--mod}$, they are split exact when considered as a complex of k-modules by 8.6.10. The unnormalized chain complex $\beta(R, M)$ associated to $\perp_* M$ is called the (unnormalized) *bar resolution* of a left R-module M. Thus $\beta(R, M)_0 = R \otimes_k M$, and $\beta(R, M)_n$ is $R^{\otimes(n+1)} \otimes_k M$. Note that $\beta(R, M) = \beta(R, R) \otimes_R M$:

$$0 \leftarrow M \overset{\varepsilon}{\longleftarrow} R \otimes_k M \leftarrow R \otimes_k R \otimes_k M \leftarrow \cdots.$$

The *normalized bar resolution* of M, written $B(R, M)$, is the normalized chain complex associated to $\perp_* M$ and is described in the following exercise.

Exercise 8.6.4 Write $\bar{R}$ for the cokernel of the k-module homomorphism $k \to R$ sending 1 to 1, and write $\otimes$ for $\otimes_k$. Show that the normalized bar resolution has $B_n(R, M) = R \otimes \bar{R} \otimes \cdots \otimes \bar{R} \otimes M$ with n factors $\bar{R}$, with (well-defined) differential

$$d(r_0 \otimes \bar{r}_1 \otimes \cdots \bar{r}_n \otimes m) = r_0 r_1 \otimes \bar{r}_2 \otimes \cdots \otimes \bar{r}_n \otimes m$$

$$+ \sum_{i=1}^{n-1} (-1)^i r_0 \otimes \cdots \otimes \bar{r}_i \bar{r}_{i+1} \otimes \cdots \otimes m$$

$$+ (-1)^n r_0 \otimes \bar{r}_1 \otimes \cdots \otimes \bar{r}_{n-1} \otimes r_n m.$$

Proposition 8.6.13 *Suppose k is commutative. If M (resp. M') is a left module over a k-algebra R (resp. R'), then there is a chain homotopy equivalence of bar resolutions of the $R \otimes R'$-module $M \otimes M'$:*

$$\text{Tot}(\beta(R, M) \otimes_k \beta(R', M')) \xrightarrow{\nabla} \beta(R \otimes_k R', M \otimes_k M').$$

Proof Let A (resp. A') denote the simplicial k-module $R^{\otimes *} \otimes M$ (resp. $R'^{\otimes *} \otimes M'$), where $\otimes$ denotes $\otimes_k$. The diagonal of the bisimplicial k-module $A \otimes A'$ is the simplicial k-module $[p] \mapsto (R^{\otimes p} \otimes M) \otimes (R'^{\otimes p} \otimes M) \cong (R \otimes R')^{\otimes p} \otimes (M \otimes M')$ whose associated chain complex is $\beta(R \otimes R', M \otimes M')$. The Eilenberg-Zilber Theorem (in the Künneth formula incarnation 8.5.3) gives a chain homotopy equivalence ∇ from the total tensor product Tot $C(A \otimes A') \cong \text{Tot } C(A) \otimes C(A') = \text{Tot } \beta(R, M) \otimes \beta(R', M')$ to C diag $(A \otimes A') \cong \beta(R \otimes R', M \otimes M')$. $\diamond$

Remark The homotopy equivalence Tot $\beta(R, R) \otimes \beta(R', R') \xrightarrow{\nabla} \beta(R \otimes R', R \otimes R')$ is fundamental; applying $\otimes_{R \otimes R'}(M \otimes M')$ to it yields the proposition.

Exercise 8.6.5 (Shuffle product) Use the explicit formula for the shuffle map ∇ of 6.5.11 and 8.5.4 to establish the explicit formula (where μ ranges over all (p, q)-shuffles):

$$\nabla((r_0 \otimes \cdots \otimes r_p \otimes m) \otimes (r'_0 \otimes \cdots \otimes r'_q \otimes m')) =$$

$$\sum_\mu (-1)^\mu (r_0 \otimes r'_0) \otimes w_{\mu(1)} \otimes \cdots \otimes w_{\mu(p+q)} \otimes (m \otimes m').$$

Here the r_i are in R, the r'_j are in R', $m \in M$, $m' \in M'$, and $w_1, \cdots, w_{p+q}$ is the ordered sequence of elements $r_1 \otimes 1, \cdots, r_p \otimes 1, 1 \otimes r'_1, \cdots, 1 \otimes r'_q$ of $R \otimes R'$.

Free Resolutions 8.6.14 Let R be a ring and FU the free module cotriple, where $U: R\text{–mod} \to \text{Sets}$ is the forgetful functor whose left adjoint $F(X)$ is the free module on X. For every R-module M, we claim that the augmented simplicial R-module $(FU)_* M \to M$ is aspherical (8.4.6). This will prove that FU_*M is a simplicial resolution of M, and that the associated chain complex $C = C(FU_*M)$ is a canonical free resolution of M because

$$H_i(C) = \pi_i(FU_*M) = \pi_i(UFU_*M) = \begin{cases} M & i = 0 \\ 0 & i \neq 0. \end{cases}$$

Indeed, the underlying augmented simplicial set $U(FU)_*M \to UM$ is fibrant and contractible by 8.6.10. If we choose $[0] = \eta(0)$ as basepoint instead of 0, then the contraction satisfies $f_n([0]) = [0]$ for all n, and therefore $U(FU)_*(M)$ is aspherical (by exercise 8.4.6). As the sets $\pi_n U(FU)_*M$ are independent of the choice of basepoint (exercise 8.3.1), the augmented simplicial R-module $FU_*(M) \to M$ is also aspherical, as claimed.

Sheaf Cohomology 8.6.15 Let X be a topological space and Sheaves(X) the category of sheaves of abelian groups on X (1.6.5). If $\mathcal{F}$ is a sheaf we can form the stalks $\mathcal{F}_x$ and take the product $\top(\mathcal{F}) = \prod_{x \in X} x_*(\mathcal{F}_x)$ of the corresponding skyscraper sheaves as in 2.3.12. As $F_x = x_*$ and $U_x(\mathcal{F}) = \mathcal{F}_x$ are adjoint, each $F_x U_x(\mathcal{F}) = x_*(\mathcal{F}_x)$ is a triple. Hence their product $\top$ is a triple on Sheaves(X). Thus we obtain a coaugmented cosimplicial sheaf $\mathcal{F} \xrightarrow{\eta} (\top^{*+1}\mathcal{F})$ and a corresponding augmented cochain complex

$$0 \longrightarrow \mathcal{F} \xrightarrow{\eta} \top(\mathcal{F}) \xrightarrow{\partial^0 - \partial^1} \top^2(\mathcal{F}) \xrightarrow{d} \cdots.$$

The resulting resolution of $\mathcal{F}$ by the Γ-acyclic sheaves $\top^{*+1}(\mathcal{F})$ is called the *Godement resolution* of $\mathcal{F}$, since it first appeared in [Gode]. (The proof that the Godement resolution is an exact sequence of sheaves involves interpreting $\prod U_x(\mathcal{F})$ as a sheaf on the disjoint union X^δ of the points of X.)

Example 8.6.16 (Commutative algebras) Let k be a commutative ring and **Commalg** the category of commutative k-algebras. Let $P_* \to R$ be an augmented simplicial object of **Commalg**; if its underlying augmented simplicial set is aspherical we say that P_* is a *simplicial resolution* of R.

The forgetful functor U: **Commalg** $\to$ **Sets** has a left adjoint taking a set X to the polynomial algebra $k[X]$ on the set X; the resulting cotriple $\bot$ on **Commalg** sends R to the polynomial algebra on the set underlying R. As with free resolutions 8.6.14, $U(\bot_* R) \to UR$ is aspherical, so $\bot_* R$ is a simplicial resolution of R. This resolution will be used in 8.8.2 to construct André-Quillen homology.

Another cotriple $\bot^S$ on arises from the left adjoint Sym of the forgetful functor U': **Commalg** $\to k$-**mod**. The *Symmetric Algebra* Sym(M) of a k-module M is defined to be the quotient of the tensor algebra $T(M)$ by the 2-sided ideal generated by all $(x \otimes y - y \otimes x)$ with $x, y \in M$ (under the identification $i: M \hookrightarrow T(M)$). From the presentation of $T(M) \cong k \oplus M \oplus \cdots \oplus M^{\otimes m} \oplus \cdots$ in 7.3.1 it follows that Sym(M) is the free commutative algebra on generators $i(x)$, $x \in M$, subject only to the two k-module relations on M:

$$\alpha\, i(x) = i(\alpha x) \quad \text{and} \quad i(x) + i(y) = i(x + y) \quad (\alpha \in k; x, y \in M).$$

Thus any k-module map $M \to R$ into a commutative k-algebra extends uniquely to an algebra map $\mathrm{Sym}(M) \to R$. This gives a natural isomorphism $\mathrm{Hom}_k(M, R) \cong \mathrm{Hom}_{\mathbf{Commalg}}(\mathrm{Sym}(M), R)$, proving that Sym is left adjoint U. The resulting cotriple on **Commalg** sends R to the symmetric algebra $\perp^S(R) = \mathrm{Sym}(U'R)$ and we have a canonical adjunction $\varepsilon \colon \mathrm{Sym}(U'R) \to R$. As the simplicial k-module $U'(\perp_* R) \to U'R$ is aspherical, $\perp^S_* R \to R$ is another simplicial resolution of R in **Commalg**, and there is a simplicial map $\perp_* R \to \perp^S_* R$, natural in R.

Exercise 8.6.6 Let X be a set and M the free k-module with basis X. Show that $\mathrm{Sym}(M)$ is the commutative polynomial ring $k[X]$. Then show that the map $\perp_* k[X] \to \perp^S_* k[X]$ is a simplicial homotopy equivalence.

Exercise 8.6.7 In general, show that $\mathrm{Sym}(M) = k \oplus M \oplus S^2(M) \oplus \cdots \oplus S^n(M) \oplus \cdots$, where $S^n(M)$ is the module $(M \otimes \cdots \otimes M)_{\Sigma_n}$ of coinvariants for the evident permutation action of the nth symmetric group Σ_n on the n-fold tensor product of M.

8.7 Cotriple Homology

Suppose that $\mathcal{A}$ is a category equipped with a cotriple $\perp = (\perp, \varepsilon, \delta)$ as described in the previous section, and suppose given a functor $E \colon \mathcal{A} \to \mathcal{M}$ with $\mathcal{M}$ some abelian category. For each object A in $\mathcal{A}$ we can apply E to the augmented simplicial object $\perp_* A \to A$ to obtain the augmented simplicial object $E(\perp_* A) \to E(A)$ in $\mathcal{M}$.

Definition 8.7.1 (Barr and Beck [BB]) The *cotriple homology of A with coefficients in E* (relative to the cotriple $\perp$) is the sequence of objects $H_n(A; E) = \pi_n E(\perp_* A)$. From the Dold-Kan correspondence, this is the same as the homology of the associated chain complex $C(E \perp_* A)$:

$$0 \leftarrow E(\perp A) \xleftarrow{\,d\,} E(\perp^2 A) \xleftarrow{\,d\,} E(\perp^3 A) \leftarrow \cdots.$$

Clearly cotriple homology is functorial with respect to maps $A \to A'$ in $\mathcal{A}$ and natural transformations of the "coefficient functors" $E \to E'$. The augmentation gives a natural transformation $\varepsilon^A_* \colon H_0(A; E) = \pi_0(E \perp_* A) \to E(A)$, but at this level of generality ε^A_* need not be an isomorphism. (Take $\perp = 0$.)

Dually, if $(\top, \eta, \mu)$ is a triple on a category $\mathcal{C}$ and $E \colon \mathcal{C} \to \mathcal{M}$ is a functor, the *triple cohomology* of an object C with coefficients in E is the sequence of objects $H^n(C; E) = \pi^n E(\mathsf{T}^{*+1}C)$, which by definition is the cohomology of the associated cochain complex

$$0 \to E(\mathsf{T}C) \xrightarrow{d} E(\mathsf{T}^2 C) \xrightarrow{d} E(\mathsf{T}^3 C) \to \cdots$$

associated to the cosimplicial object $E(\mathsf{T}^{*+1}C)$ of $\mathcal{M}$. By duality, $H^n(C; E)$ is the object $H_n(C; E^{op})$ in the opposite category $\mathcal{M}^{op}$ corresponding to $E^{op}: C^{op} \to \mathcal{M}^{op}$; we shall not belabor the dual development of triple cohomology.

Another variant occurs when we are given a *contravariant* functor E from $\mathcal{A}$ to $\mathcal{M}$. In this case $E(\bot_* A)$ is a cosimplicial object of $\mathcal{M}$. We set $H^n(A; E) = \pi^n E(\bot_* A)$ and call it the *cotriple cohomology* of A with coefficients in E. Of course if we consider $\bot$ to be a triple on $\mathcal{A}^{op}$ and take as coefficients $E: \mathcal{A}^{op} \to \mathcal{M}$, then cotriple cohomology is just triple cohomology in disguise.

Example 8.7.2 (Tor and Ext) Let R be a ring and $\bot$ the free module cotriple on **mod**–R (8.6.6). We saw in 8.6.14 that the chain complex $C(\bot_* M)$ is a free resolution of M for every R-module M. If N is a left R-module and we take $E(M) = M \otimes_R N$, then homology of the chain complex associated to $E(\bot_* M) = (\bot_* M) \otimes_R N$ computes the derived functors of E. Therefore

$$H_n(M; \otimes_R N) = \mathrm{Tor}_n^R(M, N).$$

Similarly, if N is a right R-module and $E(M) = \mathrm{Hom}_R(M, N)$, then the cohomology of the cochain complex associated to $E(\bot_* M) = \mathrm{Hom}_R(\bot_* M, N)$ computes the derived functors of E. Therefore

$$H^n(M; \mathrm{Hom}_R(-, N)) = \mathrm{Ext}_R^n(M, N).$$

Definition 8.7.3 (Barr-Beck [BB]) Let $\bot$ be a fixed cotriple on $\mathcal{A}$ and $\mathcal{M}$ an abelian category. A *theory of $\bot$-left derived functors* (L_n, λ, ∂) is the assignment to every functor $E: \mathcal{A} \to \mathcal{M}$ a sequence of functors $L_n E: \mathcal{A} \to \mathcal{M}$, natural in E, together with a natural transformation $\lambda: L_0 E \Rightarrow E$ such that

1. $\lambda : L_0(E\bot) \cong E\bot$ and $L_n(E\bot) = 0$ for $n \neq 0$ and every E.
2. Whenever $\mathcal{E}: 0 \to E' \to E \to E'' \to 0$ is an exact sequence of functors such that $0 \to E'\bot \to E\bot \to E''\bot \to 0$ is also exact, there are "connecting" maps $\partial: L_n E'' \to L_{n-1} E'$, natural in $\mathcal{E}$, such that the following sequence is exact:

$$\cdots L_n E' \to L_n E \to L_n E'' \xrightarrow{\partial} L_{n-1} E' \to L_{n-1} E \cdots.$$

Uniqueness Theorem 8.7.4 *Cotriple homology $H_*(-; E)$ is a theory of $\bot$-left derived functors. Moreover, if (L_n, λ, ∂) is any other theory of $\bot$-left derived functors then there are isomorphisms $L_n E \cong H_n(-; E)$, natural in E, under which λ corresponds to ε and ∂ corresponds to the connecting map for $H_*(-; E)$.*

Proof A theory of left derived functors is formally similar to a universal (homological) δ-functor on the functor category $\mathcal{M}^{\mathcal{A}}$, the $E \bot$ playing the role of projectives. The proof in 2.4.7 that left derived functors form a universal δ-functor formally goes through, mutatis mutandis, to prove this result as well.

$\diamond$

8.7.1 Relative Tor and Ext

8.7.5 Fix an associative ring k and let $k \to R$ be a ring map. The forgetful functor $U: \mathbf{mod}\text{-}R \to \mathbf{mod}\text{-}k$ has a left adjoint, the base-change functor $F(M) = M \otimes_k R$. If N is a left R-module, the *relative Tor groups* are defined to be the cotriple homology with coefficients in $\otimes_R N$ (relative to the cotriple $\bot = FU$):

$$\mathrm{Tor}_p^{R/k}(M, N) = H_p(M; \otimes_R N) = \pi_p((\bot_* M) \otimes_R N),$$

which is the homology of the associated chain complex $C(\bot_* M \otimes N)$ (8.3.8). Since $(\bot^{p+1} M) \otimes_R N = (\bot^p M) \otimes_k R \otimes_R N \cong \bot^p M \otimes_k N$, we can give an alternate description of this chain complex as follows. Write $\otimes$ for $\otimes_k$ and $R^{\otimes p}$ for $R \otimes R \otimes \cdots \otimes R$, so that $\bot^p M = M \otimes R^{\otimes p}$. Then $(\bot_* M \otimes N)$ is the simplicial abelian group $[p] \mapsto M \otimes R^{\otimes p} \otimes N$ with face and degeneracy operators

$$\partial_i(m \otimes r_1 \otimes \cdots \otimes r_p \otimes n) = \begin{cases} mr_1 \otimes r_2 \otimes \cdots \otimes r_p \otimes n & \text{if } i = 0 \\ m \otimes \cdots \otimes r_i r_{i+1} \otimes \cdots \otimes n & \text{if } 0 < i < p \\ m \otimes r_1 \otimes \cdots \otimes r_{p-1} \otimes r_p n & \text{if } i = p; \end{cases}$$

$$\sigma_i(m \otimes r_1 \otimes \cdots \otimes r_p \otimes n) = m \otimes \cdots \otimes r_{i-1} \otimes 1 \otimes r_i \otimes \cdots \otimes n.$$

(Check this!) Therefore $\mathrm{Tor}_*^{R/k}(M, N)$ is the homology of the chain complex

$$0 \leftarrow M \otimes N \xleftarrow{\partial_0 - \partial_1} M \otimes R \otimes N \xleftarrow{d} M \otimes R^{\otimes 2} \otimes N \leftarrow \cdots M \otimes R^{\otimes p} \otimes N \leftarrow \cdots.$$

As in 2.7.2, one could also start with left modules and form the cotriple homology of the functor $M \otimes_R: R\text{-}\mathbf{mod} \to \mathbf{Ab}$ relative to the cotriple $\bot'(N) = R \otimes_k$

N on R–**mod**. The resulting simplicial abelian group $[p] \mapsto M \otimes R^{\otimes p} \otimes N$ is just the front-to-back dual (8.2.10) of the one described above. This proves that relative Tor is a "balanced" functor in the sense that

$$\mathrm{Tor}_p^{R/k}(M, N) = H_p(M; \otimes_R N) = H_p(N; M \otimes_R).$$

If N is a right R-module we define the *relative Ext groups* to be the cotriple cohomology with coefficients in the contravariant functor $\mathrm{Hom}_R(-, N)$:

$$\mathrm{Ext}_{R/k}^p(M, N) = H^p(M; \mathrm{Hom}_R(-, N)) = \pi^p \mathrm{Hom}_R(\perp_* M, N),$$

which is the same as the cohomology of the associated cochain complex $C(\mathrm{Hom}_R(\perp_* M, N))$. Since $\mathrm{Hom}_R(M \otimes_k R, N) \cong \mathrm{Hom}_k(M, N)$ by 2.6.3, $\mathrm{Hom}_R(\perp_* M, N)$ is naturally isomorphic to the cosimplicial abelian group $[p] \mapsto \mathrm{Hom}_k(M \otimes R^{\otimes p}, N) = \{k\text{-multilinear maps } M \times R^p \to N\}$ with

$$(\partial^i f)(m, r_0, \cdots, r_p) = \begin{cases} f(mr_0, r_1, \ldots, r_p) & \text{if } i = 0 \\ f(m, \ldots, r_{i-1}r_i, \ldots) & \text{if } 0 < i < p \\ f(m, r_0, \ldots, r_{p-1})r_p & \text{if } i = p; \end{cases}$$

$$(\sigma^i f)(m, r_1, \cdots, r_{p-1}) = f(m, \cdots, r_i, 1, r_{i+1}, \cdots, r_{p-1}).$$

Exercise 8.7.1 Show that $\mathrm{Tor}_0^{R/k}(M, N) = M \otimes_R N$ and $\mathrm{Ext}_{R/k}^0(M, N) = \mathrm{Hom}_R(M, N)$.

Example 8.7.6 Suppose that $R = k/I$ for some ideal I of k. Since $\perp M \cong M$ for all M, $(\perp_* M) \otimes N$ and $\mathrm{Hom}_R(\perp_* M, N)$ are the constant simplicial groups $M \otimes N$ and $\mathrm{Hom}(M, N)$, respectively. Therefore $\mathrm{Tor}_i^{R/k}(M, N) = \mathrm{Ext}_{R/k}^i(M, N) = 0$ for $i \neq 0$. This shows one way in which the relative Tor and Ext groups differ from the absolute Tor and Ext groups of Chapter 3.

Just as with the ordinary Tor and Ext groups, the relative Tor and Ext groups can be computed from $\perp$-projective resolutions. For this, we need the following definition.

Definition 8.7.7 A chain complex P_* of R-modules is said to be *k-split* if the underlying chain complex $U(P_*)$ of k-modules is split exact (1.4.1). A resolution $P_* \to M$ is called k-split if its augmented chain complex is k-split.

Lemma 8.7.8 *If $\mathcal{E}: 0 \to M' \to M \to M'' \to 0$ is a k-split exact sequence of R-modules, there are natural long exact sequences*

$$\cdots \operatorname{Tor}_*^{R/k}(M', N) \to \operatorname{Tor}_*^{R/k}(M, N) \to \operatorname{Tor}_*^{R/k}(M'', N) \xrightarrow{\delta} \operatorname{Tor}_{*-1}^{R/k}(M', N) \cdots$$

$$\cdots \operatorname{Ext}_{R/k}^*(M'', N) \to \operatorname{Ext}_{R/k}^*(M, N) \to \operatorname{Ext}_{R/k}^*(M', N) \xrightarrow{\delta} \operatorname{Ext}_{R/k}^{*+1}(M'', N) \cdots$$

Proof Since $U(\mathcal{E})$ is split exact, for every $p \geq 1$ the complexes $(\perp^{p+1} \mathcal{E}) \otimes_R N = U(\mathcal{E}) \otimes_k (R^{\otimes p} \otimes_R N)$ and $\operatorname{Hom}_R(\perp^{p+1} \mathcal{E}, N) = \operatorname{Hom}_k(U\mathcal{E} \otimes_k R^{\otimes p}, N)$ are exact. Taking (co-) homology yields the result. $\diamond$

By combining adjectives, we see that a "k-split $\perp$-projective resolution" of an R-module M is a resolution $P_* \to M$ such that each P_i is $\perp$-projective and the augmented chain complex is k-split.

$$0 \leftarrow M \overset{\rightarrow}{\underset{\leftarrow}{}} P_0 \overset{\rightarrow}{\underset{\leftarrow}{}} P_1 \overset{\rightarrow}{\underset{\leftarrow}{}} P_2 \cdots .$$

For example, we saw in 8.6.12 that the augmented bar resolutions $B(R, M) \to M$ and $\beta(R, M) \to M$ are k-split $\perp$-projective resolutions for every R-module M.

Comparison Theorem 8.7.9 *Let $P_* \to M$ be a k-split $\perp$-projective resolution and $f': M \to N$ an R-module map. Then for every k-split resolution $Q_* \to N$ there is a map $f: P_* \to Q_*$ lifting f'. The map f is unique up to chain homotopy equivalence.*

Proof The proof of the Comparison Theorem 2.2.6 goes through. (Check this!) $\diamond$

Theorem 8.7.10 *If $P_* \to M$ is any k-split $\perp$-projective resolution of an R-module M, then there are canonical isomorphisms:*

$$\operatorname{Tor}_*^{R/k}(M, N) \cong H_*(P \otimes_R N),$$

$$\operatorname{Ext}_{R/k}^*(M, N) \cong H^* \operatorname{Hom}_R(P, N).$$

Proof Since $\otimes_R N$ is right exact and $\operatorname{Hom}_R(-, N)$ is left exact, we have isomorphisms $\operatorname{Tor}_0^{R/k}(M, N) \cong M \otimes_R N \cong H_0(P \otimes_R N)$ and $\operatorname{Ext}_{R/k}^0(M, N) \cong \operatorname{Hom}_R(M, N) \cong H^0 \operatorname{Hom}_R(P, N)$. Now the proof in 2.4.7 that derived functors form a universal δ-functor goes through to prove this result. $\diamond$

Lemma 8.7.11 *Suppose R_1 and R_2 are algebras over a commutative ring k; set $\perp_i = R_i \otimes$ and $\perp_{12} = R_1 \otimes R_2 \otimes$. If P_1 is $\perp_1$-projective and P_2 is $\perp_2$-projective, then $P_1 \otimes P_2$ is $\perp_{12}$-projective.*

Proof In general P_i is a summand of $R_i \otimes P_i$, so $P_1 \otimes P_2$ is a summand of $\perp_{12} (P_1 \otimes P_2) \cong (R_1 \otimes P_1) \otimes (R_2 \otimes P_2)$. ◇

Application 8.7.12 (External products for Tor) Suppose k is commutative, and we are given right and left R_1-modules M_1 and N_1 (resp. R_2-modules M_2 and N_2). Choose k-split $\perp_i$-projective resolutions $P_i \to N_i$; $\mathrm{Tot}(P_1 \otimes P_2)$ is therefore a k-split $\perp_{12}$-projective resolution of the $R_1 \otimes R_2$-module $N_1 \otimes N_2$. (Why?) Tensoring with $M_1 \otimes M_2$ yields an isomorphism of chain complexes

$$\mathrm{Tot}\{(M_1 \otimes_{R_1} P_1) \otimes (M_2 \otimes_{R_2} P_2)\} \cong (M_1 \otimes M_2) \otimes_{R_1 \otimes R_2} \mathrm{Tot}(P_1 \otimes P_2).$$

Applying homology yields the external product for relative Tor:

$$\mathrm{Tor}_i^{R/k}(M_1, N_1) \otimes_k \mathrm{Tor}_j^{R_2/k}(M_2, N_2) \to \mathrm{Tor}_{i+j}^{(R_1 \otimes R_2)/k}(M_1 \otimes M_2, N_1 \otimes N_2).$$

As in 2.7.8, the (porism version of the) Comparison Theorem 2.2.7 shows that this product is independent of the choice of resolution. The external product is clearly natural in M_1, N_1, M_2, N_2 and commutes with the connecting homomorphism δ in all four arguments. (Check this!) When $i = j = 0$, it is just the interchange $(M_1 \otimes_{R_1} N_1) \otimes_k (M_2 \otimes_{R_2} N_2) \cong (M_1 \otimes M_2) \otimes_{R_1 \otimes R_2} (N_1 \otimes N_2)$.

The bar resolutions $\beta(R_i, N_i)$ of 8.6.12 are concrete choices of the P_i. The shuffle map $\nabla: \mathrm{Tot}\, \beta(R_1, N_1) \otimes \beta(R_2, N_2) \to \beta(R_1 \otimes R_2, N_1 \otimes N_2)$ of 8.6.13 and exercise 8.6.5 may be used in this case to simplify the construction (cf. [MacH, X.7]).

Exercise 8.7.2 (External product for Ext) Use the notation of 8.7.12 to produce natural pairings, commuting with connecting homomorphisms:

$$\mathrm{Ext}_{R_1/k}^i(M_1, N_2) \otimes_k \mathrm{Ext}_{R_2/k}^j(M_2, N_2) \to \mathrm{Ext}_{(R_1 \otimes R_2)/k}^{i+j}(M_1 \otimes M_2, N_1 \otimes N_2).$$

If $i = j = 0$, this is just the map

$$\mathrm{Hom}(M_1, N_1) \otimes \mathrm{Hom}(M_2, N_2) \to \mathrm{Hom}(M_1 \otimes M_2, N_1 \otimes N_2).$$

Example 8.7.13 Suppose that R is a flat commutative algebra over k. If I is an ideal of R generated by a regular sequence $x = (x_1, \cdots, x_d)$, then $T = \mathrm{Tor}_1^{R/k}(R/I, R/I)$ is isomorphic to $(R/I)^d$ and

$$\mathrm{Tor}_i^{R/k}(R/I, R/I) \cong \Lambda^i T \quad \text{for } i \geq 0.$$

In particular these vanish for $i > d$. To see this, we choose the Koszul resolution $K(x) \to R/I$ (4.5.5); each $K_i(x) = \Lambda^i R^d$ is $\perp$-projective. Since every differential in $R/I \otimes_R K(x)$ is zero, we have

$$\operatorname{Tor}_i^{R/k}(R/I, R/I) \cong R/I \otimes_R K_i(x) \cong \Lambda^i T.$$

More is true: we saw in exercise 4.5.1 that $K(x)$ is a graded-commutative DG-algebra, so $\operatorname{Tor}_*^{R/k}(R/I, R/I)$ is naturally a graded-commutative R/I-algebra, namely via the exterior algebra structure. This product may also be obtained by composing the external product

$$\operatorname{Tor}_*^{R/k}(R/I, R/I) \otimes \operatorname{Tor}_*^{R/k}(R/I, R/I) \to \operatorname{Tor}_*^{R \otimes R/k}(R/I \otimes R/I, R/I \otimes R/I)$$

with multiplication arising from $R \otimes R \to R$ and $R/I \otimes R/I \to R/I$. Indeed, the external product is given by $K(x) \otimes K(x)$ and the multiplication is resolved by the Koszul product $K(x) \otimes K(x) \to K(x)$; see exercise 4.5.5.

Theorem 8.7.14 (Products of rings) *Let* $k \to R$ *and* $k \to R'$ *be ring maps. Then there are natural isomorphisms*

$$\operatorname{Tor}_*^{(R \times R')/k}(M \times M', N \times N') \cong \operatorname{Tor}_*^{R/k}(M, N) \oplus \operatorname{Tor}_*^{R'/k}(M', N'),$$

$$\operatorname{Ext}_{(R \times R')/k}^*(M \times M', N \times N') \cong \operatorname{Ext}_{R/k}^*(M, N) \oplus \operatorname{Ext}_{R'/k}^*(M', N').$$

Here M *and* N *are* R*-modules,* M' *and* N' *are* R'*-modules, and we consider* $M \times M'$ *and* $N \times N'$ *as* $(R \times R')$*-modules by taking products componentwise.*

Proof Write $\perp$ and $\perp'$ for the cotriples $\otimes R$ and $\otimes R'$, so that $\perp \oplus \perp'$ is the cotriple $\otimes(R \times R')$. Since $(\perp \oplus \perp')(M \times M') \cong (\perp M) \oplus (\perp M') \oplus (\perp' M) \oplus (\perp' M')$, both $\perp M = M \otimes R$ and $\perp' M' = M' \otimes R'$ are $(\perp \oplus \perp')$-projective $(R \times R')$-modules (exercise 8.6.3). The bar resolutions $\beta(R, M) \to M$ and $\beta(R', M') \to M'$ are therefore k-split $(\perp \oplus \perp')$-projective resolutions; so is the product $\beta(R, M) \times \beta(R', M') \to M \times M'$. Using this resolution to compute relative Tor and Ext over $R \times R'$ yields the desired isomorphisms, in view of the natural k-module isomorphisms

$$(M \times M') \otimes_{(R \times R')} (N \times N') \cong (M \otimes_R N) \oplus (M' \otimes_{R'} N'),$$

$$\operatorname{Hom}_{R \times R'}(M \times M', N \times N') \cong \operatorname{Hom}_R(M, N) \oplus \operatorname{Hom}_{R'}(M', N'). \quad \diamond$$

Call a right R-module P *relatively flat* if $P \otimes_R N_*$ is exact for every k-split exact sequence of left R-modules N_*. As in exercise 3.2.1 it is easy to see that P is relatively flat if and only if $\operatorname{Tor}_*^{R/k}(P, N) = 0$ for $* \neq 0$ and all left modules N.

Relatively Flat Resolution Lemma 8.7.15 *If $P \to M$ is a k-split resolution of M by relatively flat R-modules, then $\mathrm{Tor}_*^{R/k}(M, N) \cong H_*(P \otimes_R N)$.*

Proof The proof of the Flat Resolution Lemma 3.2.8 goes through in this relative setting. $\diamond$

Corollary 8.7.16 (Flat base change for relative Tor) *Suppose $R \to T$ is a ring map such that T is flat as an R-module. Then for all T-modules M and all R-modules N:*

$$\mathrm{Tor}_*^{R/k}(M, N) \cong \mathrm{Tor}_*^{T/k}(M, T \otimes_R N).$$

Moreover, if R is commutative and $M = L \otimes_R T$ these are isomorphic to

$$\mathrm{Tor}^{R/k}(L \otimes_R T, N) \cong T \otimes_R \mathrm{Tor}_*^{R/k}(L, N).$$

Proof This is like the Flat base change 3.2.9 for absolute Tor. Write $P \to M$ for the k-split resolution associated to $\perp_* M \to M$, with $\perp = \otimes_R T$. The right side is the homology of the chain complex $P \otimes_T (T \otimes_R N) \cong P \otimes_R N$, so it suffices to show that each $P_n = (\perp^n M) \otimes_k T$ is a relatively flat R-module. Because k is commutative there is a natural isomorphism $P \otimes_R N \cong T \otimes_R N \otimes_k (\perp^n M)$ for every N. If N_* is a k-split exact sequence of left R-modules, so is $N_* \otimes_k (\perp^n M)$; since T is flat over R, this implies that $P \otimes_R N_* \cong T \otimes_R N_* \otimes_k (\perp^n M)$ is exact. $\diamond$

Exercise 8.7.3 (Localization) Let S be a central multiplicative set in R, and M, N two R-modules. Show that

$$\mathrm{Tor}^{S^{-1}R/k}(S^{-1}M, S^{-1}N) \cong \mathrm{Tor}^{R/k}(S^{-1}M, N) \cong S^{-1}\mathrm{Tor}_*^{R/k}(M, N).$$

Vista 8.7.17 (Algebraic K-theory) Let $\mathcal{R}$ be the category of rings-without-unit. The forgetful functor $U: \mathcal{R} \to \mathbf{Sets}$ has a left adjoint functor $F: \mathbf{Sets} \to \mathcal{R}$, namely the free ring functor. The resulting cotriple $\perp: \mathcal{R} \to \mathcal{R}$ takes a ring R to the free ring-without-unit on the underlying set of R. For each ring R, the augmented simplicial ring $\perp_* R \to R$ is aspherical in the sense of 8.4.6: the underlying (based, augmented) simplicial set $U(\perp_* R) \to UR$ is aspherical. (To see this, recall from 8.6.10 that $U(\perp_* R)$ is fibrant and left contractible, hence aspherical). If $G: \mathcal{R} \to \mathbf{Groups}$ is any functor, the $\perp$-*left derived functors* of G (i.e., derived with respect to $\perp$) are defined to be $L_n G(R) = \pi_n G(\perp_* R)$, the homotopy groups of the simplicial group $G(\perp_* R)$. This is one type of non-abelian homological algebra (see 8.3.5).

Classical examples of such a functor G are the general linear groups $GL_m(R)$, defined for a ring-without-unit R as the kernel of the augmentation $GL_m(\mathbb{Z} \oplus R) \to GL_m(\mathbb{Z})$. The inclusion of $GL_m(R)$ in $GL_{m+1}(R)$ by $M \mapsto \left(\begin{smallmatrix} M & 0 \\ 0 & 1 \end{smallmatrix} \right)$ allows us to form the infinite general linear group $GL(R)$ as the union $\cup GL_m(R)$. By inspection, $L_n GL(R) = \lim_{m \to \infty} L_n GL_m(R)$.

One of the equivalent definitions of the higher K-theory of a ring R, due to Gersten and Swan, is

$$K_n(R) = L_{n-2} GL(R) = \pi_{n-2} GL(\bot_* R) \quad \text{for } n \geq 3,$$

while K_1 and K_2 are defined by the exact sequence

$$0 \to K_2(R) \to L_0 GL(R) \to GL(R) \to K_1(R) \to 0.$$

If R is a free ring, then $K_n(R) = 0$ for $n \geq 1$, because $GL(\bot_* R) \to GL(R)$ is contractible (8.6.9). If R has a unit, then $L_0 GL(R)$ is the infinite Steinberg group $St(R) = \varinjlim St_n(R)$ of 6.9.13; $St(R)$ is the universal central extension of the subgroup $E(R)$ of $GL(R)$ generated by the elementary matrices (6.9.12). For details we refer the reader to [Swan1].

8.8 André-Quillen Homology and Cohomology

In this section we fix a commutative ring k and consider the category **Commalg** $= k$-**Commalg** of commutative k-algebras R. We begin with a few definitions, which will be discussed further in Chapter 9, section 2.

8.8.1 The *Kähler differentials* of R over k is the R-module $\Omega_{R/k}$ having the following presentation: There is one generator dr for every $r \in R$, with $d\alpha = 0$ if $\alpha \in k$. For each $r, s \in R$ there are two relations:

$$d(r + s) = (dr) + (ds) \quad \text{and} \quad d(rs) = r(ds) + s(dr).$$

If M is a k-module, a k-*derivation* $D: R \to M$ is a k-module homomorphism satisfying $D(rs) = r(Ds) + s(Dr)$; the map $d: R \to \Omega_{R/k}$ (sending r to dr) is an example of a k-derivation. The set $\mathrm{Der}_k(R, M)$ of all k-derivations is an R-module in an obvious way.

Exercise 8.8.1 Show that the k-derivation $d: R \to \Omega_{R/k}$ is universal in the sense that $\mathrm{Der}_k(R, M) \cong \mathrm{Hom}_R(\Omega_{R/k}, M)$.

Exercise 8.8.2 If $R = k[X]$ is a polynomial ring on a set X, show that $\Omega_{k[X]/k}$ is the free R-module with basis $\{dx : x \in X\}$. If K is a k-algebra, conclude that $\Omega_{K[X]/K} \cong K \otimes_k \Omega_{k[X]/k}$. These results will be generalized in exercise 9.1.3 and theorem 9.1.7, using 9.2.2.

Recall from 8.6.16 that there is a cotriple $\bot$ on **Commalg**, $\bot R$ being the polynomial algebra on the set underlying R. If we take the resulting augmented simplicial k-algebra $\bot_* R \to R$, we have canonical maps from $\bot_n R = \bot^{n+1} R$ to R for every n. This makes an R-module M into a $\bot_n R$-module. The next definitions were formulated independently by M. André and D. Quillen in 1967; see [Q].

Definitions 8.8.2 The *André-Quillen cohomology* $D^n(R/k, M)$ of R with values in an R-module M is the cotriple cohomology of R with coefficients in $\mathrm{Der}_k(-, M)$:

$$D^n(R/k, M) = \pi^n \mathrm{Der}_k(\bot_* R, M) = H^n(R; \mathrm{Der}_k(-, M)).$$

The *cotangent complex* $\mathbb{L}_{R/k} = \mathbb{L}_{R/k}(\bot_* R)$ of the k-algebra R is defined to be the simplicial R-module $[n] \mapsto R \otimes_{(\bot_n R)} \Omega_{(\bot_n R)/k}$. The *André-Quillen homology* of R with values in an R-module M is the sequence of R-modules

$$D_n(R/k, M) = \pi_n(M \otimes_R \mathbb{L}_{R/k}).$$

When $M = R$, we write $D_*(R/k)$ for the R-modules $D_*(R/k, R) = \pi_* \mathbb{L}_{R/k}$.

There is a formal analogy: D_* resembles Tor_* and D^* resembles Ext^*. Indeed, the cotangent complex is constructed so that $\mathrm{Hom}_R(\mathbb{L}_{R/k}, M) \cong \mathrm{Der}_k(\bot_* R, M)$ and hence that $D^*(R/k, M) \cong \pi^* \mathrm{Hom}_R(\mathbb{L}_{R/k}, M)$. To see this, note that for each n we have

$$\mathrm{Hom}_R(R \otimes_{(\bot_n R)} \Omega_{(\bot_n R)/k}, M) \cong \mathrm{Hom}_{\bot_n R}(\Omega_{(\bot_n R)/k}, M) \cong \mathrm{Der}_k(\bot_n R, M).$$

Exercise 8.8.3 Show that $D^0(R/k, M) \cong \mathrm{Der}_k(R, M)$ and $D_0(R/k, M) \cong M \otimes_R \Omega_{R/k}$.

Exercise 8.8.4 (Algebra extensions [EGA, IV]) Let $\mathrm{Exalcomm}_k(R, M)$ denote the set of all commutative k-algebra extensions of R by M, that is, the equivalence classes of commutative algebra surjections $E \to R$ with kernel M, $M^2 = 0$. Show that

$$\mathrm{Exalcomm}_k(R, M) \cong D^1(R, M).$$

Hint: Choose a set bijection $E \cong R \times M$ and obtain an element of the module $\mathrm{Hom}_{\mathbf{Sets}}(\bot R, M) \cong \mathrm{Der}_k(\bot^2 R, M)$ by evaluating formal polynomials $f \in \bot R$ in the algebra E.

Exercise 8.8.5 Polynomial k-algebras are $\perp$-projective objects of **Commalg** (8.6.7). Show that if R is a polynomial algebra then for every M and $i \neq 0$ $D^i(R/k, M) = D_i(R/k, M) = 0$. We will see in exercise 9.4.4 that this vanishing also holds for smooth k-algebras.

Exercise 8.8.6 Show that for each M there are universal coefficient spectral sequences

$$E^2_{pq} = \mathrm{Tor}^R_p(D_q(R/k), M) \Rightarrow D_{p+q}(R/k, M);$$

$$E_2^{pq} = \mathrm{Ext}^p_R(D_q(R/k), M) \Rightarrow D^{p+q}(R/k, M).$$

If k is a field, conclude that

$$D_q(R/k, M) \cong D_q(R/k) \otimes_R M \text{ and } D^q(R/k, M) \cong \mathrm{Hom}_R(D_q(R/k), M).$$

In order to give the theory more flexibility, we need an analogue of the fact that $\perp$-projective resolutions may be used to compute cotriple homology. We say that an augmented simplicial k-algebra $P_* \to R$ is a *simplicial polynomial resolution* of R if each P_i is a polynomial k-algebra and the underlying augmented simplicial set is aspherical. The polynomial resolution $\perp_* R \to R$ is the prototype of this concept. Since polynomial k-algebras are $\perp$-projective, there is a simplicial homotopy equivalence $P_* \xrightarrow{\sim} \perp_* R$ (2.2.6, 8.6.7). Therefore $\mathrm{Der}_k(P_*, M) \simeq \mathrm{Der}_k(\perp_* R, M)$ and $D^*(R/k, M) \cong \pi^n \mathrm{Der}_k(P_*, M)$. Similarly, there is a chain homotopy equivalence between the cotangent complex $\mathbb{L}_{R/k}$ and the simplicial module $\mathbb{L}_{R/k}(P_*) : [n] \mapsto R \otimes_{P_n} \Omega_{P_n/k}$. (Exercise!) Therefore we may also compute homology using the resolution P_*.

8.8.3 Here is one useful application. Suppose that k is noetherian and that R is a finitely generated k-algebra. Then it is possible to choose a simplicial polynomial resolution $P_* \to R$ so that each P_n has finitely many variables. Consequently, if M is a finitely generated R-module, the R-modules $D^q(R/k, M)$ and $D_q(R/k, M)$ are all finitely generated.

8.8.4 (Flat base change) As another application, suppose that R and K are k-algebras such that $\mathrm{Tor}^k_i(K, R) = 0$ for $i \neq 0$. This is the case if K is flat over k. Because these Tors are the homology of the k-module chain complex $C(K \otimes_k \perp_* R)$, it follows that $K \otimes_k \perp_* R \to K \otimes_k R$ is a simplicial polynomial resolution (use 8.4.6). Therefore

$$D^*(K \otimes_k R / K, M) \cong \pi^* \mathrm{Der}_K(K \otimes_k \perp_* R, M)$$

$$\cong \pi^* \mathrm{Der}_k(\perp_* R, M) = D^*(R/k, M)$$

for every $K \otimes R$-module M. Similarly, from the fact that $\Omega_{K[X]/k} \cong K \otimes_k$ $\Omega_{k[X]/k}$ for a polynomial ring $k[X]$ it follows that $\mathbb{L}_{K \otimes R/K} \simeq K \otimes_k \mathbb{L}_{R/k}$ and hence that $D_*(K \otimes_k R /K) \cong K \otimes_k D_*(R/k)$. This family of results is called *Flat base change.*

Exercise 8.8.7 Show that $D^*(R/k, M) = D_*(R/k, M) = 0$ if R is any localization of k.

8.8.5 As a third application, suppose that R is free as a k-module. This will always be the case when k is a field. We saw in 8.6.16 that the forgetful functor U': **Commalg** $\to$ k–**mod** has a left adjoint Sym; the resulting cotriple $\bot^S(R) = \text{Sym}(U'R)$ is somewhat different than the cotriple $\bot$. Our assumption that R is free implies that $\text{Sym}(U'R)$ is a polynomial algebra, and free as a k-module. Hence $\bot^S_*(R) \to R$ is also a simplicial polynomial resolution of R. Therefore $D^*(R/k, M)$ is isomorphic to the cotriple cohomology $\pi^*(\bot^S_* R, M)$ of R with respect to the cotriple $\bot^S$. Similarly, $\mathbb{L}_{R/k}$ and $\mathbb{L}^S_{R/k} = \{R \otimes_{(\bot^s_n R)} \Omega_{(\bot^s_n R)/k}\}$ are homotopy equivalent, and $D_*(R/k, M) \cong \pi_*(M \otimes_R \mathbb{L}^S_{R/k})$.

8.8.6 (Transitivity) A fourth basic structural result, which we cite from $[Q]$, is *Transitivity*. This refers to the following exact sequences for every k-algebra map $K \to R$ and every R-module M:

$$0 \to \text{Der}_K(R, M) \to \text{Der}_k(R, M) \to \text{Der}_k(K, M) \xrightarrow{\delta} \text{Exalcomm}_K(R, M) \to$$

$$\text{Exalcomm}_k(R, M) \to \text{Exalcomm}_k(K, M) \xrightarrow{\delta} D^2(R/K, M) \to \cdots$$

$$\cdots \to D^n(R/K, M) \to D^n(R/k, M) \to D^n(K/k, M) \xrightarrow{\delta} D^{n+1}(R/K, M) \to \cdots,$$

and its homology analogue:

$$\cdots \to D_{n+1}(R/K) \xrightarrow{\partial} R \otimes_K D_n(K/k) \to D_n(R/k) \to D_n(R/K) \xrightarrow{\partial} D_{n-1}(R/K) \to \cdots.$$

The end of this sequence is the first fundamental sequence 9.2.6 for $\Omega_{R/k}$.

Exercise 8.8.8 Suppose that k is a noetherian local ring with residue field $F = R/\mathfrak{m}$. Show that $D^1(F/k) \cong D_1(F/k) \cong \mathfrak{m}/\mathfrak{m}^2$, and conclude that if R is a k/I-algebra we may have $D^*(R/k, M) \neq D^*(R/(k/I), M)$.

Exercise 8.8.9 (Barr) In this exercise we interpret André-Quillen homology as a cotriple homology. For a commutative k-algebra R, let **Commalg/R** be the "comma" category whose objects are k-algebras P equipped with an algebra map $P \to R$, and whose morphisms $P \to Q$ are algebra maps such that $P \to R$ factors as $P \to Q \to R$. Let Diff: **Commalg**/$R \to R$–**mod** be the functor $\text{Diff}(P) = \Omega_{P/k} \otimes_P R$. Show that $\perp$ induces a cotriple on **Commalg**/R, and that if we consider R as the terminal object in **Commalg**/R, then the cotriple homology groups (8.7.1) are André-Quillen homology:

$$D_n(R/k) = H_n(R; \text{Diff}) \quad \text{and} \quad D_n(R/k, M) = H_n(R; \text{Diff} \otimes_R M).$$

8.8.1 Relation to Hochschild Theory

When k is a field of characteristic zero, there is a much simpler way to calculate $D^*(R/k, M)$ and $D_*(R/k, M)$, due to M. Barr [Barr].

Barr's Theorem 8.8.7 *Suppose $C_*(R)$ is an R-module chain complex, natural in R for each R in **Commalg**, such that*

1. *$H_0(C_*(R)) \cong \Omega_{R/k}$ for each R.*
2. *If R is a polynomial algebra, $C_*(R) \to \Omega_{R/k}$ is a split exact resolution.*
3. *For each p there is a functor $F_p: k$–**mod** $\to k$–**mod** such that $C_p(R) \cong R \otimes_k F_p(UR)$, where UR is the k-module underlying R.*

Then there are natural isomorphisms

$$D^q(R/k, M) \cong H^q \text{Hom}_R(C_*(R), M) \quad \text{and}$$

$$D_q(R/k, M) \cong H_q(M \otimes_R C_*(R)).$$

Proof We give the proof for cohomology, the proof for homology being similar but more notationally involved. Form the first quadrant double complex

$$E_0^{pq} = \text{Hom}_R(C_p(\perp_q^S R), M)$$

with horizontal differentials coming from C_* and vertical differentials coming from the naturality of the C_p. We shall compute $H^* \text{Tot}(E_0)$ in two ways.

If we fix q, the ring $\perp_q^S R$ is polynomial, so by (2) $C_*(\perp_q^S R) \to \Omega_{\perp_q^S R/k}$ is split exact. Hence $H^p \text{Hom}_R(C_*(\perp_q^S R), M) = 0$ for $p \neq 0$, while

$$H^0 \text{Hom}_R(C_*(\perp_q^S R), M) \cong \text{Hom}_R(\Omega_{\perp_q^S R/k}, M) \cong \text{Der}_k(\perp_q^S R, M).$$

Thus the spectral sequence 5.6.2 associated to the row-filtration on E_0 degenerates at E_2 to yield $H^q \text{Tot}(E_0) \cong H^q \text{Der}_k(\perp_*^S R, M) = D^q(R/k, M)$.

On the other hand, if we fix p and set $G(L) = \text{Hom}_k(F_p(L), M)$ we see by condition (3) that $E_0^{p*} = G(U \perp_*^S R)$. But the augmented simplicial k-module $U \perp_*^S R \to UR$ is left contractible (8.4.6), because $\perp^S R = \text{Sym}(UR)$

(see 8.6.10). As G is a functor, $E_0^{p*} \to G(UR) = \mathrm{Hom}_R(C_p(R), M)$ is also left contractible, hence aspherical. Thus $H^q(E_0^{p*}) = 0$ for $q \neq 0$, and $H^0(E_0^{p*}) \cong \mathrm{Hom}_R(C_p(R), M)$. Thus the spectral sequence 5.6.1 associated to the column filtration degenerates at E_2 as well, yielding $H^p \mathrm{Tot}(E_0) \cong H^p \mathrm{Hom}_R(C_*(R), M)$. $\diamond$

Preview 8.8.8 In the next chapter, we will construct the Hochschild homology $H_*(R, R)$ of a commutative k-algebra R as the homology of a natural R-module chain complex $C_*^h(R)$ with $C_p^h(R) = R \otimes_k F_p(UR)$, $F_p(L)$ being the p-fold tensor product $(L \otimes_k L \otimes_k \cdots \otimes_k L)$. There is a natural isomorphism $H_1(R, R) \cong \Omega_{R/k}$ and the map $C_1^h(R) \to C_0^h(R)$ is zero. We will see in 9.4.7 that if R is a polynomial algebra, then $H_n(R, R) \cong \Omega_{R/k}^n$, so C_*^h does not quite satisfy condition (2) of Barr's Theorem.

To remedy this, we need the Hodge decomposition of Hochschild homology from 9.4.15. When $\mathbb{Q} \subseteq k$ there are natural decompositions $F_p(L) = \oplus F_p(L)^{(i)}$ such that each $C_*^h(R)^{(i)} = R \otimes_k F_*(UR)^{(i)}$ is a chain subcomplex of $C_*^h(R)$ and $C_*^h(R) = \oplus C_*^h(R)^{(i)}$. If M is an R-module (an $R-R$ bimodule via $mr = rm$), set $H_n^{(i)}(R, M) = H_n(M \otimes_R C_*^h(R)^{(i)})$ and $H_{(i)}^n(R, M) = H^n \mathrm{Hom}_R(C_*^h(R)^{(i)}, M)$. The Hodge decomposition is

$$H_n(R, M) = \oplus H_n^{(i)}(R, M) \quad \text{and} \quad H^n(R, M) = \oplus H_{(i)}^n(R, M).$$

If R is a polynomial algebra, then $H_n^{(i)}(R, R) = 0$ for $i \neq n$, and $H_n^{(n)}(R, R) \cong \Omega_{R/k}^n$ is a free R-module (exercise 9.4.4). In particular, since $C_n^h(R)^{(i)} = 0$ for $i > n$ the augmented complex $C_*^h(R)^{(i)} \to \Omega_{R/k}^i[-i]$ is split exact for all i.

If we let $C_p(R)$ be $C_{p+1}^h(R)^{(1)}$, then the above discussion show that C_* satisfies the conditions of Barr's Theorem 8.8.7. In summary, we have proven the following.

Corollary 8.8.9 *Suppose that k is a field of characteristic zero. Then André-Quillen homology is a direct summand of Hochschild homology, and André-Quillen cohomology is a direct summand of Hochschild cohomology:*

$$D_q(R/k, M) \cong H_{q+1}^{(1)}(R, M) \quad \text{and} \quad D^q(R/k, M) \cong H_{(1)}^{q+1}(R, M).$$

9

Hochschild and Cyclic Homology

In this chapter we fix a commutative ring k and construct several homology theories based on chain complexes of k-modules. For legibility, we write $\otimes$ for $\otimes_k$ and $R^{\otimes n}$ for the n-fold tensor product $R \otimes \cdots \otimes R$.

9.1 Hochschild Homology and Cohomology of Algebras

9.1.1 Let R be a k-algebra and M an $R-R$ bimodule. We obtain a simplicial k-module $M \otimes R^{\otimes *}$ with $[n] \mapsto M \otimes R^{\otimes n}$ ($M \otimes R^{\otimes 0} = M$) by declaring

$$\partial_i(m \otimes r_1 \otimes \cdots \otimes r_n) = \begin{cases} mr_1 \otimes r_2 \otimes \cdots \otimes r_n & \text{if } i = 0 \\ m \otimes r_1 \otimes \cdots \otimes r_i r_{i+1} \otimes \cdots \otimes r_n & \text{if } 0 < i < n \\ r_n m \otimes r_1 \otimes \cdots \otimes r_{n-1} & \text{if } i = n \end{cases}$$

$$\sigma_i(m \otimes r_1 \otimes \cdots \otimes r_n) = m \otimes \cdots \otimes r_i \otimes 1 \otimes r_{i+1} \otimes \cdots \otimes r_n,$$

where $m \in M$ and the r_i are elements of R. These formulas are k-multilinear, so the ∂_i and σ_i are well-defined homomorphisms, and the simplicial identities are readily verified. (Check this!) The *Hochschild homology* $H_*(R, M)$ of R with coefficients in M is defined to be the k-modules

$$H_n(R, M) = \pi_n(M \otimes R^{\otimes *}) = H_n C(M \otimes R^{\otimes *}).$$

Here $C(M \otimes R^{\otimes *})$ is the associated chain complex with $d = \sum (-1)^i \partial_i$:

$$0 \longleftarrow M \xleftarrow{\partial_0 - \partial_1} M \otimes R \xleftarrow{d} M \otimes R \otimes R \xleftarrow{d} \cdots.$$

For example, the image of $\partial_0 - \partial_1$ is the k-submodule $[M, R]$ of M that is generated by all terms $mr - rm$ ($m \in M, r \in R$). Hence $H_0(R, M) \cong M/[M, R]$.

Similarly, we obtain a cosimplicial k-module with $[n] \mapsto \mathrm{Hom}_k(R^{\otimes n}, M) =$ $\{k\text{-multilinear maps } f: R^n \to M\}$ $(\mathrm{Hom}(R^{\otimes 0}, M) = M)$ by declaring

$$(\partial^i f)(r_0, \cdots, r_n) = \begin{cases} r_0 f(r_1, \ldots, r_n) & \text{if } i = 0 \\ f(r_0, \ldots, r_{i-1}r_i, \ldots) & \text{if } 0 < i < n \\ f(r_0, \ldots, r_{n-1})r_n & \text{if } i = n \end{cases}$$

$$(\sigma^i f)(r_1, \cdots, r_{n-1}) = f(r_1, \ldots, r_i, 1, r_{i+1}, \ldots, r_n).$$

The *Hochschild cohomology* $H^*(R, M)$ of R with coefficients in M is defined to be the k-modules

$$H^n(R, M) = \pi^n(\mathrm{Hom}_k(R^{\otimes *}, M)) = H^n C(\mathrm{Hom}_k(R^{\otimes *}, M)).$$

Here $C\,\mathrm{Hom}_k(R^*, M)$ is the associated cochain complex

$$0 \longrightarrow M \xrightarrow{\partial^0 - \partial^1} \mathrm{Hom}_k(R, M) \xrightarrow{d} \mathrm{Hom}_k(R \otimes R, M) \xrightarrow{d} \cdots.$$

For example, it follows immediately that

$$H^0(R, M) = \{m \in M : rm = mr \quad \text{for all } r \in R\}.$$

Exercise 9.1.1 If R is a commutative k-algebra, show that $M \otimes R^{\otimes *}$ is a simplicial R-module via $r \cdot (m \otimes r_1 \otimes \cdots) = (rm) \otimes r_1 \otimes \cdots$. Conclude that each $H_n(R, M)$ is an R-module. Similarly, show that $\mathrm{Hom}_R(R^{\otimes *}, M)$ is a cosimplicial R-module, and conclude that each $H^n(R, M)$ is an R-module.

Exercise 9.1.2 If $0 \to M_0 \to M_1 \to M_2 \to 0$ is a k-split exact sequence of bimodules (8.7.7), show that there is a long exact sequence

$$\cdots \xrightarrow{\partial} H_i(R, M_0) \to H_i(R, M_1) \to H_i(R, M_2) \xrightarrow{\partial} H_{i-1}(R, M_0) \cdots.$$

Example 9.1.2 (Group rings) Let R be the group ring $k[G]$ of a group G, and M a right G-module. Write $_\varepsilon M$ for M considered as a $G-G$ bimodule with trivial left G-module structure ($gm = m$ for all $g \in G$, $m \in M$). If B_*^u denotes the unnormalized bar resolution of 6.5.1, then $H_*(G; M)$ is the homology of $M \otimes_{\mathbb{Z}G} B_*^u$, the chain complex that in degree i is $M \otimes (\mathbb{Z}G)^{\otimes i}$. By inspection, this is the same complex used in 9.1.1 to define the Hochschild homology of $\mathbb{Z}G$, provided that we take coefficients in the bimodule $_\varepsilon M$. Similarly, $H^*(G; M)$ is the cohomology of the chain complex $\mathrm{Hom}_G(B_*^u, M)$, which is the same as the complex $\text{Hom}_k((\mathbb{Z}G)^{\otimes *}, {}_\varepsilon M)$ used to define Hochschild cohomology. Thus

$$H_*(G; M) \cong H_*(\mathbb{Z}G; {}_\varepsilon M) \quad \text{and} \quad H^*(G; M) \cong H^*(\mathbb{Z}G; {}_\varepsilon M).$$

The above definitions, originally given by G. Hochschild in 1945, have the advantage of being completely natural in R and M. In order to put them into a homological framework, it is necessary to consider the *enveloping algebra* $R^e = R \otimes_k R^{\text{op}}$ of R. Here R^{op} is the "opposite ring"; R^{op} has the same underlying abelian group structure as R, but multiplication in R^{op} is the opposite of that in R (the product $r \cdot s$ in R^{op} is the same as the product sr in R). The main feature of R^{op} is this: A right R-module M is the same thing as a left R^{op}-module via the product $r \cdot m = mr$ because associativity requires that

$$(r \cdot s) \cdot m = (sr) \cdot m = m(sr) = (ms)r = r \cdot (ms) = r \cdot (s \cdot m).$$

Similarly a left R-module N is the same thing as a right R^{op}-module via $n \cdot r = rn$. Consequently, the main feature of R^e is that an $R{-}R$ bimodule M is the same thing as a left R^e-module via the product $(r \otimes s) \cdot m = rms$, or as a right R^e-module via the product $m \cdot (r \otimes s) = smr$. (Check this!) This gives a slick way to consider the category $R{-}\mathbf{mod}{-}R$ of $R{-}R$ bimodules as the category of left R^e-modules or as the category of right R^e-modules. In particular, the canonical $R{-}R$ bimodule structure on R makes R into both a left and right R^e-module.

Lemma 9.1.3 *Hochschild homology and cohomology are isomorphic to relative Tor and Ext for the ring map $k \to R^e = R \otimes R^{\text{op}}$:*

$$H_*(R, M) \cong \text{Tor}_*^{R^e/k}(M, R) \quad \text{and} \quad H^*(R, M) \cong \text{Ext}^*_{R^e/k}(R, M).$$

Proof Consider the unnormalized bar resolution $\beta(R, R)$ of R as a left R-module (8.6.12). Each term $\beta(R, R)_n = R^{\otimes n+1} \otimes R$ is isomorphic as an $R{-}R$ bimodule to $R \otimes R^{\otimes n} \otimes R \cong (R \otimes R^{\text{op}}) \otimes R^{\otimes n}$ and hence is $\perp$-projective (8.6.5), where $\perp = R^e \otimes$. Since $\beta(R, R)$ is a k-split $\perp$-projective resolution of the R^e-module R, 8.7.10 yields

$$\text{Tor}_*^{R^e/k}(M, R) = H_*(M \otimes_{R^e} \beta(R, R)) \quad \text{and}$$

$$\text{Ext}^*_{R^e/k}(R, M) = H^* \text{Hom}_{R^e}(\beta(R, R), M).$$

On the other hand, the isomorphism $M \otimes_{R^e} (R \otimes R^{\otimes n} \otimes R) \to M \otimes R^n$ sending $m \otimes (r_0 \otimes \cdots \otimes r_{n+1})$ to $(r_{n+1}mr_0) \otimes (r_1 \otimes \cdots \otimes r_n)$ identifies $M \otimes_{R^e}$

$\beta(R, R)$ with the chain complex $C(M \otimes R^{\otimes *})$ used to define Hochschild homology. Similarly, the isomorphism $\mathrm{Hom}_{R^e}(R \otimes R^{\otimes n} \otimes R, M) \to \mathrm{Hom}_k(R^{\otimes n}, M)$ sending f to $f(1, -, 1)$ identifies $\mathrm{Hom}_{R^e}(\beta(R, R), M)$ with the cochain complex $C(\mathrm{Hom}_k(R^{\otimes *}, M))$ used to define Hochschild cohomology.   ◇

Next we show that in good cases, such as when k is a field, we can identify Hochschild homology and cohomology with the absolute Tor and Ext over the ring R^e.

Lemma 9.1.4 *If P and Q are flat (resp. projective) k-modules, then so is $P \otimes Q$.*

Proof Let $\mathcal{E}$ be an exact sequence of k-modules. If P and Q are flat, then by definition $\mathcal{E} \otimes P$ and hence $\mathcal{E} \otimes P \otimes Q$ are exact; hence $P \otimes Q$ is flat. If P and Q are projective, then $\mathrm{Hom}(Q, \mathcal{E})$ and hence $\mathrm{Hom}(P, \mathrm{Hom}(Q, \mathcal{E})) \cong \mathrm{Hom}(P \otimes Q, \mathcal{E})$ are exact; as we saw in 2.2.3, this implies that $P \otimes Q$ is projective.   ◇

Corollary 9.1.5 *If R is flat as a k-module, then $H_*(R, M) \cong \mathrm{Tor}_*^{R^e}(M, R)$. If R is projective as a k-module, then $H^*(R, M) \cong \mathrm{Ext}_{R^e}^*(R, M)$.*

Proof If R is flat (resp. projective), then each $R^{\otimes n}$ is a flat (resp. projective) k-module, and hence each $\beta(R, R)_n \cong R^e \otimes R^{\otimes n}$ is a flat (resp. projective) R^e-module. Thus $\beta(R, R)$ is a resolution of R by flat (resp. projective) R^e-modules. It follows that the relative Tor (resp. relative Ext) modules are isomorphic to the absolute Tor (resp. absolute Ext) modules.   ◇

Here are three cases in which $H_*(R, M)$ is easy to compute. First, let us recall from 7.3.1 that the tensor algebra of a k-module V is the graded algebra

$$T(V) = k \oplus V \oplus (V \otimes V) \oplus \cdots \oplus V^{\otimes j} \oplus \cdots.$$

Proposition 9.1.6 *Let $T = T(V)$ be the tensor algebra of a k-module V, and let M be a T–T bimodule. Then $H_i(T, M) = 0$ for $i \neq 0, 1$ and there is an exact sequence*

$$0 \to H_1(T, M) \to M \otimes V \xrightarrow{b} M \to H_0(T, M) \to 0$$

where b is the usual map $b(m \otimes v) = mv - vm$. In particular, if σ denotes the cyclic permutation $\sigma(v_1 \otimes \cdots \otimes v_j) = v_j \otimes v_1 \otimes \cdots v_{j-1}$ of $V^{\otimes j}$ and we write

$(V^{\otimes j})^\sigma$ and $(V^{\otimes j})_\sigma$ for the invariants and covariants of this group action, then we have

$$H_0(T, T) = k \oplus \bigoplus_{j=1}^{\infty} (V^{\otimes j})_\sigma, \quad H_1(T, T) = \bigoplus_{j=1}^{\infty} (V^{\otimes j})^\sigma.$$

Proof The formula $d(t \otimes v \otimes t') = tv \otimes t' - t \otimes vt'$ defines a $T-T$ bimodule map from $T \otimes V \otimes T$ to $T \otimes T$. As the kernel I of the multiplication $\mu: T \otimes T \to T$ is generated by the elements $v \otimes 1 - 1 \otimes v = d(1 \otimes v \otimes 1)$ and $\mu d = 0$, the image of d is I. As d is a direct sum (over p and q) of maps from $V^{\otimes p} \otimes V \otimes V^{\otimes q}$ to $V^{\otimes p+1} \otimes V^{\otimes q}$ and to $V^{\otimes p} \otimes V^{\otimes q+1}$, each of which is an isomorphism, d is an injection. (Check this!) Hence

$$0 \to T \otimes V \otimes T \xrightarrow{d} T \otimes T \xrightarrow{\mu} T \to 0$$

is a $\perp$-projective resolution of the T^e-module T; μ is k-split by the map $id \otimes 1: T \to T \otimes T$. Hence we can compute $\text{Tor}_*^{T^e/k}(M, T)$ using this resolution. Tensoring with M yields $H_i(T, M) = 0$ for $i \neq 0, 1$ and the advertised exact sequence for H_1 and H_0. $\diamond$

Exercise 9.1.3 (Polynomials) If $R = k[x_1, \cdots, x_m]$, show that R^e is isomorphic to the polynomial ring $k[y_1, \cdots, y_n, z_1, \cdots, z_m]$ and that the kernel of $R^e \to R$ is generated by the regular sequence $x = (y_1 - z_1, \cdots, y_m - z_m)$. Using the Koszul resolution $K(x)$ of 4.5.5, show that $H_p(R, R) \cong H^p(R, R) \cong \Lambda^p(R^n)$ for $p = 0, \cdots, n$, while $H_p(R, M) = H^p(R, M) = 0$ for $p > n$ and all bimodules M. This is a special case of Theorem 9.4.7 below.

Exercise 9.1.4 (Truncated polynomials) If $R = k[x]/(x^{n+1} = 0)$, let $u = x \otimes 1 - 1 \otimes x$ and $v = x^n \otimes 1 + x^{n-1} \otimes x + \cdots + x \otimes x^{n-1} + 1 \otimes x^n$ as elements in R^e. Show that

$$0 \leftarrow R \leftarrow R^e \xleftarrow{u} R^e \xleftarrow{v} R^e \xleftarrow{u} R^e \xleftarrow{v} R^e \xleftarrow{u} \cdots$$

is a periodic R^e-resolution of R, and conclude that $H_i(R, M)$ and $H^i(R, M)$ are periodic of period 2 for $i \geq 1$. Finally, show that when $\frac{1}{n+1} \in R$ we have $H_i(R, R) \cong H^i(R, R) \cong R/(x^n R)$ for all $i \geq 1$.

Let $k \to \ell$ be a commutative ring map. If R is a k-algebra, then $R_\ell = R \otimes_k \ell$ is an ℓ-algebra. If M is an $R_\ell-R_\ell$ bimodule then via the ring map $R \to R_\ell$ ($r \mapsto r \otimes 1$) we can also consider M to be an $R-R$ bimodule. We would like to compare the Hochschild homology $H_*^k(R, M)$ of the k-algebra R with the Hochschild homology $H_*^\ell(R_\ell, M)$ of the ℓ-algebra $R_\ell = R \otimes \ell$.

Theorem 9.1.7 (Change of ground ring) *Let R be a k-algebra and $k \to \ell$ a commutative ring map. Then there are natural isomorphisms for every R_ℓ–R_ℓ bimodule M:*

$$H_*^k(R, M) \cong H_*^\ell(R_\ell, M) \quad \text{and} \quad H_k^*(R, M) \cong H_\ell^*(R_\ell, M).$$

Proof The unnormalized chain complexes used for computing homology are isomorphic by the isomorphisms $M \otimes_k R \otimes_k \cdots \otimes_k R \cong M \otimes_\ell (R \otimes_k \ell) \otimes_\ell \cdots \otimes_\ell (R \otimes_k \ell)$. Similarly, the unnormalized cochain complexes used for computing cohomology are isomorphic, by the bijection between k-multilinear maps $R^n \to M$ and ℓ-multilinear maps $(R_\ell)^n \to M$. $\diamond$

Theorem 9.1.8 (Change of rings) *Let R be a k-algebra and M an R–R bimodule.*

1. *(Product) If R' is another k-algebra and M' an R'–R' bimodule, then*

$$H_*(R \times R', M \times M') \cong H_*(R, M) \oplus H_*(R', M')$$
$$H^*(R \times R', M \times M') \cong H^*(R, M) \oplus H^*(R', M').$$

2. *(Flat base change) If R is a commutative k-algebra and $R \to T$ is a ring map such that T is flat as a (left and right) R-module, then*

$$H_*(T, T \otimes_R M \otimes_R T) \cong T \otimes_R H_*(R, M).$$

3. *(Localization) If S is a central multiplicative set in R, then*

$$H_*(S^{-1}R, S^{-1}R) \cong H_*(R, S^{-1}R) \cong S^{-1}H_*(R, R).$$

Proof For (1), note that $(R \times R')^e \cong R^e \times R'^e \times (R \otimes R'^{\mathrm{op}}) \times (R' \otimes R^{\mathrm{op}})$; since M and M' are left R^e and R'^e-modules, respectively, this is a special case of relative Tor and Ext for products of rings (8.7.14). For (2), note that $R^e \to T^e$ makes T^e flat as an R^e-module (because $T^e \otimes_{R^e} M = T \otimes_R M \otimes_R T$). By flat base change for relative Tor (8.7.16) we have

$$\mathrm{Tor}_*^{T^e/k}(T, T^e \otimes M) \cong \mathrm{Tor}_*^{R^e/k}(T, M) \cong T \otimes_R \mathrm{Tor}_*^{R^e/k}(R, M).$$

The first part of (3) is also flat base change for relative Tor 8.7.16 with $T = S^{-1}R$, and the isomorphism $H_*(R, S^{-1}R) \cong S^{-1}H_*(R, R)$ is a special case of the isomorphism $\text{Tor}^{R^e/k}(S^{-1}M, N) \cong S^{-1}\text{Tor}_*^{R^e/k}(M, N)$ for localization (3.2.10 or exercise 8.7.3). $\diamond$

Here is one way to form $R-R$ bimodules. If M and N are left R-modules, $\text{Hom}_k(M, N)$ becomes an $R-R$ bimodule by the rule $rfs : m \mapsto rf(sm)$. The Hochschild cohomology of this bimodule is just the relative Ext of 8.7.5:

Lemma 9.1.9 *Let M and N be left R-modules. Then*

$$H^n(R, \text{Hom}_k(M, N)) \cong \text{Ext}_{R/k}^n(M, N).$$

Proof Let $B = B(R, R)$ be the bar resolution of R. Thinking of M as an $R-k$ bimodule, we saw in 2.6.2 that the functor $\otimes_R M: R\text{--}\mathbf{mod}\text{--}R \to R\text{--}\mathbf{mod}\text{--}k$ is left adjoint to the functor $\text{Hom}_k(M, -)$. Naturality yields an isomorphism of chain complexes:

$$\text{Hom}_R(B \otimes_R M, N) \cong \text{Hom}_{R-R}(B, \text{Hom}_k(M, N)).$$

As $B \otimes_R M$ is the bar resolution $B(R, M)$, the homology of the left side is the relative Ext. Since the homology of the right side is the Hochschild cohomology of R with coefficients in $\text{Hom}(M, N)$, we are done. $\diamond$

9.2 Derivations, Differentials, and Separable Algebras

It is possible to give simple interpretations to the low-dimensional Hochschild homology and cohomology modules. We begin by observing that the kernel of the map $d: \text{Hom}_k(R, M) \to \text{Hom}_k(R \otimes R, M)$ is the set of all k-linear functions $f: R \to M$ satisfying the identity

$$f(r_0 r_1) = r_0 f(r_1) + f(r_0)r_1.$$

Such a function is called a *k-derivation* (or *crossed homomorphism*); the k-module of all k-derivations is written $\text{Der}_k(R, M)$ (as in 8.8.1). On the other hand, the image of the map $d: M \to \text{Hom}_k(R, M)$ is the set of all k-derivations of the form $f_m(r) = rm - mr$; call f_m a *principal derivation* and write $\text{PDer}(R, M)$ for the submodule of all principal derivations. Taking H^1, we find exactly the same situation as for the cohomology of groups (6.4.5):

Lemma 9.2.1 $H^1(R, M) = \text{Der}_k(R, M)/\text{PDer}_k(R, M)$.

Now suppose that R is commutative. Recall from 8.8.1 that the *Kähler differentials* of R over k is the R-module $\Omega_{R/k}$ defined by the presentation: There is one generator dr for every $r \in R$, with $d\alpha = 0$ if $\alpha \in k$. For each $r_1, r_2 \in R$ there are two relations:

$$d(r_0 + r_1) = d(r_0) + d(r_1) \quad \text{and} \quad d(r_0 r_1) = r_0(dr_1) + (dr_0)r_1.$$

We saw in exercise 8.8.1 that $\operatorname{Der}_k(R, M) \cong \operatorname{Hom}_R(\Omega_{R/k}, M)$ for every right R-module M. If we make M into a bimodule by setting $rm = mr$ for all $r \in R$, $m \in M$ then $H^1(R, M) \cong \operatorname{Der}_k(R, M)$. This makes the following result seem almost immediate from the Universal Coefficient Theorem (3.6.2), since the chain complex $C(M \otimes R^{\otimes *})$ is isomorphic to $M \otimes_R C(R \otimes R^{\otimes *})$.

Proposition 9.2.2 *Let R be a commutative k-algebra, and M a right R-module. Making M into an $R-R$ bimodule by the rule $rm = mr$, we have natural isomorphisms $H_0(R, M) \cong M$ and $H_1(R, M) \cong M \otimes_R \Omega_{R/k}$. In particular,*

$$H_1(R, R) \cong \Omega_{R/k}.$$

Proof Since $rm = mr$ for all m and r, the map $\partial_0 - \partial_1 \colon M \otimes R \to M$ is zero. Therefore $H_0 \cong M$ and $H_1(R, M)$ is the quotient of $M \otimes_k R$ by the relations that for all $m \in M$, $r_i \in R$ $mr_1 \otimes r_2 - m \otimes r_1 r_2 + r_2 m \otimes r_1 = 0$. It follows that there is a well-defined map $H_1(R, M) \to M \otimes_R \Omega_{R/k}$ sending $m \otimes r$ to $m \otimes dr$. Conversely, we see from the presentation of $\Omega_{R/k}$ that there is an R-bilinear map $M \times \Omega_{R/k} \to H_1(R, M)$ sending $(m, r_1 dr_2)$ to the class of $mr_1 \otimes r_2$; this induces a homomorphism $M \otimes_R \Omega_{R/k} \to H_1(R, M)$. By inspection, these maps are inverses. $\diamond$

Corollary 9.2.3 *If S is a multiplicatively closed subset of R, then*

$$\Omega_{(S^{-1}R)/k} \cong S^{-1}(\Omega_{R/k}).$$

Proof The Change of Rings Theorem (9.1.8) states that $H_1(S^{-1}R, S^{-1}R) \cong S^{-1}H_1(R, R)$. $\diamond$

Alternate Calculation 9.2.4 For any k-algebra R, let I denote the kernel of the ring map $\varepsilon \colon R \otimes R \to R$ defined by $\varepsilon(r_1 \otimes r_2) = r_1 r_2$. Since $r \mapsto r \otimes 1$ defines a k-module splitting of ε, the sequence $0 \to I \to R^e \xrightarrow{\varepsilon} R \to 0$ is k-split. As $H_1(R, R^e) = 0$, the long exact homology sequence (exercise 9.1.2) yields

$$H_1(R, M) \cong \ker(I \otimes_{R^e} M \to IM).$$

If R is commutative and $rm = mr$, then $IM = 0$ and $H_1(R, M) \cong I/I^2 \otimes_R M$. In particular, if we take $M = R$ this yields

$$\Omega_{R/k} \cong H_1(R, R) \cong I/I^2.$$

Explicitly, the generator $dr \in \Omega_{R/k}$ corresponds to $1 \otimes r - r \otimes 1 \in I/I^2$. (Check this!)

Example 9.2.5 Let k be a field and R a separable algebraic field extension of k. Then $\Omega_{R/k} = 0$. In fact, for any $r \in R$ there is a polynomial $f(x) \in k[x]$ such that $f(r) = 0$ and $f'(r) \neq 0$. Since $d: R \to \Omega_{R/k}$ is a derivation we have $f'(r)dr = d(f(r)) = 0$, and hence $dr = 0$. As $\Omega_{R/k}$ is generated by the dr's, we get $\Omega_{R/k} = 0$.

Exercise 9.2.1 Suppose that R is commutative and M is a bimodule satisfying $rm = mr$. Show that there is a spectral sequence

$$E^2_{pq} = \mathrm{Tor}^R_p(H_q(R, R), M) \Rightarrow H_{p+q}(R, M).$$

Use this to give another proof of proposition 9.2.2. Then show that if M (or every $H_*(R, R)$) is a flat R-module, then $H_n(R, M) \cong H_n(R, R) \otimes_R M$ for all n.

The following two sequences are very useful in performing calculations. They will be improved later (in 9.3.5) by adding a smoothness hypothesis.

First Fundamental Exact Sequence for Ω 9.2.6 Let $k \to R \to T$ be maps of commutative rings. Then there is an exact sequence of T-modules:

$$\Omega_{R/k} \otimes_R T \xrightarrow{\alpha} \Omega_{T/k} \xrightarrow{\beta} \Omega_{T/R} \to 0.$$

The maps in this sequence are given by $\alpha(dr \otimes t) = tdr$ and $\beta(dt) = dt$.

Proof Clearly β is onto. By the Yoneda Lemma (1.6.11), in order for this sequence of T-modules to be exact at $\Omega_{T/k}$, it is sufficient to show that for every T-module N the sequence

$$\mathrm{Hom}_T(\Omega_{R/k} \otimes_R T, N) \xleftarrow{\alpha^*} \mathrm{Hom}_T(\Omega_{T/k}, N) \xleftarrow{\beta^*} \mathrm{Hom}_T(\Omega_{T/R}, N)$$

be exact. But this is just the sequence of derivation modules

$$\mathrm{Der}_k(R, N) \leftarrow \mathrm{Der}_k(T, N) \leftarrow \mathrm{Der}_R(T, N),$$

and this is exact because any k-derivation $D: T \to N$ satisfying $D(R) = 0$ is an R-derivation. $\diamondsuit$

Second Fundamental Exact Sequence for Ω 9.2.7 Let I be an ideal of a commutative k-algebra R. Then there is an R-module map $\delta: I/I^2 \to \Omega_{R/k} \otimes_R R/I$ defined by $\delta(x) = dx \otimes 1$, fitting into an exact sequence

$$I/I^2 \xrightarrow{\delta} \Omega_{R/k} \otimes_R R/I \xrightarrow{\alpha} \Omega_{(R/I)/k} \to 0.$$

Proof If $x \in I$ and $r \in R$, then $\delta(rx) = dx \otimes r$ as $dr \otimes x = 0$; if $r \in I$ then $rx \in I^2$ and $\delta(rx) = 0$. Hence δ is well defined and R-linear. Once more we use the Yoneda Lemma 1.6.11 to take an R/I-module N and consider

$$\text{Hom}_{R/I}(I/I^2, N) \xleftarrow{\delta^*} \text{Der}_k(R, N) \xleftarrow{\alpha^*} \text{Der}_k(R/I, N) \leftarrow 0.$$

If $D: R \to N$ is a k-derivation, then $(\delta^* D)(x) = D(x)$, so if $\delta^* D = 0$, then $D(I) = 0$, and D may be considered as a k-derivation on R/I. $\diamondsuit$

9.2.1 Finite Separable Algebras

A finite-dimensional semisimple algebra R over a field k is called *separable* if for every extension field $k \subseteq \ell$ the ℓ-algebra $R_\ell = R \otimes_k \ell$ is semisimple.

Lemma 9.2.8 *If K is a finite field extension of k, this definition agrees with the usual definition of separability: every element of K is separable over k.*

Proof If $x \in K$ is not separable, its minimal polynomial $f \in k[X]$ has multiple roots in any splitting field ℓ. Then $K \otimes \ell$ contains $k(x) \otimes \ell = \ell[X]/f$, which is not reduced, so $K \otimes \ell$ is not reduced. Otherwise we can write $K = k(x)$, where the minimal polynomial f of x has distinct roots in any field extension ℓ of k. Hence $K \otimes \ell = \ell[X]/(f)$ is reduced, hence semisimple. $\diamondsuit$

Corollary 9.2.9 *A finite-dimensional commutative algebra over a field is separable if and only if it is a product of separable field extensions.*

Proof A finite commutative algebra R is semisimple if and only if it is a product of fields. R is separable if and only if these fields are separable. $\diamondsuit$

The matrix rings $M_m(k)$ form another important class of separable algebras, since $M_m(k) \otimes_k \ell \cong M_m(\ell)$. More generally, Wedderburn's Theorem states that every semisimple ring R is a finite product of simple rings, each isomorphic to $M_m(\Delta)$ for some m and some division algebra Δ; R is separable if and only if each of its simple factors is separable.

Suppose that $M_m(\Delta)$ is separable. If F is the center of Δ, then $F \otimes \ell$ is a subring of $\Delta \otimes \ell$ and $M_m(\Delta) \otimes \ell$, so F must also be a finite separable extension of k. It is easy to see that if ℓ is a splitting field of F, then $F \otimes \ell$ is a finite product of copies of ℓ, so each of the simple factors of $M_m(\Delta) \otimes \ell$ has center ℓ. As we saw in 6.6.10 (see [BAII, 8.4]), there exists a finite extension L of ℓ such that $L \otimes_k M_m(\Delta) = L \otimes_\ell (\ell \otimes_k M_m(\Delta))$ is a product of matrix rings over L. In summary, we have proven that if R is separable over k, then there is a finite extension L of k such that $R \otimes L$ is a finite product of matrix rings $M_{m_i}(L)$.

Lemma 9.2.10 *If $R = M_m(k)$, then R is a projective R^e-module.*

Proof The element $e = \sum e_{i1} \otimes e_{1i}$ of $R^e = M_m(k) \otimes M_m(k)^{\mathrm{op}}$ is idempotent ($e^2 = e$) and the product map $\varepsilon \colon R \otimes R^{\mathrm{op}} \to R$ sends e to $\sum e_{ii} = 1$. Define $\alpha \colon R \to R^e$ by $\alpha(r) = re$. Since the basis elements e_{ij} of R satisfy $e_{ij}e = e_{i1} \otimes e_{1j} = ee_{ij}$, we have $re = er$ for all $r \in R$; hence α is an R–R bimodule map. Since $\varepsilon\alpha$ is the identity on R, this shows that R is a summand of R^e. $\Diamond$

Theorem 9.2.11 *Let R be an algebra over a field k. The following are equivalent:*

1. *R is a finite-dimensional separable k-algebra.*
2. *R is projective as a left R^e-module.*
3. *$H_*(R, M) = 0$ for all $* \neq 0$ and all bimodules M.*
4. *$H^*(R, M) = 0$ for all $* \neq 0$ and all bimodules M.*

Proof From the "pd" and "fd" lemmas of 4.1.6 and 4.1.10 we see that (2), (3), and (4) are equivalent. If R is separable, choose $k \subset \ell$ so that R_ℓ is a finite product of matrix rings $R_i = M_{m_i}(\ell)$. Since every R–R bimodule is a product $M = \Pi M_i$ of R_i–R_i bimodules M_i we have $H_*(R, M) = \Pi H_*(R_i, M_i) = 0$ by 9.1.8 and the above lemma. Thus (1) $\Rightarrow$ (3).

Now assume that (2) holds for R. Then (2), (3), and (4) hold for every $R \otimes \ell$ because $R \otimes \ell$ is projective over the ring

$$(R_\ell)^e = (R \otimes \ell) \otimes_\ell (R \otimes \ell)^{\mathrm{op}} = (R \otimes R^{\mathrm{op}}) \otimes \ell = (R^e) \otimes \ell.$$

We have isolated the proof that $\dim(R) < \infty$ in lemma 9.2.12 following this proof. Now each R_ℓ is semisimple if and only if R_ℓ has global dimension 0

(4.2.2). If M and N are left R_ℓ-modules, we saw in 9.1.3 and 9.1.9 that

$$\text{Ext}^*_{R_\ell}(M, N) = \text{Ext}^*_{R_\ell/k}(M, N) \cong H^*(R_\ell, \text{Hom}_k(M, N)).$$

As (4) holds for R_ℓ, the right side is zero for $* \neq 0$ and all M, N; the Global Dimension Theorem (4.1.2) implies that R_ℓ has global dimension 0. Hence (2) $\Rightarrow$ (1). ◇

Lemma 9.2.12 (Villamayor-Zelinsky) *Let R be an algebra over a field k. If R is projective as an R^e-module, then R is finite-dimensional as a vector space over k.*

Proof Let $\{x_i\}$ be a basis for R as a vector space and $\{f^i\}$ a dual basis for $\text{Hom}_k(R, k)$. As R^e is a free left R-module on basis $\{1 \otimes x_i\}$ with dual basis $\{1 \otimes f^i\} \subseteq \text{Hom}_R(R^e, R)$, we have

$$u = \sum (1 \otimes f^i)(u) \otimes x_i \quad \text{for all } u \in R^e.$$

Now if R is a projective R^e-module, the surjection $\varepsilon: R^e \to R$ must be split. Hence there is an idempotent $e \in R^e$ such that $R^e \cdot e \cong R$ and $\varepsilon(e) = 1$. In particular, $(1 \otimes r - r \otimes 1)e = 0$ for all $r \in R$. Setting $u = (1 \otimes r)e = (r \otimes 1)e$ yields

$$(*) \qquad r = \varepsilon(u) = \sum (1 \otimes f_i)((r \otimes 1)e) \cdot x_i = r \sum (1 \otimes f_i)(e)x_i.$$

Therefore the sum in $(*)$ is over a finite indexing set independent of r. Writing $e = \sum e_{\alpha\beta}x_\alpha \otimes x_\beta$ with $e_{\alpha\beta} \in k$ allows us to rewrite $(*)$ as

$$r = \sum (1 \otimes f_i)(e_{\alpha\beta}x_\alpha \otimes rx_\beta)x_i = \sum e_{\alpha\beta} f(rx_\beta)(x_\alpha x_i).$$

Therefore the finitely many elements $x_\alpha x_i$ span R as a vector space. ◇

9.3 H² , *Extensions, and Smooth Algebras*

From the discussion in Chapter 6, section 6 about extensions and factor sets we see that $H^2(R, M)$ should have something to do with extensions. By a (square zero) *extension* of R by M we mean a k-algebra E, together with a surjective ring homomorphism $\varepsilon: E \to R$ such that $\ker(\varepsilon)$ is an ideal of square zero (so that $\ker(\varepsilon)$ has the structure of an $R-R$ bimodule), and an R-module isomorphism of M with $\ker(\varepsilon)$. We call it a *Hochschild extension* if the short exact sequence $0 \to M \to E \to R \to 0$ is k-split, that is, split as a sequence of k-modules. Choosing such a splitting $\sigma: R \to E$ yields a k-module decomposition $E \cong R \oplus M$, with multiplication given by

$$(*) \qquad (r_1, m_1)(r_2, m_2) = (r_1 r_2, r_1 m_2 + m_1 r_2 + f(r_1, r_2)).$$

We call the function $f: R \otimes R \to M$ the *factor set* of the extension corresponding to the splitting σ. Since the product $(r_0, 0)(r_1, 0)(r_2, 0)$ is associative, the factor set must satisfy the cocycle condition

$$r_0 f(r_1, r_2) - f(r_0 r_1, r_2) + f(r_0, r_1 r_2) - f(r_0, r_1)r_2 = 0.$$

Conversely, any function satisfying this cocycle condition yields a Hochschild extension with multiplication defined by $(*)$. (Check this!) A different choice $\sigma': R \to E$ of a splitting yields a factor set f', and

$$
\begin{aligned}
f'(r_1, r_2) - f(r_1, r_2) &= \sigma'(r_1)\sigma'(r_2) - \sigma'(r_1 r_2) - \sigma(r_1)\sigma(r_2) + \sigma(r_1 r_2) \\
&= \sigma'(r_1)[\sigma'(r_2) - \sigma(r_2)] - [\sigma'(r_1 r_2) - \sigma(r_1 r_2)] \\
&\quad + [\sigma'(r_1) - \sigma(r_1)]\sigma(r_2),
\end{aligned}
$$

which is the coboundary of the element $(\sigma' - \sigma) \in \mathrm{Hom}(R, M)$. Hence a Hochschild extension determines a unique cohomology class, independent of the choice of splitting σ.

The *trivial extension* is obtained by taking $E \cong R \oplus M$ with product $(r_1, m_1)(r_2, m_2) = (r_1 r_2, r_1 m_2 + m_1 r_2)$. Since its factor set is $f = 0$, the trivial extension yields the cohomology class $0 \in H^2(R, M)$.

As with group extensions, we say that two extensions E and E' are *equivalent* if there is a ring isomorphism $\varphi : E \cong E'$ making the familiar diagram commute:

$$
\begin{array}{ccccccccc}
0 & \longrightarrow & M & \longrightarrow & E & \longrightarrow & R & \longrightarrow & 0 \\
 & & \| & & {\scriptstyle \varphi}\downarrow & & \| & & \\
0 & \longrightarrow & M & \longrightarrow & E' & \longrightarrow & R & \longrightarrow & 0.
\end{array}
$$

Since E and E' share the same factor sets, they determine the same cohomology class. We have therefore proven the following result.

Classification Theorem 9.3.1 *Given a k-algebra R and an $R-R$ bimodule M, the equivalence classes of Hochschild extensions are in 1–1 correspondence with the elements of the Hochschild cohomology module $H^2(R, M)$.*

Here is a variant of the Classification Theorem 9.3.1 when R is a commutative k-algebra. If a commutative algebra E is a Hochschild extension of R by an R–R bimodule M, then M must be *symmetric* in the sense that $rm = mr$ for every $m \in M$ and $r \in R$. A moment's thought shows that symmetric bimodules are the same thing as R-modules.

If we choose a k-splitting $\sigma : R \to E$ for a commutative Hochschild extension, then the corresponding factor set f must satisfy $f(r_1, r_2) = f(r_2, r_1)$, because $\sigma(r_1)$ and $\sigma(r_2)$ must commute in E. Let us call such a factor set *symmetric*. If f is a symmetric factor set, the equation $(*)$ shows that multiplication in E is commutative.

Let us write $H_s^2(R, M)$ for the submodule of $H^2(R, M)$ consisting of the equivalence classes of symmetric factor sets. With this notation, we can summarize the above discussion as follows

Commutative Extensions 9.3.1.1 *Let R be a commutative k-algebra and M an R-module. Then the equivalence classes of commutative Hochschild extensions of R by M are in 1–1 correspondence with the elements of the module $H_s^2(R, M)$.*

Remark Let k be a field. This classification, together with Exercise 8.8.4, proves that $H_s^2(R, M)$ is just the André-Quillen cohomology $D^1(R, M)$. The characteristic zero version of this was given in 8.8.9.

9.3.2 We say that a k-algebra is *quasi-free* (over k) if for every square-zero extension $0 \to M \to E \xrightarrow{\varepsilon} T \to 0$ of a k-algebra T by a T–T bimodule M and every algebra map $\nu : R \to T$, there exists a k-algebra homomorphism $u : R \to E$ lifting ν in the sense that $\varepsilon u = \nu$. For example, it is clear that every free algebra is quasi-free over k.

$$
\begin{array}{ccccccccc}
k & \longrightarrow & & & R & & \\
\downarrow & & \swarrow & & \downarrow \nu & & \\
0 & \longrightarrow & M & \longrightarrow & E & \underset{\varepsilon}{\longrightarrow} & T & \longrightarrow & 0
\end{array}
$$

If R is quasi-free and J is a nilpotent ideal in another k-algebra E, then every algebra map $R \to E/J$ may be lifted to a map $R \to E$. In fact, we can lift it successively to $R \to E/J^2$, to $R \to E/J^3$, and so on. Since $J^m = 0$ for some m, we eventually lift it to $R \to E/J^m = E$.

Proposition 9.3.3 (J.H.C. Whitehead-Hochschild) *If k is a field, then a k-algebra R is quasi-free iff and only if $H^2(R, M) = 0$ for all R–R bimodules M.*

Proof If R is quasi-free, every extension of R by a bimodule M must be trivial, so $H^2(R, M) = 0$ by the Classification Theorem 9.3.1. Conversely, given an extension $0 \to M \to E \to T \to 0$ and $\nu : R \to T$, let D be the pullback $D = \{(r, e) \in R \times E : \nu(r) = \bar{e} \text{ in } T\}$. Then D is a subring of $R \times E$ and the kernel of $D \to R$ is a square zero ideal isomorphic to M.

$$0 \longrightarrow M \longrightarrow D \longrightarrow R \longrightarrow 0$$

$$\| \qquad \downarrow \qquad \downarrow \nu$$

$$0 \longrightarrow M \longrightarrow E \longrightarrow T \longrightarrow 0.$$

Since k is a field, D is a Hochschild extension of R and is classified by an element of $H^2(R, M)$. So if $H^2(R, M) = 0$, then there is a k-algebra splitting $\sigma: R \to D$ of $D \to R$; the composite of σ with $D \to E$ is a lifting of $R \to T$. Quantifying over all such M proves that R is quasi-free.   ◇

Corollary 9.3.3.1 *If R is an algebra over a field k and $H^2(R, M) = 0$ for every $R - R$ bimodule M, then any k-algebra surjection $E \to R$ with nilpotent kernel must be split by a k-algebra injection $\sigma : R \to E$.*

Exercise 9.3.1 (Wedderburn's Principal Theorem) Let R be a finite-dimensional algebra over a field k, with Jacobson radical $J = J(R)$. It is well known that the quotient R/J is a semisimple ring ([BAII, 4.2]). Prove that if R/J is separable, then there is a k-algebra injection $R/J \subset R$ splitting the natural surjection $R \to R/J$. *Hint:* Use the General Version 4.3.10 of Nakayama's Lemma to show that J is nilpotent.

9.3.1 Smooth Algebras

For the rest of this section, all the algebras we consider will be commutative.

We say that a commutative k-algebra is *smooth* (over k) if for every square-zero extension $0 \to M \to E \xrightarrow{\varepsilon} T \to 0$ of commutative k-algebras and every algebra map $\nu : R \to T$, there exists a k-algebra homomorphism $u : R \to E$ lifting ν in the sense that $\varepsilon u = \nu$. For example, it is clear that every polynomial algebra $R = k[x_1, \ldots, x_n]$ is smooth over k.

Proposition 9.3.4 (Whitehead-Hochschild) *Let R be an algebra over a field k. Then R is smooth if and only if $H_s^2(R, M) = 0$ for all R-modules M.*

If R is smooth, then any surjection $E \to R$ of commutative k-algebras with nilpotent kernel J must be split by a k-algebra injection $\sigma : R \to E$.

Proof The proof of the Whitehead-Hochschild result 9.3.3, and the arguments in 9.3.2, go through with no changes for commutative algebras.   ◇

Exercise 9.3.2
1. (Localization) If R is smooth over k and $S \subset R$ is a central multiplicative set, show that $S^{-1}R$ is smooth over k.
2. (Transitivity) If R is smooth over K and K is smooth over k, show that R is smooth over k.
3. (Base change) If R is smooth over k and $k \to \ell$ is any ring map, show that $R \otimes_k \ell$ is smooth over ℓ.
4. If k is a field, show that any filtered union of smooth algebras is smooth.

Exercise 9.3.3 Let $0 \to M \to E \xrightarrow{\varepsilon} T \to 0$ be a square zero algebra extension and $u: R \to E$ a k-algebra map. If $u': R \to E$ is any k-module map with $\varepsilon u' = \varepsilon u$, then $u' = u + D$ for some k-module map $D: R \to M$. Show that u' is a k-algebra map if and only if D is a k-derivation.

Fundamental Sequences for Ω with Smoothness 9.3.5 Let $k \to R \xrightarrow{f} T$ be maps of commutative rings.

1. If T is smooth over R, then the first fundamental sequence 9.2.6 becomes a *split* exact sequence by adding $0 \to$ on the left:

$$0 \to \Omega_{R/k} \otimes_R T \xrightarrow{\alpha} \Omega_{T/k} \xrightarrow{\beta} \Omega_{T/R} \to 0.$$

2. If $T = R/I$ and T is smooth over k, then the second fundamental sequence 9.2.7 becomes a *split* exact sequence by adding $0 \to$ on the left:

$$0 \to I/I^2 \xrightarrow{\delta} \Omega_{R/k} \otimes_R R/I \xrightarrow{\alpha} \Omega_{(R/I)/k} \to 0.$$

Proof For (1), let N be a T-module, and $D: R \to N$ a k-derivation. Define a ring map φ from R to the trivial extension $T \oplus N$ by $\varphi(r) = (f(r), Dr)$. By smoothness, the projection $T \oplus N \to T$ is split by an R-module homomorphism $\sigma: T \to T \oplus N$. Writing $\sigma(t) = (t, D't)$, then $D': T \to N$ is a k-derivation of T such that $D'f = D$. (Check this!) Now take N to be $\Omega_{R/k} \otimes_R T$; D' corresponds to a T-bilinear map $\gamma: \Omega_{T/k} \to \Omega_{R/k} \otimes_R T$. If D is the derivation $D(r) = dr \otimes 1$, then $\gamma\alpha$ is the identity on N and γ splits α.

For (2), note that smoothness of $T = R/I$ implies that the sequence $0 \to I/I^2 \to R/I^2 \xrightarrow{f} R/I \to 0$ is split by a k-algebra map $\sigma: R/I \to R/I^2$. The map $D = 1 - \sigma f: R \to R/I^2$ satisfies $\bar{f}D = f - (\bar{f}\sigma)f = 0$, so the image of D lies in I/I^2 and D is a derivation. Moreover the restriction of D to I is the natural projection $I \to I/I^2$. By universality, D corresponds to an R-module map $\theta: \Omega_{R/k} \to I/I^2$ sending rds to $rD(s)$. Thus θ kills $I\Omega_{R/k}$ and factors through $\Omega_{R/k} \otimes_R R/I$, with $\theta\delta$ the identity on I/I^2. $\diamond$

We are going to characterize those field extensions K that are smooth over k. For this, we recall some terminology and results from field theory [Lang, X.6]. Let k be a field and K a finitely generated extension field. We say that K is *separately generated* over k if we can find a transcendence basis $(t_1, \cdots, t_r)$ of K/k such that K is separably algebraic over the purely transcendental field $k(t_1, \cdots, t_r)$. If char$(k) = 0$, or if k is perfect, it is known that every finitely generated extension of k is separably generated.

Proposition 9.3.6 *If k is a field, every separably generated extension field K is smooth over k.*

Proof K is separably algebraic over some purely transcendental field $F = k(t_1, \cdots, t_r)$. As F is a localization of the polynomial ring $k[t_1, \cdots, t_r]$, which is smooth over k, F is smooth over k. By transitivity of smoothness, it suffices to prove that K is smooth over F. Since K is a finite separable algebraic extension of F, we may write $K = F(x)$, where $f(x) = 0$ for some irreducible polynomial f with $f'(x) \neq 0$. Suppose given a map $v \colon K \to T$ and a square zero extension $0 \to M \to E \to T \to 0$. Choosing any lift $y \in E$ of $v(x) \in T$, we have $f(y + m) = f(y) + f'(y)m$ for every $m \in M$. Since $v(f(x)) = 0$ and $v(f'(x))$ is a unit of T, $f(y) \in M$ and $f'(y)$ is a unit of E. If we put $m = -f(y)/f'(y)$, then $f(y + m) = 0$, so we may define a lift $K \to E$ by sending x to $y + m$. $\diamond$

Corollary 9.3.7 *If k is a perfect field, every extension field K is smooth over k. In particular, every extension field is smooth when $\operatorname{char}(k) = 0$.*

Proof If K_α is a finitely generated extension subfield of K, then K_α is separably generated and hence smooth. If M is a K-module, then $H_s^2(K_\alpha, M) = 0$. As tensor products and homology commute with filtered direct limits, we have $H_s^2(K, M) = \lim H_s^2(K_\alpha, M) = 0$. Hence K is smooth. $\diamond$

When $\operatorname{char}(k) \neq 0$ and k is not perfect, the situation is as follows. Call K *separable* (over k) if every finitely generated extension subfield is separably generated. The proof of the above corollary shows that separable extensions are smooth; in fact the converse is also true [Mat, 20.L]:

Theorem 9.3.8 *Let $k \subset K$ be an extension of fields. Then*

$$K \text{ is separable over } k \Leftrightarrow K \text{ is smooth over } k.$$

Remark 9.3.9 One of the major victories in field theory was the discovery that a field extension $k \subset K$ is separable if and only if for any finite field extension $k \subset \ell$ the ring $K \otimes_k \ell$ is reduced. If $\operatorname{char}(k) = p$, separability is also equivalent to *MacLane's criterion for separability*: K is linearly disjoint from the field $\ell = k^{1/p^\infty}$ obtained from k by adjoining all p-power roots of elements of k. See [Mat, 27.F] and [Lang, X.6]. Here is the most important part of this relationship.

Lemma 9.3.10 *Let K be a separably generated extension of a field k. Then for every field extension $k \subset \ell$ the ring $K \otimes_k \ell$ is reduced.*

Proof It is enough to consider the case of a purely transcendental extension and the case of a finite separable algebraic extension. If $K = k(x)$ is purely transcendental, then each $K \otimes \ell = \ell(x)$ is a field. If K is a finite separable extension, we saw that $K \otimes \ell$ is reduced for every ℓ in 9.2.8 $\diamond$

Exercise 9.3.4 A commutative algebra R over a field k is called *separable* if R is reduced and for any algebraic field extension $k \subset \ell$ the ring $R \otimes_k \ell$ is reduced. By the above remark, this agrees with the previous definition when R is a field. Show that

1. Every subalgebra of a separable algebra is again separable.
2. The filtered union of separable algebras is again separable.
3. Any localization of a separable algebra is separable.
4. If char$(k) = 0$, or more generally if k is perfect, every reduced k-algebra is separable; this completely classifies separable algebras over k.
5. An artinian k-algebra R is separable if and only if R is a finite product of separable field extensions of k (see 9.2.9).
6. A finite-dimensional algebra R is separable in the sense of this exercise if and only if it is separable in the sense of section 9.2.1.

9.3.2 Smoothness and Regularity

For the next result, we shall need the *Hilbert-Samuel function* $h_R(n) = $ length of $R/\mathfrak{m}^n$ of a d-dimensional noetherian local ring R. There is a polynomial $H_R(t)$ of degree d, called the *Hilbert-Samuel polynomial*, such that $h_R(n) = H_R(n)$ for all large n; see [Mat, 12.C&H]. For example, if R is the localization of the polynomial ring $K[x_1, \cdots, x_d]$ at the maximal ideal $M = (x_1, \cdots, x_d)$, then $h_R(n) = H_R(n) = \binom{n+d-1}{d} = \frac{n(n+1)\cdots(n+d-1)}{d!}$ for all $n \geq 1$.

Theorem 9.3.11 *Let R be a noetherian local ring containing a field k. If R is smooth over k, then R is a regular local ring.*

Proof Set $d = \dim_K(\mathfrak{m}/\mathfrak{m}^2)$, and write S for the local ring of $K[x_1, \cdots, x_d]$ at the maximal ideal $M = (x_1, \cdots, x_d)$. Note that $S/M^2 \cong K \oplus \mathfrak{m}/\mathfrak{m}^2$. By replacing k by its ground field if necessary, we may assume that the residue field $K = R/\mathfrak{m}$ is also smooth over k. This implies that the square zero extension $R/\mathfrak{m}^2 \to K$ splits, yielding an isomorphism $R/\mathfrak{m}^2 \cong K \oplus (\mathfrak{m}/\mathfrak{m}^2) \cong S/M^2$. Since R is smooth, we can lift $R \to R/\mathfrak{m}^2 \cong S/M^2$ to maps $R \to S/M^n$ for every n. By Nakayama's Lemma 4.3.9, if R maps onto S/M^n, then R maps onto S/M^{n+1} (because $\mathfrak{m}(S/M^{n+1})$ contains M^n/M^{n+1}). Inductively, this proves that $R/\mathfrak{m}^n$ maps onto S/M^n for every n and hence that $h_R(n) \geq h_S(n)$ for all n. Therefore the Hilbert polynomial $H_R(t)$ has degree $\geq d$, and hence $\dim(R) \geq d$. Since we always have $\dim(R) \leq d$ (4.4.1), this yields $\dim(R) = d$, that is, R is a regular local ring. $\diamond$

Definition 9.3.12 A commutative noetherian ring R is called *regular* if the localization of R at any prime ideal is a regular local ring (see 4.4.1). We say that R is *geometrically regular* over a field k if R contains k, and for every finite field extension $k \subset \ell$ the ring $R \otimes_k \ell$ is also regular.

Corollary 9.3.13 *Let R be a commutative noetherian ring containing a field k. If R is smooth over k, then R is geometrically regular over k.*

Proof If R is smooth over k, then so is every localization of R. Hence R is regular. For each $k \subset \ell$, $R \otimes \ell$ is smooth over ℓ, hence regular. $\diamond$

Remark In fact the converse is true: Geometrically regular k-algebras are smooth over k; see [EGA, $0_{IV}(22.5.8)$].

Theorem 9.3.14 *If R is a smooth k-algebra, then $\Omega_{R/k}$ is a projective R-module.*

Proof We will show that $\Omega_{R/k}$ satisfies the projective lifting property. Suppose given an R-module surjection $u: M \to N$ and a map $v: \Omega_{R/k} \to N$. If I is the kernel of $R^e \to R$, then the square zero algebra extension $R^e/I^2 \to R$ is trivial, that is, $R^e/I^2 \cong R \oplus I/I^2$ as a k-algebra. Moreover, $I/I^2 \cong \Omega_{R/k}$ by 9.2.4. We thus have a diagram of k-algebras

$$
\begin{array}{ccc}
R^e & \longrightarrow & R^e/I^2 \cong R \oplus \Omega_{R/k} \\
\exists \downarrow w & & \downarrow v \\
R \oplus M & \xrightarrow{\;\;(1,u)\;\;} & R \oplus N.
\end{array}
$$

The kernel of $R \oplus M \to R \oplus N$ is the square zero ideal $0 \oplus \ker(u)$. By base change (exercise 9.3.2) $R^e = R \otimes_k R$ is smooth over R, hence over k, so $R^e \to R \oplus N$ lifts to a k-algebra map $w: R^e \to R \oplus M$. Since $w(I)$ is contained in $0 \oplus M$ (why?), $w(I^2) = 0$. Thus w induces an R-module lifting $I/I^2 \to M$ of v. $\diamond$

Remark The rank of $\Omega_{R/k}$ is given in 9.4.8.

Application 9.3.15 (Jacobian criterion) Suppose that $R = k[x_1, \cdots, x_n]/J$, where J is the ideal generated by polynomials $f_1, \cdots, f_m$. The second fundamental sequence 9.2.7 is

$$
J/J^2 \xrightarrow{\;\delta\;} R^n \to \Omega_{R/k} \to 0,
$$

where R^n denotes the free R-module on basis $\{dx_1, \cdots, dx_n\}$. Since J/J^2 is generated by $f_1, \cdots, f_m$ the map δ is represented by the $m \times n$ *Jacobian matrix* $(\partial f_i/\partial x_j)$. Now suppose that R is smooth, so that this sequence is split exact and J/J^2 is also a projective R-module. If M is a maximal ideal of $k[x_1, \cdots, x_n]$ with residue field $K = R/M$, and $d = \dim(R_M)$, then J_M is generated by a regular sequence of length $n - d$, so $(J/J^2) \otimes_R K$ is a vector space of dimension $n - d$. Therefore the Jacobian matrix $(\partial f_i/\partial x_j)$ has rank $n - d$ when evaluated over $K = R/M$. This proves the necessity of the following criterion; the sufficiency is proven in [$EGA, 0_{IV}(22.6.4)$], and in [Mat, section 29].

> *Jacobian criterion:* R is smooth if and only if the Jacobian matrix $(\partial f_i/\partial x_j)$ has rank $n - \dim(R_M)$ when evaluated over R/M for every maximal ideal M.

9.4 Hochschild Products

There are external and internal products in Hochschild homology, just as there were for absolute Tor (and Ext) in 2.7.8 and exercise 2.7.5, and for relative Tor (and Ext) in 8.7.12 and exercise 8.7.2. All these external products involve two k-algebras R and R' and their tensor product algebra $R \otimes R'$. To obtain internal products in homology we need an algebra map $R \otimes R \to R$, which requires R to commutative. This situation closely resembles that of algebraic topology (pretend that R is a topological space X; the analogue of R being commutative is that X is an H-space). We shall not discuss the internal product for cohomology, since it is entirely analogous but needs an algebra map $R \to R \otimes R$, which requires R to be a Hopf algebra (or a bialgebra).

We begin with the external product for Hochschild homology. Let R and R' be k-algebras. Since the bar resolution $\beta(R, R)$ is an $R-R$ bimodule resolution of R and $\beta(R', R')$ is an $R'-R'$ bimodule resolution of R', their tensor product $\beta(R, R) \otimes \beta(R', R')$ comes from a bisimplicial object in the category **bimod** of $(R \otimes R')-(R \otimes R')$ bimodules. In 8.6.13 we showed that the shuffle product ∇ induces a chain homotopy equivalence in **bimod**:

$$\text{Tot } \beta(R, R) \otimes \beta(R', R') \xrightarrow{\nabla} \beta(R \otimes R', R \otimes R').$$

If M is an $R-R$ bimodule and M' is an $R'-R'$ bimodule, then we can tensor over $(R \otimes R')^e$ with $M \otimes M'$ to obtain a chain homotopy equivalence

$$\text{Tot}\{(M \otimes_{R^e} \beta(R, R)) \otimes (M' \otimes_{R'^e} \beta(R', R'))\} \xrightarrow{\nabla} (M \otimes M') \otimes_{(R \otimes R')^e} \beta(R \otimes R', R \otimes R').$$

Recall from 9.1.3 that the Hochschild chain complex $C(M \otimes R^{\otimes *})$ is isomorphic to $M \otimes_{R^e} \beta(R, R)$. Hence we may rewrite the latter equivalence as

$$\text{Tot}\{C(M \otimes R^{\otimes *}) \otimes C(M' \otimes R'^{\otimes *})\} \xrightarrow{\nabla} C((M \otimes M') \otimes (R \otimes R')^{\otimes *}).$$

If we apply $\text{Hom}_{\textbf{bimod}}(-, M \otimes M')$ we get an analogous cochain homotopy equivalence

$$\text{Tot Hom}_{\textbf{bimod}}(\beta(R, R) \otimes \beta(R', R'), M \otimes M') \xrightarrow{\nabla} C \text{ Hom}_k((R \otimes R')^{\otimes *}, M \otimes M'),$$

but the natural map from $\text{Hom}_R(\beta, M) \otimes \text{Hom}_{R'}(\beta', M')$ to $\text{Hom}_{\textbf{bimod}}(\beta \otimes \beta', M \otimes M')$ is not an isomorphism unless R or R' is a finite-dimensional algebra. The Künneth formula for complexes (3.6.3) yields the following result.

Proposition 9.4.1 (External products) *The shuffle product ∇ induces natural maps*

$$H_i(R, M) \otimes H_j(R', M') \xrightarrow{\nabla} H_{i+j}(R \otimes R', M \otimes M'),$$

$$H^i(R, M) \otimes H^j(R', M') \xrightarrow{\nabla} H^{i+j}(R \otimes R', M \otimes M').$$

For $i = j = 0$ these products are induced by the identity map on $M \otimes M'$. If k is a field, the direct sum of the shuffle product maps yields natural isomorphisms

$$H_n(R \otimes R', M \otimes M') \cong [H_*(R, M) \otimes H_*(R', M')]_n$$

$$= \bigoplus_{i+j=n} H_i(R, M) \otimes H_j(R', M').$$

Similarly, the shuffle product $\nabla: H^*(R, M) \otimes H^*(R', M') \to H^*(R \otimes R', M \otimes M')$ is an isomorphism when either R or R' is finite-dimensional over a field k.

Remark The explicit formula for ∇ in exercise 8.6.5 shows that the external product is associative from $H(R, M) \otimes H(R', M') \otimes H(R'', M'')$ to $H(R \otimes R' \otimes R'', M \otimes M' \otimes M'')$.

Exercise 9.4.1 Let $0 \to M_0 \to M_1 \to M_2 \to 0$ be a k-split exact sequence of $R-R$ bimodules. Show that ∇ commutes with the connecting homomorphism ∂ in the sense that there is a commutative diagram

$$
\begin{array}{ccc}
H_i(R, M_2) \otimes H_j(R', M') & \xrightarrow{\nabla} & H_{i+j}(R \otimes R', M_2 \otimes M') \\
\partial \otimes 1 \downarrow & & \downarrow \partial \\
H_{i-1}(R, M_0) \otimes H_j(R', M') & \xrightarrow{\nabla} & H_{i+j-1}(R \otimes R', M_0 \otimes M').
\end{array}
$$

9.4.1 Internal Product

Now suppose that R is a commutative k-algebra. Then the product $R \otimes R \to R$ is a k-algebra homomorphism. Composing the external products with this homomorphism yields a product in Hochschild homology

$$H_p(R, M) \otimes H_q(R, M') \to H_{p+q}(R, M \otimes_{R^e} M').$$

Here $M \otimes M'$ is an $R-R$ bimodule by $r(m \otimes m')s = (rm) \otimes (m's)$. When $M = M' = R$, the external products yield an associative product on $H_*(R, R)$.

In fact, more is true. At the chain level, the shuffle product 8.6.13 gives a map

$$\text{Tot } C(R \otimes R^{\otimes *}) \otimes C(R \otimes R^{\otimes *}) \xrightarrow{\nabla} C((R \otimes R) \otimes (R \otimes R)^{\otimes *}) \to C(R \otimes R^{\otimes *}).$$

Proposition 9.4.2 *If R is a commutative k-algebra, then*

1. $C(R \otimes R^{\otimes *}) = R \otimes_{R^e} \beta(R, R)$ *is a graded-commutative differential graded k-algebra (4.5.2).*
2. $H_*(R, R)$ *is a graded-commutative k-algebra.*

Proof It suffices to establish the first point (see exercise 4.5.1). Write C_* for $C(R \otimes R^{\otimes *}) = R \otimes_{R^e} \beta(R, R)$. The explicit formula for ∇ (exercise 8.6.5) becomes

$$(r_0 \otimes r_1 \otimes \cdots \otimes r_p)\nabla(r_0' \otimes r_{p+1} \otimes \cdots \otimes r_{p+q}) =$$
$$\sum_\mu (-1)^\mu (r_0 r_0') \otimes r_{\mu^{-1}(1)} \otimes \cdots \otimes r_{\mu^{-1}(p+q)},$$

where μ ranges over all (p, q)-shuffles. The product ∇ is associative, because an (n, p, q)-shuffle may be written uniquely either as the composition of a (p, q)-shuffle and an $(n, p + q)$-shuffle, or as the composition of an (n, p)-shuffle and an $(n + p, q)$-shuffle. Interchanging p and q amounts to precomposition with the shuffle $\nu = (p + 1, \cdots, p + q, 1, \cdots, p)$; since $(-1)^\nu = (-1)^{pq}$ the product ∇ is graded-commutative. Finally, we know that $\nabla: \text{Tot}(C_* \otimes C_*) \to C_*$ is a chain map. Therefore if we set $\rho = (r_0, r_1, \cdots, r_p)$ and $\rho' = (r_0', r_{p+1}, \cdots, r_{p+q})$ and recall the sign trick 1.2.5 for d^v we have the Leibniz formula:

$$d(\rho \nabla \rho') = \nabla(d^h + d^v)(\rho \otimes \rho') = (d\rho)\nabla\rho' + (-1)^p \rho\nabla(d\rho'). \qquad \diamond$$

Corollary 9.4.3 *If R is commutative and M is an R−R bimodule, then $H_*(R, M)$ is a graded $H_*(R, R)$-module.*

9.4.2 *The Exterior Algebra $\Omega^*_{R/k}$*

As an application, recall that $H_1(R, R)$ is isomorphic to the R-module $\Omega_{R/k}$ of Kähler differentials of R over k. If we write $\Omega^n_{R/k}$ for the nth exterior product $\wedge^n(\Omega_{R/k})$, then the exterior algebra $\Omega^*_{R/k}$ on $\Omega_{R/k}$ is

$$\Omega^*_{R/k} = R \oplus \Omega_{R/k} \oplus \Omega^2_{R/k} \oplus \cdots.$$

Note that $\Omega^0_{R/k} = R$ and $\Omega^1_{R/k} = \Omega_{R/k}$. $\Omega^*_{R/k}$ is the free graded-commutative R-algebra generated by $\Omega_{R/k}$; if K_* is a graded-commutative R-algebra, then any R-module map $\Omega_{R/k} \to K_1$ extends uniquely to an algebra map $\Omega^*_{R/k} \to K_*$.

Corollary 9.4.4 *If R is a commutative k-algebra, the isomorphism $\Omega^1_{R/k} \cong H_1(R, R)$ extends to a natural graded ring map $\psi: \Omega^*_{R/k} \to H_*(R, R)$. If $\mathbb{Q} \subset R$, this is an injection, split by a graded ring surjection $e: H_*(R, R) \to \Omega^*_{R/k}$.*

Proof Since $H_*(R, R)$ is graded-commutative, the first assertion is clear. For the second, define a map $e: R^{\otimes n+1} \to \Omega^n_{R/k}$ by the multilinear formula

$$e(r_0 \otimes r_1 \otimes \cdots \otimes r_n) = \frac{1}{n!} r_0 dr_1 \wedge \cdots \wedge dr_n.$$

The explicit formula for ∇ shows that $e(\rho \nabla \rho') = e(\rho) \wedge e(\rho')$ in $\Omega^*_{R/k}$. Therefore e is a graded R-algebra map from $R^{\otimes *+1}$ to $\Omega^*_{R/k}$. An easy calculation shows that $e(b(r_0 \otimes \cdots \otimes r_{n+1})) = 0$. (Check this!) Hence e induces an algebra map $HH_*(R, R) \to \Omega^*_{R/k}$. To see that e splits ψ, we compute that

$$e\psi(r_0 dr_1 \wedge \cdots \wedge dr_n) = e((r_0 \otimes r_1)\nabla(1 \otimes r_2)\nabla \cdots \nabla(1 \otimes r_n))$$
$$= e(r_0 \otimes r_1) \wedge e(1 \otimes r_2) \wedge \cdots \wedge e(1 \otimes r_n)$$
$$= r_0 dr_1 \wedge r_2 \wedge \cdots \wedge r_n. \qquad \diamond$$

Definition 9.4.5 We say that a commutative k-algebra R is *essentially of finite type* if it is a localization of a finitely generated k-algebra. If k is noetherian, this implies that R and $R^e = R \otimes R$ are both noetherian rings (by the Hilbert Basis Theorem).

Proposition 9.4.6 *Suppose that R is a commutative algebra, essentially of finite type over a field k. If R is smooth over k, then R^e is a regular ring.*

Proof We saw in 9.3.13 that smooth noetherian k-algebras are regular. By smooth base change and transitivity (exercise 9.3.2), $R^e = R \otimes R$ is smooth over R and hence smooth over k. Since R^e is noetherian, it is regular. $\diamond$

Theorem 9.4.7 (Hochschild-Kostant-Rosenberg) *Let R be a commutative algebra, essentially of finite type over a field k. If R is smooth over k, then ψ is an isomorphism of graded R-algebras:*

$$\psi: \Omega^*_{R/k} \xrightarrow{\cong} H_*(R, R).$$

Proof As with any R-module homomorphism, ψ is an isomorphism if and only if $\psi \otimes_R R_m$ is an isomorphism for every maximal ideal m of R. The Change of Rings Theorem (9.1.8) states that $H_*(R, R) \otimes_R R_m \cong H_*(R_m, R_m)$. Since $\Omega^*_{R/k} \otimes_R R_m = \Omega^*_{R_m/k}$, $\psi \otimes_R R_m$ is obtained by replacing R by R_m. Hence we may assume that R is a local ring.

Let I be the kernel of $R \otimes R \to R$ and M the pre-image of m in $R^e = R \otimes R$. M is a maximal ideal in the regular ring R^e, so $S = (R^e)_M$ is a regular local ring. By flat base change (8.7.16) $H_*(R, R) \cong \mathrm{Tor}^{S/k}_*(R, R)$. Since S and $R = S/I_M$ are regular local rings, I_M is generated by a regular sequence of length $d = \dim(R) = \dim(S) - \dim(R)$; see exercise 4.4.2. We also saw in 8.7.13 that the external product makes $\mathrm{Tor}^{S/k}_*(R, R)$ isomorphic to $\Lambda^*\Omega_{R/k} = \Omega^*_{R/k}$ as a graded-commutative R-algebra. Since the external product can also be computed via the bar resolution and the shuffle product (8.7.12), the above product agrees with the internal product on $H_*(R, R) \cong \mathrm{Tor}^{S/k}_*(R, R)$. $\diamond$

Remark 9.4.8 We saw in 9.3.14 and 8.7.13 that $\Omega_{R/k}$ is a projective module whose localization at a maximal ideal m of R is a free module of rank $\dim(R_m)$. Hence for $d = \dim(R) = \max\{\dim(R_m)\}$ we have $\Omega^d_{R/k} \neq 0$ and $H_n(R, R) = \Omega^n_{R/k} = 0$ for $n > d$. The converse holds: If $H_n(R, R) = 0$ for all large n, then R is smooth over k. See L. Avramov and M. Vigué-Poirrier, "Hochschild homology criteria for smoothness," *International Math. Research Notices* (1992, No.1), 17–25.

Exercise 9.4.2 Extend the Hochschild-Kostant-Rosenberg Theorem to the case in which k is a commutative noetherian ring; if R is smooth over k and essentially of finite type, then $\psi : \Omega^*_{R/k} \cong H_*(R, R)$. *Hint:* Although S and $R = S/I$ may not be regular local rings, the ideal I is still generated by a regular sequence of length d.

9.4.3 Hodge Decomposition

When $\mathbb{Q} \subset R$ and R is commutative, we shall show (in 9.4.15) that the Hochschild chain complex $C^h_*(R) = C(R \otimes R^{\otimes *})$ decomposes as the direct sum of chain complexes $C^h_*(R)^{(i)}$. The resulting decompositions $H_*(R, R) = \oplus H^{(i)}_*(R, R)$ and $H^*(R, R) = \oplus H^*_{(i)}(R, R)$ are called the Hodge decompositions of Hochschild homology and cohomology in order to reflect a relationship with the Hodge decomposition of the cohomology of complex analytic manifolds. (This relationship was noticed by Gerstenhaber and Schack [GS]; see Remark 9.8.19 for more details.) In the process, we will establish the facts needed to apply Barr's Theorem (8.8.7), showing that the summand $H_*^{(1)}(R, R)$ may be identified with the André-Quillen homology modules $D_{*-1}(R/k)$.

If R does not contain $\mathbb{Q}$, there is a filtration on $H_*(R, R)$ but need be no decomposition [Q]. This filtration may be based on certain operations λ^k; see [Loday, 4.5.15]. When $\mathbb{Q} \subset R$ the eigenspaces of the λ^k give the decomposition; λ^k acts as multiplication by $\pm k^i$ on $C_*^h(R)^{(i)}$ and hence on $H_*^{(i)}(R, R)$ and $H^*_{(i)}(R, R)$. For this reason, the Hodge decomposition is often called the λ- *decomposition*.

The symmetric group Σ_n acts on the n-fold tensor product $R^{\otimes n}$ and hence on $M \otimes R^{\otimes n}$ by permuting coordinates: $\sigma(m \otimes r_1 \otimes \cdots \otimes r_n) = m \otimes r_{\sigma^{-1}1} \otimes \cdots \otimes r_{\sigma^{-1}n}$. Consider, for example, the effect of the *signature idempotent* $\varepsilon_n = \frac{1}{n!} \sum_{\sigma \in \Sigma_n} (-1)^\sigma \sigma$ of $\mathbb{Q}\Sigma_n$; the definition of the shuffle product ∇ shows that in $R \otimes R^{\otimes n}$ we have the identity:

$$n!\varepsilon_n(r_0 \otimes r_1 \otimes \cdots \otimes r_n) = r_0(1 \otimes r_1)\nabla \cdots \nabla(1 \otimes r_n).$$

This element is an n-cycle in the Hochschild complex representing the element $\psi(r_0 dr_1 \wedge \cdots \wedge dr_n)$ of $H_n(R, R)$, where $\psi : \Omega^*_{R/k} \hookrightarrow H_*(R, R)$ is the injection discussed in 9.4.4. The formula for the chain-level splitting $e : R \otimes R^{\otimes *} \to \Omega^*_{R/k}$ of ψ is skew-symmetric, so we also have $e(r_0 \otimes r_1 \otimes \cdots \otimes r_n) = e(\varepsilon_n(r_0 \otimes r_1 \otimes \cdots \otimes r_n))$. Hence e factors through $\varepsilon_n(R \otimes R^{\otimes n})$.

The following criterion for recognizing the signature idempotent will be handy. Consider the action of Σ_n on the module $R \otimes R^{\otimes n}$.

Barr's Lemma 9.4.9 *If $u \in \mathbb{Q}\Sigma_n$ satisfies $bu(1 \otimes r_1 \otimes \cdots \otimes r_n) = 0$ for all algebras R, then $u = c\varepsilon_n$ for some $c \in \mathbb{Q}$.*

Proof Write $u = \sum c_\sigma \sigma$ with $c_\sigma \in \mathbb{Q}$. We consider its action on the element $x = (1 \otimes r_1 \otimes \cdots \otimes r_n)$ of $R \otimes R^{\otimes n}$, where R is the polynomial ring $k[r_1, \ldots, r_n]$. In $b(ux) = \sum c_\sigma b(1 \otimes r_{\sigma^{-1}1} \otimes \cdots \otimes r_{\sigma^{-1}n})$ the term

$$1 \otimes r_{\sigma^{-1}1} \otimes \cdots \otimes r_{\sigma^{-1}i}r_{\sigma^{-1}(i+1)} \otimes \cdots \otimes r_{\sigma^{-1}n}$$

occurs once with coefficient $(-1)^i c_\sigma$ and once with coefficient $(-1)^i c_{\tau\sigma}$, where τ is the transposition $(i, i+1)$. Since these terms form part of a basis for the free k-module $R \otimes R^{\otimes n}$, we must have $c_\sigma = -c_{\tau\sigma}$ for all σ and all $\tau = (i, i+1)$. Hence $c_\sigma = (-1)^\sigma c_1$ for all $\sigma \in \Sigma_n$, and therefore $u = c_1 \sum (-1)^\sigma \sigma = c_1\varepsilon_n$. $\diamond$

To fit this into a broader context, fix $n > 1$ and define the "shuffle" elements

s_{pq} of $\mathbb{Z}\Sigma_n$ to be the sum $\sum(-1)^\mu \mu$ over all (p,q)-shuffles in Σ_n (so by convention $s_{pq}=0$ unless $p+q=n$). Let s_n be the sum of the s_{pq} for $0 < p < n$.

Lemma 9.4.10 $bs_n = s_{n-1}b$ *for every* n.

Proof If $p+q=n$, $x = r_0 \otimes \cdots \otimes r_p$ and $y = 1 \otimes r_{p+1} \otimes \cdots \otimes r_n$, then $x \nabla y = s_{pq}(r_0 \otimes \cdots \otimes r_n)$. Since $R^{\otimes *+1}$ is a DG-algebra (9.4.2), we have

$$bs_{pq}(r_0 \otimes \cdots \otimes r_n) = b(x \nabla y) = (bx)\nabla y + (-1)^p x \nabla (by)$$
$$= s_{p-1,q}((bx) \otimes y) + (-1)^p s_{p,q-1}(x \otimes (by)).$$

Summing over p gives $bs_n = s_{n-1}b$. $\diamond$

Proposition 9.4.11 ([GS]) *The minimal polynomial for* $s_n \in \mathbb{Q}\Sigma_n$ *is*

$$f_n(x) = x(x - \lambda_2) \cdots (x - \lambda_n), \text{ where } \lambda_i = 2^i - 2.$$

Therefore the commutative subalgebra $\mathbb{Q}[s_n]$ *of* $\mathbb{Q}\Sigma_n$ *contains* n *uniquely determined idempotents* $e_n^{(i)}$, $i = 1, \ldots, n$ *such that* $s_n = \sum \lambda_i e_n^{(i)}$ *and* $\mathbb{Q}[s_n] = \prod \mathbb{Q}e_n^{(i)}$. *In particular,* $e_n^{(i)} e_n^{(j)} = 0$ *for* $i \neq j$. $\diamond$

Definition 9.4.12 The idempotents $e_n^{(i)}$ are called the *Eulerian idempotents* of $\mathbb{Q}\Sigma_n$. Because s_n has only n eigenvalues, $e_n^{(i)} = 0$ for $i > n$. By convention, $e_0^{(0)} = 1$ and $e_n^{(0)} = 0$ otherwise.

Proof If $n = 1$ then $s_1 = 0$, while if $\tau = (1,2)$ then $s_2 = 1 - \tau$ satisfies $x(x-2)$. For $n \geq 3$ we proceed by induction. Since $bs_n = s_{n-1}b$, we have $bf_{n-1}(s_n) = f_{n-1}(s_{n-1})b = 0$. By Barr's Lemma, $f_{n-1}(s_n) = ce_n$ for some constant c. To evaluate c, note that $\varepsilon_n s_n = \lambda_n \varepsilon_n$ because s_n has λ_n terms and $\varepsilon_n \sigma = (-1)^n \varepsilon_n$ for every $\sigma \in \Sigma_n$. Thus

$$f_{n-1}(s_n) = \varepsilon_n f_{n-1}(s_n) = f_{n-1}(\varepsilon_n s_n) = f_{n-1}(\lambda_n \varepsilon_n) = c\varepsilon_n \neq 0,$$

where $c = \lambda_n f_{n-1}(1) \neq 0$. Thus $f_n(s_n) = c\varepsilon_n(s_n - \lambda_n) = 0$. $\diamond$

Corollary 9.4.13 $e_n^{(n)}$ *is the signature idempotent* ε_n.

Proof $\mathbb{Q}[s_n]$ contains $\varepsilon_n = f_{n-1}(s_n)/c$, and $\varepsilon_n s_n = \lambda_n \varepsilon_n$. $\diamond$

Corollary 9.4.14 $be_n^{(i)} = e_{n-1}^{(i)}b$ *for* $i < n$, *and* $be_n^{(n)} = 0$.

Proof For all i, let $p_i(x)$ be the product of the terms $(x - \lambda_j)/(\lambda_i - \lambda_j)$ for $j \neq i$, $j \leq n$, so that $p_i(s_n) = e_n^{(i)}$ and $p_i(s_{n-1}) = e_{n-1}^{(i)}$; this is the Lagrange interpolation formula for diagonalizable operators and is most easily checked using $\mathbb{Q}[s_n] = \prod \mathbb{Q}e_n^{(i)}$. Since $bs_n = s_{n-1}b$, we have

$$be_n^{(i)} = bp_i(s_n) = p_i(s_{n-1})b = e_{n-1}^{(i)}b.$$

As a special case, we have the formula $be_n^{(n)} = e_{n-1}^{(n)}b = 0$.   ◇

Definition 9.4.15 Suppose that R is a commutative k-algebra containing $\mathbb{Q}$. For $i \geq 1$, let $C_n^h(R)^{(i)}$ denote the summand $e_n^{(i)} R \otimes R^{\otimes n}$ of $C_n^h(R) = R \otimes R^{\otimes n}$. By 9.4.14, each $C_*^h(R)^{(i)}$ is a chain subcomplex of $C_*^h(R)$. For $i = 0$ we let $C_*^h(R)^{(0)}$ denote the complex that is R, concentrated in degree zero, so that $C_*^h(R)$ is the direct sum of the chain subcomplexes $C_*^h(R)^{(i)}$ for $i \geq 0$. We define $H_n^{(i)}(R, R)$ to be $H_n(C_*^h(R)^{(i)})$. The resulting formula

$$H_n(R, R) = H_n^{(1)}(R, R) \oplus \cdots \oplus H_n^{(n)}(R, R), n \neq 0,$$

is called *the Hodge decomposition* of Hochschild homology. Similarly, we define $H_{(i)}^n(R, R)$ to be $H^n \operatorname{Hom}_R(C_*^h(R)^{(i)}, R)$ and call the resulting formula

$$H_n(R, R) = H_{(1)}^n(R, R) \oplus \cdots \oplus H_{(n)}^n(R, R), n \neq 0,$$

the *Hodge decomposition* of Hochschild cohomology.

The Hodge decomposition (or λ-decomposition) arose implicitly in [Barr] (via 9.4.9 and 8.8.7) and [Q] and was made explicit in [GS].

Exercise 9.4.3 Let $\bar{C}_n^h(R)^{(i)}$ denote the summand $e_n^{(i)} R \otimes (R/k)^{\otimes n}$ of the normalized Hochschild complex $R \otimes (R/k)^{\otimes n}$. Show that $H_n^{(i)}(R, R) = H_n\bar{C}_n^h(R)^{(i)}$.

Exercise 9.4.4 Show that $H_n^{(n)}(R, R) \cong \Omega_{R/R}^n$ for every R. Conclude that if R is smooth and essentially of finite type over k, then $H_n(R, R) = H_n^{(n)}(R, R)$.

9.5 Morita Invariance

Definition 9.5.1 Two rings R and S are said to be *Morita equivalent* if there is an $R-S$ bimodule P and an $S-R$ bimodule Q such that $P \otimes_S Q \cong R$ as $R-R$ bimodules and $Q \otimes_R P \cong S$ as $S-S$ bimodules. It follows that the functors $\otimes_R P\colon \mathbf{mod}\text{-}R \to \mathbf{mod}\text{-}S$ and $\otimes_S Q\colon \mathbf{mod}\text{-}S \to \mathbf{mod}\text{-}R$ are inverse equivalences, because for every right R-module M we have $(M \otimes_R P) \otimes_S Q \cong M \otimes_R (P \otimes_S Q) \cong M$ and similarly for right S-modules.

Exercise 9.5.1 Show that

1. Morita equivalence is an equivalence relation.
2. If R and S are Morita equivalent, so are R^{op} and S^{op}.
3. If R and S are Morita equivalent, then the bimodule categories $R\text{-}\mathbf{mod}\text{-}R$ and $S\text{-}\mathbf{mod}\text{-}S$ are equivalent (via $Q \otimes_R - \otimes_R P$).

Proposition 9.5.2 *The matrix rings $M_m(R)$ are Morita equivalent to R.*

Proof Let P be the module of row vectors $(r_1, \cdots, r_m)$ of length m and Q the module of column vectors of length m. The matrix ring $S = M_m(R)$ acts on the right of P and the left of Q by the usual matrix multiplication, so P is an R–S bimodule and Q is an S–R bimodule. Matrix multiplication yields bimodule maps $P \otimes_S Q \to R$ and $Q \otimes_R P \to S$: if $p = (p_1, \cdots, p_m)$ and $q = (q_1, \cdots, q_m)^T$, then $p \otimes q$ maps to $\sum p_i q_i$ and $q \otimes p$ maps to the matrix $(q_i p_j)$. It is easy to check that these maps are isomorphisms (do so!). ◇

Corollary 9.5.3 *The isomorphism $R\text{-}\mathbf{mod}\text{-}R \to M_m(R)\text{-}\mathbf{mod}\text{-}M_m(R)$ associates to an R–R bimodule M the $M_m(R)$–$M_m(R)$ bimodule $M_m(M)$ of all $m \times m$ matrices with entries in M.*

Lemma 9.5.4 *If P and Q define a Morita equivalence between R and S, then P is a finitely generated projective left R-module. P is also a finitely generated projective right S-module.*

Proof Given $p \in P$ and $q \in Q$ we write $p \cdot q$ and $q \cdot p$ for the elements of R and S corresponding to $p \otimes q \in P \otimes_S Q$ and $q \otimes p \in Q \otimes_R P$, respectively. As $Q \otimes_R P \cong S$, we can write $1 = q_1 \cdot p_1 + \cdots + q_m \cdot p_m$ for some m. Define $e\colon P \to R^m$ by $e(p) = (p \cdot q_1, \cdots, p \cdot q_m)$ and $h\colon R^m \to P$ by $h(r_1, \cdots, r_m) = \sum r_i p_i$; e and h are left R-module homomorphisms. Since $he(p) = \sum (p \cdot q_i) p_i = \sum p(q_i \cdot p_i) = p$, this expresses P as a summand of R^m in $R\text{-}\mathbf{mod}$. The proof that P is a summand of some S^n in $\mathbf{mod}\text{-}S$ is similar. ◇

Exercise 9.5.2 Show that the bimodule structures induce ring isomorphisms

$$\mathrm{End}_R(Q) \cong S \cong \mathrm{End}_R(P)^{\mathrm{op}}.$$

Conclude that if all projective R-modules are free, then any ring which is Morita equivalent to R must be a matrix ring $M_m(R)$.

Lemma 9.5.5 *If L is a left R-module and Q is a projective right R-module then $H_i(R, L \otimes Q) = 0$ for $i \neq 0$ and $H_0(R, L \otimes Q) \cong Q \otimes_R L$.*

Proof By additivity, it suffices to prove the result with $Q = R$. The standard chain complex (9.1.1) used to compute $H_*(R, L \otimes R)$ is isomorphic to the bar resolution $\beta(R, L)$ of the left R-module L (8.6.12), which has $H_i(\beta) = 0$ for $i \neq 0$ and $H_0(\beta) \cong R \otimes_R L$. $\diamond$

.

Theorem 9.5.6 (R. K. Dennis) *Hochschild homology is Morita invariant. That is, if R and S are Morita equivalent rings and M is an $R-R$ bimodule, then*

$$H_*(R, M) \cong H_*(S, Q \otimes_R M \otimes_R P).$$

Proof Let L denote the $S-R$ bimodule $Q \otimes_R M$. Consider the bisimplicial k-module $X_{ij} = S^{\otimes i} \otimes L \otimes R^{\otimes j} \otimes P$, where the jth row is the standard complex 9.1.1 for the Hochschild homology over S of the $S-S$ bimodule $L \otimes R^{\otimes j} \otimes P$ and the ith column is the standard complex for the Hochschild homology of the $R-R$ bimodule $P \otimes S^{\otimes i} \otimes L$ (with the P rotated). Using the sign trick 1.2.5, form a double complex C_{**}. We will compute the homology of $\mathrm{Tot}(C)$ in two ways.

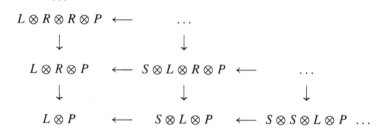

Since P is a projective right S-module, the jth row is exact except at $i = 0$, where $H_0(C_{*j}) = P \otimes_S (L \otimes R^{\otimes j}) \cong M \otimes R^{\otimes j}$ (9.5.5). The vertical differentials of the chain complex $H_0(C_{*j})$ make it isomorphic to the standard complex for the Hochschild homology of M. Thus $H_i \mathrm{Tot}(C) \cong H_i(R, M)$ for all i. On the other hand, since P is a projective left R-module, the ith column is exact except at $j = 0$, where $H_0(C_{i*}) = S^{\otimes i} \otimes L \otimes_S P$ (9.5.5). The horizontal differentials of $H_0(C_{i*})$ make it isomorphic to the standard complex for the Hochschild homology of $L \otimes_S P = Q \otimes_R M \otimes_S P$. Thus $H_i \mathrm{Tot}(C) \cong H_i(S, Q \otimes_R M \otimes_S P)$ for all i. $\diamond$

Definition 9.5.7 (Trace) The usual trace map from $M_m(R)$ to R is the map sending a matrix $g = (g_{ij})$ to its trace $\sum g_{ii}$. More generally, given an $R-R$ bimodule M we can define maps trace_n from $M_m(M) \otimes M_m(R)^{\otimes n}$ to $M \otimes R^{\otimes n}$ by the formula

$$\text{trace}_n(m \otimes g^1 \otimes \cdots \otimes g^n) = \sum_{i_0,\ldots,i_n=1}^{n} m_{i_0 i_1} \otimes g^1_{i_1 i_2} \otimes \cdots \otimes g^r_{i_r i_{r+1}} \otimes \cdots \otimes g^n_{i_n i_0}.$$

These maps are compactible with the simplicial operators ∂_i and σ_i (check this!), so they assemble to yield a simplicial module homomorphism from $M_m(M) \otimes M_m(R)^{\otimes *}$ to $M \otimes R^{\otimes *}$. They therefore induce a map on Hochschild homology, called the *trace map*.

Corollary 9.5.8 *The natural isomorphism of theorem 9.5.6 is given by the trace map* $H_*(M_m(R), M_m(M)) \to H_*(R, M)$.

Proof Let us write $F = F(R,S,P,Q,M)$ for the natural isomorphism $H_*(R, M) \to H_*(S, Q \otimes M \otimes P)$ given by the bisimplicial k-module X of theorem 9.5.6. Fixing R, set $S' = R$ and $S = M_m(R)$, $P' = R$ and $P = R^m$, $Q' = R$ and $Q = (R^m)^T$. The diagonal map $\Delta \colon R \to M_m(R)$ sending $r \in R$ to the diagonal matrix $\begin{bmatrix} r & & 0 \\ & \cdots & \\ 0 & & r \end{bmatrix}$ is compatible with the maps $P' \to P$ and $Q' \to Q$ sending $p \in P'$ and $q \in Q'$ to $(p, 0, \cdots, 0)^T$ and $(q, 0, \cdots, 0)$, respectively. It therefore yields a map $\Delta \colon X(R, S', P', Q') \to X(R, S, P, Q)$. (Check this!) This yields a commutative square

$$
\begin{array}{ccccc}
H_n(R, M) & \xrightarrow{F'} & H_n(R, R \otimes_R M \otimes_R R) & = & H_n(R, M) \\
\| & & & & \downarrow \Delta \\
H_n(R, M) & \xrightarrow{F} & H_n(M_m(R), Q \otimes M \otimes P) & = & H_n(M_m(R), M_m(M)).
\end{array}
$$

It follows that Δ is an isomorphism. At the chain level, we have

$$\Delta(m \otimes r_1 \otimes \cdots \otimes r_m) = \begin{bmatrix} m & 0 & 0 \\ 0 & \cdots & 0 \\ 0 & 0 & 0 \end{bmatrix} \otimes \begin{bmatrix} r_1 & & 0 \\ & \cdots & \\ 0 & & r_1 \end{bmatrix} \otimes \cdots \otimes \begin{bmatrix} r_n & & 0 \\ & \cdots & \\ 0 & & r_n \end{bmatrix}.$$

Clearly $\text{trace}_n(\Delta(m \otimes r_1 \otimes \cdots \otimes r_n)) = m \otimes r_1 \otimes \cdots \otimes r_n$, so the trace map $H_*(M_m(R), M_m(M)) \to H_*(R, M)$ is the inverse isomorphism to Δ. ◇

Exercise 9.5.3 For $m < n$, consider the (nonunital) inclusion $\iota: M_m(R) \hookrightarrow M_n(R)$ sending g to $\begin{bmatrix} g & 0 \\ 0 & 0 \end{bmatrix}$. Show that ι induces a chain map ι_* from the complex $M_m(M) \otimes M_m(R)^{\otimes *}$ to the complex $M_n(M) \otimes M_n(R)^{\otimes *}$ for every R-module M. Then show that this chain map is compatible with the trace maps (i.e., that trace = trace $\circ \iota_*$), and conclude that ι_* induces the Morita invariance isomorphism

$$H_*(M_m(R), M_m(M)) \cong H_*(M_n(R), M_n(M)).$$

Exercise 9.5.4 Let $e_{ij}(r)$ denote the matrix with exactly one nonzero entry, namely r, occurring in the (i, j) spot. Show that

$$\text{trace } e_{12}(r_1) \otimes e_{23}(r_2) \otimes \cdots \otimes e_{n1}(r_n) = r_1 \otimes \cdots \otimes r_n.$$

Then show that for any permutation σ of $\{1, 2, \cdots, n\}$

$$\text{trace } e_{\sigma 1, \sigma 2}(r_1) \otimes e_{\sigma n, \sigma 1}(r_n) = \begin{cases} r_1 \otimes \cdots \otimes r_n & \text{if } \sigma \in C_n \\ 0 & \text{if not,} \end{cases}$$

where C_n is the subgroup of the symmetric group generated by $(12 \cdots n)$.

9.6 Cyclic Homology

The simplicial k-module $ZR = R \otimes R^{\otimes *}$ used to construct the Hochschild homology modules $H_*(R, R)$ has a curious "cyclic" symmetry, which is suggested by writing a generator $r_0 \otimes r_1 \otimes \cdots \otimes r_n$ of $R \otimes R^{\otimes n}$ in the circular form illustrated here.

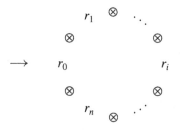

The arrow $\rightarrow$ serves as a place marker, and there are $n + 1$ of the symbols $\otimes$. The $n + 1$ face and degeneracy operators replace the appropriate symbol $\otimes$ by a product or a "$\otimes 1 \otimes$," respectively. This symmetry defines an action of the cyclic group C_{n+1} on $R \otimes R^{\otimes n}$; the generator t of C_{n+1} acts as the operator

$t(r_0 \otimes \cdots \otimes r_n) = r_n \otimes r_0 \otimes \cdots \otimes r_{n-1}$. We may visualize t as a rotation of the above circular representation (with the place marker fixed). Clearly $\partial_i t = t \partial_{i-1}$ and $\sigma_i t = t \sigma_{i-1}$ for $i > 0$; for $i = 0$ we have $\partial_0 t = \partial_n$ and $\sigma_0 t = t^2 \sigma_n$. (Check this!) This leads to the notion of an abstract cyclic k-module: a simplicial k-module with this extra cyclic symmetry. After giving the definition in this fashion, we shall construct a category ΔC such that a cyclic k-module is a contravariant functor from ΔC to k–**mod**, paralleling the definition in Chapter 8 of a simplicial object.

Definition 9.6.1 A *cyclic object* A in a category $\mathcal{A}$ is a simplicial object together with an automorphism t_n of order $n+1$ on each A_n such that $\partial_i t = t \partial_{i-1}$ and $\sigma_i t = t \sigma_{i-1}$ for $i \neq 0$, $\partial_0 t_n = \partial_n$ and $\sigma_0 t_n = t_{n+1}^2 \sigma_n$. (Writing t instead of t_n is an abuse of notation we shall often employ for legibility.)

We will use the term "cyclic module" for a cyclic object in the category of modules. For example, there is a cyclic k-module ZR associated to every k-algebra R; $Z_n R$ is $R^{\otimes n+1}$ and the rest of the structure was described above.

Example 9.6.2 We will also use the term "cyclic set" for a cyclic object in the category of sets. For example, let G be a group. The simplicial set BG (8.1.7) may be considered as a cyclic set by defining t on $BG_n = G^n$ to be $t(g_1, \cdots, g_n) = (g_0, g_1, \cdots, g_{n-1})$, where $g_0 = (g_1 \cdots g_n)^{-1}$. Another cyclic set is ZG, which has $(ZG)_n = G^{n+1}$,

$$\partial_i(g_0, \cdots, g_n) = \begin{cases} (g_0, \cdots, g_i g_{i+1}, \cdots, g_n) & \text{if } i < n \\ (g_n g_0, g_1, \cdots, g_{n-1}) & \text{if } i = n \end{cases}$$

$$\sigma_i(g_0, \cdots, g_n) = (g_0, \cdots, g_i, 1, g_{i+1}, \cdots)$$

$$t(g_0, \cdots, g_n) = (g_n, g_0, \cdots, g_{n-1}).$$

As the notation suggests, there is a natural inclusion $BG \subset ZG$ and the free k-modules $k(ZG)_n$ fit together to form the cyclic k-module $Z(kG)$.

We now propose to construct a category ΔC containing Δ such that a cyclic object in $\mathcal{A}$ is the same thing as a contravariant functor from ΔC to $\mathcal{A}$. Recall from Chapter 8, section 1 that the simplicial category Δ has for its objects the finite (ordered) sets $[n] = \{0, 1, \cdots, n\}$, morphisms being nondecreasing monotone functions. Let t_n be the "cyclic" automorphism of the set $[n]$ defined by $t_n(0) = n$ and $t_n(j) = j - 1$ for $j \neq 0$.

Definition 9.6.3 Let $\text{Hom}_{\Delta C}([n], [p])$ denote the family of formal pairs (α, t^i), where $0 \leq i \leq n$ and $\alpha: [n] \to [p]$ is a nondecreasing monotone function. Let $\text{Hom}_C([n], [p])$ denote the family of all set maps $\varphi: [n] \to [p]$

which factor as $\varphi = \alpha t_n^i$ for some pair (α, t^i) in $\text{Hom}_{\Delta C}([n], [p])$. Note that $\varphi(i) \le \varphi(i+1) \le \cdots \le \varphi(i-1)$ in this case. Therefore the obvious surjection from $\text{Hom}_{\Delta C}([n], [p])$ to $\text{Hom}_C([n], [p])$ is almost a bijection—that is, φ uniquely determines (α, t^i) such that $\varphi = \alpha t^i$ unless φ is a constant map, in which case φ determines α ($\alpha = \varphi$) but all $n+1$ of the pairs (φ, t^i) yield the set map φ. We identify $\text{Hom}_\Delta([n], [p])$ as the subset of all pairs $(\alpha, 1)$ in $\text{Hom}_{\Delta C}([n], [p])$.

There is a subcategory C of **Sets**, containing Δ, whose objects are the sets $[n]$, $n \ge 0$, and whose morphisms are the functions in $\text{Hom}_C([n], [p])$. To see this we need only check that the composition of $\psi = \beta t_m^j$ and $\varphi = \alpha t_n^i$ is in C, and this follows from the following identities of set functions for the functions $\varepsilon_i : [n-1] \to [n]$ and $\eta_j : [n+1] \to [n]$ generating Δ (see exercise 8.1.1)

$$t_n \varepsilon_i = \begin{Bmatrix} \varepsilon_n & i = 0 \\ \varepsilon_{i-1} t_{n-1} & i > 0 \end{Bmatrix} \quad \text{and} \quad t_n \eta_i = \begin{Bmatrix} \eta_n t_{n+1}^2 & i = 0 \\ \eta_{i-1} t_{n+1} & i > 0 \end{Bmatrix}.$$

Proposition 9.6.4 (A. Connes) *The formal pairs in* $\text{Hom}_{\Delta C}([n], [p])$ *form the morphisms of a category* ΔC *containing* Δ, *the objects being the sets* $[n]$ *for* $n \ge 0$. *Moreover, a cyclic object in a category* $\mathcal{A}$ *is the same thing as a contravariant functor from* ΔC *to* $\mathcal{A}$.

Proof We need to define the composition (γ, t^k) of $(\beta, t^j) \in \text{Hom}_{\Delta C}([m], [n])$ and $(\alpha, t^i) \in \text{Hom}_{\Delta C}([n], [p])$ in such a way that if $i = j = 0$, then $(\gamma, t^k) = (\alpha\beta, 1)$. If β is not a constant set map, then the composition $t^i \beta t^j$ in C is not constant, so there is a unique (β', t^k) such that $t^i \beta t^j = \beta' t^k$; we set $(\gamma, t^k) = (\alpha\beta', t^k)$. If β is constant, we set $(\gamma, t^k) = (\alpha\beta, t^j)$. By construction, the projections from $\text{Hom}_{\Delta C}$ to Hom_C are compatible with composition; as C is a category, it follows that the $(\text{id}, 1)$ are 2-sided identity maps and that composition in ΔC is associative (except possibly for the identity $(\varphi \circ (\beta, t^j)) \circ \psi = \varphi \circ ((\beta, t^j) \circ \psi)$ when β is constant, which is easily checked). Thus ΔC is a category and $\Delta \to \Delta C \to C$ are functors. The final assertion is easily checked using the above identities for $t\varepsilon_i$ and $t\eta_j$. $\diamond$

Remark The original definition given by A. Connes in [Connes] is that $\text{Hom}_{\Delta C}([n], [p])$ is the set of equivalence classes of continuous increasing maps of degree 1 from $S^1 = \{z \in \mathbb{C} : |z| = 1\}$ to S^1 sending the $(n+1)^{st}$ roots of unity to $(p+1)^{st}$ roots of unity. Connes also observed that ΔC is isomorphic to its opposite category $(\Delta C)^{\text{op}}$. See [Loday] for more details.

Exercise 9.6.1 Show that the automorphisms of $[n]$ in ΔC form the cyclic group C_{n+1} of order $n + 1$.

Definitions 9.6.5 Let A be a cyclic object in an abelian category $\mathcal{A}$. The chain complex $C_*^h(A)$ associated to the underlying simplicial object of A (8.2.1) is called the *Hochschild complex* of A. It is traditional to write b for the differential of $C_*^h(A)$, so that $b = \partial_0 - \partial_1 + \cdots \pm \partial_n$ goes from $C_n^h(A) = A_n$ to $C_{n-1}^h(A) = A_{n-1}$. The *Hochschild homology* $HH_*(A)$ of A is the homology of $C_*^h(A)$; when $A = ZR$ (9.6.1) we will write $HH_*(R)$ for $HH_*(ZR) = H_*(R, R)$. The *acyclic complex* of A, $C_*^a(A)$, is the complex obtained from $C_*^h(A)$ by omitting the last face operator. Thus $C_n^a(A) = A_n$, and we write b' for the resulting differential $\partial_0 - \partial_1 + \cdots \mp \partial_{n-1}$ from A_n to A_{n-1}.

Exercise 9.6.2 Show the "acyclic" complex $C_*^a(A)$ is indeed acyclic. *Hint:* The path space PA (8.3.14) is a simplicial resolution of A_0.

Definition 9.6.6 (Tsygan's double complex) If A is a cyclic object in an abelian category, there is an associated first quadrant double complex $CC_{**}(A)$, first found by B. Tsygan in [Tsy], and independently by Loday and Quillen in [LQ]. The columns are periodic of order two: If p is even, the pth column is the Hochschild complex C_*^h of A; if p is odd, the pth column is the acyclic complex C_*^a of A with differential $-b'$. (The minus sign comes from the sign trick of 1.2.5.) Thus $CC_{pq}(A)$ is A_q, independently of p. The qth row of $CC_{**}(A)$ is the periodic complex associated to the action of the cyclic group C_{q+1} on A_q, in which the generator acts as multiplication by $(-1)^q t$. Thus the differential $A_q \to A_q$ is multiplication by $1 - (-1)^q t$ when p is odd; when p is even it is multiplication by the norm operator

$$N = 1 + (-1)^q t + \cdots + (-1)^{iq} t^i + \cdots + (-1)^q t^q.$$

$$\downarrow b \qquad \downarrow -b' \qquad \downarrow b \qquad \downarrow -b'$$

$$A_2 \xleftarrow{1-t} A_2 \xleftarrow{N} A_2 \xleftarrow{1-t} A_2 \xleftarrow{N}$$

$$\downarrow b \qquad \downarrow -b' \qquad \downarrow b \qquad \downarrow -b'$$

$$A_1 \xleftarrow{1+t} A_1 \xleftarrow{N} A_1 \xleftarrow{1+t} A_1 \xleftarrow{N}$$

$$\downarrow b \qquad \downarrow -b' \qquad \downarrow b \qquad \downarrow -b'$$

$$A_0 \xleftarrow{1-t} A_0 \xleftarrow{N} A_0 \xleftarrow{1-t} A_0 \xleftarrow{N}$$

Tsygan's double complex $CC_{**}(A)$

Definition 9.6.7 The *cyclic homology* $HC_*(A)$ of a cyclic object A is the homology of Tot $CC_{**}(A)$. The cyclic homology $HC_*(R)$ of an k-algebra R is the cyclic homology of the cyclic object ZR $(= R \otimes R^{\otimes *})$ of 9.6.1. In particular, $HC_0(A) = HH_0(A)$ and $HC_0(R) = R/[R, R]$.

One of the advantages of generalizing from algebras to cyclic objects is that a short exact sequence $0 \to A \to B \to C \to 0$ of cyclic objects gives rise to short exact sequences of Hochschild complexes as well as Tsygan complexes, which in turn give rise to long exact sequences

$$\cdots HH_n(A) \to HH_n(B) \to HH_n(C) \to HH_{n+1}(A) \cdots$$

$$\cdots HC_n(A) \to HC_n(B) \to HC_n(C) \to HC_{n-1}(A) \cdots.$$

Lemma 9.6.8 $CC_{**}(A)$ *is a double complex.*

Proof Set $\eta = (-1)^q$. We have to see that $b(1 - \eta t) = (1 + \eta t)b'$ and $Nb = b'N$ as maps from A_q to A_{q-1}. Now $b - b' = \eta\partial_q$ and the cyclic relations imply that $bt = \partial_q - tb'$, yielding the first relation. The cyclic relations also imply that

$$b' = \sum_{i=0}^{q-1}(-1)^i\partial_q t^{q-i} \quad \text{and} \quad b = \sum_{i=0}^{q}(-1)^{q-i}\partial_q t^i.$$

(Check this!) Since $(1 - \eta t)N = 0$, we have $t^i N = \eta^i N$ on A_q. Since $N(1 + \eta t) = 0$, we have $Nt^i = (-\eta)^i N$ on A_{q-1}. Thus

$$\eta Nb = \eta \sum_{i=0}^{q} N(\eta)^{q-i}\partial_q t^i = \eta^{q+1}N\partial_q \sum(\eta t)^i = N\partial_q N,$$

$$\eta b'N = \eta \sum_{i=0}^{q-1}(-1)^i\partial_q\eta^{q-i}N = \eta^{q+1}\sum(-\eta t)^i\partial_q N = N\partial_q N.$$

This yields the second relation, $Nb = b'N$. $\diamond$

Corollary 9.6.9 *Let* $A_n/\sim$ *denote the quotient of* A_n *by the action of the cyclic group. These form a quotient chain complex* $A_*/\sim$ *of the Hochschild complex* $C_*^h(A)$:

$$0 \leftarrow A_0 \xleftarrow{b} A_1/\sim \xleftarrow{b} A_2/\sim \xleftarrow{b} \cdots.$$

Indeed, $A_*/\sim$ *is the cokernel of the chain map* $CC_{1*} \to CC_{0*}$, *so there is a natural map from* $H_n(A_*/\sim)$ *to* $HC_n(A)$.

Remark Some authors define the cyclic homology of R to be $H_n(R^{\otimes *+1}/\sim)$, especially when $k = \mathbb{C}$. The following lemma states that their definition is equivalent to ours.

Lemma 9.6.10 *If k contains $\mathbb{Q}$, then $HC_*(A)$ may be computed as the homology of the quotient complex $A_*/\sim$ of the Hochschild complex.*

Proof Filtering Tsygan's double complex 9.6.6 by rows yields a spectral sequence starting with group homology of the cyclic groups:

$$E^1_{pq} = H_p(C_{q+1}; A_q) \Rightarrow HC_{p+q}(A).$$

The edge map from $HC_*(A)$ to the homology of $E^1_{0q} = H_0(C_{q+1}; A_q) = A_q/\sim$ arises from the augmentation $CC_{0q} \to A_q/\sim$, so the E^2 edge map maps $H_n(A_*/\sim)$ to $HC_n(A)$. In characteristic zero the group homology vanishes (6.1.10) and the spectral sequence degenerates at E^2. $\diamond$

Remark Filtering Tsygan's double complex by columns yields the even more interesting spectral sequence 9.8.6 (see exercise 9.8.2).

The three basic homomorphisms S, B, and I relating cyclic and Hochschild homology are obtained as follows. The inclusion of $C^h_*(A)$ as the column $p = 0$ in $CC_{**}(A)$ yields a map $I: HH_n(A) \to HC_n(A)$. Now let CC^{01}_{**} denote the double subcomplex of $CC_{**}(A)$ consisting of the columns $p = 0, 1$; the inclusion of $C^h_*(A)$ into CC^{01}_{**} induces an isomorphism $HH_n(A) \cong H_n \text{Tot}(CC^{01}_{**})$ because the quotient is the acyclic complex $C^a_*(A)$. The quotient double complex $CC[-2] = CC/CC^{01}$, which consists of the columns $p \geq 2$, is isomorphic to CC_{**} except that it has been translated 2 columns to the right. The quotient map $\text{Tot}(CC_{**}) \to \text{Tot}(CC[-2])$ therefore yields a map $S: HC_n(A) \to HC_{n-2}(A)$. The short exact sequence of double complexes

$$0 \to CC^{01} \xrightarrow{I} CC(A) \xrightarrow{S} CC[-2] \to 0$$

yields the map $B: HC_{n-1}(A) \to HH_n(A)$ and the following "SBI" sequence.

Proposition 9.6.11 (SBI sequence) *For any cyclic object A there is a long exact "SBI" sequence*

$$\cdots HC_{n+1}(A) \xrightarrow{S} HC_{n-1}(A) \xrightarrow{B} HH_n(A) \xrightarrow{I} HC_n(A) \xrightarrow{S} HC_{n-2}(A) \cdots.$$

In particular, there is a long exact sequence for every algebra R:

$$\cdots HC_{n+1}(R) \xrightarrow{S} HC_{n-1}(R) \xrightarrow{B} H_n(R, R) \xrightarrow{I} HC_n(R) \xrightarrow{S} HC_{n-2}(R) \cdots.$$

Remark In the literature the "SBI" sequence is also called "Connes' sequence" and the "Gysin" sequence. See exercise 9.7.4 for an explanation.

Corollary 9.6.12 *If $A \to A'$ is a morphism of cyclic objects with $HH_n(A) \overset{\cong}{\longrightarrow} HH_n(A')$, then the induced maps $HC_n(A) \to HC_n(A')$ are all isomorphisms.*

Proof This follows from induction on n via the 5-lemma and 9.6.7.   ◇

Application 9.6.13 Let R be a k-algebra. The explicit formula in 9.5.7 for the trace map $Z(M_m R) \to Z(R)$ shows that it is actually a map of cyclic k-modules. Since it induces isomorphisms on Hochschild homology, it also induces isomorphisms

$$HC_*(M_m R) \overset{\cong}{\longrightarrow} HC_*(R).$$

Exercise 9.6.3 For $m < n$, show that the nonunital inclusion $\iota \colon M_m(R) \hookrightarrow M_n(R)$ of exercise 9.5.3 induces a cyclic map $ZM_m(R) \to ZM_n(R)$, which in turn induces isomorphisms

$$\iota_* \colon HC_* M_m(R) \cong HC_* M_n(R).$$

Example 9.6.14 Since $H_n(k, k) = 0$ for $n \neq 0$, the SBI sequence quickly yields

$$HC_n(k) = \begin{cases} k & \text{if } n \text{ is even} \\ 0 & \text{if } n \text{ is odd,} \end{cases}$$

with the maps $S \colon HC_{n+2}(k) \to HC_n(k)$ all isomorphisms. The same calculation applies for any finite separable algebra R over a field k because we saw in 9.2.11 that $H_n(R, R) = 0$ for all $n \neq 0$.

HC₁ 9.6.15 The SBI sequence interprets $HC_1(R)$ as a quotient of $H_1(R, R)$:

$$H_0(R, R) \overset{B}{\longrightarrow} H_1(R, R) \to HC_1(R) \to 0.$$

Now suppose that R is commutative, so that $H_0(R, R) = R$ and $H_1(R, R) = \Omega_{R/k}$. The map $B \colon R \to \Omega_{R/k}$ maps $r \in R$ to dr. (Check this!) Therefore we may identify B with d and make the identification

$$HC_1(R) \cong \Omega_{R/k}/(dR).$$

Example 9.6.16 Since $H_n(k[x], k[x]) = 0$ for $n \geq 2$, the $S: HC_{n+2}(k[x]) \to HC_n(k[x])$ are isomorphisms for all $n \geq 1$ and there is an exact sequence

$$0 \to HC_2(k[x]) \xrightarrow{S} k[x] \xrightarrow{d} \Omega_{k[x]/k} \xrightarrow{I} HC_1(k[x]) \to 0.$$

If k contains $\mathbb{Q}$, then $x^n dx = d(x^{n+1}/n + 1)$ for all $n \geq 0$, so d is onto and $HC_1(k[x]) = 0$. This yields the calculation

$$HC_n(k[x]) = \begin{cases} k[x] & \text{if } n = 0 \\ k & \text{if } n \geq 2 \text{ is even} \\ 0 & \text{if } n \geq 1 \text{ is odd}. \end{cases}$$

Similar remarks pertain to the Laurent polynomial ring $k[x, x^{-1}]$, except that the map $d: k[x, x^{-1}] \to \Omega_{k[x,x^{-1}]/k} \cong k[x, x^{-1}]$ has cokernel k (on dx/x) when $\mathbb{Q} \subseteq k$. Thus when $\mathbb{Q} \subseteq k$ we have

$$HC_n(k[x, x^{-1}]) \cong k \quad \text{for all} \quad n \geq 1.$$

Remark We will compute $HC_*(R)$ for a smooth algebra R in 9.8.11 and 9.8.12 in terms of de Rham cohomology.

Exercise 9.6.4 Consider the truncated polynomial ring $R = k[x]/(x^{n+1})$ over a field k of characteristic 0. We saw in exercise 9.1.4 that $\dim_k H_i(R, R) = n$ for $i > 0$. Show explicitly that $HC_1(R) = 0$. Then use the SBI sequence to show that $HC_i(R) = 0$ for all odd i, while for even $i \neq 0$ $HC_i(R) \cong HC_i(k) \oplus H_i(R, R) \cong k^{n+1}$. Another approach will be given in exercise 9.9.2.

9.6.1 Variations: HP and HN

9.6.17 We may use the periodicity of Tsygan's first quadrant double complex $CC_{**}(A)$ to extend it to the left, obtaining an upper half-plane double complex $CC_{**}^P(A)$. (See 9.6.6.) The *periodic cyclic homology* of A is the homology of the product total complex

$$HP_*(A) = H_* \text{Tot}^\Pi(CC_{**}^P(A)).$$

If we truncate CC_{**}^P to the left of the $2p^{th}$ column, we obtain Tsygan's double complex 9.6.6 translated $2p$ times. These truncations $\{CC_{**}[-2p]\}$ form a tower of double chain complexes in the sense of Chapter 3, section 5. The homology of this tower of double complexes is the tower of k-modules

$$\cdots \xrightarrow{S} HC_{n+4}(A) \xrightarrow{S} HC_{n+2}(A) \xrightarrow{S} HC_n(A).$$

As we saw in 3.5.8, this means that there is an exact sequence

$$0 \to \varprojlim{}^1 HC_{n+2p+1}(A) \to HP_n(A) \to \varprojlim HC_{n+2p}(A) \to 0.$$

Moreover, it is visually clear from the periodicity of $CC^P_{**}(A)$ that each map $S: HP_{n+2}(A) \to HP_n(A)$ is an isomorphism. This accounts for the name "periodic cyclic homology": the modules $HP_n(A)$ are periodic of order 2.

Similarly, we can consider the "negative" subcomplex $CC^N_{**}(A)$ of the periodic complex $CC^P_{**}(A)$ consisting of the columns with $p \leq 0$. This is a second quadrant double complex. The *negative cyclic homology* of A is defined to be the homology of the product total complex of $CC^N_{**}(A)$:

$$HN_*(A) = H_* \operatorname{Tot}^\Pi(CC^N_{**}(A)).$$

We leave it to the reader to check that there is an SBI exact sequence 9.6.11 for $I: HN_* \to HP_*$ fitting into the following commutative diagram:

$$
\begin{array}{ccccccccc}
HP_{n+1}(A) & \xrightarrow{S} & HC_{n-1}(A) & \xrightarrow{B} & HN_n(A) & \xrightarrow{I} & HP_n(A) & \xrightarrow{S} & HC_{n-2}(A) \quad \cdots \\
\downarrow{\scriptstyle S} & & \downarrow{\scriptstyle =} & & \downarrow & & \downarrow{\scriptstyle S} & & \downarrow{\scriptstyle =} \\
HC_{n+1}(A) & \xrightarrow{S} & HC_{n-1}(A) & \xrightarrow{B} & HH_n(A) & \xrightarrow{I} & HC_n(A) & \xrightarrow{S} & HC_{n-2}(A) \quad \cdots \\
\downarrow{\scriptstyle B} & & & & \downarrow{\scriptstyle I} & & \downarrow{\scriptstyle B} & & \\
HN_n(A) & & & & HN_{n-1}(A) & = & HN_{n-1}(A) & &
\end{array}
$$

9.7 Group Rings

In this section we fix a commutative ring k and a group G. Our goal is to compute HH_* and HC_* of the group ring kG (9.7.5 and 9.7.9). To prepare for this we calculate HC_* of kBG, which we call $HC_*(G)$.

In 9.6.2 we saw that BG could be regarded as a cyclic set by defining $t(g_1, \cdots, g_n) = ((g_1 \cdots g_n)^{-1}, g_1, \cdots, g_{n-1})$. Applying the free k-module functor to BG therefore yields a cyclic k-module kBG. If we adopt the notation $HH_*(G) = HH_*(kBG)$, $HC_*(G) = HC_*(kBG)$, and so on, then we see (using 8.2.3) that

$$HH_n(G) = \pi_n(kBG) = H_n(BG; k) = H_n(G; k).$$

Theorem 9.7.1 (Karoubi) *For each group G,*

$$HC_n(G) \cong H_n(G; k) \oplus H_{n-2}(G; k) \oplus H_{n-4}(G; k) \oplus \cdots.$$

Moreover, the maps $S: HC_n(G) \to HC_{n-2}(G)$ are the natural projections with kernel $H_n(G; k)$, and the maps B are zero.

Remark It is suggested to write Karoubi's Theorem in the form $HC_*(G) \cong H_*(G; k) \otimes HC_*(k)$.

Proof Consider the path space $EG = P(BG)$ of BG (8.3.14 and exercise 8.3.8), which as a simplicial set has $(EG)_n = G^{n+1}$ and $\partial_i(g_0, \cdots, g_n) = (\cdots, g_i g_{i+1}, \cdots)$ for $i \neq n$ and $\partial_n(g_0, \cdots, g_n) = (g_0, \cdots, g_{n-1})$. If we define

$$t(g_0, \cdots, g_n) = (g_0 \cdots g_n, (g_1 \cdots g_n)^{-1}, g_1, g_2, \cdots, g_{n-1}),$$

then the cyclic identities ($t^{n+1} = 1$, $\partial_i t = t \partial_{i-1}$, etc.) are readily verified. (Do so!) Therefore EG is also a cyclic set, and the projection $\pi: EG \to BG$, which forgets g_0, is a morphism of cyclic sets. Applying the free k-module functor, $\pi: kEG \to kBG$ is a morphism of cyclic k-modules. More is true: The group G acts on EG by $g(g_0, g_1, \cdots) = (gg_0, g_1, \cdots)$ in a way that makes kEG into a cyclic left kG-module, and $kBG = k \otimes_{kG} kEG$. In particular, Tsygan's double complex $CC_{**}(kEG)$ is a double complex of free kG-modules and $CC_{**}(kBG) = k \otimes_{kG} CC_{**}(kEG)$. It follows that $HC_*(G) = H_* \mathrm{Tot}(CC_{**}(kBG))$ is the hyperhomology $\mathbb{H}_*(G; \mathrm{Tot} CC_{**}(kEG))$ of the group G (6.1.15), because each summand $CC_{pq}(kEG)$ of $\mathrm{Tot} CC_{**}(kEG)$ is a free (hence flat) kG-module.

We saw in exercise 8.3.7 that the augmentation $EG \to 1$ is a simplicial homotopy equivalence. Applying the free module functor, the augmentation $kEG \to k$ is a simplicial homotopy equivalence. Hence $C_*^h(kEG)$ is a resolution of the trivial kG-module k, just as $C_*^a(kEG)$ is a resolution of the kG-module 0. Fitting these together, Tsygan's double complex $CC_{**}(kEG)$ is a "resolution" (in the sense of hyperhomology) of the trivial chain complex

$$K_*: 0 \leftarrow k \leftarrow 0 \leftarrow k \leftarrow 0 \leftarrow k \leftarrow \cdots$$

which has $K_i = 0$ for $i < 0$ or i odd and $K_i = k$ for i even, $i \geq 0$. But the hyperhomology of K_* is easy to compute:

$$HC_n(G) = \mathbb{H}_n(G; K_*) = \bigoplus_{i=0}^{\infty} \mathbb{H}_{n-2i}(G; k) = \bigoplus H_{n-2i}(G; k).$$

The assertions that the maps $S: HC_n(G) \to HC_{n-2}(G)$ are the natural projections with kernel $HH_n(G) = H_n(G; k)$, and that the maps $B: HC_{n-1}(G) \to HH_n(G)$ are thus all zero, follow from a visual inspection of $\mathbb{H}_*(G; K_*)$. $\diamond$

Corollary 9.7.2

$$HP_n(G) = \varprojlim HC_{n+2i}(G) = \begin{cases} \prod_{i=0}^{\infty} H_{2i}(G; k), & n \text{ even} \\ \prod_{i=0}^{\infty} H_{2i+1}(G; k), & n \text{ odd} \end{cases}$$

Exercise 9.7.1 When $\mathbb{Q} \subset k$, use $kBG/\sim$ to compute $HC_*(G)$.

We now turn to the Hochschild homology of the group ring kG. Let $<G>$ denote the set of conjugacy classes of elements of G. Our first step is to find a decomposition of the cyclic set ZG of 9.6.2 and the cyclic module $Z(kG) = k(ZG)$ which is indexed by $<G>$. There is a cyclic set map from ZG to the trivial cyclic set $<G>$, which sends $(g_0, g_1, \cdots, g_n) \in (ZG)_n = G^{n+1}$ to the conjugacy class of the product $g_0 \cdots g_n$ in $<G>$. (Check this!) For $n = 0$ this yields an isomorphism

$$HC_0(kG) = HH_0(kG) \xrightarrow{\cong} k <G> = \bigoplus_{<G>} k.$$

Indeed, the kernel of the surjection $kG \to k<G>$ is generated by the elements $x - gxg^{-1} = g^{-1}(gx) - (gx)g^{-1} = b(g^{-1} \otimes gx)$, and $HC_0(k<G>) = k<G>$.

Definition 9.7.3 For $x \in G$, let $Z_n(G, x)$ denote the subset of $G^{n+1} = Z_n G$ consisting of all $(g_0, \cdots, g_n)$ such that $g_0 \cdots g_n$ is conjugate to x, that is, $Z_n(G, x)$ is the inverse image of $<x> \in <G>$. As n varies, these form a cyclic subset $Z(G, x)$ of ZG. Note that $Z(G, 1)$ is isomorphic to the cyclic set BG (forget g_0). Applying the free k-module functor gives cyclic k-submodules $kZ(G, x)$ of $kZ(G)$, one for each conjugacy class. We shall write $HH_*(G, x)$ for $HH_*(kZ(G, x))$, $HC_*(G, x)$ for $HC_*(kZ(G, x))$, etc. for simplicity. As $Z(G)$ is the disjoint union of the cyclic sets $Z(G, x)$, $kZ(G)$ is the direct sum of the $kZ(G, x)$. Therefore $HH_*(kG) \cong \bigoplus_x HH_*(G, x)$ and $HC_*(kG) \cong \bigoplus_x HC_*(G, x)$.

To describe $HH_*(G, x)$ etc. we recall that the *centralizer subgroup* of $x \in G$ is the subgroup $C_G(x) = \{g \in G : gxg^{-1} = x\}$. If x' is conjugate to x, then $C_G(x')$ and $C_G(x)$ are conjugate subgroups of G. In fact, if we let G act on itself by conjugation, then $C_G(x)$ is the stabilizer subgroup of x; if we choose a set $\{y\}$ of coset representatives for $G/C_G(x)$, then for each x' conjugate to x there is a unique coset representative y such that $yx'y^{-1} = x$.

Proposition 9.7.4 *For each $x \in G$ the inclusion $C_G(x) \subseteq G$ induces isomorphisms $HH_*(C_G(x), x) \cong HH_*(G, x)$ and $HC_*(C_G(x), x) \cong HC_*(G, x)$.*

Proof Write H for $C_G(x)$, and choose a set $\{y\}$ of coset representatives for G/H, the coset of H being represented by $y = 1$. Given $(g_0, \cdots, g_n) \in Z_n(G, x)$, let y_i be the (unique) coset representative such that $y_i(g_{i+1} \cdots g_n g_0 \cdots g_i)y_i^{-1} = x$ and set

$$\rho(g_0, \cdots, g_n) = (y_n g_0 y_0^{-1}, y_0 g_1 y_1^{-1}, \cdots, y_{i-1} g_i y_i^{-1}, \cdots, y_{n-1} g_n y_n^{-1}).$$

Each $y_{i-1} g_i y_i^{-1}$ is in H (check this!), so $\rho(g_0, \cdots, g_n) \in Z_n(H, x)$. By inspection, $\rho: Z(G, x) \to Z(H, x)$ is a cyclic morphism splitting the inclusion $\iota: Z(H, x) \hookrightarrow Z(G, x)$. There is a simplicial homotopy h from the identity map of $Z(G, x)$ to $\iota\rho$ defined by

$$h_j(g_0, \cdots, g_n) = (g_0 y_0^{-1}, y_0 g_1 y_1^{-1}, \cdots, y_{j-1} g_j y_j^{-1}, y_j, g_{j+1}, \cdots, g_n),$$

$j = 0, \cdots, n$. (Check this!) Hence the inclusion $Z(H, x) \subseteq Z(G, x)$ is a simplicial homotopy equivalence. This implies that $kZ(H, x) \subseteq kZ(G, x)$ is also a homotopy equivalence. Hence $HH_*(H, x) = \pi_* kZ(H, x)$ is isomorphic to $HH_*(G, x) = \pi_* kZ(G, x)$, which in turn implies that $HC_*(H, x) \cong HC_*(G, x)$. $\diamond$

Corollary 9.7.5 *For each $x \in G$, $HH_*(G, x) \cong H_*(C_G(x); k)$. Hence*

$$HH_*(kG) = \bigoplus_{x \in <G>} H_*(C_G(x); k).$$

Proof We have to show that $HH_*(C_G(x), x)$ is isomorphic to $H_*(C_G(x); k)$ for each x, so suppose x is in the center of G. There is an isomorphism $Z(G, 1) \to Z(G, x)$ of simplicial sets given by $(g_0, \cdots, g_n) \mapsto (xg_0, g_1, \cdots, g_n)$. Therefore $H_*(G; k) = HH_*(kBG) \cong HH_*(G, 1)$ is isomorphic to $HH_*(G, x)$. $\diamond$

Remark One might naively guess from the above calculation that $HC_*(kG)$ would be the sum of the modules $HC_*(C_G(x)) = H_*(C_G(x); k) \otimes HC_*(k)$. However, when G is the infinite cyclic group T and $\mathbb{Q} \subseteq k$, we saw in 9.6.16 that for $n \geq 1$

$$HC_n(kT) = HC_n(k[t, t^{-1}]) \cong k \cong HC_n(T).$$

Therefore if $\mathbb{Q} \subseteq k$, then for all $x \neq 1$ in T we have $HC_n(T, x) = 0, n \neq 0$.

Exercise 9.7.2 Show that $t^{n-1} \otimes t \in Z_1(kT, t^n)$ represents the differential $t^{n-1}dt$ in $HH_1(kT) \cong \Omega_{kT/k}$, and use this to conclude that for general k.

$$HC_i(T, t^n) \cong \begin{cases} k, & i = 0 \\ k/nk, & i \geq 1 \text{ odd} \\ \mathrm{Tor}(k, \mathbb{Z}/n), & i \geq 2 \text{ even.} \end{cases}$$

Lemma 9.7.6 *If $\mathbb{Q} \subseteq k$ and $x \in G$ is a central element of finite order, then*

$$HC_*(G, x) \cong HC_*(G) \cong H_*(G; k) \otimes HC_*(k).$$

Proof Let $\bar{G}$ denote the quotient of G by the subgroup $\{x\}$ generated by x, and write $\bar{g}$ for the image of $g \in G$ in $\bar{G}$. The map of cyclic sets $Z(G, x) \to Z(\bar{G}, 1)$ sending $(g_0, \cdots, g_n)$ to $(\bar{g}_0, \cdots, \bar{g}_n)$ induces the natural map from $H_*(G; k) \cong HH_*(G, x)$ to $H_*(\bar{G}; k) \cong HH_*(\bar{G}, 1)$, because its composition with the simplicial isomorphism $Z(G, 1) \to Z(G, x)$ is the natural quotient map. The Hochschild-Serre spectral sequence $E_{pq}^2 = H_p(\bar{G}; H_q(\{x\}; k)) \Rightarrow H_{p+q}(G; k)$ degenerates since $\mathbb{Q} \subset k$ (6.1.10) to show that the natural map $H_p(G; k) \to H_p(\bar{G}; k)$ is in fact an isomorphism. This yields $HC_*(G) \cong HC_*(\bar{G})$ by Karoubi's Theorem 9.7.1, as well as $HC_*(G, x) \cong HC_*(\bar{G}, 1) \cong HC_*(\bar{G})$. $\diamond$

Corollary 9.7.7 *If $\mathbb{Q} \subseteq k$ and G is a finite group, then*

$$HC_*(kG) \cong \bigoplus_{x \in <G>} HC_*(C_G(x)) \cong k < G > \otimes HC_*(k).$$

Remark When k is a field of characteristic zero, Maschke's Theorem states that kG is a semisimple (hence separable) k-algebra. In 9.2.11 we saw that this implied that $HH_n(kG) = 0$ for $n \neq 0$, so the SBI sequence yields an alternate proof of this corollary.

Example 9.7.8 $(G = C_2)$ Things are more complicated for general k, even when G is the cyclic group $C_2 = \{1, x\}$ of order 2. For example, when $k = \mathbb{Z}$ the group $HC_n(C_2, x)$ is $\mathbb{Z}$ for n even and 0 for n odd, which together with Karoubi's Theorem for $HC_*(C_2)$ yields

$$HC_n(\mathbb{Z}C_2) = \begin{cases} \mathbb{Z} \oplus \mathbb{Z}, & n \text{ even} \\ (\mathbb{Z}/2)^{(n+1)/2}, & n \text{ odd .} \end{cases}$$

This calculation may be found in {G. Cortiñas, J. Guccione, and O. Villamayor, "Cyclic homology of $K[\mathbb{Z}/p\mathbb{Z}]$," *K-theory* 2 (1989), 603–616}.

Exercise 9.7.3 (Kassel) Set $k = \mathbb{Z}$ and show that $HP_n(\mathbb{Z}C_2)$ is not the inverse limit of the groups $HC_{n+2i}(\mathbb{Z}C_2)$ by showing that

$$HP_0(C_2, x) \cong \varprojlim{}^1 HC_{2i+1}(C_2, x) \cong \hat{\mathbb{Z}}_2/\mathbb{Z},$$

where $\hat{\mathbb{Z}}_2$ denotes the 2-adic integers. *Hint:* Show that the SBI sequence breaks up, conclude that S is multiplication by 2, and use 3.5.5.

Theorem 9.7.9 (Burghelea) *Suppose that $\mathbb{Q} \subset k$. Then $HC_*(kG)$ is the direct sum of*

$$\bigoplus_{\substack{x \in <G> \\ \text{finite order}}} HC_*(C_G(x)) \cong \bigoplus_{\substack{x \in <G> \\ \text{finite order}}} H_*(C_G(x); k) \otimes HC_*(k)$$

and

$$\bigoplus_{\substack{x \in <G> \\ \text{infinite order}}} H_*(W(x); k).$$

Here $W(x)$ denotes the quotient group $C_G(x)/\{x^n\}$.

Proof We have already seen that $HC_*(kG)$ is the direct sum over all x in $<G>$ of the groups $HC_*(C_G(x), x)$, and that if x has finite order this equals $HC_*(C_G(x))$. Therefore it remains to suppose that $x \in G$ is a central element of infinite order and prove that $HC_*(G, x) \cong H_*(G/T; k)$, where T is the subgroup of G generated by x. For this, we pull back the path space $E(G/T)$ of 9.7.1 to $Z(G, x)$.

Let E be the cyclic subset of $E(G/T) \times Z(G, x)$ consisting of all pairs (e, z) which agree in $B(G/T)$. Forgetting the redundant first coordinates of e and z, we may identify E_n with $(G/T) \times G^n$ in such a way that (for $\bar{g}_0 \in G/T, g_1 \in G$):

$$\partial_i(\bar{g}_0, g_1, \cdots, g_n) = \begin{cases} (\bar{g}_0\bar{g}_1, g_2, \ldots, g_n), & i = 0 \\ (\bar{g}_0, \ldots, g_i g_{i+1}, \ldots), & 0 < i < n \\ (\bar{g}_0, g_1, \ldots, g_{n-1}), & i = n \end{cases}$$

$$t(\bar{g}_0, g_1, \cdots, g_n) = (\bar{g}_0 \cdots \bar{g}_n, (g_1 \cdots g_n)^{-1}, g_1, \cdots, g_{n-1}).$$

As in the proof of Karoubi's theorem 9.7.1, the action of G/T on the $\bar{g}_0$ coordinate makes E into a cyclic G/T-set and makes the morphism of cyclic sets $\pi: E \to Z(G, x)$ into a principal G/T-fibration (exercise 8.2.6). Therefore $kZ(G, x) = k \otimes_{kG/T} kE$, Tsygan's double complex $CC_{**}(kE)$ consists of free kG/T-modules and $CC_{**}kZ(G, x) = k \otimes_{kG/T} CC_{**}(kE)$. We will prove that $\mathrm{Tot} CC_{**}(kE)$ is a free kG/T-module resolution of k, so that

$$HC_*(G, x) = H_*(k \otimes_{kG/T} \mathrm{Tot} CC_{**}(kE)) \cong H_*(G/T; k).$$

The homotopy sequence for the principal G/T-fibration $E \to Z(G, x)$ (exercise 8.2.6 and 8.3.5) shows that $\pi_i(E) = 0$ for $i \neq 1$ and $\pi_1(E) \cong T$. The natural cyclic map $Z(T, x) \to E$, which sends $(t_0, \cdots, t_n) \in T^{n+1}$ to $(1, t_1, \cdots, t_n) \in E_n = (G/T) \times G^n$ induces isomorphisms on simplicial homotopy groups and therefore on simplicial homology (see 8.2.3). That is, $HH_*(T, x) \cong HH_*(kE)$. It follows that if $\mathbb{Q} \subseteq k$, then

$$HC_n(kE) \cong HC_n(T, x) = \begin{cases} k & n = 0 \\ 0 & n \neq 0. \end{cases}$$

Hence the natural map from $CC_{00}(kE) = kG/T$ to $k = HC_0(kE)$ provides the augmentation making $\mathrm{Tot}\, CC_{**}(kE) \to k$ into a free kG/T-resolution of k, as claimed. $\diamond$

Exercise 9.7.4 Show that the SBI sequence for $Z(G, x)$ may be identified with the Gysin sequence of 6.8.6:

$$\cdots H_n(G; k) \xrightarrow{\mathrm{coinf}} H_n(G/T; k) \to H_{n-2}(G/T; k) \to H_{n-1}(G; k) \cdots.$$

Hint: Compare $C^h_*(G, x) \to CC_{**}(G, x)$ to the coinflation map for $G \to G/T$.

9.8 Mixed Complexes

We can eliminate the odd (acyclic) columns in Tsygan's double complex 9.6.6 $CC_{**}(A)$, and obtain a double complex $\mathcal{B}_{**}(A)$ due to A. Connes. To do this, fix the chain contraction $s_n = t\sigma_n : A_n \to A_{n+1}$ of the acyclic complex $C^a_*(A)$ and define $B : A_n \to A_{n+1}$ to be the composite $(1 + (-1)^n t) s N$, where N is the norm operator on A_n. (Exercise: Show that s is a chain contraction.) Setting $\eta = (-1)^n$, we have

$$B^2 = (1 - \eta t) s N (1 + \eta t) s N = 0$$

$$bB + Bb = b(1 + \eta t) s N + (1 - \eta t) s N b = (1 - \eta t)(b's + sb')N$$

$$= (1 - \eta t) N = 0.$$

Connes' double complex $\mathcal{B}_{**}(A)$ is formed using b and B as vertical and horizontal differentials, with $\mathcal{B}_{pq} = A_{q-p}$ for $p \geq 0$. We can formalize this construction as follows.

$$
\begin{array}{ccccccc}
\cdots & & & & \cdots & & \\
M_3 & \xleftarrow{B} & M_2 & \xleftarrow{B} & M_1 & \xleftarrow{B} & M_0 \\
\downarrow{\scriptstyle b} & & \downarrow{\scriptstyle b} & & \downarrow{\scriptstyle b} & & \\
M_2 & \xleftarrow{B} & M_1 & \xleftarrow{B} & M_0 & & \\
\downarrow{\scriptstyle b} & & \downarrow{\scriptstyle b} & & & & \\
M_1 & \xleftarrow{B} & M_0 & & & & \\
\downarrow{\scriptstyle b} & & & & & & \\
M_0 & & & & & &
\end{array}
$$

Definition 9.8.1 (Kassel) A *mixed complex* (M, b, B) in an abelian category $\mathcal{A}$ is a graded object $\{M_m : m \geq 0\}$ endowed with two families of morphisms $b : M_m \to M_{m-1}$ and $B : M_m \to M_{m+1}$ such that $b^2 = B^2 = bB + Bb = 0$. Thus a mixed complex is both a chain and a cochain complex.

The above calculation shows that every cyclic object A gives rise to a mixed complex (A, b, B), where A is considered as a graded object, b is the Hochschild differential on A and B is the map constructed as above.

Definition 9.8.2 (Connes' double complex) Let (M, b, B) be a mixed complex. Define a first quadrant double chain complex $\mathcal{B}_{**}(M)$ as follows. $\mathcal{B}_{pq}$ is M_{q-p} if $0 \leq p \leq q$ and zero otherwise. The vertical differentials are the b maps, and the horizontal differentials are the B maps.

We write $H_*(M)$ for the homology of the chain complex (M, b), and $HC_*(M)$ for the homology of the total complex $\mathrm{Tot}(\mathcal{B}_{**}(M))$. $HC_*(M)$ is called the *cyclic homology* of the mixed complex (M, b, B), a terminology which is justified by the following result.

Proposition 9.8.3 *If A is a cyclic object, then $HC_*(A)$ is naturally isomorphic to the cyclic homology of the mixed complex (A, b, B).*

Proof For each $0 \leq p \leq q$, set $t = q - p$ and map $\mathcal{B}_{pq} = A_t$ to $CC_{2p,t} \oplus CC_{2p-1,t+1} = A_t \oplus A_{t+1}$ by the map $(1, sN)$. The direct sum over p, q gives a morphism of chain complexes $\mathrm{Tot}(\mathcal{B}_{**}) \to \mathrm{Tot}(CC_{**})$. (Check this!) These two complexes compute $HC_*(A, b, B)$ and $HC_*(A)$, respectively by 9.8.2 and 9.6.6; we have to see that this morphism is a quasi-isomorphism. For this we filter $\mathcal{B}_{**}$ by columns and select the "double column" filtration for $CC_{**}: F_p CC = \oplus\{CC_{st} : t \le 2p\}$. The morphism $\mathrm{Tot}(\mathcal{B}_{**}) \to \mathrm{Tot}(CC_{**})$ is filtration-preserving, so it induces a morphism of the corresponding spectral sequences 5.4.1. To compare these spectral sequences we must compute the E^1 terms. Clearly $E^1_{pq}(\mathcal{B}) = H_{q-p}(A)$. Let T_p denote the total complex of the 2-column double complex obtained from the $(2p-1)^{st}$ and $(2p)^{th}$ columns of CC_{**}; the degree $p+q$ part of T_p is $CC_{2p,q-p} \oplus CC_{2p-1,q-p+1}$. The translates (1.2.8) of $C^a_*(A)$ and $C^h_*(A)$ fit into a short exact sequence $0 \to C^a_*(A)[1-2p] \to T_p \to C^h_*(A)[-2p] \to 0$, so the spectral sequence 5.4.1 associated to the double column filtration of CC has $E^0_{pq} = (T_p)_{p+q}$ and

$$E^1_{pq}(CC) = H_{p+q}(T_p) \cong H_{p+q}(C^h_*(A)[-2p]) \cong H_{q-p}(A).$$

By inspection, the map $E^1_{pq}(\mathcal{B}) \to E^1_{pq}(CC)$ is an isomorphism for all p and q. By the Comparison Theorem (5.2.12), $\mathrm{Tot}(\mathcal{B}) \to \mathrm{Tot}(CC)$ is a quasi-isomorphism. $\diamond$

Remark If A is a cyclic object, any other choice of the chain contraction s, such as $s_n = (-1)^n \sigma_n$, will yield a slightly different mixed complex $M = (A, b, B')$. The proof of the above proposition shows that we would still have $HC_*(M) \cong HC_*(A)$. Our choice is dictated by the next application and by the historical selection $s(r_0 \otimes \cdots \otimes r_n) = 1 \otimes r_0 \otimes \cdots \otimes r_n$ for $A = ZR$ in [LQ].

Application 9.8.4 (Normalized mixed complex) By the Dold-Kan Theorem 8.4.1, the Hochschild homology of a cyclic k-module A may be computed using either the unnormalized chain complex $C^h_*(A)$ or the normalized chain complex $\bar{C}_*(A) = C^h_*(A)/D_*(A)$, obtained by modding out by the degenerate subcomplex $D_*(A)$. Since $D_*(A)$ is preserved by t (why?) as well as our choice of s, it is preserved by $B = (1 \pm t)s(\sum \pm t^i)$. Hence B passes to the quotient complex $\bar{C}_*(A)$, yielding a mixed complex $(\bar{C}_*(A), b, B)$. Since the morphism of mixed complexes from (A, b, B) to $(\bar{C}_*(A), b, B)$ induces an isomorphism on homology, it follows (say from the SBI sequence 9.8.7 below) that it also induces an isomorphism on cyclic homology: $HC_*(A) \cong HC_*(\bar{C}_*(A))$.

One advantage of the normalized mixed complex is that it simplifies the expression for $B = (1 \pm t)sN$. Since $ts = t^2\sigma_n = \sigma_0 t = 0$ on $\bar{C}_n(A)$, we have

$$B = t\sigma_n N = t\sigma_n + (-1)^n t^2 \sigma_{n-1} + \cdots + (-1)^{ni} t^{i+1} \sigma_{n-i} + \cdots + (-1)^n t^{n+1} \sigma_0.$$

In particular, if R is a k-algebra and $A = ZR$, then in $\bar{C}_n(A) = B_n(R, R)$:

$$B(r_0 \otimes \cdots \otimes r_n) = \sum_{i=0}^{n} (-1)^{in} \otimes r_i \otimes \cdots \otimes r_n \otimes r_0 \otimes \cdots \otimes r_{i-1}.$$

Example 9.8.5 (Tensor algebra) Let $T = T(V)$ be the tensor algebra (7.3.1) of a k-module V. If $v_1, \cdots, v_j \in V$, write $(v_1 \cdots v_j)$ for their product in the degree j part $V^{\otimes j}$ of T; the generator σ of the cyclic group C_j acts on $V^{\otimes j}$ by $\sigma(v_1 \cdots v_j) = (v_j v_1 \cdots v_{j-1})$. In 9.1.6 we saw that $H_i(T, T) = 0$ for $i \neq 0$, so to use Connes' double complex 9.8.2 it suffices to describe the map

$$B: H_0(T, T) = \bigoplus (V^{\otimes j})_\sigma \to \bigoplus (V^{\otimes j})^\sigma = H_1(T, T).$$

Of course the definition of $B: T \to T \otimes T$ yields $B(r) = 1 \otimes r + r \otimes 1$ for every $r \in R$. If we modify this by elements of the form $b(r_0 \otimes r_1 \otimes r_2) = r_0 r_1 \otimes r_2 - r_0 \otimes r_1 r_2 + r_2 r_0 \otimes r_1$ we obtain a different representative of the same element of $H_1(T, T)$. Thus for $r = (v_1 \cdots v_j)$ we have

$$B(r) = r \otimes 1 + 1 \otimes r \sim v_1 \otimes (v_2 \cdots v_j) + (v_2 \cdots v_j) \otimes v_1 + r \otimes 1$$

$$\sim (v_1 v_2) \otimes (v_3 \cdots v_j) + (v_3 \cdots v_j v_1) \otimes v_2$$

$$+ (v_2 \cdots v_j) \otimes v_1 + r \otimes 1$$

$$\sim \sum (v_{i+1} \cdots v_j v_1 \cdots v_{i-1}) \otimes v_i + r \otimes 1.$$

Upon identifying the degree j part of $T \otimes V$ with $V^{\otimes j}$ and ignoring the degenerate term $r \otimes 1$ by passing to $\bar{C}_*$, we see that $B(r) = (1 + \sigma + \cdots + \sigma^{j-1})r$ as a map from $(V^{\otimes j})_\sigma$ to $(V^{\otimes j})^\sigma$. Identifying B with the norm map for the action of C_j on $V^{\otimes j}$, we see from Connes' complex and 6.2.2 that

$$HC_n(T) = HC_n(k) \oplus \bigoplus_{j=1}^{\infty} H_n(C_j; V^{\otimes j}).$$

In particular, if $\mathbb{Q} \subseteq k$, then $HC_n(T) = HC_n(k)$ for all $n \neq 0$.

Exercise 9.8.1 If R has an ideal I with $I^2 = 0$ and $R/I \cong k$, show that

$$HC_n(R) = HC_n(k) \oplus \bigoplus_{j=1}^{n+1} H_{n+1-j}(C_j; I^{\otimes j}).$$

Connes' Spectral Sequence 9.8.6 The increasing filtration by columns on $\mathcal{B}_{**}(M)$ gives a spectral sequence converging to $HC_*(M)$, as in 5.6.1. Since the pth column is the translate $M[-p]$ of $(M, \pm b)$, we have

$$E^1_{pq} = H_{q-p}(M) \Rightarrow HC_{p+q}(M)$$

with d^1 differential $H_i(M) \to H_{i+1}(M)$ induced by Connes' operator B. This quickly yields $HC_0(M) = H_0(M)$, $HC_1(M) = H_1(M)/B(M_0)$ and a sequence of low degree terms

$$H_1(M) \xrightarrow{B} H_2(M) \xrightarrow{I} HC_2(M) \to H_0(M) \xrightarrow{B} H_1(M) \xrightarrow{I} HC_1(M) \to 0.$$

In order to extend this sequence to the left, it is convenient to proceed as follows. The inclusion of M_* as the column $p = 0$ of $\mathcal{B} = \mathcal{B}_{**}(M)$ yields a short exact sequence of chain complexes

$$0 \to M_* \xrightarrow{I} \mathrm{Tot}(\mathcal{B}) \xrightarrow{S} \mathrm{Tot}(\mathcal{B})[-2] \to 0,$$

since $\mathcal{B}/M_*$ is the double complex obtained by translating $\mathcal{B}$ up and to the right. The associated long exact sequence in homology is what we sought:

$$\cdots HC_{n+1}(M) \xrightarrow{S} HC_{n-1}(M) \xrightarrow{B} H_n(M) \xrightarrow{I} HC_n(M) \xrightarrow{S} HC_{n-2}(M) \cdots .$$
(9.8.7)

We call this the "SBI sequence" of the mixed complex M, since the proof of 9.8.3 above shows that when $M = (A, b, B)$ is the mixed complex of a cyclic object A this sequence is naturally isomorphic to the SBI sequence of A constructed in 9.6.11. As in *loc. cit.*, if $M \to M'$ is a morphism of mixed complexes such that $H_*(M) \cong H_*(M')$, then $HC_*(M) \cong HC_*(M')$ as well.

Exercise 9.8.2 Show that the spectral sequence 5.6.1 arising from Tsygan's double complex $CC_{**}(A)$, which has $E^2_{2p,q} = HH_q(A)$, has for its d^2 differential the map $HH_q(A) \to HH_{q+1}(A)$ induced by Connes' operator B. Then show that this spectral sequence is isomorphic (after reindexing) to Connes' spectral sequence 9.8.6. *Hint:* Show that the exact couple 5.9.3 of the filtration on $\mathcal{B}_{**}$ is the derived couple of the exact couple associated to $CC_{**}(A)$.

Notational consistency   Our uses of the letter "B" are compatible. The map $B: M_m \to M_{m+1}$ defining the mixed complex M induces the d^1 differentials $B: H_m(M) \to H_{m+1}(M)$ in Connes' spectral sequence because it is used for the horizontal arrows in Connes' double complex 9.8.2. This is the same map as the composition $BI: H_m(M) \to HC_m(M) \to H_{m+1}(M)$ in the SBI sequence (9.8.7). (Exercise!)

Trivial Mixed Complexes 9.8.8 If (C_*, b) is any chain complex, we can regard it as a trivial mixed complex $(C_*, b, 0)$ by taking $B = 0$. Since the horizontal differentials vanish in Connes' double complex we have

$$HC_n(C_*, b, 0) = H_n(C) \oplus H_{n-2}(C) \oplus H_{n-4}(C) \oplus \cdots.$$

Similarly, if (C^*, B) is any cochain complex, we can regard it as the trivial mixed complex $(C^*, 0, B)$. Since the rows of Connes' double complex are the various brutal truncations (1.2.7) of C, we have

$$HC_n(C^*, 0, B) = C^n/B(C^{n-1}) \oplus H^{n-2}(C) \oplus H^{n-4}(C) \oplus \cdots.$$

The de Rham complex 9.8.9 provides us with an important example of this phenomenon.

9.8.1 de Rham Cohomology

9.8.9 Let R be a commutative k-algebra and $\Omega^*_{R/k}$ the exterior algebra of Kähler differentials discussed in sections 9.2 and 9.4. The *de Rham differential* $d: \Omega^n_{R/k} \to \Omega^{n+1}_{R/k}$ is characterized by the formula

$$d(r_0 dr_1 \wedge \cdots \wedge dr_n) = dr_0 \wedge dr_1 \wedge \cdots \wedge dr_n \quad (r_i \in R).$$

We leave it to the reader to check (using the presentation of $\Omega_{R/k}$ in 8.8.1; see [EGA, IV.16.6.2]) that d is well defined. Since $d^2 = 0$, we have a cochain complex $(\Omega^*_{R/k}, d)$ called the *de Rham complex*; the cohomology modules $H^*_{dR}(R) = H^*(\Omega^*_{R/k})$ are called the (algebraic) *de Rham cohomology* of R. All this is an algebraic parallel to the usual construction of de Rham cohomology for manifolds in differential geometry and has applications to algebraic geometry that we will not pursue here. The material here is based on [LQ].

Exercise 9.8.3 Show that d makes $\Omega^*_{R/k}$ into a differential graded algebra (4.5.2), and conclude that $H^*_{dR}(R)$ is a graded-commutative k-algebra.

If we consider $(\Omega^*_{R/k}, d)$ as a trivial mixed complex with $b = 0$, then by 9.8.8

$$HC_n(\Omega^*_{R/k}, 0, d) = \Omega^n_{R/k}/d\Omega^{n-1}_{R/k} \oplus H^{n-2}_{dR}(R) \oplus \cdots.$$

In many ways, this serves as a model for the cyclic homology of R. For example, in 9.4.4 we constructed a ring homomorphism $\psi: \Omega^*_{R/k} \to H_*(R, R)$, which was an isomorphism if R is smooth over k (9.4.7). The following result allows us to interpret the d^1 differentials in Connes' spectral sequence.

Lemma 9.8.10 *The following square commutes:*

$$
\begin{array}{ccc}
\Omega^n_{R/k} & \xrightarrow{\psi} & H_n(R, R) \\
d \downarrow & & \downarrow B \\
\Omega^{n+1}_{R/k} & \xrightarrow{\psi} & H_{n+1}(R, R).
\end{array}
$$

Proof Given a generator $\omega = r_0 dr_1 \wedge \cdots \wedge dr_n$ of $\Omega^n_{R/k}$, $\psi(\omega)$ is the class of

$$(r_0 \otimes r_1)\nabla(1 \otimes r_2)\nabla \cdots \nabla(1 \otimes r_n) = n!\varepsilon_n(r_0 \otimes \cdots \otimes r_n)$$
$$= \sum_\sigma (-1)^\sigma r_0 \otimes r_{\sigma^{-1}(1)} \otimes \cdots \otimes r_{\sigma^{-1}(n)}$$

where σ ranges over all permutations of $\{1, \cdots, n\}$ and ∇ denotes the shuffle product on $\beta(R, R)$ given in 9.4.2. Passing to the normalized complex $B_n(R, R)$, defining $\sigma(0) = 0$ and applying B, the description in 9.8.4 gives us

$$\sum_\sigma (-1)^\sigma \sum_t (-1)^t \otimes r_{\sigma^{-1}t^{-1}(0)} \otimes r_{\sigma^{-1}t^{-1}(1)} \otimes \cdots \otimes r_{\sigma^{-1}t^{-1}(n)}$$

where t ranges over the cyclic permutations $p \mapsto p + i$ of $\{0, 1, \cdots, n\}$. Since every permutation μ of $\{0, 1, \cdots, n\}$ can be written uniquely as a composite $t\sigma$, this expression equals the representative of $\psi(dr_0 \wedge dr_1 \wedge \cdots \wedge dr_n)$:

$$(n+1)!\varepsilon_n(1 \otimes r_0 \otimes \cdots \otimes r_n) = \sum_\mu (-1)^\mu \otimes r_{\mu^{-1}(0)} \otimes \cdots \otimes r_{\mu^{-1}(n)}. \quad \diamond$$

Porism Suppose that $1/(n+1)! \in R$. The above proof shows that

$$B(n!\varepsilon_n)(r_0 \otimes \cdots \otimes r_n) = (n+1)!\varepsilon_{n+1}(1 \otimes r_0 \otimes \cdots \otimes r_n)$$
$$= n!\varepsilon_{n+1}B(r_0 \otimes \cdots \otimes r_n).$$

Dividing by $n!$ gives the identity $B\varepsilon_n = \varepsilon_{n+1}B$.

Corollary 9.8.11 *If R is smooth over k, the E^2 terms of Connes' spectral sequence are*

$$E_{pq}^2 = \begin{cases} \Omega_{R/k}^q/d\Omega_{R/k}^{q-1} & \text{if } p = 0 \\ H_{dR}^{q-p}(R) & \text{if } p > 0. \end{cases}$$

We will now show that in characteristic zero this spectral sequence collapses at E^2; we do not know if it collapses in general. Of course, when R is smooth, the sequence of low-degree terms always yields the extension (split if $1/2 \in R$):

$$0 \to \Omega_{R/k}^2/d\Omega_{R/k} \to HC_2(R) \to H_{dR}^0(R) \to 0.$$

9.8.12 Assuming that R is commutative and $\mathbb{Q} \subset R$, we saw in 9.4.4 that the maps $e: R^{\otimes n+1} \to \Omega_{R/k}^n$ defined by $e(r_0 \otimes \cdots) = r_0 dr_1 \wedge \cdots \wedge dr_n/n!$ satisfied $eb = 0$ and $e\psi = $ identity. In fact, e is a morphism of mixed complexes from $(R^{\otimes *+1}, b, B)$ to $(\Omega_{R/k}^*, 0, d)$ because by 9.8.4

$$eB(r_0 \otimes \cdots) = \sum \frac{(-1)^{in}}{(n+1)!} dr_i \wedge \cdots \wedge dr_n \wedge dr_0 \wedge \cdots \wedge dr_{i-1} = de(r_0 \otimes \cdots).$$

Therefore e induces natural maps

$$HC_n(R) \to HC_n(\Omega_{R/k}^*) = \Omega_{R/k}^n/d\Omega_{R/k}^{n-1} \oplus H_{dR}^{n-2}(R) \oplus H_{dR}^{n-4}(R) \oplus \cdots.$$

Theorem 9.8.13 *If R is a smooth commutative algebra, essentially of finite type over a field k of characteristic 0, then e induces natural isomorphisms*

$$HC_n(R) \cong \Omega_{R/k}^n/d\Omega_{R/k}^{n-1} \oplus H_{dR}^{n-2}(R) \oplus H_{dR}^{n-4}(R) \oplus \cdots,$$

$$HP_n(R) \cong \prod_{i \in \mathbb{Z}} H_{dR}^{n+2i}(R).$$

Proof On Hochschild homology, e induces maps $H_n(R, R) \to HH_n(\Omega_{R/k}^*) = \Omega_{R/k}^n$. When R is smooth, the Hochschild-Kostant-Rosenberg Theorem 9.4.7 states that these are isomorphisms. It follows (9.8.7) that e induces isomorphisms on HC_* and HP_* as well. ◇

Exercise 9.8.4 When R is commutative and $\mathbb{Q} \subset R$, show that $\Omega_{R/k}^n/d\Omega_{R/k}^{n-1}$ and $H_{dR}^{n-2}(R)$ are always direct summands of $HC_n(R)$. I do not know if the other $H_{dR}^{n-2i}(R)$ are direct summands.

Exercise 9.8.5 Show that the SBI sequence for a trivial mixed complex $(C^*, 0, B)$ is not split in general. Conclude that the SBI sequence of a smooth algebra R need not split in low degrees. Of course, if R is smooth and finitely generated, we observed in 9.4.8 that $H_n(R, R) = 0$ for $n > d = \dim(R)$, so the first possible non-split map is $S: HC_{d+1}(R) \to HC_{d-1}(R)$.

9.8.2 Hodge Decomposition

There is a decomposition for cyclic homology analogous to that for Hochschild homology. To construct it we consider Connes' double complex $\mathcal{B}_{**}$ (9.8.2) for the normalized mixed complex $(\bar{C}_*^h(R), b, B)$. Lemma 9.8.15 below shows that B sends $\bar{C}_n^h(R)^{(i)}$ to $\bar{C}_{n+1}^h(R)^{(i+1)}$. Therefore there is a double subcomplex $\mathcal{B}_{**}^{(i)}$ of $\mathcal{B}_{**}$ whose p^{th} column is the complex $\bar{C}_*^h(R)^{(i-p)}$ shifted p places vertically.

$$
\begin{array}{ccccccccc}
\cdots & & \cdots & & & & \cdots & & \\
\bar{C}_n^{(i)} & \xleftarrow{B} & \bar{C}_{n-1}^{(i-1)} & \xleftarrow{B} & \cdots & \xleftarrow{B} & \bar{C}_{n-i+1}^{(1)} & \longleftarrow & 0 \\
\cdots & & \cdots & & & & \cdots & & \\
\downarrow b & & \downarrow b & & & & \downarrow b & & \\
\bar{C}_{i+1}^{(i)} & \xleftarrow{B} & \bar{C}_i^{(i-1)} & \xleftarrow{B} & \cdots & \xleftarrow{B} & \bar{C}_2^{(1)} & \longleftarrow & 0 \\
\downarrow b & & \downarrow b & & & & \downarrow b & & \\
\bar{C}_i^{(i)} & \xleftarrow{B} & \bar{C}_{i-1}^{(i-1)} & \xleftarrow{B} & \cdots & \xleftarrow{B} & \bar{C}_1^{(1)} & \xleftarrow{B} & R \longleftarrow 0 \\
\downarrow b & & \downarrow b & & & & \downarrow b & & \\
0 & & 0 & & & & 0 & &
\end{array}
$$

Definition 9.8.14 (Loday) If $i \geq 1$, then $HC_n^{(i)}(R) = H_n \operatorname{Tot} \mathcal{B}_{**}^{(i)}$. Because $e_n^{(0)} = 0$ for $n \neq 0$, $HC_*^{(0)}(R) = HC_0^{(0)}(R) = R$. The *Hodge decomposition* of HC_n for $n \geq 1$ is

$$HC_n(R) = HC_n^{(1)}(R) \oplus HC_n^{(2)} \oplus \cdots \oplus HC_n^{(n)}(R).$$

Lemma 9.8.15 $e_{n+1}^{(i+1)}B = Be_n^{(i)}$ *for every n and $i \leq n$.*

Proof When $n = i = 1$ we have $Be_1^{(1)}(r_0 \otimes r_1) = B(r_0 \otimes r_1) = 1 \otimes r_0 \otimes r_1 - 1 \otimes r_1 \otimes r_0$, which is $\varepsilon_2 B(r_0 \otimes r_1)$. More generally, if $i = n$, the equality

$\varepsilon_{n+1}B = B\varepsilon_n$ was established in the porism to lemma 9.8.10. For $i < n$, we proceed by induction. Set $F = e_{n+1}^{(i+1)}B - Be_n^{(i)}$. The following calculation shows that $b(F) = 0$:

$$be_{n+1}^{(i+1)}B = e_n^{(i+1)}bB = -e_n^{(i+1)}Bb = -Be_{n-1}^{(i)}b = -Bbe_n^{(i)} = +bBe_n^{(i)}.$$

Now observe that there is an element u of $\mathbb{Q}\Sigma_{n+1}$ such that

$$u(1 \otimes r_0 \otimes \cdots \otimes r_n) = e_{n+1}^{(i+1)}B(r_0 \otimes \cdots \otimes r_n) - Be_n^{(i)}(r_0 \otimes \cdots \otimes r_n).$$

By Barr's Lemma 9.4.9, $u = c\varepsilon_n$ and it suffices to evaluate the constant c. Because $i < n$ we have $\varepsilon_{n+1}e_{n+1}^{(i+1)} = 0$ and $\varepsilon_n e_n^{(i)} = 0$. Therefore

$$\varepsilon_{n+1}u(1 \otimes r_0 \otimes \cdots \otimes r_n) = -\varepsilon_{n+1}Be_n^{(i)}(r_0 \otimes \cdots \otimes r_n)$$
$$= -B\varepsilon_n e_n^{(i)}(r_0 \otimes \cdots \otimes r_n)$$
$$= 0.$$

This gives the desired relation $u = \varepsilon_{n+1}u = 0$. $\qquad\diamond$

Corollary 9.8.16 $HC_n^{(n)}(R) = \Omega_{R/k}^n / d\Omega_{R/k}^{n-1}.$

Proof Filtering $\mathcal{B}_{**}^{(i)}$ by columns and looking in the lower left-hand corner, we see that $HC_n^{(n)}(R)$ is the cokernel of the map $B = d : H_{n-1}^{(n-1)}(R, R) \to H_n^{(n)}(R, R)$. $\qquad\diamond$

Theorem 9.8.17 *When $\mathbb{Q} \subseteq R$, the SBI sequence breaks up into the direct sum of exact sequences*

$$\cdots HC_{n+1}^{(i)}(R) \xrightarrow{S} HC_{n-1}^{(i-1)}(R) \xrightarrow{B} H_n^{(i)}(R, R) \xrightarrow{I} HC_n^{(i)}(R) \xrightarrow{S} HC_{n-1}^{(i-1)}(R) \cdots.$$

Proof The quotient double complex $\mathcal{B}_{**}^{(i)} / \bar{C}_*^h(R)^{(i)}$ is a translate of $\mathcal{B}_{**}^{(i-1)}$. $\diamond$

Corollary 9.8.18 *Let k be a field of characteristic zero. Then*

$$HC_n^{(1)}(R) \cong H_n^{(1)}(R, R) \cong D_{n-1}(R/k)$$

(André-Quillen homology) for $n \geq 3$, while for $n = 2$ there is an exact sequence

$$0 \to D_1(R/k) \to HC_2^{(1)} \to H_{dR}^0(R/k) \to 0.$$

Exercise 9.8.6 Show that if R is smooth over k, then $HC_n^{(i)}(R) = 0$ for $i < n/2$, while if $n/2 \leq i < n$ we have $HC_n^{(i)}(R) \cong H_{dR}^{2i-n}(R/k)$.

Exercise 9.8.7 Show that there is also a Hodge decomposition for $HP_*(R)$:

$$HP_*(R) = \Pi HP_*^{(i)}(R).$$

If R is smooth, show that $HP_*^{(i)}(R) \cong H_{dR}^{2i-n}(R/k)$.

Remark 9.8.19 (Schemes) It is possible to extend Hochschild and cyclic homology to schemes over k by formally replacing R by $\mathcal{O}_X$ and $R^{\otimes n}$ by $\mathcal{O}_X^{\otimes n}$ to get chain complexes of sheaves on X, and then taking hypercohomology (Chapter 5, section 7). For details, see [G-W]. If X is smooth over k and contains $\mathbb{Q}$, it turns out that $HH_n^{(i)}(X) \cong H^{i-n}(X, \Omega_X^i)$ and $HP_n^{(i)}(X) = H_{dR}^{2i-n}(X)$. If X is a smooth projective scheme and $p = i - n$, then $HC_n^{(i)}(X)$ is the pth level $F^p H_{dR}^{2i-n}(X)$ of the classical Hodge filtration on $H_{dR}^*(X) \cong H^*(X(\mathbb{C}); k)$. This direct connection to the classical Hodge filtration of $H_{dR}^*(X)$ justifies our use of the term "Hodge decomposition."

9.9 Graded Algebras

Let $R = \oplus R_i$ be a graded k-algebra. If $r_0, \cdots, r_p$ are homogeneous elements, define the *weight* of $r_0 \otimes \cdots \otimes r_p \in R^{\otimes p+1}$ to be $w = \sum |r_i|$, where $|r_i| = j$ means that $r_i \in R_j$. This makes the tensor product $R^{\otimes p+1}$ into a graded k-module, $(R^{\otimes p+1})_w$ being generated by elements of weight w. Since the face and degeneracy maps, as well as the cyclic operator t, all preserve weight, the $\{(R^{\otimes p+1})_w\}$ form a cyclic submodule $(ZR)_w$ of $ZR = R^{\otimes *+1}$ and allows us to view $ZR = \oplus(ZR)_w$ as a graded cyclic module or cyclic object in the abelian category of graded k-modules (9.6.1). As our definitions work in any abelian category, this provides each $HH_p(R) = HH_p(ZR)$ and $HC_p(R) = HC_p(ZR)$ with the structure of graded k-modules: $HH_p(R)_w = HH_p((ZR)_w)$ and $HC_p(R)_w = HC_p((ZR)_w)$. We are going to prove the following theorem, due to T. Goodwillie [Gw].

Goodwillie's Theorem 9.9.1 *If R is a graded k-algebra, then the image of $S: HC_p(R)_w \to HC_{p-2}(R)_w$ is annihilated by multiplication by w. In particular, if $\mathbb{Q} \subset R$, then $S = 0$ on $HC_*(R)_w$ for $w \neq 0$, and the SBI sequence splits up into short exact sequences*

$$0 \to HC_{p-1}(R)_w \xrightarrow{B} HH_p(R)_w \xrightarrow{I} HC_p(R)_w \to 0.$$

If R is positively graded ($R = R_0 \oplus R_1 \oplus \cdots$), then clearly $(ZR)_0 = Z(R_0)$, so that the missing piece $w = 0$ of the theorem has $HC(R)_0 = HC(R_0)$.

Corollary 9.9.2 *If R is positively graded and $\mathbb{Q} \subset R$, then $HP_*(R) \cong HP_*(R_0)$.*

Corollary 9.9.3 (Poincaré Lemma) *If R is commutative, positively graded, and $\mathbb{Q} \subset R$, then*

$$H^*_{dR}(R) \cong H_{dR}(R_0).$$

Proof It suffices to show that the weight w part of the de Rham complex $(\Omega^*_{R/k}, d)$ of 9.8.9 is zero for $w \neq 0$. This is a direct summand (by 9.4.4, exercise 9.4.4) of the chain complex $(HH_*(R)_w, BI)$, which is exact because the kernel of $BI \colon HH_p(R)_w \to HH_{p+1}(R)_w$ is $HC_{p-1}(R)_w$. $\diamond$

Example 9.9.4 The tensor algebra $T = T(V)$ of a k-module V may be graded by setting $T_i = V^{\otimes i}$. We saw in 9.1.6 that $HH_n(T) = 0$ for $n \neq 0, 1$. If $\mathbb{Q} \subseteq k$, this immediately yields $HC_n(T)_w = 0$ for $n \neq 0$ and $w \neq 0$, and hence we have $HC_n(T) = HC_n(k)$ for $n \neq 0$. If $\mathbb{Q} \not\subset k$, the explicit calculation in 9.8.5 shows that $HC_n(T)_w \cong H_n(C_w; V^{\otimes w})$, which is a group of exponent w as the cyclic group C_w has order w.

Exercise 9.9.1 Given a k-module V we can form the ring $R = k \oplus V$ with $V^2 = 0$. If we grade R with $R_1 = V$ and fix $w \neq 0$, show that

$$HC_n(R)_w \cong H_{n+1-w}(C_w; V^{\otimes w}).$$

Exercise 9.9.2 Let R be the truncated polynomial ring $k[x]/(x^{m+1})$, and suppose that $\mathbb{Q} \subset k$. We saw that $HH_n(R) \cong k^m$ for all $n \neq 0$ in exercise 9.1.4. Show that $HC_n(R) = 0$ for n odd, while for n even $HC_n(R) \cong k^{m+1}$. Compare this approach with that of exercise 9.6.4.

Exercise 9.9.3 (Generating functions) Let k be a field of characteristic zero, and suppose that R is a positively graded k-algebra with each R_i finite-dimensional. Show that $h(n, w) = \dim HH_n(R)_w$ is finite and that for every $w \neq 0$ we have

$$\dim HC_n(R)_w = (-1)^n \sum_{i=0}^{n} (-1)^i h(i, w).$$

Now set $h_w(t) = \sum h(n, w)t^n$, $f_w(t) = \sum \dim HC_n(R)_w t^n$, and show that $h_w(t) = (1 + t)f_w(t)$.

In order to prove Goodwillie's Theorem, we work with the normalized mixed complex $\bar{C}_*(R)$ of R. First we describe those maps $F: R^{\otimes m+1} \to \bar{C}_n(R)$ which are natural with respect to the graded ring R (and k). For each sequence of weights $w = (w_0, \cdots, w_m)$ we must give a map F_w from $R_{w_0} \otimes \cdots \otimes R_{w_m}$ to $\bar{C}_n(R)$. Let T_w denote the free k-algebra on elements $x_0, \cdots, x_m$, graded so that x_i has weight w_i. Given $r_i \in R_{w_i}$ there is a graded algebra map $T_w \to R$ sending x_i to r_i; the map $\bar{C}_n(T_w) \to \bar{C}_n(R)$ must send $y = F_w(x_0 \otimes \cdots \otimes x_m)$ to $F_w(r_0 \otimes \cdots \otimes r_m)$. Thus F_w is determined by the element $y = y(x_0, \cdots, x_m)$ of $\bar{C}_n(T_w) = T_w \otimes \bar{T}_w \otimes \cdots$, that is, by a k-linear combination of terms $M_0 \otimes \cdots \otimes M_n$, where the M_j are noncommutative monomials in the x_i, and $M_j \ne 1$ for $i \ne 0$. In order for y to induce a natural map F_w we must have multilinearity:

$$\lambda y(x_0, \cdots, x_m) = y(x_0, \cdots, \lambda x_i, \cdots, x_m)$$

for all i and all $\lambda \in k$. Changing k if necessary (so that for each j there is a $\lambda \in k$ such that $\lambda^j \ne \lambda$), this means there can be at most one occurrence of each x_i in each monomial $M_0 \otimes \cdots \otimes M_n$ in $y(x_0, \cdots, x_m)$.

If $n \ge m + 2$, then at least two of the monomials M_i must be one in each term $M_0 \otimes \cdots \otimes M_n$ of y. This is impossible unless $y = 0$. If $n = m + 1$, then we must have $M_0 = 1$ in each term, and y must be a linear combination of the monomials $1 \otimes x_{\sigma 0} \otimes \cdots \otimes x_{\sigma m}$ as σ runs over all permutations of $\{0, \cdots, m\}$. An example of such a natural map is B; the universal formula in this case is given by $y = B(x_0 \otimes \cdots \otimes x_m)$, where only cyclic permutations are used. From this we make the following deduction.

Lemma 9.9.5 *Any natural map* $F: R^{\otimes m+1} \to \bar{C}_{m+1}(R)$ *must satisfy* $FB = BF = 0$, *and induces a map* $\bar{F}: \bar{C}_m(R) \to \bar{C}_{m+1}(R)$.

Examples 9.9.6 If $m = n$, there is a natural map $D: \bar{C}_m(R) \to \bar{C}_m(R)$ which is multiplication by $w = \sum w_i$ on $R_{w_0} \otimes \cdots \otimes R_{w_m}$. When $m = 0$, D is the map from $R = \bar{C}_0(R)$ to itself sending $r \in R_w$ to wr. The formula

$$e(r_0 \otimes \cdots \otimes r_m) = (-1)^{m-1}(Dr_m)r_0 \otimes r_1 \otimes \cdots \otimes r_{m-1}$$

gives a natural map $e: \bar{C}_m(R) \to \bar{C}_{m-1}(R)$. This map is of interest because $eb + be = 0$ (check this!), and also because of its resemblance to the face map ∂_n (which is natural on $R^{\otimes m+1}$ but does not induce a natural map $\bar{C}_m \to \bar{C}_{m-1}$).

Proof of Theorem 9.9.1 Since D commutes with B and b, it is a map of mixed complexes and induces an endomorphism of $HC_*(R)$ — namely, it is multiplication by w on $HC_*(R)_w$. We must show that $DS = 0$. To do this we construct a chain contraction $Se + SE$ of DS: $\mathrm{Tot}_n \mathcal{B}_{**} \to \mathrm{Tot}_{n-2} \mathcal{B}_{**}$, where $\mathcal{B}_{**}$ is Connes' double complex for the normalized complex $\bar{C}_*(R)$ and S is the periodicity map $\mathcal{B}_{pq} \to \mathcal{B}_{p-1,q-1}$. The map $e: \mathcal{B}_{pq} \to \mathcal{B}_{p+1,q}$ is the map $\bar{C}_m \to \bar{C}_{m-1}$ given in 9.9.6, and E will be a map $\mathcal{B}_{pq} \to \mathcal{B}_{p,q+1}$ induced by natural maps $E_m: \bar{C}_m \to \bar{C}_{m+1}$. If we choose E so that D equals

$$(*) \qquad (e+E)(B+b) + (B+b)(e+E) = eB + Be + Eb + bE$$

$$\bar{C}_{m+1}$$
$$\uparrow E$$
$$\bar{C}_{m+1} \xleftarrow{\ B\ } \bar{C}_m = \mathcal{B}_{pq} \xrightarrow{\ e\ } \bar{C}_{m-1}$$
$$\downarrow b$$
$$\bar{C}_{m-1}$$

on $\bar{C}_m(R)$, then $S(e+E)$ will be a chain contraction of DS. Note that the term eB of $(*)$ does not make sense on $\mathcal{B}_{0q}$, but the term SeB does.

All that remains is to construct $E_m: \bar{C}_m(R) \to \bar{C}_{m+1}(R)$, and we do this by induction on m, starting with $E_0 = 0$ and $E_1(r_0 \otimes r_1) = 1 \otimes Dr_1 \otimes r_0$. Because

$$(eB + Be)(r_0) = e(1 \otimes r_0) = Dr_0,$$
$$(eB + Be + bE_1)(r_0 \otimes r_1) = e(1 \otimes r_0 \otimes r_1 - 1 \otimes r_1 \otimes r_0)$$
$$+ B(Dr_1)r_0 + bE_1(r_0 \otimes r_1)$$
$$= -Dr_1 \otimes r_0 + Dr_0 \otimes r_1$$
$$+ 1 \otimes (Dr_1)r_0 + b(1 \otimes Dr_1 \otimes r_0)$$
$$= Dr_0 \otimes r_1 + r_0 \otimes Dr_1$$
$$= D(r_0 \otimes r_1),$$

the expression $(*)$ equals D on $\bar{C}_0(R)$ and $\bar{C}_1(R)$. For $m \geq 2$, we assume E_{m-1}, E_{m-2} constructed; for each w we need to find elements $y \in \bar{C}_{m+1}(T_w)$ such that

$$by + (eB + Be + E_{m-1}b)(x_0 \otimes \cdots \otimes x_m) = D(x_0 \otimes \cdots \otimes x_m)$$

in $\bar{C}_m(T_w)$. Set $z = (D - eB - Be - E_{m-1}b)(x_0 \otimes \cdots \otimes x_m)$; by induction and $(*)$,

$$bz = (Db + ebB + Bbe - bE_{m-1}b - E_{m-2}b^2)(x_0 \otimes \cdots \otimes x_m)$$
$$= (D - eB - Be - bE_{m-1} - E_{m-2}b)b(x_0 \otimes \cdots \otimes x_m)$$
$$= 0.$$

We saw in 9.1.6 that $H_m(T_w, T_w) = 0$ for $m \geq 2$, so the normalized complex $\bar{C}_*(T_w)$ and hence its summand $\bar{C}_*(T_w)_w$ of weight w are exact at m. Thus there is an element y in $\bar{C}_{m+1}(T_w)_w$ such that $by = z$. Since y has weight w_i with respect to each x_i, there can be at most one occurrence of each x_i in each monomial in $y(x_0, \cdots, x_m)$. Hence if we define

$$E_m(r_0 \otimes \cdots \otimes r_m) = y(r_0, \cdots, r_m),$$

then E_m is a natural map from $\bar{C}_m(R)$ to $\bar{C}_{m+1}(R)$ such that $(*)$ equals D on $\bar{C}_m(R)$. This finishes the construction of E and hence the proof of Goodwillie's Theorem. $\diamond$

Remark 9.9.7 The "weight" map $D: R \to R$ is a derivation, and Goodwillie's Theorem 9.9.1 holds more generally for any derivation acting on a k-algebra R; see [Gw]. All the basic formulas in the proof—such as the formula $(*)$ for D—were discovered by G. Rinehart 20 years earlier; see sections 9, 10 of "Differential forms on general commutative algebras, *Trans. AMS* 108 (1963), 195–222.

As an application of Goodwillie's Theorem, suppose that I is an ideal in a k-algebra R. Let $Z(R, I)$ denote the kernel of the surjection $Z(R) \to Z(R/I)$; we define the cyclic homology modules $HC_*(R, I)$ to be the cyclic homology of the cyclic module $Z(R, I)$. Since cyclic homology takes short exact sequences of cyclic modules to long exact sequences, we have a long exact sequence

$$\cdots HC_{n+1}(R) \to HC_{n+1}(R/I) \to HC_n(R, I) \to HC_n(R) \to HC_n(R/I) \cdots.$$

Thus $HC_*(R, I)$ measures the difference between $HC_*(R)$ and $HC_*(R/I)$.

We can filter each module $Z_p R = R^{\otimes p+1}$ by the submodules F_p^i generated by all the $I^{i_0} \otimes \cdots \otimes I^{i_p}$ with $i_0 + \cdots + i_p = i$. Since the structure maps ∂_i, σ_i, t preserve this filtration, the F_*^i are cyclic submodules of ZR. As F_*^1 is $Z(R, I)$, we have $F_*^0/F_*^1 = Z(R/I)$.

Exercise 9.9.4 If k is a field, show that the graded cyclic vector spaces $\oplus F_*^i/F_*^{i+1}$ and $Z(gr R)$ are isomorphic, where $gr(R) = R/I \oplus I/I^2 \oplus \cdots \oplus I^m/I^{m+1} \oplus \cdots$ is the associated graded algebra of $I \subset R$.

Proposition 9.9.8 *Let k be a field of characteristic zero. If $I^{m+1} = 0$, then the maps $S^i: HC_{p+2i}(R, I) \to HC_p(R, I)$ are zero for $i \geq m(p + 1)$.*

Proof By the above exercise, $HC_*(gr R)_w \cong HC_*(F_*^w / F_*^{w+1})$. Since $gr(R)$ is graded, the map S is zero on all but the degree zero part of $HC_*(gr R)$. Hence $S^i = 0$ on $HC_*(F^1 / F^{i+1})$. Since $F_p^{i+1} = 0$ for $i \geq m(p + 1)$, the map S^i factors as

$$HC_{p+2i}(R, I) \to HC_{p+2i}(F_*^1 / F_*^{i+1}) \xrightarrow{S^i} HC_p(F_*^1 / F_*^{i+1}) = HC_p(R, I),$$

which is the zero map. $\diamond$

Corollary 9.9.9 *If I is a nilpotent ideal of R, then $HP_*(R, I) = 0$ and $HP_*(R) \cong HP_*(R/I)$.*

Proof The tower $\{HC_{*+2i}(R, I)\}$ satisfies the trivial Mittag-Leffler condition. $\diamond$

Exercise 9.9.5 If I is a nilpotent ideal of R and k is a field with $\mathrm{char}(k) = 0$, show that $H_{dR}^*(R) \cong H_{dR}^*(R/I)$. *Hint:* Study the complex $(HH_*(R), BI)$.

9.9.1 Homology of DG-Algebras

9.9.10 It is not hard to extend Hochschild and cyclic homology to DG-algebras, that is, graded algebras with a differential $d: R_n \to R_{n-1}$ satisfying the Leibnitz identity $d(r_0 r_1) = (dr_0)r_1 + (-1)^{|r_0|} r_0(dr_1)$; see 4.5.2. (Here $|r_0| = j$ if $r_0 \in R_j$.) If we forget the differential, we can consider ZR (9.6.1) as a graded cyclic module as in Goodwillie's Theorem 9.9.1. If we lay out the Hochschild complex in the plane with $(R^{\otimes q+1})_p$ in the (p, q) spot, then there is also a "horizontal" differential given by

$$d(r_0 \otimes \cdots \otimes r_q) = \sum_{i=0}^{q} (-1)^{|r_0|+\cdots+|r_{i-1}|} r_0 \otimes \cdots \otimes dr_i \otimes \cdots \otimes r_q.$$

Thus the Hochschild complex becomes a double complex $C_*^h(R, d)_*$; we define the Hochschild homology $HH_*^{DG}(R)$ to be the homology of $\mathrm{Tot}^\oplus C_*^h(R)_*$. If R is positively graded, then $C^h(R, d)$ lies in the first quadrant and there is a spectral sequence converging to $HH_*^{DG}(R)$ with $E_{pq}^2 = H_p^h(HH_q(R)_*)$. *Warning:* If R is a graded algebra endowed with differential $d = 0$, then $HH_n^{DG}(R)$ is the sum of the $HH_q(R)_p$ with $p + q = n$ and not $HH_n(R)$.

In the literature (e.g., in [MacH, X]) one often considers DG-algebras to have a differential $d\colon R^n \to R^{n+1}$ and $R^n = 0$ for $n < 0$. If we reindex R^n as R_{-n} this is a negatively graded DG-algebra. It is more natural to convert $C_*^h(R, d)_*$ into a cochain double complex in the fourth quadrant and to write $HH^n_{DG}(R)$ for $HH^{DG}_{-n}(R)$.

Exercise 9.9.6 If $R^0 = k$ and $R^1 = 0$, construct a convergent fourth quadrant spectral sequence converging to $HH^*_{DG}(R)$ with $E_2^{pq} = H^p HH_{-q}(R)$.

Exercise 9.9.7 Let (R_*, d) be a DG-algebra and M a chain complex that is also a graded R-module in such a way that the Leibnitz identity holds with $r_0 \in M$, $r_1 \in R$. Define $H_*^{DG}(R, M)$ to be the homology of the total complex $(M \otimes R^{\otimes q})_p$ obtained by taking $r_0 \in M$ in 9.9.10. If M and R are positively graded, show that there is a spectral sequence

$$E_{pq}^2 = H_p^h H_q(R, M) \Rightarrow H_{p+q}^{DG}(R, M).$$

We now return to the cyclic viewpoint. The chain complexes $Z_q(R)_* = (R^{\otimes q+1})_*$ fit together to form a cyclic object $Z(R, d)$ in $\mathbf{Ch}(k\text{–}\mathbf{mod})$, the abelian category of chain complexes, provided that we use the sign trick to insert a sign of $(-1)^{|r_q|(|r_0| + \cdots + |r_{q-1}|)}$ in the formulas for ∂_q and t. (Check this!) As in any abelian category, we can form HH_* and HC_* in $\mathbf{Ch}(k\text{–}\mathbf{mod})$. However, since $C_*^h(Z(R, d))$ is really a double complex whose total complex yields $HH_*^{DG}(R)$ it makes good sense to imitate 9.6.7 and define $HC_*^{DG}(R)$ as $H_*\operatorname{Tot}^\oplus CC_{**}Z(R, d)$. If R is positively graded, then we can define $HP_*^{DG}(R)$ using the product total complex of $CC_{**}^P Z(R, d)$. All the major structural results for ordinary cyclic homology clearly carry over to this DG-setting.

Proposition 9.9.11 *If $f\colon (R, d) \to (R', d')$ is a homomorphism of flat DG-algebras such that $H_*(R) \cong H_*(R')$, then f induces isomorphisms*

$$HH_*^{DG}(R) \cong HH_*^{DG}(R') \quad \text{and} \quad HC_*^{DG}(R) \cong HC_*^{DG}(R').$$

Proof As each $R^{\otimes n}$ is also flat as a k-module, the chain maps

$$f^{\otimes n+1}\colon R^{\otimes n+1} \to R^{\otimes n} \otimes R' \to (R')^{\otimes n+1}$$

are quasi-isomorphisms for all n. Filtering by rows 5.6.2 yields a convergent spectral sequence

$$E_{pq}^1 = H_q(R^{\otimes p+1}) \Rightarrow HH_{p+q}^{DG}(R).$$

By the Comparison Theorem 5.2.12, we have $HH_*^{DG}(R,d) \cong HH_*^{DG}(R',d')$. The isomorphism on HC_*^{DG} follows formally using the 5-lemma and the SBI sequence 9.6.11. ⬦

Vista 9.9.12 (Free loop spaces) Suppose that X is a fixed simply connected topological space, and write $C^*(X)$ for the DG-algebra of singular chains on X with coefficients in a field k; the singular cohomology $H^*(X)$ of X is the cohomology of $C^*(X)$. Let X^I denote the space of all maps $f: I \to X$, I denoting the interval $[0,1]$; the *free loop space* ΛX is $\{f \in X^I : f(0) = f(1)\}$ and if $* \in X$ is fixed, the *loop space* ΩX is $\{f \in X^I : f(0) = f(1) = *\}$. The general machinery of the "Eilenberg-Moore spectral sequence" [Smith] for the diagram

$$
\begin{array}{ccccc}
\Omega X & \longrightarrow & \Lambda X & \longrightarrow & X^I \\
\downarrow & & \downarrow & & \downarrow \\
* & \longrightarrow & X & \xrightarrow{\Delta} & X \times X
\end{array}
$$

yields isomorphisms:

$$H^n(\Omega X) \cong HH_{DG}^n(C^*(X), k) \cong HH_{-n}^{DG}(C^*(X), k);$$

$$H^n(\Lambda X) \cong HH_{DG}^n(C^*(X)) \cong HH_{-n}^{DG}(C^*(X)).$$

We say that a space X is *formal* (over k) if there are DG-algebra homomorphisms $C^*(X) \leftarrow R \to H^*(X)$ that induce isomorphisms in cohomology. Here we regard the graded ring $H^*(X)$ as a DG-algebra with $d = 0$, either positively graded as a cochain complex or negatively graded as a chain complex. Proposition 9.9.11 above states that for formal spaces we may replace $C^*(X)$ by $H^*(X)$ in the above formulas for $H^n(\Omega X)$ and $H^n(\Lambda X)$.

All this has an analogue for cyclic homology, using the fact that the topological group S^1 acts on ΛX by rotating loops. The equivariant homology $H_*^{S^1}(\Lambda X)$ of the S^1-space ΛX is defined to be $H_*(\Lambda X \times_{S^1} ES^1)$, the singular homology of the topological space $\Lambda X \times_{S^1} ES^1 = \{(\lambda, e) \in \Lambda X \times ES^1 : \lambda(1) = \pi(e)\}$. Several authors (see [Gw], for example) have identified $H_*^{S^1}(\Lambda X)$ with the cyclic homology $HC_*^{DG}(R_*)$ of the DG-algebra R_* whose homology is $H_*(\Omega X)$.

9.10 Lie Algebras of Matrices

In this section we fix a field k of characteristic zero and an associative k-algebra with unit R. Our goal is to relate the homology of the Lie algebra $\mathfrak{gl}_m(R) = \mathrm{Lie}(M_m(R))$ of $m \times m$ matrices, described in Chapter 7, to the cyclic homology of R. This relationship was discovered in 1983 by J.-L. Loday and D. Quillen [LQ], and independently by B. Feigin and B. Tsygan. We shall follow the exposition in [LQ].

The key to this relationship is the map

$$H_*^{Lie}(\mathfrak{gl}_m(R); k) \xrightarrow{\lambda_*} HC_*(M_m(R)) \cong HC_*(R)$$

constructed as follows. Recall from 7.7.3 that the homology of a Lie algebra $\mathfrak{g}$ can be computed as the homology of the Chevalley-Eilenberg complex $\Lambda^* \mathfrak{g} = k \otimes_{U\mathfrak{g}} V_*(\mathfrak{g})$, with differential

$$d(x_1 \wedge \cdots \wedge x_p) = \sum_{i<j}(-1)^{i+j}[x_i, x_j] \wedge x_1 \wedge \cdots \wedge \hat{x}_i \wedge \cdots \wedge \hat{x}_j \wedge \cdots \wedge x_p.$$

On the other hand, we saw in 9.6.10 that the cyclic homology of R may be computed using the quotient complex $C_*(R) = C_*^h(R)/\sim$ of the Hochschild complex $C_*^h(R)$. Define $\lambda \colon \Lambda^{p+1}\mathfrak{gl}_m(R) \to C_*(M_m(R))$ by

$$\lambda(x_0 \wedge \cdots \wedge x_p) = (-1)^p \sum_{\sigma}(-1)^{\sigma} x_0 \otimes x_{\sigma 1} \otimes \cdots \otimes x_{\sigma p},$$

where the sum is over all possible permutations σ of $\{1, \cdots, p\}$. (Exercise: Why is λ well defined?)

Lemma 9.10.1 λ *is a morphism of chain complexes, and induces maps*

$$\lambda_* \colon H_{p+1}(\mathfrak{gl}_m(R); k) \to HC_p(R).$$

Moreover λ is compatible with the usual nonunital inclusion $\iota\colon M_m(R) \hookrightarrow M_{m+1}(R)$, $\iota(g) = \begin{bmatrix} g & 0 \\ 0 & 0 \end{bmatrix}$, in the sense that the following diagram commutes.

$$
\begin{array}{ccccc}
\Lambda^{*+1}\mathfrak{gl}_m(R) & \xrightarrow{\lambda} & C_*(M_m(R)) & \xrightarrow{\text{trace}} & C_*(R) \\
{\scriptstyle \mathrm{Lie}(\iota)}\downarrow & & \downarrow{\scriptstyle \iota_*} & & \| \\
\Lambda^{*+1}\mathfrak{gl}_{m+1}(R) & \xrightarrow{\lambda} & C_*(M_{m+1}(R)) & \xrightarrow{\text{trace}} & C_*(R).
\end{array}
$$

Proof Commutativity of the right square amounts to the assertion that ι_* is compatible with the trace maps, and was established in exercise 9.5.3. Now set $\omega = x_0 \wedge \cdots \wedge x_p$ with $x_i \in \mathfrak{gl}_m(R)$. The formula for λ shows that $\iota_*(\lambda\omega) = \lambda(\iota x_0 \wedge \cdots \wedge \iota x_p) = \lambda(\iota\omega)$, which gives commutativity of the left square. It also shows that

$$b\lambda(\omega) = (-1)^n \sum (-1)^\nu (x_{\nu 0} x_{\nu 1}) \otimes x_{\nu 2} \otimes \cdots \otimes x_{\nu p},$$

the sum being over all permutations ν of $\{0, 1, \cdots, p\}$. Since

$$\omega = (-1)^{i+j+1} x_i \wedge x_j \wedge x_0 \wedge \cdots \wedge \hat{x}_i \wedge \cdots \wedge \hat{x}_j \wedge \cdots \wedge x_p$$

for $i < j$, it is readily verified (do so!) that $\lambda(d\omega) = b(\lambda\omega)$. This proves that λ is a morphism of complexes. $\diamond$

Primitive Elements 9.10.2 An element x in a coalgebra H (6.7.13) is called *primitive* if $\Delta(x) = x \otimes 1 + 1 \otimes x$. The primitive elements form a submodule $\mathrm{Prim}(H)$ of the k-module underlying H. If H is a graded coalgebra and Δ is a graded map, the homogeneous components of any primitive element must be primitive, so $\mathrm{Prim}(H)$ is a graded submodule of H.

We saw in exercise 7.3.8 that the homology $H = H_*(\mathfrak{g}; k)$ of any Lie algebra $\mathfrak{g}$ is a graded coalgebra with coproduct $\Delta \colon H \to H \otimes H$ induced by the diagonal $\mathfrak{g} \to \mathfrak{g} \times \mathfrak{g}$. When $\mathfrak{g}$ is the Lie algebra $\mathfrak{gl}(R) = \cup \mathfrak{gl}_m(R)$, we are going to prove in 9.10.10 that $\mathrm{Prim}\, H_i(\mathfrak{g}; k) \cong HC_{i-1}(R)$.

The first step in the proof is to recall from exercise 7.7.6 that any Lie group $\mathfrak{g}$ acts on $\Lambda^n \mathfrak{g}$ by the formula $[x_1 \wedge \cdots \wedge x_n, g] = \sum x_1 \wedge \cdots \wedge [x_i g] \wedge \cdots \wedge x_n$. This makes the Chevalley-Eilenberg complex $\Lambda^* \mathfrak{g}$ into a chain complex of right $\mathfrak{g}$-modules, and $\mathfrak{g}$ acts trivially on $H_*(\mathfrak{g}; k) = H_*(\Lambda^* \mathfrak{g})$, again by exercise 7.7.6. Applying this to $\mathfrak{gl}_m(R)$, we observe that $\Lambda^* \mathfrak{gl}_m(R)$ is a chain complex of modules over $\mathfrak{gl}_m(R)$ and hence over the simple Lie algebra $\mathfrak{sl}_m = \mathfrak{sl}_m(k)$ of matrices over k with trace 0 (7.1.3, 7.8.1). Therefore we may take coinvariants to form the chain complex $H_0(\mathfrak{sl}_m; \Lambda^* \mathfrak{gl}_m(R))$.

Proposition 9.10.3 *Taking coinvariants gives a quasi-isomorphism of complexes*

$$\Lambda^* \mathfrak{gl}_m(R) \to H_0(\mathfrak{sl}_m; \Lambda^* \mathfrak{gl}_m(R)).$$

Proof Weyl's Theorem 7.8.11 states that, like every finite-dimensional $\mathfrak{sl}_m$-module, $\Lambda^n \mathfrak{gl}_m(k)$ is a direct sum of simple modules. As R is a free k-module, each $\Lambda^n \mathfrak{gl}_m(R) = \Lambda^n \mathfrak{gl}_m(k) \otimes R$ is also a direct sum of simple modules. Write

Q^n for the direct sum of the simple modules on which $\mathfrak{sl}_m$ acts non-trivially, so that $\Lambda^*\mathfrak{gl}_m(R) = Q^* \oplus H_0(\mathfrak{sl}_m; \Lambda^*\mathfrak{gl}_m(R))$ as an $\mathfrak{sl}_m$-module complex. As $\mathfrak{sl}_m$ acts trivially on the homology of $\Lambda^*\mathfrak{gl}_m(R)$ by exercise 7.7.6, the complex Q^* has to be acyclic, proving the proposition. $\diamond$

Corollary 9.10.4 *If $m \geq n$ the maps $H_n(\mathfrak{gl}_m(R); k) \to HC_{n-1}(R)$ are split surjections.*

Proof Let $e_{ij}(r)$ denote the matrix which is r in the (i, j) spot and zero elsewhere. Exercise 9.5.4 showed that if we set

$$\omega = \omega(r_1, \cdots, r_n) = e_{12}(r_1) \wedge e_{23}(r_2) \wedge \cdots \wedge e_{n-1,n}(r_{n-1}) \wedge e_{n1}(r_n),$$

then $\omega \in \Lambda^n\mathfrak{gl}_m(R)$ satisfies trace$(\lambda\omega) = (-1)^{n-1}r_1 \otimes \cdots \otimes r_n$. Moreover

$$-d\omega = e_{13}(r_1r_2) \wedge \cdots + e_{12}(r_1) \wedge e_{24}(r_2r_3) \wedge \cdots$$
$$+ (-1)^{n+1}e_{n2}(r_nr_1) \wedge e_{23}(r_2) \wedge \cdots .$$

Modulo coinvariants this equals $-\omega(b(r_1 \otimes \cdots \otimes r_n))$. Therefore ω defines a chain complex homomorphism from the translated cyclic complex $R^{\otimes *}/ \sim = (R^{\otimes *+1}/ \sim)[-1]$ to $H_0(\mathfrak{sl}_m; \Lambda^*\mathfrak{gl}_m(R))$. As ω is split by trace(λ), the result follows upon taking homology. $\diamond$

Invariant Theory Calculation 9.10.5 Let Σ_n be the symmetric group of permutations of $\{1, \cdots, n\}$ and (sgn) the 1-dimensional Σ_n-module on which $\sigma \in \Sigma_n$ acts as multiplication by its signature $(-1)^\sigma$. If Σ_n acts on $V^{\otimes n}$ by permuting coordinates, then $\Lambda^n V = V^{\otimes n} \otimes_{k\Sigma_n} (sgn)$. In particular,

$$\Lambda^n\mathfrak{gl}_m(R) = (\mathfrak{gl}_m(k) \otimes R)^{\otimes n} \otimes_{k\Sigma_n} (sgn) = (\mathfrak{gl}_m(k)^{\otimes n} \otimes R^{\otimes n}) \otimes_{k\Sigma_n} (sgn).$$

To compute the coinvariants, we pull a rabbit out of the "hat" of classical invariant theory. The action of Σ_n on $V^{\otimes n}$ gives a homomorphism from $k\Sigma_n$ to End$(V^{\otimes n}) = $ End$(V)^{\otimes n}$; the Lie algebra $\mathfrak{g}$ associated (7.1.2) to the associative algebra End(V) also acts on $V^{\otimes n}$ and the action of Σ_n is $\mathfrak{g}$-invariant, so the image of $k\Sigma_n$ belongs to the invariant submodule $(\text{End}(V)^{\otimes n})^\mathfrak{g} = (\mathfrak{g}^{\otimes n})^\mathfrak{g}$. The classical invariant theory of [Weyl] asserts that $k\Sigma_n \cong (\mathfrak{g}^{\otimes n})^\mathfrak{g}$ whenever $\dim(V) \geq n$. If $\dim(V) = m$, then $\mathfrak{g} \cong \mathfrak{gl}_m(k) \cong k \times \mathfrak{sl}_m(k)$ and the abelian Lie algebra k acts trivially on $(\mathfrak{g}^{\otimes n})$. By Weyl's Theorem (7.8.11), $\mathfrak{g}^{\otimes n}$ is a direct sum of simple $\mathfrak{sl}_m(k)$-modules, so

$$k\Sigma_n \cong (\mathfrak{g}^{\otimes n})^{\mathfrak{sl}_m} \cong (\mathfrak{g}^{\otimes n})_{\mathfrak{sl}_m(k)}, \quad m \geq n.$$

Tensoring with the trivial $\mathfrak{g}$-module $R^{\otimes n}$ therefore yields (for $m \geq n$):

$$H_0(\mathfrak{sl}_m; \wedge^n \mathfrak{gl}_m(k)) = H_0(\mathfrak{sl}_m; (\mathfrak{gl}_m^{\otimes n} \otimes R^{\otimes n}) \otimes_{k\Sigma_n} (sgn))$$

$$= (H_0(\mathfrak{sl}_m; \mathfrak{gl}_m^{\otimes n}) \otimes R^{\otimes n}) \otimes_{k\Sigma_n} (sgn)$$

$$= (k\Sigma_n \otimes R^{\otimes n}) \otimes_{k\Sigma_n} (sgn).$$

The action of Σ_n on $k\Sigma_n$ in the final term is by conjugation.

Corollary 9.10.6 (Stabilization) *For every associative k-algebra R and every n the following stabilization homomorphisms are isomorphisms:*

$$H_n(\mathfrak{gl}_n(R); k) \cong H_n(\mathfrak{gl}_{n+1}(R); k) \cong \cdots \cong H_n(\mathfrak{gl}(R); k).$$

Proof The invariant theory calculation shows that the first $n + 1$ terms (resp. n terms) of the chain complex $H_0(\mathfrak{sl}_m; \wedge^* \mathfrak{gl}_m(R))$ are independent of m, as long as $m \geq n + 1$ (resp. $m \geq n$). This yields a surjection $H_n(\mathfrak{gl}_n(R); k) \to H_n(\mathfrak{gl}_{n+1}(R); k)$ and stability for $m \geq n + 1$. For the more subtle invariant theory needed to establish stability for $m = n$, we cite [Loday, 10.3.5]. $\diamond$

Remark 9.10.7 (Loday-Quillen) It is possible to describe the obstruction to improving the stability result to $m = n - 1$. If R is commutative, we have a naturally split exact sequence

$$H_n(\mathfrak{gl}_{n-1}(R); k) \to H_n(\mathfrak{gl}_n(R); k) \xrightarrow{\lambda} \Omega_{R/k}^{n-1}/d\Omega_{R/k}^{n-2} \to 0.$$

The right-hand map is the composite of $\lambda_*: H_n(\mathfrak{gl}_n(R); k) \to HC_{n-1}(R)$, defined in 9.10.1, and the projection $HC_i(R) \to \Omega_{R/k}^i/d\Omega_{R/k}^{i-1}$ of 9.8.12. The proof of this assertion uses slightly more invariant theory and proposition 9.10.9 below; see [LQ, 6.9]. If R is not commutative, we only need to replace $\Omega_{R/k}^{n-1}/d\Omega_{R/k}^{n-2}$ by a suitable quotient of $\wedge^n R$; see [Loday, 10.3.3 and 10.3.7] for details.

9.10.8 In order to state our next proposition, we need to introduce some standard facts about DG-coalgebras, expanding upon the discussion of graded coalgebras in 6.7.13 and 9.10.2.

If V is any vector space, the exterior algebra $\wedge^*(V)$ is a graded coalgebra with counit $\varepsilon: \wedge^*(V) \to \wedge^*(0) = k$ induced by $V \to 0$ and coproduct

$$\Delta: \wedge^*(V) \to \wedge^*(V \times V) \cong (\wedge^* V) \otimes (\wedge^* V)$$

induced by the diagonal $V \to V \times V$. (Check this!) In particular, $\wedge^* \mathfrak{g}$ is a graded coalgebra for every Lie algebra $\mathfrak{g}$. Since $\mathfrak{g} \to 0$ and $\mathfrak{g} \to \mathfrak{g} \times \mathfrak{g}$ are

Lie algebra maps, $H_0(\mathfrak{h}; \Lambda^*\mathfrak{g})$ is a coalgebra for every Lie subalgebra $\mathfrak{h}$ of $\mathfrak{g}$. (Check this!) In particular, $H_0(\mathfrak{sl}_m(k); \Lambda^*\mathfrak{gl}_m(R))$ is a graded coalgebra for each m.

A *differential graded coalgebra* (or *DG-coalgebra*) C is a graded coalgebra endowed with a differential d making it into a chain complex in such a way that $\varepsilon: C_* \to k$ and $\Delta: C \to C \otimes C$ are morphisms of complexes. For example, $\Lambda^*\mathfrak{g}$ and $H_0(\mathfrak{sl}_m(k); \Lambda^*\mathfrak{gl}_m(R))$ are *DG*-coalgebras because ε and Δ arise from Lie algebra homomorphisms. By the Künneth formula 3.6.3, Δ induces a map

$$H_*(C) \to H_*(C \otimes C) \cong H_*(C) \otimes H_*(C),$$

making the homology of a *DG*-coalgebra C again into a graded coalgebra. Moreover, if $x \in C_n$ is primitive (9.10.2), then $dx \in C_{n-1}$ is primitive, because

$$\Delta(dx) = d\Delta(x) = d(x \otimes 1 + 1 \otimes x) = (dx) \otimes 1 + 1 \otimes (dx).$$

Therefore the graded submodule Prim(C) is a chain subcomplex of C.

Proposition 9.10.9 *The chain complex* $L_* = H_0(\mathfrak{sl}(k); \Lambda^*\mathfrak{gl}(R))$ *is a DG-coalgebra whose primitive part* Prim(L_*) *is the translate* $C_{*-1}(R) = R^{\otimes *}/\sim$ *of the chain complex for cyclic homology.*

Proof Recall from the discussion 9.10.5 on invariant theory that we have

$$L_n \cong (k\Sigma_n \otimes R^{\otimes n}) \otimes_{k\Sigma_n} (sgn).$$

This Σ_n-module splits into a direct sum of modules, one for each conjugacy class of elements of Σ_n. Let U_n be the conjugacy class of the cyclic permutation $\tau = (12 \cdots n)$; we first prove that Prim(L_n) is $(kU_n \otimes R^n) \otimes_{k\Sigma_n} (sgn)$. If $\sigma \in \Sigma_n$ and $r_i \in R$, then consider the element $x = \sigma \otimes (r_1 \otimes \cdots \otimes r_n)$ of L_n. We have

$$\Delta(x) = \sum_{I,J}(\sigma_I \otimes (\cdots \otimes r_i \otimes \cdots)) \otimes (\sigma_J \otimes (\cdots \otimes r_j \otimes \cdots)),$$

where the sum is over all partitions (I, J) of $\{1, \cdots, n\}$ such that $\sigma(I) = I$ and $\sigma(J) = J$, and where σ_I (resp. σ_J) denotes the restriction of σ to I (resp. to J). (Check this!) By inspection, x is primitive if and only if σ admits no nontrivial partitions (I, J), that is, if and only if $\sigma \in U_n$.

Now Σ_n acts on U_n by conjugation, the stabilizer of τ being the cyclic group C_n generated by τ. Hence U_n is isomorphic to the coset space $\Sigma_n/C_n = \{C_n\sigma\}$

and $k[\Sigma_n/C_n] = k \otimes_{kC_n} k\Sigma_n$. From this we deduce the following sequence of isomorphisms:

$$\mathrm{Prim}(L_n) \cong (kU_n \otimes R^{\otimes n}) \otimes_{k\Sigma_n} (sgn)$$

$$\cong (k[\Sigma_n/C_n] \otimes R^{\otimes n}) \otimes_{k\Sigma_n} (sgn)$$

$$\cong R^{\otimes n} \otimes_{kC_n} (sgn)$$

$$\cong R^{\otimes n}/\sim,$$

because $R^{\otimes n} \otimes_{kC_n} (sgn)$ is the quotient of $R^{\otimes n}$ by $1 - (-1)^n \tau$. Note that this sequence of isomorphism sends the class of

$$\omega = e_{12}(r_1) \wedge e_{23}(r_2) \wedge \cdots \wedge e_{n1}(r_n) \in \Lambda^n \mathfrak{gl}_n(R)$$

to $(-1)^{n-1} r_1 \otimes \cdots \otimes r_n$. We leave it as an exercise for the reader to show that the class of $d\omega \in \Lambda^{n-1}\mathfrak{gl}_n(R)$ is sent to $b(r_1 \otimes \cdots \otimes r_n)$. This identifies the differential d on $\mathrm{Prim}(L_*)$ with the differential b of $R^{\otimes *}/\sim$ up to a sign. $\diamond$

Theorem 9.10.10 (Loday-Quillen, Tsygan) *Let k be a field of characteristic zero and R an associative k-algebra. Then*

1. *The restriction of trace(λ) to primitive elements is an isomorphism*

$$\mathrm{Prim}\, H_n(\mathfrak{gl}(R); k) \cong HC_{n-1}(R).$$

2. *$H_*(\mathfrak{gl}(R); k)$ is a graded Hopf algebra, isomorphic to the tensor product*

$$\mathrm{Sym}\left(\bigoplus_{i=1}^{\infty} HC_{2i-1}(R)\right) \otimes_k \Lambda^*\left(\bigoplus_{i=0}^{\infty} HC_{2i}(R)\right).$$

Proof The direct sums $\oplus: \mathfrak{gl}_m(R) \times \mathfrak{gl}_n(R) \to \mathfrak{gl}_{m+n}(R)$ sending (x, y) to $x \oplus y = \begin{pmatrix} x & 0 \\ 0 & y \end{pmatrix}$ yield chain complex homomorphisms

$$\mu_{mn}: H_0(\mathfrak{sl}_m; \Lambda^*\mathfrak{gl}_m(R)) \otimes H_0(\mathfrak{sl}_n; \Lambda^*\mathfrak{gl}_n(R)) \to H_0(\mathfrak{sl}_{m+n}; \Lambda^*\mathfrak{gl}_{m+n}(R)).$$

Because we have taken coinvariants, which allow us to move the indices of $\mathfrak{gl}_{m+n}$ around inside $\mathfrak{gl}_{m+n+1}$, the maps μ_{mn}, $\mu_{m,n+1}$, and $\mu_{m+1,n}$ are compatible. Taking the limit as $m, n \to \infty$ yields an associative product μ on $L_* = H_0(\mathfrak{sl}; \Lambda^*\mathfrak{gl}(R))$. This makes L_* into a DG-algebra as well as a

DG-coalgebra. In fact L_* is a graded Hopf algebra (6.7.15) because the formula $(x, x) \oplus (y, y) \sim (x \oplus y, x \oplus y)$ in $\mathfrak{gl}_{m+n}(R) \times \mathfrak{gl}_{m+n}(R)$ shows that $\Delta: L_* \to L_* \otimes L_*$ is an algebra map. It follows that $H_*(\mathfrak{gl}(R); k) = H_*(L_*)$ is also a Hopf algebra.

The classification of graded-commutative Hopf algebras H_* over a field k of characteristic zero is known [MM]. If $H_0 = k$, then $H_* = \mathrm{Sym}(P_e) \otimes \Lambda^*(P_o)$, where P_e (resp. P_o) is the sum of the $\mathrm{Prim}(H_i)$ with i even (resp. i odd). Thus (1) implies (2). Applying this classification to L_*, a simple calculation (exercise!) shows that $\mathrm{Prim}\, H_n(L_*) \cong H_n \mathrm{Prim}(L_*)$. But $H_n \mathrm{Prim}(L_*) = HC_{n-1}(R)$ by Proposition 9.10.9. $\diamond$

Exercise 9.10.11 (Bloch, Kassel-Loday) Use the Hochschild-Serre spectral sequence (7.5.2) for $\mathfrak{sl} \subset \mathfrak{gl}$ to show that $H_2(\mathfrak{sl}_2(R); k) \cong HC_1(R)$.

10

The Derived Category

There are many formal similarities between homological algebra and algebraic topology. The Dold-Kan correspondence, for example, provides a dictionary between positive complexes and simplicial theory. The algebraic notions of chain homotopy, mapping cones, and mapping cylinders have their historical origins in simplicial topology.

The *derived category* $\mathbf{D}(\mathcal{A})$ of an abelian category is the algebraic analogue of the homotopy category of topological spaces. $\mathbf{D}(\mathcal{A})$ is obtained from the category $\mathbf{Ch}(\mathcal{A})$ of (cochain) complexes in two stages. First one constructs a quotient $\mathbf{K}(\mathcal{A})$ of $\mathbf{Ch}(\mathcal{A})$ by equating chain homotopy equivalent maps between complexes. Then one "localizes" $\mathbf{K}(\mathcal{A})$ by inverting quasi-isomorphisms via a calculus of fractions. These steps will be explained below in sections 10.1 and 10.3. The topological analogue is given in section 10.9.

10.1 The Category $\mathbf{K}(\mathcal{A})$

Let $\mathcal{A}$ be an abelian category, and consider the category $\mathbf{Ch} = \mathbf{Ch}(\mathcal{A})$ of cochain complexes in $\mathcal{A}$. The quotient category $\mathbf{K} = \mathbf{K}(\mathcal{A})$ of $\mathbf{Ch}$ is defined as follows: The objects of $\mathbf{K}$ are cochain complexes (the objects of $\mathbf{Ch}$) and the morphisms of $\mathbf{K}$ are the chain homotopy equivalence classes of maps in $\mathbf{Ch}$. That is, $\mathrm{Hom}_{\mathbf{K}}(A, B)$ is the set $\mathrm{Hom}_{\mathbf{Ch}}(A, B)/\sim$ of equivalence classes of maps in $\mathbf{Ch}$. We saw in exercise 1.4.5 that $\mathbf{K}$ is well defined as a category and that $\mathbf{K}$ is an additive category in such a way that the quotient $\mathbf{Ch}(\mathcal{A}) \to \mathbf{K}(\mathcal{A})$ is an additive functor.

It is useful to consider categories of complexes having special properties. If $\mathcal{C}$ is any full subcategory of $\mathbf{Ch}(\mathcal{A})$, let $\mathcal{K}$ denote the full subcategory of $\mathbf{K}(\mathcal{A})$ whose objects are the cochain complexes in $\mathcal{C}$. $\mathcal{K}$ is a "quotient category" of $\mathcal{C}$

in the sense that

$$\mathrm{Hom}_{\mathcal{K}}(A, B) = \mathrm{Hom}_{\mathbf{K}}(A, B) = \mathrm{Hom}_{\mathbf{Ch}}(A, B)/\sim \; = \mathrm{Hom}_{\mathcal{C}}(A, B)/\sim .$$

If $\mathcal{C}$ is closed under $\oplus$ and contains the zero object, then by 1.6.2 both $\mathcal{C}$ and $\mathcal{K}$ are additive categories and $\mathcal{C} \to \mathcal{K}$ is also an additive functor.

We write $\mathbf{K}^b(\mathcal{A})$, $\mathbf{K}^-(\mathcal{A})$, and $\mathbf{K}^+(\mathcal{A})$ for the full subcategories of $\mathbf{K}(\mathcal{A})$ corresponding to the full subcategories $\mathbf{Ch}^b$, $\mathbf{Ch}^-$, and $\mathbf{Ch}^+$ of bounded, bounded above, and bounded below cochain complexes described in section 1.1. These will be useful in section 5 below.

Of course, we could have equally well considered chain complexes instead of cochain complexes when constructing $\mathbf{K}$. However, the historical origins of derived categories were in Grothendieck's study of sheaf cohomology [HartRD], and the choice to use cochains is fixed in the literature.

Having introduced the cast of categories, we turn to their properties.

Lemma 10.1.1 *The cohomology $H^*(C)$ of a cochain complex C induces a family of well-defined functors H^i from the category $\mathbf{K}(\mathcal{A})$ to $\mathcal{A}$.*

Proof As we saw in 1.4.5, the map $u^*: H^i(A) \to H^i(B)$ induced by $u: A \to B$ is independent of the chain homotopy equivalence class of u. $\diamond$

Proposition 10.1.2 (Universal property) *Let $F: \mathbf{Ch}(\mathcal{A}) \to \mathcal{D}$ be any functor that sends chain homotopy equivalences to isomorphisms. Then F factors uniquely through $\mathbf{K}(\mathcal{A})$.*

$$\mathbf{Ch}(\mathcal{A}) \overset{F}{\longrightarrow} \mathcal{D}$$
$$\downarrow \quad \nearrow \exists!$$
$$\mathbf{K}(\mathcal{A})$$

Proof Let $\mathrm{cyl}(B)$ denote the mapping cylinder of the identity map of B; it has $B^n \oplus B^{n+1} \oplus B^n$ in degree n. We saw in exercise 1.5.4 that the inclusion $\alpha(b) = (0, 0, b)$ of B into $\mathrm{cyl}(B)$ is a chain homotopy equivalence with homopy inverse $\beta(b', b,'' b) = b' + b$; $\beta\alpha = \mathrm{id}_B$ and $\alpha\beta \sim \mathrm{id}_{\mathrm{cyl}(B)}$. By assumption, $F(\alpha): F(B) \to F(\mathrm{cyl}(B))$ is an isomorphism with inverse $F(\beta)$. Now the map $\alpha': B \to \mathrm{cyl}(B)$ defined by $\alpha'(b) = (b, 0, 0)$ has $\beta\alpha' = \mathrm{id}_B$, so

$$F(\alpha') = F(\alpha)F(\beta)F(\alpha') = F(\alpha)F(\beta\alpha') = F(\alpha).$$

Now suppose there is a chain homotopy s between two maps $f, g: B \to C$. Then $\gamma = (f, s, g): \mathrm{cyl}(B) \to C$ is a chain complex map (exercise 1.5.3).

Moreover, $\gamma\alpha' = f$ and $\gamma\alpha = g$. Hence in $\mathcal{D}$ we have

$$F(f) = F(\gamma)F(\alpha') = F(\gamma)F(\alpha) = F(g).$$

It follows that F factors through the quotient **K**(𝒜) of **Ch**(𝒜).   ◇

Exercise 10.1.1 Taking F to be **Ch**(𝒜) → **K**(𝒜), the proof shows that $\alpha': B \to \mathrm{cyl}(B)$ is a chain homotopy equivalence. Use an involution on $\mathrm{cyl}(B)$ to produce an explicit chain homotopy $\beta\alpha' \sim \mathrm{id}_{\mathrm{cyl}(B)}$.

Definition 10.1.3 (Triangles in **K**(𝒜)) Let $u: A \to B$ be a morphism in **Ch**. Recall from 1.5.2 that the mapping cone of u fits into an exact sequence

$$0 \to B \xrightarrow{v} \mathrm{cone}(u) \xrightarrow{\delta} A[-1] \to 0$$

in **Ch**. (The degree n part of $\mathrm{cone}(u)$ is $A^{n+1} \oplus B^n$ and A^{n+1} is the degree n part of $A[-1]$; see 1.2.8.) The *strict triangle* on u is the triple (u, v, δ) of maps in **K**; this data is usually written in the form

$$
\begin{array}{ccc}
 & \mathrm{cone}(u) & \\
\delta\nearrow & & \nwarrow v \\
A & \xrightarrow{\ u\ } & B.
\end{array}
$$

Now consider three fixed cochain complexes A, B and C. Suppose we are given three maps $u: A \to B$, $v: B \to C$, and $w: C \to A[-1]$ in **K**. We say that (u, v, w) is an *exact triangle* on (A, B, C) if it is "isomorphic" to a strict triangle (u', v', δ) on $u': A' \to B'$ in the sense that there is a diagram of chain complexes,

$$
\begin{array}{ccccccc}
A & \xrightarrow{\ u\ } & B & \xrightarrow{\ v\ } & C & \xrightarrow{\ w\ } & A[-1] \\
f\downarrow & & \downarrow g & & \downarrow h & & \downarrow f[-1] \\
A' & \xrightarrow{\ u'\ } & B' & \xrightarrow{\ v'\ } & \mathrm{cone}(u') & \xrightarrow{\ \delta\ } & A'[-1],
\end{array}
$$

commuting in **K** (i.e., commuting in **Ch** up to chain homotopy equivalences) and such that the maps f, g, h are isomorphisms in **K** (i.e., chain homotopy equivalences). If we replace u, v, and w by chain homotopy equivalent maps, we get the same diagram in **K**. This allows us to think of (u, v, w) as a triangle in the category **K**. A triangle is usually written as follows:

$$
\begin{array}{ccc}
 & C & \\
w \swarrow & & \nwarrow v \\
A & \xrightarrow{\;u\;} & B.
\end{array}
$$

Corollary 10.1.4 *Given an exact triangle* (u, v, w) *on* (A, B, C), *the cohomology sequence*

$$
\cdots \xrightarrow{w^*} H^i(A) \xrightarrow{u^*} H^i(B) \xrightarrow{v^*} H^i(C) \xrightarrow{w^*} H^{i+1}(A) \xrightarrow{u^*} \cdots
$$

is exact. Here we have identified $H^i(A[-1])$ *and* $H^{i+1}(A)$.

Proof For a strict triangle, this is precisely the long exact cohomology sequence of 1.5.2. Exactness for any exact triangle follows from this by the definition of a triangle and the fact that each H^i is a functor on **K**.   ◇

Example 10.1.5 The endomorphisms 0 and 1 of A fit into the exact triangles

$$
\begin{array}{ccccccc}
& A \oplus A[-1] & & & & 0 & \\
\swarrow & & \nwarrow & & \swarrow & & \nwarrow \\
A & \xrightarrow{\;\;0\;\;} & A & & A & \xrightarrow{\;1\;} & A.
\end{array}
$$

Indeed, $\mathrm{cone}(0) = A \oplus A[-1]$ and we saw in exercise 1.5.1 that $\mathrm{cone}(1)$ is a split exact complex, that is, $\mathrm{cone}(1)$ is isomorphic to zero in **K**.

Example 10.1.6 (Rotation) If (u, v, w) is an exact triangle, then so are its "rotates"

$$
\begin{array}{ccccccc}
& A[-1] & & & & B & \\
-u[-1] \swarrow & & \nwarrow w & & v \swarrow & & \nwarrow u \\
B & \xrightarrow{\;v\;} & C & \text{and} & C[+1] & \xrightarrow{-w[1]} & A.
\end{array}
$$

To see this, we may suppose that $C = \mathrm{cone}(u)$. In this case, the assertions amount to saying that the maps $\mathrm{cone}(v) \to A[-1]$ and $B[-1] \to \mathrm{cone}(\delta)$ are chain homotopy equivalences. The first was verified in exercises 1.5.6 and 1.5.8, and the second assertion follows from the observation that $\mathrm{cone}(\delta) = \mathrm{cyl}(-u)[-1]$.

Remark 10.1.7 Given a short exact sequence $0 \to A \xrightarrow{u} B \xrightarrow{v} C \to 0$ of complexes, there may be no map $C \xrightarrow{w} A[-1]$ making (u, v, w) into an exact triangle in $\mathbf{K}(\mathcal{A})$, even though there is a long exact cohomology sequence begging to be seen as coming from an exact triangle (but see 10.4.9 below). This cohomology sequence does arise from the mapping cylinder triangle

$$\text{cone}(u)$$
$$w \swarrow \qquad \nwarrow$$
$$A \longrightarrow \text{cyl}(u)$$

and the quasi-isomorphisms $\beta: \text{cyl}(u) \to B$ and $\varphi: \text{cone}(u) \to C$ of exercises 1.5.4 and 1.5.8.

Exercise 10.1.2 Regard the abelian groups $\mathbb{Z}/2$ and $\mathbb{Z}/4$ as cochain complexes concentrated in degree zero, and show that the short exact sequence $0 \to \mathbb{Z}/2 \xrightarrow{2} \mathbb{Z}/4 \xrightarrow{1} \mathbb{Z}/2 \to 0$ cannot be made into an exact triangle $(2, 1, w)$ on $(\mathbb{Z}/2, \mathbb{Z}/4, \mathbb{Z}/2)$ in the category $\mathbf{K}(\mathcal{A})$.

10.2 Triangulated Categories

The notion of triangulated category generalizes the structure that exact triangles give to $\mathbf{K}(\mathcal{A})$. One should think of exact triangles as substitutes for short exact sequences.

Suppose given a category $\mathbf{K}$ equipped with an automorphism T. A *triangle* on an ordered triple (A, B, C) of objects of $\mathbf{K}$ is a triple (u, v, w) of morphisms, where $u: A \to B$, $v: B \to C$, and $w: C \to T(A)$. A triangle is usually displayed as follows:

$$C$$
$$w \swarrow \qquad \nwarrow v$$
$$A \xrightarrow{u} B$$

A *morphism of triangles* is a triple (f, g, h) forming a commutative diagram in $\mathbf{K}$:

$$
\begin{array}{ccccccc}
A & \xrightarrow{u} & B & \xrightarrow{v} & C & \xrightarrow{w} & TA \\
\downarrow{f} & & \downarrow{g} & & \downarrow{h} & & \downarrow{Tf} \\
A' & \xrightarrow{u'} & B' & \xrightarrow{v'} & C' & \xrightarrow{w'} & TA'.
\end{array}
$$

Definition 10.2.1 (Verdier) An additive category $\mathbf{K}$ is called a *triangulated category* if it is equipped with an automorphism $T: \mathbf{K} \to \mathbf{K}$ (called the *translation functor*) and with a distinguished family of triangles (u, v, w) (called the *exact triangles* in $\mathbf{K}$), which are subject to the following four axioms:

(TR1) Every morphism $u: A \to B$ can be embedded in an exact triangle (u, v, w). If $A = B$ and $C = 0$, then the triangle $(\mathrm{id}_A, 0, 0)$ is exact. If (u, v, w) is a triangle on (A, B, C), isomorphic to an exact triangle (u', v', w') on (A', B', C'), then (u, v, w) is also exact.

$$
\begin{array}{ccccccc}
A & \xrightarrow{u} & B & \xrightarrow{v} & C & \xrightarrow{w} & TA \\
\downarrow{\scriptstyle\cong} & & \downarrow{\scriptstyle\cong} & & \downarrow{\scriptstyle\cong} & & \downarrow{\scriptstyle\cong} \\
A' & \xrightarrow{u'} & B' & \xrightarrow{v'} & C' & \xrightarrow{w'} & TA'
\end{array}
$$

(TR2) (Rotation). If (u, v, w) is an exact triangle on (A, B, C), then both its "rotates" $(v, w, -Tu)$ and $(-T^{-1}w, u, v)$ are exact triangles on (B, C, TA) and $(T^{-1}C, A, B)$, respectively.

(TR3) (Morphisms). Given two exact triangles

$$
\begin{array}{ccc}
& C & \\
{\scriptstyle w}\nearrow & & \nwarrow{\scriptstyle v} \\
A & \xrightarrow{\;u\;} & B
\end{array}
\qquad \text{and} \qquad
\begin{array}{ccc}
& C' & \\
{\scriptstyle w'}\nearrow & & \nwarrow{\scriptstyle v'} \\
A & \xrightarrow{\;u'\;} & B'
\end{array}
$$

with morphisms $f: A \to A'$, $g: B \to B'$ such that $gu = u'f$, there exists a morphism $h: C \to C'$ so that (f, g, h) is a morphism of triangles.

$$
\begin{array}{ccccccc}
A & \xrightarrow{u} & B & \xrightarrow{v} & C & \xrightarrow{w} & TA \\
\downarrow{\scriptstyle f} & & \downarrow{\scriptstyle g} & & \exists\downarrow{\scriptstyle h} & & \downarrow{\scriptstyle Tf} \\
A' & \xrightarrow{u'} & B' & \xrightarrow{v'} & C' & \xrightarrow{w'} & TA'
\end{array}
$$

(TR4) (The octahedral axiom). Given objects A, B, C, A', B', C' in $\mathbf{K}$, suppose there are three exact triangles: (u, j, ∂) on (A, B, C'); (v, x, i) on (B, C, A'); (vu, y, δ) on (A, C, B'). Then there is a fourth exact triangle $(f, g, (Tj)i)$ on (C', B', A')

$$
\begin{array}{ccc}
& A' & \\
{\scriptstyle (Tj)i}\nearrow & & \nwarrow{\scriptstyle g} \\
C' & \xrightarrow[f]{} & B'
\end{array}
$$

such that in the following octahedron we have (1) the four exact triangles form four of the faces; (2) the remaining four faces commute (that is, $\partial = \delta f\colon C' \to B' \to TA$ and $x = gy\colon C \to B' \to A'$); (3) $yv = fj\colon B \to B'$; and (4) $u\delta = ig\colon B' \to B$.

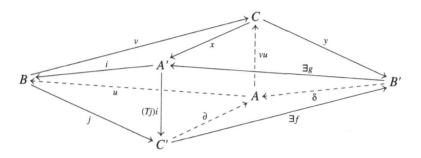

Exercise 10.2.1 If (u, v, w) is an exact triangle, show that the compositions vu, wv, and $(Tu)w$ are zero in **K**. *Hint*: Compare the triangles $(\mathrm{id}_A, 0, 0)$ and (u, v, w).

Exercise 10.2.2 (5-lemma) If (f, g, h) is a morphism of exact triangles, and both f and g are isomorphisms, show that h is also an isomorphism.

$$
\begin{array}{ccccccc}
A & \longrightarrow & B & \longrightarrow & C & \longrightarrow & TA \\
\downarrow{f} & & \downarrow{g} & & \exists\downarrow{h} & & \downarrow{Tf} \\
A' & \longrightarrow & B' & \longrightarrow & C' & \longrightarrow & TA'
\end{array}
$$

Remark 10.2.2 Every exact triangle is determined up to isomorphism by any one of its maps. Indeed, (TR3) gives a morphism between any two exact triangles (u, v, w) on (A, B, C) and (u, v', w') on (A, B, C'), and the 5-lemma shows that it is an isomorphism. In particular, the data of the octahedral axiom are completely determined by the two maps $A \xrightarrow{u} B \xrightarrow{v} C$.

Exegesis 10.2.3 The octahehral axiom (TR4) is sufficiently confusing that it is worth giving another visualization of this axiom, following [BBD]. Write the triangles as straight lines (ignoring the morphism $C \to T(A)$), and form the diagram

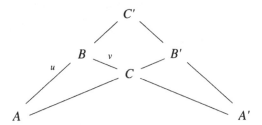

The octahedral axiom states that the three lines through A, B, and C determine the fourth line through (C', B', A'). This visualization omits the identity $\partial = \delta f$.

Proposition 10.2.4 $K(\mathcal{A})$ *is a triangulated category.*

Proof The translation $TA = A[-1]$ is defined in 1.2.8. We have already seen that axioms (TR1) and (TR2) hold. For (TR3) we may suppose that $C = \mathrm{cone}(u)$ and $C' = \mathrm{cone}(u')$; the map h is given by the naturality of the mapping cone construction.

It remains to check the octahedral axiom (TR4). For this we may assume that the given triangles are strict, that is, that $C' = \mathrm{cone}(u)$, $A' = \mathrm{cone}(v)$, and $B' = \mathrm{cone}(vu)$. Define f^n from $(C')^n = B^n \oplus A^{n+1}$ to $(B')^n = C^n \oplus A^{n+1}$ by $f^n(b, a) = (v(b), a)$, and define g^n from $(B')^n = C^n \oplus A^{n+1}$ to $(A')^n = C^n \oplus B^{n+1}$ by $g^n(c, a) = (c, u(a))$. Manifestly, these are chain maps, $\partial = \delta f$ and $x = gy$. Since the degree n part of $\mathrm{cone}(f)$ is $(C^n \oplus A^{n+1}) \oplus (B^{n+1} \oplus A^{n+2})$, there is a natural inclusion γ of A' into $\mathrm{cone}(f)$ such that the following diagram of chain complexes commutes.

$$
\begin{array}{ccccccc}
C' & \xrightarrow{f} & B' & \xrightarrow{g} & A' & \xrightarrow{(Tj)i} & C'[-1] \\
\| & & \| & & \downarrow{\gamma} & & \| \\
C' & \xrightarrow{f} & B' & \longrightarrow & \mathrm{cone}(f) & \longrightarrow & C'[-1]
\end{array}
$$

To see that γ is a chain homotopy equivalence, define $\varphi : \mathrm{cone}(f) \to A'$ by $\varphi(c, a_{n+1}, b, a_{n+2}) = (c, b + u(a_{n+1}))$. We leave it to the reader to check that φ is a chain map, that $\varphi\gamma = \mathrm{id}_{A'}$ and that $\gamma\varphi$ is chain homotopic to the identity map on $\mathrm{cone}(f)$. (Exercise!) This shows that $(f, g, (Tj)i)$ is an exact triangle, because it is isomorphic to the strict triangle of f. ◇

Corollary 10.2.5 *Let C be a full subcategory of* **Ch**$(\mathcal{A})$ *and K its corresponding quotient category. Suppose that C is an additive category and is closed* *under translation and the formation of mapping cones. Then* $\mathcal{K}$ *is a triangulated category.*

In particular, $\mathbf{K}^b(\mathcal{A})$, $\mathbf{K}^-(\mathcal{A})$, *and* $\mathbf{K}^+(\mathcal{A})$ *are triangulated categories.*

Definition 10.2.6 A *morphism* $F: \mathbf{K}' \to \mathbf{K}$ of triangulated categories is an additive functor that commutes with the translation functor T and sends exact triangles to exact triangles. There is a category of triangulated categories and their morphisms. We say that $\mathbf{K}'$ is a *triangulated subcategory* of $\mathbf{K}$ if $\mathbf{K}'$ is a full subcategory of $\mathbf{K}$, the inclusion is a morphism of triangulated categories, and if every exact triangle in $\mathbf{K}$ is exact in $\mathbf{K}'$.

For example, $\mathbf{K}^b$, $\mathbf{K}^+$, and $\mathbf{K}^-$ are triangulated subcategories of $\mathbf{K}(\mathcal{A})$. More generally, $\mathcal{K}$ is a triangulated subcategory of $\mathbf{K}$ in the above corollary.

Definition 10.2.7 Let $\mathbf{K}$ be a triangulated category and $\mathcal{A}$ an abelian category. An additive functor $H: \mathbf{K} \to \mathcal{A}$ is called a (covariant) *cohomological functor* if whenever (u, v, w) is an exact triangle on (A, B, C) the long sequence

$$\cdots \xrightarrow{w^*} H(T^iA) \xrightarrow{u^*} H(T^iB) \xrightarrow{v^*} H(T^iC) \xrightarrow{w^*} H(T^{i+1}A) \xrightarrow{u^*} \cdots$$

is exact in $\mathcal{A}$. We often write $H^i(A)$ for $H(T^iA)$ and $H^0(A)$ for $H(A)$ because, as we saw in 10.1.1, the zeroth cohomology $H^0: \mathbf{K}(\mathcal{A}) \to \mathcal{A}$ is the eponymous example of a cohomological functor. Here is another important cohomological functor:

Example 10.2.8 (Hom) If X is an object of a triangulated category $\mathbf{K}$, then $\text{Hom}_\mathbf{K}(X, -)$ is a cohomological functor from $\mathbf{K}$ to $\mathbf{Ab}$. To see this, we have to see that for every exact triangle (u, v, w) on (A, B, C) that the sequence

$$\text{Hom}_\mathbf{K}(X, A) \xrightarrow{u} \text{Hom}_\mathbf{K}(X, B) \xrightarrow{v} \text{Hom}_\mathbf{K}(X, C)$$

is exact; exactness elsewhere will follow from (TR2). The composition is zero since $vu = 0$. Given $g \in \text{Hom}_\mathbf{K}(X, B)$ such that $vg = 0$ we apply (TR3) and (TR2) to

$$
\begin{array}{ccccccc}
X & = & X & \longrightarrow & 0 & \longrightarrow & TX \\
{\scriptstyle\exists}\downarrow{\scriptstyle f} & & \downarrow{\scriptstyle g} & & \downarrow{\scriptstyle 0} & & {\scriptstyle\exists}\downarrow{\scriptstyle Tf} \\
A & \xrightarrow{u} & B & \xrightarrow{v} & C & \xrightarrow{w} & TA
\end{array}
$$

and conclude that there exists an $f \in \text{Hom}_\mathbf{K}(X, A)$ so that $uf = g$.

Exercise 10.2.3 If K is triangulated, show that the opposite category K^{op} is also triangulated. A covariant cohomological functor H from K^{op} to $\mathcal{A}$ is sometimes called a *contravariant cohomological functor* on K. If Y is any object of K, show that $\mathrm{Hom}_K(-, Y)$ is a contravariant cohomological functor on K.

Exercise 10.2.4 Let $\mathcal{A}^{\mathbb{Z}}$ be the category of graded objects in $\mathcal{A}$, a morphism from $A = \{A_n\}$ to $B = \{B_n\}$ being a family of morphisms $f_n: A_n \to B_n$. Define TA to be the translated graded object $A[-1]$, and call (u, v, w) an exact triangle on (A, B, C) if for all n the sequence

$$A_n \xrightarrow{u} B_n \xrightarrow{v} C_n \xrightarrow{w} A_{n-1} \xrightarrow{u} B_{n-1}$$

is exact. Show that axioms (TR1) and (TR2) hold, but that (TR3) fails for $\mathcal{A} = \mathbf{Ab}$. If $\mathcal{A}$ is the category of vector spaces over a field, show that $\mathcal{A}^{\mathbb{Z}}$ is a triangulated category, and that cohomology $H^*: \mathbf{K}(\mathcal{A}) \to \mathcal{A}^{\mathbb{Z}}$ is a morphism of triangulated categories.

Exercise 10.2.5 Let H be a cohomological functor on a triangulated category K, and let K_H denote the full subcategory of K consisting of those objects A such that $H^i(A) = 0$ for all i. Show that K_H is a triangulated subcategory of K.

Exercise 10.2.6 (Verdier) Show that every commutative square on the left in the diagram below can be completed to the diagram on the right, in which all the rows and columns are exact triangles and all the squares commute, except the one marked "–" which anticommutes. *Hint:* Use (TR1) to construct everything except the third column, and construct an exact triangle on (A, B', D). Then use the octahedral axiom to construct exact triangles on (C, D, B''), $(A,'' D, C')$, and finally (C', C'', C).

$$
\begin{array}{ccccccc}
A & \xrightarrow{i} & B & & & & \\
\scriptstyle u\downarrow & & \downarrow & & & & \\
A' & \longrightarrow & B' & & & &
\end{array}
\qquad
\begin{array}{ccccccccc}
A & \xrightarrow{i} & B & \xrightarrow{j} & C & \xrightarrow{k} & T(A) \\
\scriptstyle u\downarrow & & \downarrow & & \downarrow & & \downarrow \scriptstyle Tu \\
A' & \longrightarrow & B' & \longrightarrow & C' & \longrightarrow & T(A') \\
\scriptstyle v\downarrow & & \downarrow & & \downarrow & & \downarrow \scriptstyle Tv \\
A'' & \longrightarrow & B'' & \longrightarrow & C'' & \longrightarrow & T(A'') \\
\scriptstyle w\downarrow & & \downarrow & & \downarrow & \scriptstyle - & \downarrow \scriptstyle Tw \\
T(A) & \xrightarrow{Ti} & T(B) & \xrightarrow{Tj} & T(C) & \xrightarrow{Tk} & T^2(A)
\end{array}
$$

10.3 Localization and the Calculus of Fractions

The derived category $\mathbf{D}(\mathcal{A})$ is defined to be the localization $Q^{-1}\mathbf{K}(\mathcal{A})$ of category $\mathbf{K}(\mathcal{A})$ at the collection Q of quasi-isomorphisms, in the sense of the following definition.

Definition 10.3.1 Let S be a collection of morphisms in a category $\mathcal{C}$. A *localization of $\mathcal{C}$ with respect to S* is a category $S^{-1}\mathcal{C}$, together with a functor $q:\mathcal{C} \to S^{-1}\mathcal{C}$ such that

1. $q(s)$ is a isomorphism in $S^{-1}\mathcal{C}$ for every $s \in S$.
2. Any functor $F:\mathcal{C} \to \mathcal{D}$ such that $F(s)$ is an isomorphism for all $s \in S$ factors in a unique way through q. (It follows that $S^{-1}\mathcal{C}$ is unique up to equivalence.)

Examples 10.3.2

1. Let S be the collection of chain homotopy equivalences in $\mathbf{Ch}(\mathcal{A})$. The universal property 10.1.2 for $\mathbf{Ch}(\mathcal{A}) \to \mathbf{K}(\mathcal{A})$ shows that $\mathbf{K}(\mathcal{A})$ is the localization $S^{-1}\mathbf{Ch}(\mathcal{A})$.
2. Let $\widetilde{Q}$ be the collection of all quasi-isomorphisms in $\mathbf{Ch}(\mathcal{A})$. Since $\widetilde{Q}$ contains the S of part (1), it follows that

$$\widetilde{Q}^{-1}\mathbf{Ch}(\mathcal{A}) = Q^{-1}(S^{-1}\mathbf{Ch}(\mathcal{A})) = Q^{-1}\mathbf{K}(\mathcal{A}) = \mathbf{D}(\mathcal{A}).$$

Therefore we could have defined the derived category to be the localization $\widetilde{Q}^{-1}\mathbf{Ch}(\mathcal{A})$. However, in order to prove that $\widetilde{Q}^{-1}\mathbf{Ch}(\mathcal{A})$ exists we must first prove that $Q^{-1}\mathbf{K}(\mathcal{A})$ exists, by giving an explicit description of the morphisms.

Set-Theoretic Remark 10.3.3 If $\mathcal{C}$ is a small category, every localization $S^{-1}\mathcal{C}$ of $\mathcal{C}$ exists. (Add inverses to the presentation of $\mathcal{C}$ by generators and relations; see [MacH, II.8].) It is also not hard to see that $S^{-1}\mathcal{C}$ exists when the class S is a set. However, when the class S is not a set, the existence of localizations is a delicate set-theoretic question.

The standard references [Verd], [HarRD], [GZ] all ignore these set-theoretic problems. Some adherents of the Grothendieck school avoid these difficulties by imagining the existence of a larger universe in which $\mathcal{C}$ is small and constructing the localization in that universe. Nevertheless, the issue of whether or not $S^{-1}\mathcal{C}$ exists in our universe is important to other schools of thought, and in particular to topologists who need to localize with respect to homology theories; see [A, III.14].

In this section we shall consider a special case in which localizations $S^{-1}C$ may be constructed within our universe, the case in which S is a "locally small multiplicative system." This is due to the presence of a kind of calculus of fractions.

In section 10.4 we will see that the multiplicative system Q of quasi-isomorphisms in $\mathbf{K}(\mathcal{A})$ is locally small when $\mathcal{A}$ is either **mod–R** or **Sheaves**(X). This will prove that $\mathbf{D}(\mathcal{A})$ exists within our universe. We will also see that if $\mathcal{A}$ has enough injectives (resp. projectives), the existence of Cartan-Eilenberg resolutions 5.7.1 allows us to forget about the set-theoretical difficulties in asserting that $\mathbf{D}^+(\mathcal{A})$ exists (resp. that $\mathbf{D}^-(\mathcal{A})$ exists).

Definition 10.3.4 A collection S of morphisms in a category C is called a *multiplicative system* in C if it satisfies the following three self-dual axioms:

1. S is closed under composition (if $s, t \in S$ are composable, then $st \in S$) and contains all identity morphisms (id$_X \in S$ for all objects X in C).
2. (Ore condition) If $t: Z \to Y$ is in S, then for every $g: X \to Y$ in C there is a commutative diagram "$gs = tf$" in C with s in S.

$$
\begin{array}{ccc}
W & \xrightarrow{f} & Z \\
s\downarrow & & \downarrow t \\
X & \xrightarrow{g} & Y
\end{array}
$$

   (The slogan is "$t^{-1}g = fs^{-1}$ for some f and s.") Moreover, the symmetric statement (whose slogan is "$fs^{-1} = t^{-1}g$ for some t and g") is also valid.
3. (Cancellation) If $f, g: X \to Y$ are parallel morphisms in C, then the following two conditions are equivalent:
   (a) $sf = sg$ for some $s \in S$ with source Y.
   (b) $ft = gt$ for some $t \in S$ with target X.

Prototype 10.3.5 (Localizations of rings) An associative ring R with unit may be considered as an additive category $\mathcal{R}$ with one object $\cdot$ via $R = \mathrm{End}_{\mathcal{R}}(\cdot)$. Let S be a subset of R closed under multiplication and containing 1. If R is commutative, or more generally if S is in the center of R, then S is always a multiplicative system in $\mathcal{R}$; the usual ring of fractions $S^{-1}R$ is also the localization $S^{-1}\mathcal{R}$ of the category $\mathcal{R}$.

If S is not central, then S is a multiplicative system in $\mathcal{R}$ if and only if S is a "2-sided denominator set" in R in the sense of [Faith]. The classical ring of fractions $S^{-1}R$ is easy to construct in this case, each element being represented as either fs^{-1} or $t^{-1}g$ ($f, g \in R$ and $s, t \in S$), and again $S^{-1}R$ is the localization of the category $\mathcal{R}$.

The construction of the ring of fractions $S^{-1}R$ serves as the prototype for the construction of the localization $S^{-1}\mathcal{C}$. We call a chain in $\mathcal{C}$ of the form

$$fs^{-1}: X \xleftarrow{s} X_1 \xrightarrow{f} Y$$

a (left) "fraction" if s is in S. Call fs^{-1} *equivalent* to $X \xleftarrow{t} X_2 \xrightarrow{g} Y$ just in case there is a fraction $X \leftarrow X_3 \to Y$ fitting into a commutative diagram in $\mathcal{C}$:

$$
\begin{array}{ccccc}
 & & X_1 & & \\
 & {}^{s}\swarrow & \uparrow & \searrow^{f} & \\
X & \longleftarrow & X_3 & \longrightarrow & Y \, . \\
 & {}^{t}\nwarrow & \downarrow & \nearrow^{g} & \\
 & & X_2 & &
\end{array}
$$

It is easy to see that this is an equivalence relation. Write $\operatorname{Hom}_S(X, Y)$ for the family of equivalence classes of such fractions. Unfortunately, there is no *a priori* reason for this to be a set, unless S is "locally small" in the following sense.

Set-Theoretic Considerations 10.3.6 A multiplicative system S is called *locally small* (on the left) if for each X there exists a set S_X of morphisms in S, all having target X, such that for every $X_1 \to X$ in S there is a map $X_2 \to X_1$ in $\mathcal{C}$ so that the composite $X_2 \to X_1 \to X$ is in S_X.

If S is locally small, then $\operatorname{Hom}_S(X, Y)$ is a set for every X and Y. To see this, we make S_X the objects of a small category, a morphism from $X_1 \xrightarrow{s} X$ to $X_2 \xrightarrow{t} X$ being a map $X_2 \to X_1$ in $\mathcal{C}$ so that t is $X_2 \to X_1 \xrightarrow{s} X$. The Øre condition says that by enlarging S_X slightly we can make it a filtered category (2.6.13). There is a functor $\operatorname{Hom}_{\mathcal{C}}(-, Y)$ from S_X to **Sets** sending s to the set of all fractions fs^{-1}, and $\operatorname{Hom}_S(X, Y)$ is the colimit of this functor.

Composition of fractions is defined as follows. To compose $X \leftarrow X' \xrightarrow{g} Y$ with $Y \xleftarrow{t} Y' \to Z$ we use the Ore condition to find a diagram

$$
\begin{array}{ccccc}
W & \xrightarrow{f} & Y' & \longrightarrow & Z \\
\downarrow^{s} & & \downarrow^{t} & & \\
X & \longleftarrow & X' & \xrightarrow{g} & Y
\end{array}
$$

with s in S; the composite is the class of the fraction $X \leftarrow W \rightarrow Z$ in $\mathrm{Hom}_S(X, Z)$. The slogan for the Ore condition, $t^{-1}g = fs^{-1}$, is a symbolic description of composition. It is not hard to see that the equivalence class of the composite is independent of the choice of X' and Y', so that we have defined a pairing

$$\mathrm{Hom}_S(X, Y) \times \mathrm{Hom}_S(Y, Z) \rightarrow \mathrm{Hom}_S(X, Z).$$

(Check this!) It is clear from the construction that composition is associative, and that $X = X = X$ is a 2-sided identity element. Hence the $\mathrm{Hom}_S(X, Y)$ (if they are sets) form the morphisms of a category having the same objects as C; it will be our localization $S^{-1}C$.

Gabriel-Zisman Theorem 10.3.7 ([GZ]) *Let S be a locally small multiplicative system of morphisms in a category C. Then the category $S^{-1}C$ constructed above exists and is a localization of C with respect to S. The universal functor $q: C \rightarrow S^{-1}C$ sends $f: X \rightarrow Y$ to the sequence $X = X \xrightarrow{f} Y$.*

Proof To see that $q: C \rightarrow S^{-1}C$ is a functor, observe that the composition of $X = X \xrightarrow{f} Y$ and $Y = Y \xrightarrow{h} Z$ is $X = X \xrightarrow{hf} Z$ since we can choose $t = \mathrm{id}_X$ and $f = g$. If s is in S, then $q(s)$ is an isomorphism because the composition of $X = X \xrightarrow{s} Y$ and $Y \xleftarrow{s} X = X$ is $X = X = X$ (take $W = X$). Finally, suppose that $F: C \rightarrow \mathcal{D}$ is another functor sending S to isomorphisms. Define $S^{-1}F: S^{-1}C \rightarrow \mathcal{D}$ by sending the fraction fs^{-1} to $F(f)F(s)^{-1}$. Given g and t, the equality $gs = tf$ in C shows that $F(g)F(s) = F(t)F(f)$, or $F(t^{-1}g) = F(fs^{-1})$; it follows that $S^{-1}F$ respects composition and is a functor. It is clear that $F = (S^{-1}F) \circ q$ and that this factorization is unique. $\diamond$

Corollary 10.3.8 $S^{-1}C$ *can be constructed using equivalence classes of "right fractions" $t^{-1}g: X \xrightarrow{g} Y' \xleftarrow{t} Y$, provided that S is "locally small on the right" (the dual notion to locally small, involving maps $Y \rightarrow Y'$ in S).*

Proof S^{op} is a multiplicative system in C^{op}. Since $C^{\mathrm{op}} \rightarrow (S^{\mathrm{op}})^{-1}C^{\mathrm{op}}$ is a localization, so is its dual $C \rightarrow [(S^{\mathrm{op}})^{-1}(C^{\mathrm{op}})]^{\mathrm{op}}$. But this is constructed using the fractions $t^{-1}g$. $\diamond$

Corollary 10.3.9 *Two parallel maps $f, g: X \rightarrow Y$ in C become identified in $S^{-1}C$ if and only if $sf = sg$ for some $s: X_3 \rightarrow X$ in S.*

Exercise 10.3.1

1. If Z is a zero object (resp. an initial object, a terminal object) in C, show that $q(Z)$ is a zero object (resp. an initial object, a terminal object) in $S^{-1}C$.

2. If the product $X \times Y$ exists in C, show that $q(X \times Y) \cong q(X) \times q(Y)$ in $S^{-1}C$.

Corollary 10.3.10 *Suppose that C has a zero object. Then for every X in C:*

$$q(X) \cong 0 \text{ in } S^{-1}C \Leftrightarrow S \text{ contains the zero map } X \xrightarrow{0} X.$$

Proof Since $q(0)$ is a zero object in $S^{-1}C$, $q(X) \cong 0$ if and only if the parallel maps $0, \mathrm{id}_X : X \to X$ become identified in $S^{-1}C$, that is, iff $0 = s0 = s$ for some s. $\diamond$

Corollary 10.3.11 *If C is an additive category, then so is $S^{-1}C$, and q is an additive functor.*

Proof If C is an additive category, we can add fractions from X to Y as follows. Given fractions $f_1 s_1^{-1}$ and $f_2 s_2^{-1}$, we use the Ore condition to find an $s: X_2 \to X$ in S and $f_1', f_2': X_2 \to Y$ so that $f_1 s_1^{-1} \sim f_1' s^{-1}$ and $f_2 s_2^{-1} \sim f_2' s^{-1}$; the sum $(f_1' + f_2') s^{-1}$ is well defined up to equivalence. (Check this!) Since $q(X \times Y) \cong q(X) \times q(Y)$ in $S^{-1}C$ (exercise 10.3.1), it follows that $S^{-1}C$ is an additive category (A.4.1) and that q is an additive functor. $\diamond$

It is often useful to compare the localizations of subcategories with $S^{-1}C$. For this we introduce the following definition.

Definition 10.3.12 (Localizing subcategories) Let B be a full subcategory of C, and let S be a locally small multiplicative system in C whose restriction $S \cap B$ to B is also a multiplicative system. For legibility, we will write $S^{-1}B$ for $(S \cap B)^{-1}B$. B is called a *localizing subcategory* of C (for S) if the natural functor $S^{-1}B \to S^{-1}C$ is fully faithful. That is, if it identifies $S^{-1}B$ with the full subcategory of $S^{-1}C$ on the objects of B.

Lemma 10.3.13 *A full subcategory B of C is localizing for S iff (1) holds. Condition (2) implies that B is localizing if S is locally small on the left, and condition (3) implies that B is localizing if S is locally small on the right.*

1. *For each B and B' in B, the colimit $\mathrm{Hom}_{S \cap B}(B, B')$ (taken in B) maps bijectively to the colimit $\mathrm{Hom}_S(B, B')$ (taken in C).*

2. *Whenever $C \to B$ is a morphism in S with B in $\mathcal{B}$, there is a morphism $B' \to C$ in $\mathcal{C}$ with B' in $\mathcal{B}$ such that the composite $B' \to B$ is in S.*

3. *Whenever $B \to C$ is a morphism in S with B in $\mathcal{B}$, there is a morphism $C \to B'$ in $\mathcal{C}$ with B' in $\mathcal{B}$ such that the composite $B \to B'$ is in S.*

Proof The statement that $S^{-1}\mathcal{B} \to S^{-1}\mathcal{C}$ is fully faithful means that the morphisms coincide (A.2.3), which by the Gabriel-Zisman Theorem 10.3.7 is assertion (1). Part (2) states that every left fraction $B \leftarrow C \to B''$ is equivalent to a fraction $B \leftarrow B' \to B''$, which must lie in the full subcategory $\mathcal{B}$. In particular, if two left fractions $\mathcal{B}$ are equivalent via a fraction $B \leftarrow C \to B''$ with C in $\mathcal{C}$, they are equivalent via a fraction with C in $\mathcal{B}$. Thus (2) implies (1) when S is locally small on the left. Replacing 'left' by 'right' and citing 10.3.8 proves that (3) implies (1) when S is locally small on the right. $\diamond$

Corollary 10.3.14 *If $\mathcal{B}$ is a localizing subcategory of $\mathcal{C}$, and for every object C in $\mathcal{C}$ there is a morphism $C \to B$ in S with B in $\mathcal{B}$, then $S^{-1}\mathcal{B} \cong S^{-1}\mathcal{C}$.*

Suppose in addition that $S \cap \mathcal{B}$ consists of isomorphisms. Then

$$\mathcal{B} \cong S^{-1}\mathcal{B} \cong S^{-1}\mathcal{C}.$$

Example 10.3.15 Assume $\mathbf{D}(\mathcal{A})$ exists. The subcategories $\mathbf{K}^b(\mathcal{A})$, $\mathbf{K}^+(\mathcal{A})$, and $\mathbf{K}^-(\mathcal{A})$ of $\mathbf{K}(\mathcal{A})$ are localizing for Q (check this). Thus their localizations exist and are the full subcategories $\mathbf{D}^b(\mathcal{A})$, $\mathbf{D}^+(\mathcal{A})$, and $\mathbf{D}^-(\mathcal{A})$ of $\mathbf{D}(\mathcal{A})$ whose objects are the cochain complexes which are bounded, bounded below, and bounded above, respectively.

Example 10.3.16 Let S be a multiplicative system in a ring, and let Σ be the collection of all morphisms $A \to B$ in **mod**–R such that $S^{-1}A \to S^{-1}B$ is an isomorphism. It is not hard to see that Σ is a multiplicative system in **mod**–R. The subcategory **mod**–$S^{-1}R$ is localizing, because the natural map $A \to S^{-1}A$ is in Σ for every R-module A. Since $\Sigma \cap$ **mod**–$S^{-1}R$ consists of isomorphisms, we therefore have

$$\mathbf{mod}\text{–}S^{-1}R \cong \Sigma^{-1}\mathbf{mod}\text{–}R.$$

Exercise 10.3.2 (Serre subcategories) Let $\mathcal{A}$ be an abelian category. An abelian subcategory $\mathcal{B}$ is called a *Serre subcategory* if it is closed under subobjects, quotients, and extensions. Suppose that $\mathcal{B}$ is a Serre subcategory of $\mathcal{A}$, and let Σ be the family of all morphisms f in $\mathcal{A}$ with $\ker(f)$ and $\operatorname{coker}(f)$ in $\mathcal{B}$.

1. Show that Σ is a multiplicative system in $\mathcal{A}$. We write $\mathcal{A}/\mathcal{B}$ for the localization $\Sigma^{-1}\mathcal{A}$ (provided that it exists).

2. Show that $q(X) \cong 0$ in $\mathcal{A}/\mathcal{B}$ if and only if X is in $\mathcal{B}$.

3. Assume that $\mathcal{B}$ is a small category, and show that Σ is locally small. This is one case in which $\mathcal{A}/\mathcal{B} = \Sigma^{-1}\mathcal{A}$ exists. More generally, $\mathcal{A}/\mathcal{B}$ exists whenever $\mathcal{A}$ is *well-powered*, that is, whenever the family of subobjects of any object of $\mathcal{A}$ is a set; see [Swan, pp.44ff].

4. Show that $\mathcal{A}/\mathcal{B}$ is an abelian category, and that $q: \mathcal{A} \to \mathcal{A}/\mathcal{B}$ is an exact functor.

5. Let S be a multiplicative system in a ring R, and let $\mathbf{mod}_S R$ denote the full subcategory of R-modules A such that $S^{-1}A \cong 0$. Show that $\mathbf{mod}_S R$ is a Serre subcategory of $\mathbf{mod}{-}R$. Conclude that $\mathbf{mod}{-}S^{-1}R \cong \mathbf{mod}{-}R/\mathbf{mod}_S R$.

10.4 The Derived Category

In this section we show that $\mathbf{D}(\mathcal{A})$ is a triangulated category and that $\mathbf{D}^+(\mathcal{A})$ is determined by maps between bounded below complexes of injectives. We also show that $\mathbf{D}(\mathcal{A})$ exists within our universe, at least if $\mathcal{A}$ is $\mathbf{mod}{-}R$ or $\mathbf{Sheaves}(X)$.

For this we generalize slightly. Let $\mathbf{K}$ be a triangulated category. The system S *arising from* a cohomological functor $H: \mathbf{K} \to \mathcal{A}$ is the collection of all morphisms s in $\mathbf{K}$ such that $H^i(s)$ is an isomorphism for all integers i. For example, the quasi-isomorphisms Q arise from the cohomological functor H^0.

Proposition 10.4.1 *If S arises from a cohomological functor, then*

1. *S is a multiplicative system.*
2. *$S^{-1}\mathbf{K}$ is a triangulated category, and $\mathbf{K} \to S^{-1}\mathbf{K}$ is a morphism of triangulated categories (in any universe containing $S^{-1}\mathbf{K}$).*

Proof We first show that the system S is multiplicative (10.3.4). Axiom (1) is trivial. To prove (2), let $f: X \to Y$ and $s: Z \to Y$ be given. Embed s in an exact triangle (s, u, δ) on (Z, Y, C) using (TR1). Complete $uf: X \to C$ into an exact triangle (t, uf, v) on (W, X, C). By axiom (TR3) there is a morphism g such that

$$
\begin{array}{ccccccc}
W & \xrightarrow{t} & X & \xrightarrow{uf} & C & \xrightarrow{v} & W[-1] \\
{\scriptstyle g}\downarrow & & \downarrow{\scriptstyle f} & & \| & & \downarrow{\scriptstyle Tg} \\
Z & \xrightarrow{s} & Y & \xrightarrow{u} & C & \xrightarrow{\delta} & Z[-1]
\end{array}
$$

is a morphism of triangles. If $H^*(s)$ is a isomorphism, then $H^*(C) = 0$. Applying this to the long exact sequence of the other triangle, we see that

$H^*(t)$ is also an isomorphism. The symmetric assertion may be proven similarly, or by appeal to $\mathbf{K}^{\mathrm{op}} \to \mathcal{A}^{\mathrm{op}}$.

To verify axiom (3), we consider the difference $h = f - g$. Given $s\colon Y \to Y'$ in S with $sf = sg$, embed s in an exact triangle (u, s, δ) on (Z, Y, Y'). Note that $H^*(Z) = 0$. Since $\mathrm{Hom}_{\mathbf{K}}(X, -)$ is a cohomological functor (by 10.2.8),

$$\mathrm{Hom}_{\mathbf{K}}(X, Z) \xrightarrow{\ u\ } \mathrm{Hom}_{\mathbf{K}}(X, Y) \xrightarrow{\ s\ } \mathrm{Hom}_{\mathbf{K}}(X, Y')$$

is exact. Since $s(f - g) = 0$, there is a $g\colon X \to Z$ in $\mathbf{K}$ such that $f - g = ug$. Embed g in an exact triangle (t, g, w) on (X', X, Z). Since $gt = 0$, $(f - g)t = ugt = 0$, whence $ft = gt$. And since $H^*(Z) = 0$, the long exact sequence for H shows that $H^*(X') \cong H^*(X)$, that is, $t \in S$. The other implication of axiom (3) is analogous and may be deduced from the above by appeal to $\mathbf{K}^{\mathrm{op}} \to \mathcal{A}^{\mathrm{op}}$.

Now suppose that $S^{-1}\mathbf{K}$ exists. The formula $T(fs^{-1}) = T(f)T(s)^{-1}$ defines a translation functor T on $S^{-1}\mathbf{K}$. To show that $S^{-1}\mathbf{K}$ is triangulated, we need to define exact triangles. Given $us_1^{-1}\colon A \to B$, $vs_2^{-1}\colon B \to C$, and $ws_3^{-1}\colon C \leftarrow C' \to T(A)$, the Ore condition for S yields morphisms $t_1\colon A' \to A$ and $t_2\colon B' \to B$ in S and $u'\colon A' \to B'$, $v'\colon B' \to C'$ in $\mathcal{C}$ so that $us_1^{-1} \cong t_2 u' t_1^{-1}$ and $vs_2^{-1} \cong s_3 v' t_2^{-1}$. We say that $(us_1^{-1}, vs_2^{-1}, ws_3^{-1})$ is an *exact triangle* in $S^{-1}\mathbf{K}$ just in case (u', v', w) is an exact triangle in $\mathbf{K}$. The verification that $S^{-1}\mathbf{K}$ is triangulated is left to the reader as an exercise, being straightforward but lengthy; one uses the fact that $\mathrm{Hom}_S(X, Y)$ may also be calculated using fractions of the form $t^{-1}g$. $\Diamond$

Corollary 10.4.2 (Universal property) *Let $F\colon \mathbf{K} \to \mathbf{L}$ be a morphism of triangulated categories such that $F(s)$ is an isomorphism for all s in S, where S arises from a cohomological functor. Since $q\colon \mathbf{K} \to S^{-1}\mathbf{K}$ is a localization, there is a unique functor $F'\colon S^{-1}\mathbf{K} \to \mathbf{L}$ such that $F = F' \circ q$. In fact, F' is a morphism of triangulated categories.*

Corollary 10.4.3 $\mathbf{D}(\mathcal{A})$, $\mathbf{D}^b(\mathcal{A})$, $\mathbf{D}^+(\mathcal{A})$ *and* $\mathbf{D}^-(\mathcal{A})$ *are triangulated categories (in any universe containing them).*

Proposition 10.4.4 *Let R be a ring. Then $\mathbf{D}(\mathcal{A})$ exists and is a triangulated category if $\mathcal{A}$ is $\mathbf{mod}\text{–}R$, or either of*

- *Presheaves(X), presheaves of R-modules on a topological space X, or*
- *Sheaves(X), sheaves of R-modules on a topological space X.*

Proof We have to prove that the multiplicative system Q is locally small (10.3.6). Given a fixed cochain complex of R-modules A, choose an infinite cardinal number κ larger than the cardinality of the sets underlying the A^i and R. Call a cochain complex B *petite* if its underlying sets have cardinality $< \kappa$; there is a set of isomorphism classes of petite cochain complexes, hence a set S_X of isomorphism classes of quasi-isomorphisms $A' \to A$ with A' petite.

Given a quasi-isomorphism $B \to A$, it suffices to show that B contains a petite subcomplex B' quasi-isomorphic to A. Since $H^*(A)$ has cardinality $< \kappa$, there is a petite subcomplex B_0 of B such that the map $f_0^*: H^*(B_0) \to H^*(A)$ is onto. Since $\ker(f_0^*)$ has cardinality $< \kappa$, we can enlarge B_0 to a petite subcomplex B_1 such that $\ker(f_0^*)$ vanishes in $H^*(B_1)$. Inductively, we can construct an increasing sequence of petite subcomplexes B_n of B such that the kernel of $H^*(B_n) \to H^*(A)$ vanishes in $H^*(B_{n+1})$. But then their union $B' = \cup B_n$ is a petite subcomplex of B with

$$H^*(B') \cong \varinjlim H^*(B_n) \cong H^*(A).$$

The proof for presheaves is identical, except that κ must bound the number of open subsets U as well as the cardinality of $A(U)$ for every open subset U of X. The proof for sheaves is similar, using the following three additional facts, which may be found in [Hart] or [Gode]: (1) if κ bounds card $A(U)$ for all U and the number of such U, then κ also bounds the cardinality of the stalks A_x for $x \in X$ (2.3.12); (2) a map $B \to A$ is a quasi-isomorphism in **Sheaves**(X) iff every map of stalks $B_x \to A_x$ is a quasi-isomorphism; and (3) for every directed system of sheaves we have $H^*(\varinjlim B_n) = \varinjlim H^*(B_n)$. $\Diamond$

Remark 10.4.5 (Gabber) The proof shows that $\mathbf{D}(\mathcal{A})$ exists within our universe for every well-powered abelian category $\mathcal{A}$ that satisfies (AB5) and has a set of generators.

We conclude with a discussion of the derived category $\mathbf{D}^+(\mathcal{A})$. Assuming that $\mathcal{A}$ has enough injectives and we are willing to always pass to complexes of injectives, there is no need to leave the homotopy category $\mathbf{K}^+(\mathcal{A})$. In particular, $\mathbf{D}^+(\mathcal{A})$ will exist in our universe even if $\mathbf{D}(\mathcal{A})$ may not.

Lemma 10.4.6 *Let Y be a bounded below cochain complex of injectives. Every quasi-isomorphism $t: Y \to Z$ of complexes is a split injection in $\mathbf{K}(\mathcal{A})$.*

Proof The mapping cone cone$(t) = T(Y) \oplus Z$ is exact (1.5.4), and there is a natural map $\varphi: \text{cone}(t) \to T(Y)$. The Comparison Theorem of 2.3.7 (or rather its proof; see 2.2.6) shows that φ is null-homotopic, say, by a chain homotopy $v = (k, s)$ from $T(Y) \oplus Z$ to Y. The first coordinate of the equation $-y =$

$\varphi(y, z) = (vd + dv)(y, z)$ yields the equation

$$y = (kdy + sty - dky) + (dsz - sdz).$$

Thus $ds = sd$ (i.e., s is a morphism of complexes) and $st = \mathrm{id}_Y + dk - kd$, that is, k is a chain homotopy equivalence $st \simeq \mathrm{id}_Y$. Hence $st = \mathrm{id}_Y$ in $\mathbf{K}^+(\mathcal{A})$.
                                                                                    $\diamond$

Corollary 10.4.7 *If I is a bounded below cochain complex of injectives, then*

$$\mathrm{Hom}_{\mathbf{D}(\mathcal{A})}(X, I) \cong \mathrm{Hom}_{\mathbf{K}(\mathcal{A})}(X, I)$$

for every X. Dually, if P is a bounded above cochain complex of projectives, then

$$\mathrm{Hom}_{\mathbf{D}(\mathcal{A})}(P, X) \cong \mathrm{Hom}_{\mathbf{K}(\mathcal{A})}(P, X).$$

Proof We prove the assertion for $Y = I$, using the notation of the lemma. Every right fraction $t^{-1}g\colon X \xrightarrow{g} Z \xleftarrow{t} Y$ is equivalent to $sg = (st)t^{-1}g\colon X \to Y$. Conversely, if two parallel arrows $f, g\colon X \to Y$ in $\mathbf{K}(\mathcal{A})$ become identified in $\mathbf{D}(\mathcal{A}) = Q^{-1}\mathbf{K}(\mathcal{A})$, then $tf = tg$ for some quasi-isomorphism $t\colon Y \to Z$ by 10.3.9, which implies that $f = stf = stg = g$ in $\mathbf{K}(\mathcal{A})$. $\diamond$

Exercise 10.4.1 In the situation of the lemma, show that $(tk, 1)\colon \mathrm{cone}(t) \to Z$ induces an isomorphism $Z \cong Y \oplus \mathrm{cone}(t)$ in $\mathbf{K}(\mathcal{A})$.

Theorem 10.4.8 *Suppose that $\mathcal{A}$ has enough injectives. Then $\mathbf{D}^+(\mathcal{A})$ exists in our universe because it is equivalent to the full subcategory $\mathbf{K}^+(\mathcal{I})$ of $\mathbf{K}^+(\mathcal{A})$ whose objects are bounded below cochain complexes of injectives*

$$\mathbf{D}^+(\mathcal{A}) \cong \mathbf{K}^+(\mathcal{I}).$$

Dually, if $\mathcal{A}$ has enough projectives, then the localization $\mathbf{D}^-(\mathcal{A})$ of $\mathbf{K}^-(\mathcal{A})$ exists and is equivalent to the full subcategory $\mathbf{K}^-(\mathcal{P})$ of bounded above cochain complexes of projectives in $\mathbf{K}^-(\mathcal{A})$:

$$\mathbf{D}^-(\mathcal{A}) \cong \mathbf{K}^-(\mathcal{P}).$$

Proof Recall from 5.7.2 that every X in $\mathbf{Ch}^+(\mathcal{A})$ has a Cartan-Eilenberg resolution $X \to I$ with $\mathrm{Tot}(I)$ in $\mathbf{K}^+(\mathcal{I})$; since X is bounded below, this is a quasi-isomorphism (exercise 5.7.1). If $Y \to X$ is a quasi-isomorphism, then so is

$Y \to \mathrm{Tot}(I)$; by 10.3.13(3), $\mathbf{K}^+(\mathcal{I})$ is a localizing subcategory of $\mathbf{K}^+(\mathcal{A})$. This proves that $\mathbf{D}^+(\mathcal{A}) \cong S^{-1}\mathbf{K}^+(\mathcal{I})$, and by 10.3.14 it suffices to show that every quasi-isomorphism in $\mathbf{K}^+(\mathcal{I})$ is an isomorphism. Let Y and X be bounded below cochain complexes of injectives and $t\colon Y \to X$ a quasi-isomorphism. By lemma 10.4.6, there is a map $s\colon X \to Y$ so that $st = \mathrm{id}_Y$ in $\mathbf{K}^+(\mathcal{A})$. Interchanging the roles of X and Y, s and t, we see that $us = \mathrm{id}_X$ for some u. Hence t is an isomorphism in $\mathbf{K}^+(\mathcal{I})$ with $t^{-1} = s$.

Dually, if $\mathcal{A}$ has enough projectives, then $\mathcal{A}^{\mathrm{op}}$ has enough injectives and $\mathbf{D}^-(\mathcal{A}) \cong \mathbf{D}^+(\mathcal{A}^{\mathrm{op}})^{\mathrm{op}} \cong \mathbf{K}^+(\mathcal{P}^{\mathrm{op}})^{\mathrm{op}} \cong \mathbf{K}^-(\mathcal{P})$.   ◇

Example 10.4.9 Every short exact sequence $0 \to A \xrightarrow{u} B \xrightarrow{v} C \to 0$ of cochain complexes fits into an exact triangle in $\mathbf{D}(\mathcal{A})$, isomorphic to the strict triangle on u. Indeed, the quasi-isomorphism $\varphi\colon \mathrm{cone}(u) \to C$ of 1.5.8 allows us to form the exact triangle $(u, v, \delta\varphi^{-1})$ on (A, B, C). This construction should be contrasted with the observation in 10.1.7 that there may be no similar exact triangle (u, v, w) in $\mathbf{K}(\mathcal{A})$.

Note that the construction of $\mathbf{D}(\mathcal{A})$ implies the following two useful criteria. A chain complex X is isomorphic to 0 in $\mathbf{D}(\mathcal{A})$ iff it is exact. A morphism $f\colon X \to Y$ in $\mathbf{Ch}(\mathcal{A})$ becomes the zero map in $\mathbf{D}(\mathcal{A})$ iff there is a quasi-isomorphism $s\colon Y \to Y'$ such that sf is null homotopic (chain homotopic to zero). The following exercise shows the subtlety of being zero.

Exercise 10.4.2 Give examples of maps f, g in $\mathbf{Ch}(\mathcal{A})$ such that (1) $f = 0$ in $\mathbf{D}(\mathcal{A})$, but f is not null homotopic, and (2) g induces the zero map on cohomology, but $g \neq 0$ in $\mathbf{D}(\mathcal{A})$. *Hint:* For (2) try $X\colon 0 \to \mathbb{Z} \xrightarrow{2} \mathbb{Z} \to 0$, $Y\colon 0 \to \mathbb{Z} \xrightarrow{1} \mathbb{Z}/3 \to 0$, $g = (1, 2)$.

Exercise 10.4.3 ($\mathbf{K}_{\mathcal{B}}(\mathcal{A})$ and $\mathbf{D}_{\mathcal{B}}(\mathcal{A})$) Let $\mathcal{B}$ be a Serre subcategory of $\mathcal{A}$, and let $\pi\colon \mathcal{A} \to \mathcal{A}/\mathcal{B}$ be the quotient map constructed in exercise 10.3.2.

1. Show that $H = \pi H^0\colon \mathbf{K}(\mathcal{A}) \to \mathcal{A} \to \mathcal{A}/\mathcal{B}$ is a cohomological functor, so that $\mathbf{K}_H(\mathcal{A})$ is a triangulated category by exercise 10.2.5. The notation $\mathbf{K}_{\mathcal{B}}(\mathcal{A})$ is often used for $\mathbf{K}_H(\mathcal{A})$, because of the description in part (2).
2. Show that X is in $\mathbf{K}_{\mathcal{B}}(\mathcal{A})$ iff the cohomology $H^i(X)$ is in $\mathcal{B}$ for all i.
3. Show that $\mathbf{K}_{\mathcal{B}}(\mathcal{A})$ is a localizing subcategory of $\mathbf{K}(\mathcal{A})$, and conclude that its localization $\mathbf{D}_{\mathcal{B}}(\mathcal{A})$ is a triangulated subcategory of $\mathbf{D}(\mathcal{A})$ (10.2.6).
4. Suppose that $\mathcal{B}$ has enough injectives and that every injective object of $\mathcal{B}$ is also injective in $\mathcal{A}$. Show that there is an equivalence $\mathbf{D}^+(\mathcal{B}) \cong \mathbf{D}_{\mathcal{B}}^+(\mathcal{A})$.

Exercise 10.4.4 (Change of Universe) This is a continuation of the previous exercise. Suppose that our universe is contained in a larger universe $\mathcal{U}$, and that **mod–R** and **Sheaves**(X) are small categories in $\mathcal{U}$. Let **MOD–R** and **SHEAVES**(X) denote the categories of modules and sheaves in $\mathcal{U}$, respectively. Show that **mod–R** and **Sheaves**(X) are Serre subcategories of **MOD–R** and **SHEAVES**(X), respectively. Conclude that **D(mod–R)** $\cong$ **D$_{mod–R}$(MOD–R)** and **D(Sheaves**$(X)) \cong$ **D$_{Sheaves(X)}$(SHEAVES**$(X))$.

Exercise 10.4.5 Here is a construction of **D**$(\mathcal{A})$ when $\mathcal{A}$ is **mod–R**, valid whenever $\mathcal{A}$ has enough projectives and satisfies (AB5). It is based on the construction of CW spectra in algebraic topology [LMS]. Call a chain complex C *cellular* if it is the increasing union of subcomplexes C_n, with $C_0 = 0$, such that each quotient C_n/C_{n-1} is a complex of projectives with all differentials zero. Let $\mathbf{K}_{cell}$ denote the full subcategory of $\mathbf{K}(\mathcal{A})$ consisting of cellular complexes. Show that

1. For every X there is a quasi-isomorphism $C \to X$ with C cellular.
2. If C is cellular and X is acyclic, then every map $C \to X$ is null-homotopic.
3. If C is cellular and $f: X \to Y$ is a quasi-isomorphism, then

$$f_* : \mathrm{Hom}_{\mathbf{K}(\mathcal{A})}(C, X) \cong \mathrm{Hom}_{\mathbf{K}(\mathcal{A})}(C, Y).$$

4. (Whitehead's Theorem) If $f: C \to D$ is a quasi-isomorphism of cellular complexes, then f is a homotopy equivalence, that is, $C \cong D$ in $\mathbf{K}(\mathcal{A})$.
5. $\mathbf{K}_{cell}$ is a localizing triangulated subcategory of $\mathbf{K}(\mathcal{A})$.
6. The natural map is an equivalence: $\mathbf{K}_{cell} \cong \mathbf{D}(\mathcal{A})$.

Exercise 10.4.6 Let R be a noetherian ring, and let $\mathbf{M}(R)$ denote the category of all finitely generated R-modules. Let $\mathbf{D}_{fg}(R)$ denote the full subcategory of $\mathbf{D}(\mathbf{mod}–R)$ consisting of complexes A whose cohomology modules $H^i(A)$ are all finitely generated, that is, the category $\mathbf{D}_{\mathbf{M}(R)}(\mathbf{mod}–R)$ of exercise 10.4.3. Show that $\mathbf{D}_{fg}(R)$ is a triangulated category and that there is an equivalence $\mathbf{D}^-(\mathbf{M}(R)) \cong \mathbf{D}_{fg}^-(R)$. *Hint:* $\mathbf{M}(R)$ is a Serre subcategory of **mod–R** (exercise 10.3.2).

10.5 Derived Functors

There is a category of triangulated categories; a *morphism* $F: \mathbf{K} \to \mathbf{K}'$ of triangulated categories is a (covariant) additive functor that commutes with the translation functor T and sends exact triangles to exact triangles. Morphisms are sometimes called *covariant ∂-functors*; a morphism $\mathbf{K}^{op} \to \mathbf{K}'$ is of course a contravariant ∂-functor.

For example, suppose given an additive functor $F: \mathcal{A} \to \mathcal{B}$ between two abelian categories. Since F preserves chain homotopy equivalences, it extends to additive functors $\mathbf{Ch}(\mathcal{A}) \to \mathbf{Ch}(\mathcal{B})$ and $\mathbf{K}(\mathcal{A}) \to \mathbf{K}(\mathcal{B})$. Since F commutes with translation of chain complexes, it even preserves mapping cones and exact triangles. Thus $F: \mathbf{K}(\mathcal{A}) \to \mathbf{K}(\mathcal{B})$ is a morphism of triangulated categories.

We would like to extend F to a functor $\mathbf{D}(\mathcal{A}) \to \mathbf{D}(\mathcal{B})$. If $F: \mathcal{A} \to \mathcal{B}$ is exact, this is easy. However, if F is not exact, then the functor $\mathbf{K}(\mathcal{A}) \to \mathbf{K}(\mathcal{B})$ will not preserve quasi-isomorphisms, and this may not be possible. The thing to expect is that if F is left or right exact, then the derived functors of F will be needed to extend something like the hyper-derived functors of F.

Our experience in Chapter 5, section 7 tells us that the right hyper-derived functors $\mathbb{R}^i F$ work best if we restrict attention to bounded below cochain complexes. With this in mind, let $\mathbf{K}$ denote $\mathbf{K}^+(\mathcal{A})$ or any other localizing triangulated subcategory of $\mathbf{K}(\mathcal{A})$, and let $\mathbf{D}$ denote the full subcategory of the derived category $\mathbf{D}(\mathcal{A})$ corresponding to $\mathbf{K}$.

Definition 10.5.1 Let $F: \mathbf{K} \to \mathbf{K}(\mathcal{B})$ be a morphism of triangulated categories. A *(total) right derived functor* of F on $\mathbf{K}$ is a morphism $\mathbf{R}F: \mathbf{D} \to \mathbf{D}(\mathcal{B})$ of triangulated categories, together with a natural transformation ξ from $qF: \mathbf{K} \to \mathbf{K}(\mathcal{B}) \to \mathbf{D}(\mathcal{B})$ to $(\mathbf{R}F)q: \mathbf{K} \to \mathbf{D} \to \mathbf{D}(\mathcal{B})$ which is universal in the sense that if $G: \mathbf{D} \to \mathbf{D}(\mathcal{B})$ is another morphism equipped with a natural transformation $\zeta: qF \Rightarrow Gq$, then there exists a unique natural transformation $\eta: \mathbf{R}F \Rightarrow G$ so that $\zeta_A = \eta_{qA} \circ \xi_A$ for every A in $\mathbf{D}$.

This universal property guarantees that if $\mathbf{R}F$ exists, then it is unique up to natural isomorphism, and that if $\mathbf{K}' \subset \mathbf{K}$, then there is a natural transformation from the right derived functor $\mathbf{R}'F$ on $\mathbf{D}'$ to the restriction of $\mathbf{R}F$ to $\mathbf{D}'$. If there is a chance of confusion, we will write $\mathbf{R}^b F$, $\mathbf{R}^+ F$, $\mathbf{R}_\mathcal{B} F$, and so on for the derived functors of F on $\mathbf{K}^b(\mathcal{A})$, $\mathbf{K}^+(\mathcal{A})$, $\mathbf{K}_\mathcal{B}(\mathcal{A})$, etc.

Similarly, a *(total) left derived functor* of F is a morphism $\mathbf{L}F: \mathbf{D} \to \mathbf{D}(\mathcal{B})$ together with a natural transformation $\xi: (\mathbf{L}F)q \Rightarrow qF$ satisfying the dual universal property (G factors through $\eta: G \Rightarrow \mathbf{L}F$). Since $\mathbf{L}F$ is $\mathbf{R}(F^{op})^{op}$, where $F^{op}: \mathbf{K}^{op} \to \mathbf{K}(\mathcal{B}^{op})$, we can translate any statement about $\mathbf{R}F$ into a dual statement about $\mathbf{L}F$.

Exact Functors 10.5.2 If $F: \mathcal{A} \to \mathcal{B}$ is an exact functor, F preserves quasi-isomorphisms. Hence F extends trivially to $F: \mathbf{D}(\mathcal{A}) \to \mathbf{D}(\mathcal{B})$. In effect, F is its own left and derived functor. The following two examples generalize this observation.

Example 10.5.3 Let $\mathbf{K}^+(\mathcal{I})$ denote the triangulated category of bounded below complexes of injectives. We saw in 10.4.8 that every quasi-isomorphism
in $\mathbf{K}^+(\mathcal{I})$ is an isomorphism, so $\mathbf{K}^+(\mathcal{I})$ is isomorphic to its derived category
$\mathbf{D}^+(\mathcal{I})$. The functor $qFq^{-1}\colon \mathbf{D}^+(\mathcal{I}) \cong \mathbf{K}^+(\mathcal{I}) \xrightarrow{F} \mathbf{K}^+(\mathcal{B}) \to \mathbf{D}^+(\mathcal{B})$ satisfies
$qF \cong (qFq^{-1})q$, so it is both the left and right total derived functor of F.

Similarly, for the category $\mathbf{K}^-(\mathcal{P})$ of bounded above cochain complexes of
projectives, we have $\mathbf{K}^-(\mathcal{P}) \cong \mathbf{D}^-(\mathcal{P})$. Again, qFq^{-1} is both the left and right
derived functor of F.

Definition 10.5.4 Let $F\colon \mathbf{K} \to \mathbf{K}(\mathcal{B})$ be a morphism of triangulated categories. A complex X in $\mathbf{K}$ is called F-*acyclic* if $F(X)$ is acyclic, that is, if
$H^i(FX) \cong 0$ for all i. (Compare with 2.4.3.)

Example 10.5.5 (F-*acyclic complexes*) Suppose that $\mathbf{K}$ is a triangulated
subcategory of $\mathbf{K}(\mathcal{A})$ such that every acyclic complex in $\mathbf{K}$ is F-acyclic. If
$s\colon X \to Y$ is a quasi-isomorphism in $\mathbf{K}$, then cone(s) and hence $F(\text{cone}(s))$
is acyclic. Since F preserves exact triangles, the cohomology sequence shows
that $F(s)^*\colon H^*(FX) \cong H^*(FY)$, that is, that $F(s)$ is a quasi-isomorphism.
By the universal property of the localization $\mathbf{D} = Q^{-1}\mathbf{K}$ there is a unique
functor $Q^{-1}F$ from $\mathbf{D}$ to $\mathbf{D}(\mathcal{B})$ such that $qF = (Q^{-1}F)q$. Once again, $Q^{-1}F$
is both the left and right derived functor of F.

Existence Theorem 10.5.6 *Let* $F\colon \mathbf{K}^+(\mathcal{A}) \to \mathbf{K}(\mathcal{B})$ *be a morphism of triangulated categories. If* $\mathcal{A}$ *has enough injectives, then the right derived functor*
$\mathbf{R}^+F$ *exists on* $\mathbf{D}^+(\mathcal{A})$, *and if I is a bounded below complex of injectives, then*

$$\mathbf{R}^+F(I) \cong qF(I).$$

Dually, if $\mathcal{A}$ *has enough projectives, then the left derived functor* $\mathbf{L}^-F$ *exists
on* $\mathbf{D}^-(\mathcal{A})$, *and if P is a bounded above cochain complex of projectives, then*

$$\mathbf{L}^-F(P) \cong qF(P).$$

Proof Choose an equivalence $U\colon \mathbf{D}^+(\mathcal{A}) \xrightarrow{\cong} \mathbf{K}^+(\mathcal{I})$ inverse to the natural
map $T\colon \mathbf{K}^+(\mathcal{I}) \xrightarrow{\cong} \mathbf{D}^+(\mathcal{A})$ of 10.4.8, and define $\mathbf{R}F$ to be the composite
qFU:

$$\mathbf{D}^+(\mathcal{A}) \xrightarrow{\cong} \mathbf{K}^+(\mathcal{I}) \xrightarrow{F} \mathbf{K}^+(\mathcal{B}) \xrightarrow{q} \mathbf{D}^+(\mathcal{B}).$$

To construct ξ we use the natural isomorphism of 10.4.7

$$\text{Hom}_{\mathbf{D}^+(\mathcal{A})}(qX, TUqX) \cong \text{Hom}_{\mathbf{K}^+(\mathcal{A})}(X, UqX).$$

Under this isomorphism there is a natural map $f_X : X \to UqX$ in $\mathbf{K}^+(\mathcal{A})$ corresponding to the augmentation $\eta : qX \to TUqX$ in $\mathbf{D}^+(\mathcal{A})$. We define ξ_X to be the natural transformation $qF(f_X) : qF(X) \to qF(UqX) \cong (qFU)(qX)$. It is not hard to see that ξ has the required universal property, making $(\mathbf{R}F, \xi)$ into a right derived functor of F. As usual, the dual assertion that the composite

$$\mathbf{D}^-(\mathcal{A}) \xrightarrow{\cong} \mathbf{K}^-(\mathcal{P}) \xrightarrow{F} \mathbf{K}^-(\mathcal{B}) \xrightarrow{q} \mathbf{D}^-(\mathcal{B})$$

is a left derived functor of F follows by passage to F^{op}. $\diamond$

Corollary 10.5.7 *Let $F : \mathcal{A} \to \mathcal{B}$ be an additive functor between abelian categories.*

1. *If $\mathcal{A}$ has enough injectives, the hyper-derived functors $\mathbb{R}^i F(X)$ are the cohomology of $\mathbf{R}F(X)$: $\mathbb{R}^i F(X) \cong H^i \mathbf{R}^+ F(X)$ for all i.*
2. *If $\mathcal{A}$ has enough projectives, the hyper-derived functors $\mathbb{L}_i F(X)$ are the cohomology of $\mathbf{L}F(X)$: $\mathbb{L}_i F(X) \cong H^{-i} \mathbf{L}^- F(X)$ for all i.*

Remark 10.5.8 The assumption in 5.7.4 that F be left or right exact was not necessary to define $\mathbb{R}^i F$ or $\mathbb{L}_i F$; it was made to retain the connection with F. Suppose that we consider an object A of $\mathcal{A}$ as a complex concentrated in degree zero. The assumption that F be left exact is needed to ensure that the $\mathbb{R}^i F(A)$ are the ordinary derived functors $R^i F(A)$ and in particular that $\mathbb{R}^0 F(A) = F(A)$. Similarly, the assumption that F be right exact is needed to ensure that $\mathbb{L}_i F(A)$ is the ordinary derived functor $L_i F(A)$, and that $\mathbb{L}_0 F(A) = F(A)$.

Exercise 10.5.1 Suppose that $F : \mathbf{K}^+(\mathcal{A}) \to \mathbf{K}(\mathcal{C})$ is a morphism of triangulated categories and that $\mathcal{B}$ is a Serre subcategory of $\mathcal{A}$. If $\mathcal{A}$ has enough injectives, show that the restriction of $\mathbf{R}^+ F$ to $\mathbf{D}_{\mathcal{B}}^+(\mathcal{A})$ is the derived functor $\mathbf{R}_{\mathcal{B}}^+ F$. If in addition $\mathcal{B}$ has enough injectives, which are also injective in $\mathcal{A}$, this proves that the composition $\mathbf{D}^+(\mathcal{B}) \to \mathbf{D}^+(\mathcal{A}) \xrightarrow{\mathbf{R}F} \mathbf{D}^+(\mathcal{C})$ is the derived functor $\mathbf{R}^+ F|_{\mathcal{B}}$ of the restriction $F|_{\mathcal{B}}$ of F to $\mathcal{B}$, since we saw in exercise 10.4.3 that in this case $\mathbf{D}^+(\mathcal{B}) \cong \mathbf{D}_{\mathcal{B}}^+(\mathcal{A})$.

Generalized Existence Theorem 10.5.9 ([HartRD, I.5.1]) *Suppose that $\mathbf{K}'$ is a triangulated subcategory of $\mathbf{K}$ such that*

1. *Every X in $\mathbf{K}$ has a quasi-isomorphism $X \to X'$ to an object of $\mathbf{K}'$.*

2. *Every exact complex in* $\mathbf{K}'$ *is F-acyclic (10.5.4).*

Then $\mathbf{D}' \overset{\cong}{\longrightarrow} \mathbf{D}$ *and* $\mathbf{R}F\colon \mathbf{D} \cong \mathbf{D}' \overset{\mathbf{R}'F}{\longrightarrow} \mathbf{D}(\mathcal{B})$ *is a right derived functor of F.*

Proof By (1) and 10.3.14, $\mathbf{K}'$ is localizing and $\mathbf{D}' \overset{\cong}{\longrightarrow} \mathbf{D}$. Now modify the proof of the Existence Theorem 10.5.6, using F-acyclic complexes. $\diamond$

Definition 10.5.10 Let $F\colon \mathcal{A} \to \mathcal{B}$ be an additive functor between abelian categories. When $\mathcal{A}$ has enough injectives, so that the usual derived functors $R^i F$ (of Chapter 2) exist, we say that F has *cohomological dimension n* if $R^n F = 0$ for all $i > n$, yet $R^n F \neq 0$. Dually, when $\mathcal{A}$ has enough projectives, so that the usual derived functors $L_i F$ exist, we say that F has *homological dimension n* if $L_i F = 0$ for all $i > n$, yet $L_n F \neq 0$.

Exercise 10.5.2 If F has finite cohomological dimension, show that every exact complex of F-acyclic objects (2.4.3) is an F-acyclic complex in the sense of 10.5.4.

Corollary 10.5.11 *Let* $F\colon \mathcal{A} \to \mathcal{B}$ *be an additive functor. If F has finite cohomological dimension n, then* $\mathbf{R}F$ *exists on* $\mathbf{D}(\mathcal{A})$, *and its restriction to* $\mathbf{D}^+(\mathcal{A})$ *is* $\mathbf{R}^+ F$. *Dually, if F has finite homological dimension n, then* $\mathbf{L}F$ *exists on* $\mathbf{D}(\mathcal{A})$, *and its restriction to* $\mathbf{D}^-(\mathcal{A})$ *is* $\mathbf{L}^- F$.

Proof Let $\mathbf{K}'$ be the full subcategory of $\mathbf{K}(\mathcal{A})$ consisting of complexes of F-acyclic objects in $\mathcal{A}$ (2.4.3). We need to show that every complex X has a quasi-isomorphism $X \to X'$ with X' a complex of F-acyclic objects. To see this, choose a Cartan-Eilenberg resolution $X^{\cdot} \to I^{\cdot\cdot}$ and let τI be the double subcomplex of I obtained by taking the good truncation $\tau_{\leq n}(I^{p\cdot})$ of each column (1.2.7). Since each $X^p \to I^{p\cdot}$ is an injective resolution, each $\tau_{\leq n}(I^{p\cdot})$ is a finite resolution of X^p by F-acyclic objects. Therefore $X' = \mathrm{Tot}(\tau I)$ is a chain complex of F-acyclic objects. The bounded spectral sequence $H^p H^q(\tau I) \Rightarrow H^{p+q}(X')$ degenerates to yield $H^*(X) \overset{\cong}{\longrightarrow} H^*(X')$, that is, $X \to X'$ is a quasi-isomorphism. $\diamond$

10.6 The Total Tensor Product

Let R be a ring. In order to avoid notational problems, we shall use the letters A, B, and so on to denote cochain complexes of R-modules. For each cochain complex A of right R-modules the total tensor product complex 2.7.1 is a functor $F(B) = \text{Tot}^{\oplus}(A \otimes_R B)$ from $\mathbf{K}(R\text{–mod})$ to $\mathbf{K}(\mathbf{Ab})$. Since $R\text{–mod}$ has enough projectives, its derived functor $\mathbf{L}^- F: \mathbf{D}^-(R\text{–mod}) \to \mathbf{D}(\mathbf{Ab})$ exists by 10.5.6.

Definition 10.6.1 The *total tensor product* of A and B is

$$A \otimes_R^{\mathbf{L}} B = \mathbf{L}^- \text{Tot}^{\oplus}(A \otimes_R -)B.$$

Lemma 10.6.2 *If A, A', and B are bounded above cochain complexes and $A \to A'$ is a quasi-isomorphism, then $A \otimes_R^{\mathbf{L}} B \cong A' \otimes_R^{\mathbf{L}} B$.*

Proof We may change B up to quasi-isomorphism to suppose that B is a complex of flat modules. In this case $A \otimes_R^{\mathbf{L}} B$ is $\text{Tot}^{\oplus}(A \otimes_R B)$ and $A' \otimes_R^{\mathbf{L}} B$ is $\text{Tot}^{\oplus}(A' \otimes_R B)$ by 10.5.5. Now apply the Comparison Theorem 5.2.12 to $E_1^{pq}(A) \to E_1^{pq}(A')$, where

$$E_1^{pq}(A) = H^q(A) \otimes_R B^p \Rightarrow H^{p+q}(A \otimes_R^{\mathbf{L}} B).$$

The spectral sequences converge when A, A', and B are bounded above 5.6.2. $\diamond$

Theorem 10.6.3 *The total tensor product is a functor*

$$\otimes_R^{\mathbf{L}}: \mathbf{D}^-(\mathbf{mod}\text{–}R) \times \mathbf{D}^-(R\text{–mod}) \to \mathbf{D}^-(\mathbf{Ab}).$$

Its cohomology is the hypertor of 5.7.8:

$$\mathbf{Tor}_i^R(A, B) \cong H^{-i}(A \otimes_R^{\mathbf{L}} B).$$

Proof For each fixed B, the functor $F(A) = A \otimes_R^{\mathbf{L}} B$ from $\mathbf{K}^-(\mathbf{mod}\text{–}R)$ to $\mathbf{D}^-(\mathbf{Ab})$ sends quasi-isomorphisms to isomorphisms, so F factors through the localization $\mathbf{D}^-(\mathbf{mod}\text{–}R)$ of $\mathbf{K}^-(\mathbf{mod}\text{–}R)$. If P and Q are chain complexes of flat modules, then by definition the hypertor groups $\mathbf{Tor}_i^R(P, Q)$ are the homology of $\text{Tot}^{\oplus} P \otimes_R Q$. Reindexing the chain complexes as cochain complexes, the cochain complex $\text{Tot}^{\oplus}(P \otimes_R Q)$ is isomorphic to $P \otimes_R^{\mathbf{L}} Q$. $\diamond$

Corollary 10.6.4 *If A and B are R-modules, the usual* Tor*-group* $\text{Tor}_i^R(A, B)$ *of Chapter 3 is* $H^{-i}(A \otimes_R^{\mathbf{L}} B)$, *where A and B are considered as cochain complexes in degree zero.*

Exercise 10.6.1 Form the derived functor $\mathbf{L} \text{Tot}^{\oplus}(- \otimes_R B)$ and show that $A \otimes_R^{\mathbf{L}} B$ is naturally isomorphic to $\mathbf{L}^- \text{Tot}^{\oplus}(- \otimes_R B)A$ in $\mathbf{D}(\mathbf{Ab})$. This isomorphism underlies the fact that hypertor is a balanced functor (2.7.7).

Exercise 10.6.2 If A is a complex of R_1–R bimodules, and B is a complex of R–R_2 bimodules, $A \otimes_R B$ is a double complex of R_1–R_2 bimodules. Show that the total tensor product may be refined to a functor

$$\otimes_R^{\mathbf{L}} : \mathbf{D}^-(R_1\text{–}\mathbf{mod}\text{–}R) \times \mathbf{D}^-(R\text{–}\mathbf{mod}\text{–}R_2) \to \mathbf{D}^-(R_1\text{–}\mathbf{mod}\text{–}R_2).$$

By "refine" we mean that the composition to $\mathbf{D}(\mathbf{Ab})$ induced by the usual forgetful functor is the total tensor product in $\mathbf{D}(\mathbf{Ab})$. Then show that if R is a commutative ring, we may refine it to a functor

$$\otimes_R^{\mathbf{L}} : \mathbf{D}^-(R\text{–}\mathbf{mod}) \times \mathbf{D}^-(R\text{–}\mathbf{mod}) \to \mathbf{D}^-(R\text{–}\mathbf{mod}),$$

and that there is a natural isomorphism $A \otimes_R^{\mathbf{L}} B \cong B \otimes_R^{\mathbf{L}} A$.

Remark 10.6.5 (see [HartRD, II.4]) If X is a topological space with a sheaf $\mathcal{O}_X$ of rings, there is a category of $\mathcal{O}_X$-modules [Hart]. This category has enough flat modules (see [Hart, exercise III.6.4]), even though it may not have enough projectives, and this suffices to construct the total tensor product $\mathcal{E} \otimes_{\mathcal{O}_X}^{\mathbf{L}} \mathcal{F}$ of complexes of $\mathcal{O}_X$-modules.

10.6.1 Ring Homomorphisms and $\mathbf{L}f^*$

10.6.6 Let $f: R \to S$ be a ring homomorphism. By the Existence Theorem 10.5.6, the functor $f^* = -\otimes_R S$ from R-modules to S-modules has a left-derived functor

$$\mathbf{L}f^* = \mathbf{L}(-\otimes_R S) : \mathbf{D}^-(\mathbf{mod}\text{–}R) \to \mathbf{D}^-(\mathbf{mod}\text{–}S).$$

The discussion in 5.7.8 shows that the hypertor groups are

$$\mathbf{Tor}_i^R(A, S) = \mathbb{L}_i f^*(A) \cong H^{-i}(\mathbf{L}f^*A).$$

If S has finite flat dimension n (4.1.1), then f^* has homological dimension n, and we may extend the derived functor $\mathbf{L}f^*$ using 10.5.11 to $\mathbf{L}f^*$: $\mathbf{D}(\mathbf{mod}\text{–}R) \to \mathbf{D}(\mathbf{mod}\text{–}S)$.

The forgetful functor $f_*: \mathbf{mod}\text{–}S \to \mathbf{mod}\text{–}R$ is exact, so it "is" its own derived functor $f_*: \mathbf{D}(\mathbf{mod}\text{–}S) \to \mathbf{D}(\mathbf{mod}\text{–}R)$. The composite $f_*(\mathbf{L}f^*)A$ is the total tensor product $A \otimes_R^{\mathbf{L}} S$ because, when A is a bounded above complex of flat modules, both objects of the derived category are represented by $A \otimes_R S$. We will see in the next section that $f_* (= \mathbf{R}f_*)$ and $\mathbf{L}f^*$ are adjoint functors in a suitable sense.

Remark If we pass from rings to schemes, the map f reverses direction, going from $\operatorname{Spec}(S)$ to $\operatorname{Spec}(R)$. This explains the use of the notation f_*, which suggests a covariant functor on $\operatorname{Spec}(R)$. Of course f_* is not always exact when we pass to more general schemes, and one needs to replace f_* by $\mathbf{R}f_*$; see [HartRD, II.5.5].

Lemma 10.6.7 *If $f: R \to S$ is a commutative ring homomorphism, there is a natural isomorphism in $\mathbf{D}^-(\text{mod}–S)$ for every A, B in $\mathbf{D}^-(\text{mod}–R)$:*

$$\mathbf{L}f^*(A) \otimes_S^{\mathbf{L}} \mathbf{L}f^*(B) \xrightarrow{\cong} \mathbf{L}f^*(A \otimes_R^{\mathbf{L}} B).$$

Proof Replacing A and B by complexes of flat R-modules, this is just the natural isomorphism $(A \otimes_R S) \otimes_S (S \otimes_R B) \cong (A \otimes_R B) \otimes_R S$. ◇

Exercise 10.6.3 (finite Tor-dimension) The *Tor-dimension* of a bounded complex A of right R-modules is the smallest n such that the hypertor $\mathbf{Tor}_i^R(A, B)$ vanish for all modules B when $i > n$. If A is a module, the Tor-dimension is just the flat dimension of 4.1.1.

1. Show that A has finite Tor-dimension if and only if there is a quasi-isomorphism $P \to A$ with P a bounded complex of flat R-modules.
2. If A has finite Tor-dimension, show that the derived functor $A \otimes_R^{\mathbf{L}} -$ on $\mathbf{D}^-(R\text{–mod})$ extends to a functor

$$\mathbf{L}(A \otimes_R): \mathbf{D}(R\text{–mod}) \to \mathbf{D}(\mathbf{Ab}).$$

3. Let $f: R \to S$ be a ring map, with S of finite flat dimension over R. Show that the forgetful functor $f_*: \mathbf{D}^b(\text{mod}–S) \to \mathbf{D}^b(\text{mod}–R)$ sends complexes of finite Tor-dimension over S to complexes of finite Tor-dimension over R.

10.6.2 The Derived Functors of Γ and f_*

10.6.8 Let X be a topological space, and Γ the global sections functor from **Sheaves**(X) (sheaves of abelian groups) to **Ab**; see 2.5.4. For simplicity, we shall write $\mathbf{D}(X)$, $\mathbf{D}^+(X)$, and so on for the derived categories $\mathbf{D}(\mathbf{Sheaves}(X))$, $\mathbf{D}^+(\mathbf{Sheaves}(X))$, and so on. By 2.3.12 the category **Sheaves**(X) has enough injectives. Therefore Γ has a right-derived functor $\mathbf{R}^+\Gamma: \mathbf{D}^+(X) \to \mathbf{D}^+(\mathbf{Ab})$, and for every sheaf $\mathcal{F}$ the usual cohomology functors $H^i(X, \mathcal{F})$ of 2.5.4 are the groups $H^i(\mathbf{R}^+\Gamma(\mathcal{F}))$. More generally, if $\mathcal{F}^*$ is a bounded below complex of sheaves on X, then the hypercohomology groups of 5.7.10 are given by:

$$\mathbb{H}^i(X, \mathcal{F}^*) \cong H^i \mathbf{R}^+ \Gamma(\mathcal{F}^*).$$

In algebraic geometry, one usually works with topological spaces that are noetherian (the closed subspaces satisfy the descending chain condition) and have finite Krull dimension n (the longest chain of irreducible closed subsets has length n). Grothendieck proved in [Tohuku, 3.6.5] (see [Hart, III.2.7]) that for such a space the functors $H^i(X, -)$ vanish for $i > n$, that is, that Γ has cohomological dimension n. As we have seen in 10.5.11, this permits us to extend $\mathbf{R}^+\Gamma$ to a functor

$$\mathbf{R}\Gamma \colon \mathbf{D}(X) \to \mathbf{D}(\mathbf{Ab}).$$

Now let $f \colon X \to Y$ be a continuous map of topological spaces. Just as for Γ, the direct image sheaf functor f_* (2.6.6) has a derived functor

$$\mathbf{R} f_* \colon \mathbf{D}^+(X) \to \mathbf{D}^+(Y).$$

If $\mathcal{F}$ is a sheaf on X, its higher direct image sheaves (2.6.6) are the sheaves

$$R^i f_*(\mathcal{F}) = H^i \mathbf{R} f_*(\mathcal{F}).$$

When X is noetherian of finite Krull dimension, the functor f_* has finite cohomological dimension because, by [Hart, III.8.1], $R^i f_*(\mathcal{F})$ is the sheaf on Y associated to the presheaf sending U to $H^i(f^{-1}(U), \mathcal{F})$. Once again, we can extend $\mathbf{R} f_*$ from $\mathbf{D}^+(X)$ to a functor $\mathbf{R} f_* \colon \mathbf{D}(X) \to \mathbf{D}(Y)$.

$\mathbf{R}\Gamma$ is just a special case of $\mathbf{R} f_*$. Indeed, if Y is a point, then $\mathbf{Sheaves}(Y) = \mathbf{Ab}$ and Γ is f_*; it follows that $\mathbf{R}\Gamma$ is $\mathbf{R} f_*$.

10.7 Ext and RHom

Let A and B be cochain complexes. In 2.7.4 we constructed the total Hom cochain complex $\mathrm{Hom}^{\cdot}(A, B)$, and observed that $H^n \mathrm{Hom}^{\cdot}(A, B)$ is the group of chain homotopy equivalence classes of morphisms $A \to B[-n]$. That is,

$$\mathrm{Hom}_{\mathbf{K}(\mathcal{A})}(A, T^n B) = H^n(\mathrm{Hom}^{\cdot}(A, B)).$$

Both $\mathrm{Hom}^{\cdot}(A, -)$ and $\mathrm{Hom}^{\cdot}(-, B)$ are morphisms of triangulated functors, from $\mathbf{K}(\mathcal{A})$ and $\mathbf{K}(\mathcal{A})^{\mathrm{op}}$ to $\mathbf{K}(\mathbf{Ab})$, respectively. In fact, $\mathrm{Hom}^{\cdot}$ is a bimorphism

$$\mathrm{Hom}^{\cdot} \colon \mathbf{K}(\mathcal{A})^{\mathrm{op}} \times \mathbf{K}(\mathcal{A}) \to \mathbf{K}(\mathbf{Ab}).$$

(Exercise!) In this section we construct an object $\mathbf{R}\text{Hom}(A, B)$ in the derived category $\mathbf{D}(\mathcal{A})$ and prove that if A and B are bounded below, then

$$\text{Hom}_{\mathbf{D}(\mathcal{A})}(A, T^n B) = H^n(\mathbf{R}\text{Hom}(A, B)).$$

Since $\mathbf{D}^+(\mathcal{A})$ is a full subcategory of $\mathbf{D}(\mathcal{A})$, this motivates the following.

Definition 10.7.1 Let A and B be cochain complexes in an abelian category $\mathcal{A}$. The nth *hyperext* of A and B is the abelian group

$$\text{Ext}^n(A, B) = \text{Hom}_{\mathbf{D}(\mathcal{A})}(A, T^n B).$$

Note that since $\mathbf{D}(\mathcal{A})$ is a triangulated category, its Hom-functors $\text{Ext}^n(A, -)$ and $\text{Ext}^n(-, B)$ are cohomological functors, that is, they convert exact triangles into long exact sequences (10.2.8). Since $\mathbf{K}(\mathcal{A})$ is a triangulated category, its Hom-functors $H^n \text{Hom}^{\cdot}(A, -)$ and $H^n \text{Hom}^{\cdot}(-, B)$ are also cohomological functors, and there are canonical morphisms

$$H^n \text{Hom}^{\cdot}(A, B) = \text{Hom}_{\mathbf{K}(\mathcal{A})}(A, T^n B) \to \text{Hom}_{\mathbf{D}(\mathcal{A})}(A, T^n B) = \text{Ext}^n(A, B).$$

Definition 10.7.2 Suppose that $\mathcal{A}$ has enough injectives, so that the derived functor $\mathbf{R}^+ \text{Hom}^{\cdot}(A, -): \mathbf{D}^+(\mathcal{A}) \to \mathbf{D}(\mathbf{Ab})$ exists for every cochain complex A. We write $\mathbf{R}\text{Hom}(A, B)$ for the object $\mathbf{R}^+ \text{Hom}^{\cdot}(A, -)B$ of $\mathbf{D}(\mathbf{Ab})$.

Lemma 10.7.3 *If* $A \to A'$ *is a quasi-isomorphism, then* $\mathbf{R}\text{Hom}(A', B) \xrightarrow{\cong} \mathbf{R}\text{Hom}(A, B)$.

Proof We may change B up to quasi-isomorphism to suppose that B is a bounded below cochain complex of injectives. But then $\mathbf{R}\text{Hom}(A', B) \cong \text{Hom}^{\cdot}(A', B)$ is quasi-isomorphic to $\mathbf{R}\text{Hom}(A, B) \cong \text{Hom}^{\cdot}(A, B)$, because we saw in 10.4.7 that

$$H^n \text{Hom}^{\cdot}(A', B) = \text{Hom}_{\mathbf{K}(\mathcal{A})}(A', T^n B)$$
$$\cong \text{Hom}_{\mathbf{D}(\mathcal{A})}(A', T^n B) \cong \text{Hom}_{\mathbf{D}(\mathcal{A})}(A, T^n B)$$
$$\cong \text{Hom}_{\mathbf{K}(\mathcal{A})}(A, T^n B) = H^n \text{Hom}^{\cdot}(A, B). \qquad \diamond$$

Theorem 10.7.4 *If* $\mathcal{A}$ *has enough injectives, then* $\mathbf{R}Hom$ *is a bifunctor*

$$\mathbf{R}\text{Hom}: \mathbf{D}(\mathcal{A})^{\text{op}} \times \mathbf{D}^+(\mathcal{A}) \to \mathbf{D}(\mathbf{Ab}).$$

Dually, if $\mathcal{A}$ *has enough projectives, then* $\mathbf{R}\text{Hom}$ *is a bifunctor*

$$\mathbf{R}\text{Hom}: \mathbf{D}^-(\mathcal{A})^{\text{op}} \times \mathbf{D}(\mathcal{A}) \to \mathbf{D}(\mathbf{Ab}).$$

In both cases, we have $\text{Ext}^n(A, B) \cong H^n(\mathbf{R}\text{Hom}(A, B))$.

Proof The lemma shows that, for each fixed B, the functor $F(A) = \mathbf{R}\text{Hom}(A, B)$ from $\mathbf{K}(\mathcal{A})^{\text{op}}$ to $\mathbf{D}(\mathbf{Ab})$ sends quasi-isomorphisms to isomorphisms, so F factors through the localization $\mathbf{D}(\mathcal{A})^{\text{op}}$ of $\mathbf{K}(\mathcal{A})^{\text{op}}$. Therefore, to compute $H^n(\mathbf{R}\text{Hom}(A, B))$ we may suppose that B is a bounded below cochain complex of injectives. But then by the construction of $\mathbf{R}\text{Hom}(A, B)$ as $\text{Hom}^{\cdot}(A, B)$ we have

$$H^n \mathbf{R}\text{Hom}(A, B) = H^n \text{Hom}^{\cdot}(A, B) = \text{Hom}_{\mathbf{K}(\mathcal{A})}(A, B) = \text{Hom}_{\mathbf{D}(\mathcal{A})}(A, B). \quad \diamond$$

Corollary 10.7.5 *If $\mathcal{A}$ has enough injectives, or enough projectives, then for any A and B in $\mathcal{A}$ the group $\text{Ext}^n(A, B)$ is the usual Ext-group of Chapter 3.*

Proof If $B \to I$ is an injective resolution, then the usual definition of $\text{Ext}^n(A, B)$ is $H^n \text{Hom}(A, I) = H^n \text{Tot} \text{Hom}(A, I) \cong H^n \mathbf{R}\text{Hom}(A, B)$. Similarly, if $P \to A$ is a projective resolution, the usual $\text{Ext}^n(A, B)$ is $H^n \text{Hom}(P, B) = H^n \mathbf{R}\text{Hom}(A, B)$. $\diamond$

Exercise 10.7.1 (balancing $\mathbf{R}\text{Hom}$) Suppose that $\mathcal{A}$ has both enough injectives and enough projectives. Show that the two ways of defining the functor $\mathbf{R}\text{Hom}: \mathbf{D}^-(\mathcal{A})^{\text{op}} \times \mathbf{D}^+(\mathcal{A}) \to \mathbf{D}^+(\mathbf{Ab})$ are canonically isomorphic.

Exercise 10.7.2 Suppose that $\mathcal{A}$ has enough injectives. We say that a bounded below complex B has *injective dimension n* if $\text{Ext}^i(A, B) = 0$ for all $i > n$ and all A in $\mathcal{A}$, and $\text{Ext}^n(A, B) \neq 0$ for some A.

1. Show that B has finite injective dimension $\Leftrightarrow$ there is a quasi-isomorphism $B \to I$ into a bounded complex I of injectives.
2. If B has finite injective dimension, show that $\mathbf{R}\text{Hom}(-, B): \mathbf{D}(\mathcal{A})^{\text{op}} \to \mathbf{D}(\mathbf{Ab})$ of 10.7.4 is the derived functor 10.5.1 of $\text{Hom}(-, B)$.

10.7.1 Adjointness of $\mathbf{L}f^*$ and f_*

We can refine the above construction slightly when $\mathcal{A}$ is the category $R\text{–}\mathbf{mod}$ of modules over a commutative ring R. For simplicity we shall write $\mathbf{D}(R)$, $\mathbf{D}^+(R)$, and so on for the derived categories $\mathbf{D}(R\text{–}\mathbf{mod})$, $\mathbf{D}^+(R\text{–}\mathbf{mod})$, and so on. Write $\text{Hom}_R(A, B)$ for $\text{Hom}^{\cdot}(A, B)$, considered as a complex of R-modules. If we replace $\mathbf{D}(\mathbf{Ab})$ by $\mathbf{D}(R)$ in the above construction, we obtain an object $\mathbf{RHom}_R(A, B)$ in $\mathbf{D}(R)$ whose image under $\mathbf{D}(R) \to \mathbf{D}(\mathbf{Ab})$ is the unrefined $\mathbf{RHom}(A, B)$ of 10.7.2.

Suppose now that $f: R \to S$ is a map of commutative rings. The forgetful functor $f_*: \mathbf{mod}\text{–}S \to \mathbf{mod}\text{–}R$ is exact, so it is its own derived functor $f_*: \mathbf{D}(S) \to \mathbf{D}(R)$. If A is in $\mathbf{D}(S)$, the functor $f_* \mathbf{RHom}_S(A, -): \mathbf{D}^+(S) \to \mathbf{D}(R)$ is the right derived functor of $f_* \mathrm{Hom}_S(A, -)$ because if I is a complex of injectives, then $f_* \mathbf{RHom}_S(A, I) = f_* \mathrm{Hom}_S(A, I)$. The universal property of derived functors yields a natural map:

$$(\dagger) \qquad \zeta: f_* \mathbf{RHom}_S(A, B) \to \mathbf{RHom}_R(f_*A, f_*B).$$

Theorem 10.7.6 *If $f: R \to S$ is a map of commutative rings, then the functor $\mathbf{L}f^*: \mathbf{D}^-(R) \to \mathbf{D}^-(S)$ is left adjoint to $f_*: \mathbf{D}^+(S) \to \mathbf{D}^+(R)$. That is, for A in $\mathbf{D}^-(R)$ and B in $\mathbf{D}^+(S)$ there is a natural isomorphism*

$$(*) \qquad \mathrm{Hom}_{\mathbf{D}(S)}(\mathbf{L}f^*A, B) \xrightarrow{\cong} \mathrm{Hom}_{\mathbf{D}(R)}(A, f_*B).$$

The adjunction morphisms are $\eta_A: A \to f_*\mathbf{L}f^*A$ and $\varepsilon_B: \mathbf{L}f^*(f_*B) \to B$, respectively. Moreover, the isomorphism $(*)$ comes from a natural isomorphism

$$\tau: f_* \mathbf{RHom}_S(\mathbf{L}f^*A, B) \xrightarrow{\cong} \mathbf{RHom}_R(A, f_*B).$$

Proof Since f_* is exact, $f_*\mathbf{L}f^*$ is the left derived functor of f_*f^*; the universal property gives a map $\eta_A: A \to \mathbf{L}(f_*f^*)A = f_*\mathbf{L}f^*A$. Using $(\dagger)$, this gives the map

$$\tau: f_* \mathbf{RHom}_S(\mathbf{L}f^*A, B) \xrightarrow{\zeta} \mathbf{RHom}_R(f_*\mathbf{L}f^*A, f_*B) \xrightarrow{\eta^*} \mathbf{RHom}_R(A, f_*B).$$

To evaluate this map, we suppose that A is a bounded above complex of projective R-modules. In this case the map τ is the isomorphism

$$\mathrm{Tot}(f_* \mathrm{Hom}_S(A \otimes_R S, B)) \cong \mathrm{Tot}(\mathrm{Hom}_R(A, \mathrm{Hom}_S(S, B)))$$
$$= \mathrm{Tot}(\mathrm{Hom}_R(A, f_*B)).$$

Passing to cohomology, τ induces the adjoint isomorphism $(*)$. $\qquad\qquad \diamond$

Remark For schemes one needs to be able to localize the above data to form the $\mathcal{O}_X$-module analogue of $\mathbf{RHom}_R$. By 3.3.8 one needs A to be finitely presented in order to have an isomorphism $S^{-1} \mathrm{Hom}_R(A, B) \cong \mathrm{Hom}_{S^{-1}R}(S^{-1}A, S^{-1}B)$. Thus one needs to restrict A to a subcategory of $\mathbf{D}(X)$ which is locally the $\mathbf{D}_{\mathrm{fg}}(R)$ of exercise 10.4.6; see [HartRD, II.5.10] for details.

Exercise 10.7.3 Let X be a topological space. Given two sheaves $\mathcal{E}, \mathcal{F}$ on X, the *sheaf hom* is the sheaf $\mathcal{H}om(\mathcal{E}, \mathcal{F})$ is the sheaf on X associated to the presheaf sending U to $\mathrm{Hom}(\mathcal{E}|U, \mathcal{F}|U)$; see [Hart, exercise II.1.15]. Mimic the construction of $\mathbf{R}\mathrm{Hom}$ to obtain a functor

$$\mathbf{R}\mathcal{H}om: \mathbf{D}(X)^{\mathrm{op}} \times \mathbf{D}^+(X) \to \mathbf{D}(X).$$

Now suppose that $f: X \to Y$ is a continuous map, and that X is noetherian of finite Krull dimension. Generalize (†) for $\mathcal{E}$ in $\mathbf{D}^-(X)$, $\mathcal{F}$ in $\mathbf{D}^+(X)$ to obtain a natural map in $\mathbf{D}^+(Y)$:

$$\zeta: \mathbf{R}f_* (\mathbf{R}\mathcal{H}om_X(\mathcal{E}, \mathcal{F})) \to \mathbf{R}\mathcal{H}om_Y(\mathbf{R}f_*\mathcal{E}, \mathbf{R}f_*\mathcal{F}).$$

10.8 Replacing Spectral Sequences

We have seen that the objects $\mathbf{R}F(A)$ in the derived category are more flexible than their cohomology groups, the hyper-derived functors $\mathbb{R}^i F(A) = H^i \mathbf{R}F(A)$. Of course, if we are interested in the groups themselves, we can use the spectral sequence $E_2^{pq} = (R^p F)(H^q A) \Rightarrow \mathbb{R}^{p+q} F(A)$ of 5.7.9. Things get more complicated when we compose two or more functors, because then we need spectral sequences to compute the E_2-terms of other spectral sequences.

Example 10.8.1 Consider the problem of comparing the two ways of forming the total tensor product of three bounded below cochain complexes $A \in \mathbf{D}^-(\mathbf{mod}\text{–}R)$, $B \in \mathbf{D}^-(R\text{–}\mathbf{mod}\text{–}S)$, and $C \in \mathbf{D}^-(S\text{–}\mathbf{mod})$. Replacing A and C by complexes of projectives, we immediately see that there is a natural isomorphism

(∗) $$A \otimes_R^{\mathbf{L}} (B \otimes_S^{\mathbf{L}} C) \cong (A \otimes_R^{\mathbf{L}} B) \otimes_S^{\mathbf{L}} C.$$

However, it is quite a different matter to try to establish this quasi-isomorphism by studying the two hypertor modules $\mathbf{Tor}_i^R(A, B)$ and $\mathbf{Tor}_j^S(B, C)$! Cf. [EGA, III.6.8.3]. Another way to establish the isomorphism (∗) is to set $F = \mathrm{Tot}(A \otimes_R)$ and $G = \mathrm{Tot}(\otimes_S C)$. Since $FG \cong GF$, (∗) follows immediately from the following result.

Composition Theorem 10.8.2 *Let* $\mathbf{K} \subset \mathbf{K}(\mathcal{A})$ *and* $\mathbf{K}' \subset \mathbf{K}(\mathcal{B})$ *be localizing triangulated subcategories, and suppose given two morphisms of triangulated categories* $G: \mathbf{K} \to \mathbf{K}'$, $F: \mathbf{K}' \to \mathbf{K}(\mathcal{C})$. *Assume that* $\mathbf{R}F$, $\mathbf{R}G$, *and* $\mathbf{R}(FG)$ *exist, with* $\mathbf{R}F(\mathbf{D}) \subseteq \mathbf{D}'$. *Then:*

1. *There is a unique natural transformation* $\zeta = \zeta_{F,G} \colon \mathbf{R}(FG) \Rightarrow \mathbf{R}F \circ \mathbf{R}G$, *such that the following diagram commutes in* $\mathbf{D}(\mathcal{C})$ *for each A in* $\mathbf{K}$.

$$
\begin{array}{ccc}
q\,FG(A) & \xrightarrow{\;\xi_F\;} & (\mathbf{R}F)(qGA) \\[4pt]
\Big\downarrow{\scriptstyle \xi_{FG}} & & \Big\downarrow{\scriptstyle \xi_G} \\[6pt]
\mathbf{R}(FG)(qA) & \xrightarrow{\;\zeta_{qA}\;} & (\mathbf{R}F)(\mathbf{R}G)(qA)
\end{array}
$$

2. *Suppose that there are triangulated subcategories* $\mathbf{K}_0 \subseteq \mathbf{K}$, $\mathbf{K}_0' \subseteq \mathbf{K}'$ *satisfying the hypotheses of the Generalized Existence Theorem 10.5.9 for G and F, and suppose that G sends* $\mathbf{K}_0$ *to* $\mathbf{K}_0'$. *Then* ζ *is an isomorphism*

$$
\zeta : \mathbf{R}(FG) \cong (\mathbf{R}F) \circ (\mathbf{R}G).
$$

Proof Part (1) follows from the universal property 10.5.1 of $\mathbf{R}(FG)$. For (2) it suffices to observe that if A is in $\mathbf{K}_0$, then

$$
\mathbf{R}(FG)(qA) = q\,FG(A) \cong \mathbf{R}F(q(GA)) \cong \mathbf{R}F(\mathbf{R}G(qA)). \qquad \Diamond
$$

Corollary 10.8.3 (Grothendieck spectral sequences) *Let* $\mathcal{A}$, $\mathcal{B}$, *and* $\mathcal{C}$ *be abelian categories such that both* $\mathcal{A}$ *and* $\mathcal{B}$ *have enough injectives, and suppose given left exact functors* $G \colon \mathcal{A} \to \mathcal{B}$ *and* $F \colon \mathcal{B} \to \mathcal{C}$.

$$
\begin{array}{ccc}
\mathcal{A} & \xrightarrow{\;G\;} & \mathcal{B} \\[4pt]
& {\scriptstyle FG}\searrow \quad \swarrow{\scriptstyle F} & \\[4pt]
& \mathcal{C} &
\end{array}
$$

If G sends injective objects of $\mathcal{A}$ *to F-acyclic objects of* $\mathcal{B}$, *then*

$$
\zeta : \mathbf{R}^+(FG) \cong (\mathbf{R}^+F) \circ (\mathbf{R}^+G).
$$

If in addition G sends acyclic complexes to F-acyclic complexes, and both F and G have finite cohomological dimension, then $\mathbf{R}(FG) \colon \mathbf{D}(\mathcal{A}) \to \mathbf{D}(\mathcal{C})$ *exists, and*

$$
\zeta : \mathbf{R}(FG) \cong (\mathbf{R}F) \circ (\mathbf{R}G).
$$

In both cases, there is a convergent spectral sequence for all A:

$$
E_2^{pq} = (R^p F)(\mathbb{R}^q G)(A) \Rightarrow \mathbb{R}^{p+q}(FG)(A).
$$

If A is an object of $\mathcal{A}$, *this is the Grothendieck spectral sequence of 5.8.3.*

Proof The hypercohomology spectral sequence 5.7.9 converging to $(\mathbb{R}^{p+q}F)(\mathbf{R}G(A))$ has E_2^{pq} term $(R^pF)H^q(\mathbf{R}G(A)) = (R^pF)(\mathbb{R}^qG(A))$.

$\diamond$

Remark 10.8.4 Conceptually, the composition of functors $\mathbf{R}(FG) \cong (\mathbf{R}F) \circ (\mathbf{R}G)$ is much simpler than the original spectral sequence. The reader having some familiarity with algebraic geometry may wish to glance at [EGA, III.6], and especially at the "six spectral sequences" of III.6.6 or III.6.7.3, to appreciate the convenience of the derived category.

Exercise 10.8.1 If F, G, H are three consecutive morphisms, show that as natural transformations from $\mathbf{R}(FGH)$ to $\mathbf{R}F \circ \mathbf{R}G \circ \mathbf{R}H$ we have

$$\zeta_{G,H} \circ \zeta_{F,GH} = \zeta_{F,G} \circ \zeta_{FG,H}.$$

In the rest of this section, we shall enumerate three consequences of the Composition Theorem 10.8.2, usually replacing a spectral sequence with an isomorphism in the derived category. We will implicitly use the dual formulation $LF \circ LG \cong L(FG)$ of the Composition Theorem without comment.

10.8.1 The Projection Formula

10.8.5 Let $f: R \to S$ be a ring homomorphism, A a bounded above complex of right R-modules, and B a complex of left S-modules. The functor $f^*: \mathbf{mod}\text{-}R \to \mathbf{mod}\text{-}S$ sends A to $A \otimes_R S$, so it preserves projectives. Since $f^*(A) \otimes_S B = (A \otimes_R S) \otimes_S B \cong A \otimes_R f_*B$, the Composition Theorem 10.8.2 yields

$$(*) \qquad\qquad Lf^*(A) \otimes_S^{\mathbf{L}} B \xrightarrow{\cong} A \otimes_R^{\mathbf{L}} (f_*B)$$

in $\mathbf{D}(\mathbf{Ab})$. If S is commutative, we may regard B as an S–S bimodule and f_*B as an R–S bimodule. As we saw in exercise 10.6.2, this allows us to interpret $(*)$ as an isomorphism in $\mathbf{D}(S)$. From the standpoint of algebraic geometry, however, it is better to apply f_* to obtain the following isomorphism in $\mathbf{D}(R)$:

$$f_*(Lf^*(A) \otimes_S^{\mathbf{L}} B) \cong A \otimes_R^{\mathbf{L}} (f_*B).$$

This is sometimes called the "projection formula"; see [HartRD, II.5.6] for the generalization to schemes. The projection formula underlies the "Base change for Tor" spectral sequence 5.6.6.

Exercise 10.8.2 Use the universal property of $\otimes_R^{\mathbf{L}}$ to construct the natural map $\mathbf{L}f^*(A) \otimes_S^{\mathbf{L}} B \to A \otimes_R^{\mathbf{L}} (f_* B)$.

10.8.6 Similarly, if $g: S \to T$ is another ring homomorphism, we have $(gf)^* \cong g^* f^*$. The Composition Theorem 10.8.2 yields a natural isomorphism

$$(\mathbf{L}g^*)(\mathbf{L}f^*)A \cong \mathbf{L}(gf)^*A.$$

This underlies the spectral sequence $\mathrm{Tor}_p^S(\mathrm{Tor}_q^R(A, S), T) \Rightarrow \mathrm{Tor}_{p+q}^R(A, T)$.

10.8.2 Adjointness of $\otimes^{\mathbf{L}}$ and $\mathbf{R}\mathrm{Hom}$

Theorem 10.8.7 *If R is a commutative ring and B is a bounded above complex of R-modules, then $\otimes_R^{\mathbf{L}} B: \mathbf{D}^-(R) \to \mathbf{D}^-(R)$ is left adjoint to the functor $\mathbf{R}\mathrm{Hom}_R(B, -): \mathbf{D}^+(R) \to \mathbf{D}^+(R)$. That is, for A in $\mathbf{D}^-(R)$ and C in $\mathbf{D}^+(R)$ there is a natural isomorphism*

$$\mathrm{Hom}_{\mathbf{D}(R)}(A, \mathbf{R}\mathrm{Hom}_R(B, C)) \cong \mathrm{Hom}_{\mathbf{D}(R)}(A \otimes_R^{\mathbf{L}} B, C).$$

This isomorphism arises by applying H^0 to the isomorphism

$$(\dagger) \qquad \mathbf{R}\mathrm{Hom}_R(A, \mathbf{R}\mathrm{Hom}_R(B, C)) \xrightarrow{\cong} \mathbf{R}\mathrm{Hom}_R(A \otimes_R^{\mathbf{L}} B, C)$$

in $\mathbf{D}^+(R)$. The adjunction morphisms are $\eta_A: A \to \mathbf{R}\mathrm{Hom}_R(B, A \otimes_R^{\mathbf{L}} B)$ and $\varepsilon_C: \mathbf{R}\mathrm{Hom}_R(B, C) \otimes_R^{\mathbf{L}} B \to C$.

Proof Fix a projective complex A and an injective complex C. The functor $A \otimes_R^{\mathbf{L}} -$ preserves projectives, while the functor $\mathrm{Hom}_R(-, C)$ sends projectives to injectives. By the Composition Theorem 10.8.2, the two sides of ($\dagger$) are both isomorphic to the derived functors of the composite functor $\mathrm{Hom}(A, \mathrm{Hom}(B, C)) \cong \mathrm{Hom}(A \otimes_R B, C)$. $\diamond$

Exercise 10.8.3 Let R be a commutative ring and C a bounded complex of finite Tor dimension over R (exercise 10.6.3). Show that there is a natural isomorphism in $\mathbf{D}(R)$:

$$\mathbf{R}\mathrm{Hom}_R(A, B) \otimes_R^{\mathbf{L}} C \xrightarrow{\cong} \mathbf{R}\mathrm{Hom}_R(A, B \otimes_R^{\mathbf{L}} C).$$

Here A is in $\mathbf{D}(R)$ and B is in $\mathbf{D}^+(R)$. For the scheme version of this result, see [HartRD, II.5.14].

We now consider the effect of a ring homomorphism $f: R \to S$ upon **RHom**. We saw in 2.3.10 that $\mathrm{Hom}_R(S, -): \mathbf{mod}\text{--}R \to \mathbf{mod}\text{--}S$ preserves injectives. Therefore for every S-module complex A, and every bounded below R-module complex B, we have

$$\mathbf{RHom}_S(A, \mathbf{RHom}_R(S, B)) \cong \mathbf{RHom}_R(f_*A, B).$$

This isomorphism underlies the "Base change for Ext" spectral sequence of exercise 5.6.3.

Exercise 10.8.4 Suppose that S is a flat R-module, so that f^* is exact and $\mathbf{L}f^* \cong f^*$. Suppose that A is quasi-isomorphic to a bounded above complex of finitely generated projective modules. Show that we have a natural isomorphism for every B in $\mathbf{D}^+(R)$:

$$\mathbf{L}f^* \mathbf{RHom}_R(A, B) \to \mathbf{RHom}_S(\mathbf{L}f^*A, \mathbf{L}f^*B).$$

Exercise 10.8.5 (Lyndon/Hochschild-Serre) Let H be a normal subgroup of a group G. Show that the functors $A_H = A \otimes_{\mathbb{Z}H} \mathbb{Z}$ and $A^H = \mathrm{Hom}_H(\mathbb{Z}, A)$ of Chapter 6 have derived functors $A \otimes_H^{\mathbf{L}} \mathbb{Z}: \mathbf{D}(G\text{--}\mathbf{mod}) \to \mathbf{D}(G/H\text{--}\mathbf{mod})$ and $\mathbf{RHom}_H(\mathbb{Z}, A): \mathbf{D}(G\text{--}\mathbf{mod}) \to \mathbf{D}(G/H\text{--}\mathbf{mod})$ such that

$$A \otimes_G^{\mathbf{L}} \mathbb{Z} \cong (A \otimes_H^{\mathbf{L}} \mathbb{Z}) \otimes_{G/H}^{\mathbf{L}} \mathbb{Z} \quad \text{and}$$

$$\mathbf{RHom}_G(\mathbb{Z}, A) \cong \mathbf{RHom}_{G/H}(\mathbb{Z}, \mathbf{RHom}_H(\mathbb{Z}, A)).$$

Use these to obtain the Lyndon/Hochschild-Serre spectral sequences 6.8.2.

10.8.3 Leray Spectral Sequences

10.8.8 Suppose that $f: X \to Y$ is a continuous map of topological spaces. We saw in 5.8.6 that f_* preserves injectives and that the Leray spectral sequence

$$E_2^{pq} = H^p(Y; R^q f_* \mathcal{F}) \Rightarrow H^{p+q}(X; \mathcal{F})$$

arose from the fact that $\Gamma(X, \mathcal{F})$ is the composite $\Gamma(Y, f_*\mathcal{F})$. The Composition Theorem 10.8.2 promotes this into an isomorphism for every $\mathcal{F}$ in $\mathbf{D}^+(X)$:

$$\mathbf{R}\Gamma(X, \mathcal{F}) \cong \mathbf{R}\Gamma(Y, \mathbf{R}f_*\mathcal{F}).$$

Of course, if X and Y are noetherian spaces of finite Krull dimension, then this isomorphism is valid for every $\mathcal{F}$ in $\mathbf{D}(X)$.

We can generalize this by replacing $\Gamma(Y, -)$ by g_*, where $g: Y \to Z$ is another continuous map. For this, we need the following standard identity.

Lemma 10.8.9 $(gf)_*\mathcal{F} = g_*(f_*\mathcal{F})$ *for every sheaf $\mathcal{F}$ on X.*

Proof By its very definition (2.6.6), for every open subset U of X we have

$$(gf)_*\mathcal{F}(U) = \mathcal{F}((gf)^{-1}U)$$
$$= \mathcal{F}(f^{-1}g^{-1}U) = (f_*\mathcal{F})(g^{-1}U) = g_*(f_*\mathcal{F})(U). \qquad \diamond$$

Corollary 10.8.10 *For every $\mathcal{F}$ in $\mathbf{D}^+(X)$ there is a natural isomorphism*

$$\mathbf{R}(gf)_*(\mathcal{F}) \cong \mathbf{R}g_*(\mathbf{R}f_*(\mathcal{F}))$$

in $\mathbf{D}(Z)$. If moreover X and Y are noetherian of finite Krull dimension, then this isomorphism holds for every $\mathcal{F}$ in $\mathbf{D}(X)$.

Exercise 10.8.6 If $\mathcal{F}$ is an injective sheaf, the sheaf hom $\mathcal{H}om(\mathcal{E}, \mathcal{F})$ is Γ-acyclic ("flasque") by [Gode, II.7.3.2]. For any two sheaves $\mathcal{E}$ and $\mathcal{F}$, show that $\mathcal{H}om_X(\mathcal{E}, \mathcal{F}) \cong \Gamma(X, \mathcal{H}om(\mathcal{E}, \mathcal{F}))$. Then use the Composition Theorem 10.8.2 to conclude that there is a natural isomorphism

$$\mathbf{R}\mathrm{Hom}(\mathcal{E}, \mathcal{F}) \cong (\mathbf{R}\Gamma) \circ \mathbf{R}\mathcal{H}om(\mathcal{E}, \mathcal{F})$$

of bifunctors from $\mathbf{D}^-(X)^{\mathrm{op}} \times \mathbf{D}^+(X)$ to $\mathbf{D}(\mathbf{Ab})$.

10.9 The Topological Derived Category

At the same time (1962–1963) as Verdier was inventing the algebraic notion of the derived category [Verd], topologists (e.g., D. Puppe) were discovering that the stable homotopy category $\mathbf{D}(\mathcal{S})$ was indeed a triangulated category. In this last section we show how to construct this structure with a minimum of topology, mimicking the passage from chain complexes to the homotopy category $\mathbf{K(Ab)}$ in section 10.1 and the localization from $\mathbf{K(Ab)}$ to the derived category $\mathbf{D(Ab)}$. This provides a rich analogy between derived categories and stable homotopy theory, which has only recently been exploited (see [Th] and [Rob], for example).

Our first task is to define the category of spectra $\mathcal{S}$. Here is the "modern" (coordinatized) definition, following [LMS].

Definition 10.9.1 A *spectrum* E is a sequence of based topological spaces E_n and based homeomorphisms $\alpha_n \colon E_n \xrightarrow{\cong} \Omega E_{n+1}$. A *map of spectra* $f \colon E \to F$ is a sequence of based continuous maps $f_n \colon E_n \to F_n$ strictly compatible with the given structural homeomorphisms. As these maps are closed under composition, the spectra and their maps form a category $\mathcal{S}$. The sequence of 1-point spaces forms a spectrum $*$, which is the zero object in $\mathcal{S}$, because $\operatorname{Hom}_{\mathcal{S}}(*, E) = \operatorname{Hom}_{\mathcal{S}}(E, *) = \{\text{point}\}$ for all E. The product $E \times F$ of two spectra is the spectrum whose n^{th} space is $E_n \times F_n$.

Historically, spectra arose from the study of "infinite loop spaces;" E_0 is an infinite loop space, because we have described it as the p-fold loop space $E_0 \cong \Omega^p E_p$ for all p. The most readable reference for this is part III of Adams' book [A], although it is far from optimal on the foundations, which had not yet been worked out in 1974.

Looping and Delooping 10.9.2 If E is a spectrum, we can form its loop spectrum ΩE by setting $(\Omega E)_n = \Omega(E_n)$, the structural maps being the $\Omega(\alpha_n)$. More subtly, we can form the delooping $\Omega^{-1}E$ by reindexing and forgetting $E_0 \colon (\Omega^{-1}E)_n = E_{n+1}$. Clearly $\Omega\Omega^{-1}E = \Omega^{-1}\Omega E \cong E$, so Ω is an automorphism of the category $\mathcal{S}$. When we construct a triangulated structure on the stable homotopy category, Ω^{-1} will become our "translation functor."

Example 10.9.3 (Sphere spectra) There is a standard map from the m-sphere S^m to the ΩS^{m+1} (put S^m at the equator of S^{m+1} and use the longitudes). The *n-sphere spectrum* $\mathbf{S}^n$ is obtained by applying Ω^i and taking the colimit

$$(\mathbf{S}^n)_p = \operatorname*{colim}_{i \to \infty} \Omega^i S^{n+p+i}.$$

Of course, to define the negative sphere spectrum $\mathbf{S}^n$ we only use $i \geq -n$. The zero-th space of the sphere spectrum $\mathbf{S}^0$ is often written as $\Omega^\infty S^\infty$. Note that our notational conventions are such that for all integers n and p we have $\Omega^p \mathbf{S}^n = \mathbf{S}^{n-p}$.

Definition 10.9.4 (The stable category) The homotopy groups of a spectrum E are:

$$\pi_n E = \pi_{n+i}(E_i) \quad \text{for } i \geq 0, n + i \geq 0.$$

These groups are independent of the choice of i, because for all m $\pi_{i+1}E_m \cong \pi_i(\Omega E_m)$. We say that $f: E \to F$ is a *weak homotopy equivalence* if f induces an isomorphism on homotopy groups. Let $\widetilde{W}$ denote the family of all weak homotopy equivalences in $\mathcal{S}$. The *stable homotopy category*, or *topological derived category* $\mathbf{D}(\mathcal{S})$, is the localization $\widetilde{W}^{-1}\mathcal{S}$ of $\mathcal{S}$ at $\widetilde{W}$.

Of course, in order to see that the stable category exists within our universe we need to prove something. Mimicking the procedure of section 1 and section 3, we shall first construct a homotopy category $\mathbf{K}(\mathcal{S})$ and prove that the system W of weak homotopy equivalences form a locally small multiplicative system in $\mathbf{K}(\mathcal{S})$ (10.3.6). Then we shall show that the homotopy category of "CW spectra" forms a localizing subcategory $\mathbf{K}(\mathcal{S}_{CW})$ of $\mathbf{K}(\mathcal{S})$ (10.3.12), and that we may take the topological derived category to be $\mathbf{K}(\mathcal{S}_{CW})$. This parallels theorem 10.4.8, that the category $\mathbf{D}^+(\mathbf{Ab})$ is equivalent to the homotopy category of bounded below complexes of injective abelian groups.

For this program, we need the notion of homotopy in $\mathcal{S}$ and the notion of a CW spectrum, both of which are constructed using prespectra and the "spectrification" functor Ω^∞. Let SX denote the usual based suspension of a topological space X, and recall that maps $SX \to Y$ are in 1–1 correspondence with maps $X \to \Omega Y$.

Definition 10.9.5 A *prespectrum* D is a sequence of based topological spaces D_n and based continuous maps $S(D_n) \to D_{n+1}$, or equivalently, maps $D_n \to \Omega D_{n+1}$. If C and D are prespectra, a *function* $f: C \to$ D is a sequence of based continuous maps $f_n: C_n \to D_n$ which are strictly compatible with the given structural maps. There is a category $\mathcal{P}$ of prespectra and functions, as well as a forgetful functor $\mathcal{S} \to \mathcal{P}$. A *CW prespectrum* is a prespectrum D in which all the spaces D_n are CW complexes and all the structure maps $SD_n \to D_{n+1}$ are cellular inclusions.

Warning: Terminology has changed considerably over the years, even since the 1970s. A prespectrum used to be called a "suspension spectrum," and the present notion of spectrum is slightly stronger than the notion of "Ω-spectrum," in which the structural maps were only required to be weak equivalences. Our use of "function" agrees with [A], but the category of CW prespectra in [A] has more morphisms than just the functions; see [A, *p*.140] or [LMS, *p*.2] for details.

10.9.6   There is a functor $\Omega^\infty \colon \mathcal{P} \to \mathcal{S}$, called "spectrification." It sends a CW prespectrum D to the spectrum $\Omega^\infty D$ whose n^{th} space is

$$(\Omega^\infty D)_n = \operatorname*{colim}_{i \to \infty} \Omega^i D_{n+i},$$

where the colimit is taken with respect to the iterated loops on the maps $D_j \to \Omega D_{j+1}$. The structure maps $(\Omega^\infty D)_n \to (\Omega^\infty D)_{n+1}$ are obtained by shifting the indices, using the fact that Ω commutes with colimits. The effect of Ω^∞ on functions should be clear.

A *CW spectrum* is a spectrum of the form $E = \Omega^\infty D$ for some CW prespectrum D. The full subcategory of $\mathcal{S}$ consisting of CW spectra is written as $\mathcal{S}_{CW}$. Although the topological spaces E_n of a CW spectrum are obviously not CW complexes themselves, they do have the homotopy type of CW complexes.

Exercise 10.9.1   Show that $\Omega^\infty E \cong E$ in $\mathcal{S}$ for every spectrum E.

Topology Exercise 10.9.2   If D is a CW prespectrum, show that the structure maps $D_n \to \Omega D_{n+1}$ are closed embeddings. Use this to show that

$$\pi_n(\Omega^\infty D) = \operatorname*{colim}_{i \to \infty} \pi_{n+i}(D_{n+i}).$$

Analogy 10.9.7   There is a formal analogy between the theory of spectra and the theory of (chain complexes of) sheaves. The analogue of a presheaf is a prespectrum. Just as the forgetful functor from sheaves to presheaves has a left adjoint (sheafification), the forgetful functor from spectra to prespectra has Ω^∞ as its left adjoint. The reader is referred to the Appendix of [LMS] for the extension of Ω^∞ to general spectra, as well as the verification that Ω^∞ is indeed the left adjoint of the forgetful functor.

Just as many standard operations on sheaves (inverse image, direct sum, cokernels) are defined by sheafification, many standard operations on spectra (cylinders, wedges, mapping cones) are defined on spectra by applying Ω^∞ to the corresponding operation on prespectra. This is not surprising, since both are right adjoint functors and therefore must preserve coproducts and colimits by 2.6.10.

Example 10.9.8 (Coproduct)   Recall that the coproduct in the category of based topological spaces is the *wedge* $\vee_\alpha X_\alpha$, obtained from the disjoint union by identifying the basepoints. If $\{D_\alpha\}$ is a family of prespectra, their wedge is the prespectrum whose n^{th} space is $(\vee D_\alpha)_n = \vee (D_\alpha)_n$; it is the coproduct in the category of prespectra. (Why?) Since Ω^∞ preserves coproducts, $\vee D_\alpha = \Omega^\infty \{\vee (D_\alpha)_n\}$ is the coproduct in the category of spectra.

Example 10.9.9 (Suspension) The *suspension* SE of a spectrum E is Ω^∞ applied to the prespectrum whose n^{th} space is SE_n and whose structure maps are the suspensions of the structure maps $SE_n \to E_{n+1}$. Adams proves in [A, III.3.7] that the natural maps $E_n \to \Omega S(E_n)$ induce a weak homotopy equivalence $E \to \Omega SE$, and hence a weak homotopy equivalence

$$\Omega^{-1} E \xrightarrow{\sim} SE.$$

Definition 10.9.10 (Homotopy category) The *cylinder spectrum* $\mathrm{cyl}(E)$ of a spectrum E is obtained by applying Ω^∞ to the prespectrum $(I_+ \wedge E)_n = [0, 1] \times E_n/[0, 1] \times \{*\}$. Just as in ordinary topology, we say that two maps of spectra $f_0, f_1: E \to F$ are *homotopic* if there is a map $h: \mathrm{cyl}(E) \to F$ such that the f_i are the composites $E \cong \{i\} \times E \hookrightarrow \mathrm{cyl}(E) \to F$. It is not hard to see that this is an equivalence relation (exercise!).

We write $[E, F]$ for the set of homotopy classes of maps of spectra; these form the morphisms of the homotopy category $\mathbf{K}(\mathcal{S})$ of spectra. The full subcategory of $\mathbf{K}(\mathcal{S})$ consisting of the CW spectra is written as $\mathbf{K}(\mathcal{S}_{CW})$.

Exercise 10.9.3 Show that $E \times F$ and $E \vee F$ are also the product and coproduct in $\mathbf{K}(\mathcal{S})$.

Proposition 10.9.11 $\mathbf{K}(\mathcal{S})$ *is an additive category.*

Proof Since $\mathbf{K}(\mathcal{S})$ has a zero object $*$ and a product $E \times F$, we need only show that it is an **Ab**-category (Appendix, A.4.1), that is, that every Hom-set $[E, F]$ has the structure of an abelian group in such a way that composition distributes over addition. The standard proof in topology that homotopy classes of maps into any loop space form an abelian group proves this; one splits $\mathrm{cyl}(F)$ into $[0, \frac{1}{2}] \times F/\sim$ and $[\frac{1}{2}, 1] \times F/\sim$ and concatenates loops. We leave the verification of this to readers familiar with the standard proof. $\diamond$

Corollary 10.9.12 *The natural map* $E \vee F \to E \times F$ *is an isomorphism in* $\mathbf{K}(\mathcal{S})$.

The role of CW spectra is based primarily upon the two following fundamental results.

Proposition 10.9.13 *For each spectrum E there is a natural weak homotopy equivalence $C \to E$, with C a CW spectrum. In particular, $\mathbf{K}(\mathcal{S}_{CW})$ is a localizing subcategory of $\mathbf{K}(\mathcal{S})$ in the sense of 10.3.12.*

Proof Let $\mathrm{Sing}(X)$ denote the singular simplicial set (8.2.4) of a topological space X, and $|\mathrm{Sing}(X)| \to X$ the natural map. Since $|\mathrm{Sing}(X)|$ is a CW complex, the cellular inclusions $S|\mathrm{Sing}(E_n)| \hookrightarrow |\mathrm{Sing}(SE_n)| \hookrightarrow |\mathrm{Sing}(E_{n+1})|$ make $|\mathrm{Sing}(E)|$ into a CW prespectrum and give us a function of prespectra $|\mathrm{Sing}(E)| \to E$. Taking adjoints gives a map of spectra $C \to E$, where $C = \Omega^\infty |\mathrm{Sing}(E)|$. Since $\pi_*|\mathrm{Sing}(X)| \cong \pi_*(X)$ for every topological space X, we have

$$\pi_i|\mathrm{Sing}(E_m)| \cong \pi_i(E_m) \cong \pi_{i+1}(E_{m+1}) \cong \pi_{i+1}|\mathrm{Sing}(E_{m+1})|$$

for all m and i. Since $\pi_n(C) \cong \mathrm{colim}_{i\to\infty} \pi_{n+i}(|\mathrm{Sing}(E_{n+i})|)$ by the topology exercise 10.9.2, it follows that $C \to E$ is a weak homotopy equivalence.   ◇

Whitehead's Theorem 10.9.14

1. *If C is a CW spectrum, then for every weak homotopy equivalence $f \colon E \to F$ of spectra (10.9.4) we have $f_* \colon [C, E] \cong [C, F]$.*
2. *Every weak homotopy equivalence of CW spectra is a homotopy equivalence (10.9.10), that is, an isomorphism in $\mathbf{K}(\mathcal{S})$.*

Proof See [A, pp.149–150] or [LMS, p.30]. Note that (1) implies (2), by setting $C = F$.   ◇

Corollary 10.9.15 *The stable homotopy category $\mathbf{D}(\mathcal{S})$ exists and is equivalent to the homotopy category of CW spectra*

$$\mathbf{D}(\mathcal{S}) \cong \mathbf{K}(\mathcal{S}_{CW}).$$

Proof The generalities on localizing subcategories in section 3 show that $\mathbf{D}(\mathcal{S}) \cong W^{-1}\mathbf{K}(\mathcal{S}_{CW})$. But by Whitehead's Theorem we have $\mathbf{K}(\mathcal{S}_{CW}) = W^{-1}\mathbf{K}(\mathcal{S}_{CW})$.   ◇

We are going to show in 10.9.18 that the topological derived category $\mathbf{D}(\mathcal{S}) \cong \mathbf{K}(\mathcal{S}_{CW})$ is a triangulated category in the sense of 10.2.1. For this we need to define exact triangles. The exact triangles will be the cofibration sequences, a term that we must now define. In order to avoid explaining a technical hypothesis ("cofibrant") we shall restrict our attention to CW spectra.

Mapping Cones 10.9.16 Suppose that $u: E \to F$ is a map of spectra. The sequence of topological mapping cones $\mathrm{cone}(u_n) = \mathrm{cone}(E_n) \cup_u F_n$ form a prespectrum (why?), and the *mapping cone of f* is defined to be the spectrum $\Omega^\infty\{\mathrm{cone}(f_n)\}$. Applying Ω^∞ to the prespectrum functions $i_n: F_n \to \mathrm{cone}(f_n)$ and $\mathrm{cone}(f_n) \to SE_n$ give maps of spectra $i: F \to \mathrm{cone}(f)$ and $j: \mathrm{cone}(f) \to SE$. The triangle determined by this data is called the *Puppe sequence* associated to f:

$$E \xrightarrow{\ u\ } F \xrightarrow{\ i\ } \mathrm{cone}(u) \xrightarrow{\ j\ } SE.$$

A *cofibration sequence* in $\mathbf{K}(\mathcal{S}_{CW})$ is any triangle isomorphic to a Puppe sequence. Since $* \to E \xrightarrow{\mathrm{id}} E \to *$ is a Puppe sequence, the following elementary exercise shows that cofibration sequences satisfy axioms (TR1) and (TR2).

Exercise 10.9.4 (Rotation) Use the fact that SE_n is homotopy equivalent to the cone of $i_n: F_n \to \mathrm{cone}(f_n)$ to show that $SE \cong \mathrm{cone}(i)$. Then show that

$$F \xrightarrow{\ i\ } \mathrm{cone}(u) \xrightarrow{\ j\ } SE \xrightarrow{\ -Su\ } SF$$

is a cofibration sequence.

We say that a diagram of spectra is *homotopy commutative* if it commutes in the homotopy category $\mathbf{K}(\mathcal{S})$.

Proposition 10.9.17 *Every homotopy commutative square of spectra*

$$
\begin{array}{ccc}
E & \xrightarrow{\ u\ } & F \\
\downarrow{\scriptstyle f} & & \downarrow{\scriptstyle g} \\
E' & \xrightarrow{\ u'\ } & F'
\end{array}
$$

can be made to commute. That is, there is a homotopy commutative diagram

$$
\begin{array}{ccc}
E & \xrightarrow{\ u\ } & F \\
\| & & \downarrow{\scriptstyle \simeq} \\
E & \longrightarrow & \mathrm{cyl}(u) \\
\downarrow{\scriptstyle f} & & \downarrow{\scriptstyle g'} \\
E' & \xrightarrow{\ u'\ } & F'
\end{array}
$$

in which the bottom square strictly commutes in S and the map $\simeq$ is a homotopy equivalence.

Proof Let cyl(u_n) denote the topological mapping cylinder of u_n (Chapter 1, section 5). The mapping cylinder spectrum cyl(u) is Ω^∞ of the prespectrum $\{$cyl(u_n)$\}$. It is homotopy equivalent to F because the homotopy equivalences $F_n \xrightarrow{\simeq}$ cyl(u_n) are canonical. The map cyl(E) $\to F'$ expressing the homotopy commutativity of the square corresponds to a prespectrum function from $\{$cyl(E_n)$\}$ to F'; together with g they define a prespectrum function from $\{$cyl(u_n)$\}$ to F' and hence a spectrum map g': cyl(u) $\to F'$. The inclusions of E_n into the top of cyl(u_n) give the middle row after applying Ω^∞. It is now a straightforward exercise to check that the diagram homotopy commutes and that the bottom square commutes. $\diamond$

Theorem 10.9.18 $\mathbf{K}(\mathcal{S}_{CW})$ *is a triangulated category.*

Proof We have already seen that axioms (TR1) and (TR2) hold. For (TR3) we may suppose that $C = $ cone(u) and $C' = $ cone(u') and that $gu = u'f$ in S; the map h is given by the naturality of the mapping cone construction.

It remains to check the octahedral axiom (TR4). For this we may assume that the given triangles are Puppe sequences, that is, that $C' = $ cone(u), $A' = $ cone(v), and $B' = $ cone(vu). We shall mimic the proof in 10.2.4 that the octahedral axiom holds in $\mathbf{K}(\mathcal{A})$. Define a prespectrum function $\{f_n\}$ from $\{$cone(u_n)$\}$ to $\{$cone($v_n u_n$)$\}$ by letting f_n be the identity on cone(A_n) and v_n on B_n. Define a prespectrum function $\{g_n\}$ from $\{$cone($v_n u_n$)$\}$ to $\{$cone(v_n)$\}$ by letting g_n be cone(u_n): cone(A_n) $\to$ cone(B_n) and the identity on C. Manifestly, these are prespectrum functions; we define f and g by applying Ω^∞ to $\{f_n\}$ and $\{g_n\}$. Since it is true at the prespectrum level, ∂ is the composite cone(u) $\xrightarrow{f}$ cone(vu) $\xrightarrow{\delta}$ SA and x is the composite $C \xrightarrow{y}$ cone(vu) $\xrightarrow{g}$ cone(v). (Check this!)

Since cone(f_n) is a quotient of the disjoint union of cone(cone(A_n)), cone(B_n), and C_n, the natural maps from cone(B_n) and C_n to cone(f_n) induce an injection cone(v_n) $\hookrightarrow$ cone(f_n). As n varies, this forms a function of prespectra. Applying Ω^∞ gives a natural map of spectra γ: cone(v) $\to$ cone(f) such that the following diagram of spectra commutes in S:

$$
\begin{array}{ccccccc}
C' & \xrightarrow{\;f\;} & B' & \xrightarrow{\;g\;} & \text{cone}(v) & \xrightarrow{\;(Tj)i\;} & SC' \\
\| & & \| & & \downarrow{\gamma} & & \| \\
C' & \xrightarrow{\;f\;} & B' & \longrightarrow & \text{cone}(f) & \longrightarrow & SC'.
\end{array}
$$

To see that γ is a homotopy equivalence, define $\varphi_n: \text{cone}(f_n) \to \text{cone}(v_n)$ by sending $\text{cone}(B_n)$ and C_n to themselves via the identity, and composing the natural retract $\text{cone}(\text{cone}(A_n)) \to \text{cone}(0 \times A_n)$ with $\text{cone}(u_n): \text{cone}(A_n) \to \text{cone}(B_n)$. Since the φ_n are natural, they form a function of prespectra; applying Ω^∞ gives a map of spectra $\varphi: \text{cone}(f) \to \text{cone}(v)$. We leave it to the reader to check that $\varphi\gamma$ is the identity on $\text{cone}(v)$ and that $\gamma\varphi$ is homotopic to the identity map on $\text{cone}(f)$. (Exercise!). This shows that $(f, g, (Tj)i)$ is a cofibration sequence (exact triangle), because it is isomorphic to the Puppe sequence of f. $\diamond$

Geometric Realization 10.9.19 By the Dold-Kan correspondence (8.4.1), there is a geometric realization functor from **Ch(Ab)** to $\mathcal{S}_{CW}$. Indeed, if A is a chain complex of abelian groups, then the good truncation $\tau A = \tau_{\geq 0}(A)$ corresponds to a simplicial abelian group, and its realization $|\tau A|$ is a CW complex. In the sequence

$$\tau A \longrightarrow \tau \, \text{cone}(A) \overset{\delta}{\longrightarrow} \tau(A[-1]),$$

the map δ is a Kan fibration (8.2.9, exercise 8.2.5). Since the mapping cone is contractible (exercise 1.5.1), there is a weak homotopy equivalence $|\tau A| \to \Omega|\tau A[-1]|$, and its adjoint $S|\tau A| \to |\tau A[-1]|$ is a cellular inclusion. (Check this!) Thus the sequence of spaces $|\tau A[-n]|$ form a CW prespectrum; applying Ω^∞ gives a spectrum. This construction makes it clear that the functor $|\tau|: \text{Ch(Ab)} \to \mathcal{S}_{CW}$ sends quasi-isomorphisms to weak equivalences and sends the translated chain complex $A[n]$ to $\Omega^n|\tau A|$. In particular, it induces a functor on the localized categories $|\tau|: \text{D(Ab)} \to \text{D}(\mathcal{S})$.

Vista 10.9.20 Let $H\mathbb{Z}$ denote the geometric realization $|\tau\mathbb{Z}|$ of the abelian group $\mathbb{Z}$, regarded as a chain complex concentrated in degree zero. It turns out that $H\mathbb{Z}$ is a "ring spectrum" and that **D(Ab)** is equivalent to the stable category of "module spectra" over $H\mathbb{Z}$. This equivalence takes the total tensor product $\otimes_{\mathbb{Z}}^{\text{L}}$ in **D(Ab)** to smash products of module spectra over $H\mathbb{Z}$. See [Rob] and {A. Elmendorf, I. Kriz, and J. P. May, "E_∞ Modules Over E_∞ Ring Spectra," preprint (1993)}.

Appendix A

Category Theory Language

This Appendix provides a swift summary of some of the basic notions of category theory used in this book. Many of the terms are defined in Chapters 1 and 2, but we repeat them here for the convenience of the reader.

A.1 Categories

Definition A.1.1 A *category* $\mathcal{C}$ consists of the following: a class obj($\mathcal{C}$) of *objects*, a set $\mathrm{Hom}_{\mathcal{C}}(A, B)$ of *morphisms* for every ordered pair (A, B) of objects, an *identity morphism* $\mathrm{id}_A \in \mathrm{Hom}_{\mathcal{C}}(A, A)$ for each object A, and a *composition function* $\mathrm{Hom}_{\mathcal{C}}(A, B) \times \mathrm{Hom}_{\mathcal{C}}(B, C) \to \mathrm{Hom}_{\mathcal{C}}(A, C)$ for every ordered triple (A, B, C) of objects. We write $f: A \to B$ to indicate that f is a morphism in $\mathrm{Hom}_{\mathcal{C}}(A, B)$, and we write gf or $g \circ f$ for the composition of $f: A \to B$ with $g: B \to C$. The above data is subject to two axioms:

> *Associativity Axiom:* $(hg)f = h(gf)$ for $f: A \to B$, $g: B \to C$, $h: C \to D$
> *Unit Axiom:* $\mathrm{id}_B \circ f = f = f \circ \mathrm{id}_A$ for $f: A \to B$.

Paradigm A.1.2 The fundamental category to keep in mind is the category **Sets** of sets. The objects are sets and the morphisms are (set) functions, that is, the elements of $\mathrm{Hom}_{\mathbf{Sets}}(A, B)$ are the functions from A to B. Composition of morphisms is just composition of functions, and id_A is the function $\mathrm{id}_A(a) = a$ for all $a \in A$. Note that the objects of **Sets** do not form a set (or else we would encounter Russell's paradox of a set belonging to itself!); this explains the pedantic insistence that obj($\mathcal{C}$) be a class and not a set. Nevertheless, we shall often use the notation $C \in \mathcal{C}$ to indicate that C is an object of $\mathcal{C}$.

Examples A.1.3 Another fundamental category is the category **Ab** of abelian groups. The objects are abelian groups, and the morphisms are group homomorphisms. Composition is just ordinary composition of homomorphisms.

The categories **Groups** of groups (and group maps) and **Rings** of rings (and ring maps) are defined similarly.

If R is a ring, R–**mod** is the category of left R-modules. Here the objects are left R-modules, the morphisms are R-module homomorphisms, and composition has its usual meaning. The category **mod**–R of right R-modules is defined similarly, and it is the same as R–**mod** when R is a commutative ring.

A *discrete* category is one in which every morphism is an identity morphism. Every set (or class!) may be regarded as a discrete category, since composition is forced by discreteness.

Small categories A.1.4 A category C is *small* if obj(C) is a set (not just a class). **Sets**, **Ab** and R–**mod** are not small, but a poset or a group may be thought of as a small category as follows.

A *partially ordered set*, or *poset*, is a set P with a reflexive, transitive antisymmetric relation $\leq$. We regard a poset as a small category as follows. Given $p, q \in P$ the set $\mathrm{Hom}_P(p, q)$ is the empty set unless $p \leq q$, in which case there is exactly one morphism from p to q (denoted $p \leq q$ of course). Composition is given by transitivity and the reflexive axiom ($p \leq p$) yields identity morphisms.

A category with exactly one object $*$ is the same thing as a *monoid*, that is, a set M (which will be $\mathrm{Hom}(*, *)$) equipped with an associative law of composition and an identity element. In this way we may consider a group as a category with one object.

The word "category" is due to Eilenberg and MacLane (1947) but was taken from Aristotle and Kant. It is chiefly used as an organizing principle for familiar notions. It is also useful to have other words to describe familiar types of morphisms that we encounter in many different categories; here are a few.

A morphism $f: B \to C$ is called an *isomorphism* in C if there is a morphism $g: C \to B$ such that $gf = \mathrm{id}_B$ and $fg = \mathrm{id}_C$. The usual proof shows that if g exists it is unique, and we often write $g = f^{-1}$. An isomorphism in **Sets** is a set bijection; an isomorphism in the category **Top** of topological spaces and continuous maps is a homeomorphism; an isomorphism in the category of smooth manifolds and smooth maps is called a diffeomorphism. In most algebraic categories, isomorphism has its usual meaning. In a group (considered as a category), every morphism is an isomorphism.

A.1.5 A morphism $f: B \to C$ is called *monic* in C if for any two distinct morphisms $e_1, e_2: A \to B$ we have $fe_1 \neq fe_2$; in other words, we can cancel f on the left. In **Sets**, **Ab**, R–**mod**, ..., in which objects have an underlying set ("concrete" categories; see A.2.3), the monic morphisms are precisely the morphisms that are set injections (monomorphisms) in the usual sense. If $B \to C$ is monic, we will sometimes say that B is a *subobject* of C. (Technically a subobject is an equivalence class of monics, two monics being equivalent if they factor through each other.)

A morphism $f: B \to C$ is called *epi* in C if for any two distinct morphisms $g_1, g_2: C \to D$ we have $g_1 f \neq g_2 f$; in other words, we can cancel f on the right. In **Sets**, **Ab**, and R–**mod** the epi morphisms are precisely the onto maps (epimorphisms). In other concrete categories such as **Rings** or **Top** this fails; the morphisms whose underlying set map is onto are epi, but there are other epis.

Exercise A.1.1 Show that $\mathbb{Z} \subset \mathbb{Q}$ is epi in **Rings**. Show that $\mathbb{Q} \subset \mathbb{R}$ is epi in the category of Hausdorff topological spaces.

A.1.6 An *initial object* (if it exists) in C is an object I such that for every C in C there is exactly one morphism from I to C. A *terminal object* in C (if it exists) is an object T such that for every C in C there is exactly one morphism from C to T. All initial objects must be isomorphic, and all terminal objects must be isomorphic. For example, in **Sets** the empty set ϕ is the initial object and any 1-point set is a terminal object. An object that is both initial and terminal is called a *zero object*. There is no zero object in **Sets**, but 0 is a zero object in **Ab** and in R–**mod**.

Suppose that C has a zero object 0. Then there is a distinguished element in each set $\text{Hom}_C(B, C)$, namely the composite $B \to 0 \to C$; by abuse we shall write 0 for this map. A *kernel* of a morphism $f: B \to C$ is a morphism $i: A \to B$ such that $fi = 0$ and that satisfies the following universal property: Every morphism $e: A' \to B$ in C such that $fe = 0$ factors through A as $e = ie'$ for a unique $e': A' \to A$. Every kernel is monic, and any two kernels of f are isomorphic in an evident sense; we often identify a kernel of f with the corresponding subobject of B. Similarly, a *cokernel* of $f: B \to C$ is a morphism $p: C \to D$ such that $pf = 0$ and that satisfies the following universal property: Every morphism $g: C \to D'$ such that $gf = 0$ factors through D as $g = g'p$ for a unique $g': D \to D'$. Every cokernel is an epi, and any two cokernels are isomorphic. In **Ab** and R–**mod**, kernel and cokernel have their usual meanings.

Exercise A.1.2 In **Groups**, show that monics are just injective set maps, and kernels are monics whose image is a normal subgroup.

Opposite Category A.1.7 Every category C has an *opposite category* C^{op}. The objects of C^{op} are the same as the objects in C, but the morphisms (and composition) are reversed, so that there is a 1–1 correspondence $f \mapsto f^{op}$ between morphisms $f: B \to C$ in C and morphisms $f^{op}: C \to B$ in C^{op}. If f is monic, then f^{op} is epi; if f is epi, then f^{op} is monic. Similarly, taking opposites interchanges kernels and cokernels, as well as initial and terminal objects. Because of this duality, C^{op} is also called the *dual category* of C.

Example A.1.8 If R is a ring (a category with one object), R^{op} is the ring with the same underlying set, but in which multiplication is reversed. The category (R^{op})–**mod** of left R^{op}-modules is isomorphic to the category **mod**–R of right R-modules. However, $(R$–**mod**$)^{op}$ cannot be S–**mod** for any ring S (see A.4.7).

Exercise A.1.3 (Pontrjagin duality) Show that the category C of finite abelian groups is isomorphic to its opposite category C^{op}, but that this fails for the category T of torsion abelian groups. We will see in exercise 6.11.4 that T^{op} is the category of profinite abelian groups.

Products and Coproducts A.1.9 If $\{C_i: i \in I\}$ is a set of objects of C, a *product* $\prod_{i \in I} C_i$ (if it exists) is an object of C, together with maps $\pi_j: \prod C_i \to C_j$ ($j \in I$) such that for every $A \in C$, and every family of morphisms $\alpha_i: A \to C_i$ ($i \in I$), there is a unique morphism $\alpha: A \to \prod C_i$ in C such that $\pi_i \alpha = \alpha_i$ for all $i \in I$. *Warning:* Any object of C isomorphic to a product is also a product, so $\prod C_i$ is not a well-defined object of C. Of course, if $\prod C_i$ exists, then it is unique up to isomorphism. If $I = \{1, 2\}$, then we write $C_1 \times C_2$ for $\prod_{i \in I} C_i$. Many concrete categories (**Sets**, **Groups**, **Rings**, R–**mod**, ... A.2.3) have arbitrary products, but others (e.g., **Fields**) have no products at all.

Dually, a *coproduct* $\coprod_{i \in I} C_i$ of a set of objects in C (if it exists) is an object of C, together with maps $\iota_j: C_j \to \coprod C_i$ ($j \in I$) such that for every family of morphisms $\alpha_i: C_i \to A$ there is a unique morphism $\alpha: \coprod C_i \to A$ such that $\alpha \iota_j = \alpha_j$ for all $j \in I$. That is, a coproduct in C is a product in C^{op}. If $I = \{1, 2\}$, then we write $C_1 \amalg C_2$ for $\coprod_{i \in I} C_i$. In **Sets**, the coproduct is disjoint union; in **Groups**, the coproduct is the free product; in R–**mod**, the coproduct is direct sum.

Exercise A.1.4 Show that $\text{Hom}_C(A, \prod C_i) \cong \prod_{i \in I} \text{Hom}_C(A, C_i)$ and that $\text{Hom}_C(\coprod C_i, A) \cong \prod_{i \in I} \text{Hom}_C(C_i, A)$.

Exercise A.1.5 Let $\{\alpha_i: A_i \to C_i\}$ be a family of maps in C. Show that

1. If $\prod A_i$ and $\prod C_i$ exist, there is a unique map $\alpha: \prod A_i \to \prod C_i$ such that $\pi_i \alpha = \alpha_i \pi_i$ for all i. If every α_i is monic, so is α.

2. If $\coprod A_i$ and $\coprod C_i$ exist, there is a unique map $\alpha: \coprod A_i \to \coprod C_i$ such that $\iota_i \alpha_i = \alpha \iota_i$ for all i. If every α_i is an epi, so is α.

A.2 Functors

By a *functor* $F: \mathcal{C} \to \mathcal{D}$ from a category $\mathcal{C}$ to a category $\mathcal{D}$ we mean a rule that associates an object $F(C)$ (or FC or even F_C) of $\mathcal{D}$ to every object C of $\mathcal{C}$, and a morphism $F(f): F(C_1) \to F(C_2)$ in $\mathcal{D}$ to every morphism $f: C_1 \to C_2$ in $\mathcal{C}$. We require F to preserve identity morphisms ($F(\mathrm{id}_C) = \mathrm{id}_{FC}$) and composition ($F(gf) = F(g)F(f)$). Note that F induces set maps

$$\mathrm{Hom}_{\mathcal{C}}(C_1, C_2) \to \mathrm{Hom}_{\mathcal{D}}(FC_1, FC_2)$$

for every C_1, C_2 in $\mathcal{C}$. If $G: \mathcal{D} \to \mathcal{E}$ is another functor, the composite $GF: \mathcal{C} \to \mathcal{E}$ is defined in the obvious way: $(GF)(C) = G(F(C))$ and $(GF)(f) = G(F(f))$.

The identity functor $\mathrm{id}_{\mathcal{C}}: \mathcal{C} \to \mathcal{C}$ is the rule fixing all objects and morphisms, that is, $\mathrm{id}_{\mathcal{C}}(C) = C$, $\mathrm{id}_{\mathcal{C}}(f) = f$. Clearly, for a functor $F: \mathcal{C} \to \mathcal{D}$ we have $F \circ \mathrm{id}_{\mathcal{C}} = F = \mathrm{id}_{\mathcal{D}} \circ F$. Except for set-theoretic difficulties, we could form a category **CAT** whose objects are categories and whose morphisms are functors. Instead, we form **Cat**, whose objects are small categories; $\mathrm{Hom}_{\mathbf{Cat}}(\mathcal{C}, \mathcal{D})$ is the set (!) of all functors from $\mathcal{C}$ to $\mathcal{D}$, the identity of $\mathcal{C}$ is $\mathrm{id}_{\mathcal{C}}$, and composition is composition of functors.

Hom and Tensor Product A.2.1 Let R be a ring and M a right R-module. For every left R-module N the tensor product $M \otimes_R N$ is an abelian group and $M \otimes_R -$ is a functor from R–**mod** to **Ab**. For every right R-module N, $\mathrm{Hom}_R(M, N)$ is an abelian group and $\mathrm{Hom}_R(M, -)$ is a functor from **mod**–R to **Ab**. These two functors are discussed in Chapter 3.

Forgetful Functors A.2.2 A functor that does nothing more than forget some of the structure of a category is commonly called a *forgetful functor*, and written with a U (for "underlying"). For example, there is a forgetful functor from R–**mod** to **Ab** (forget the R-module structure), one from **Ab** to **Sets** (forget the group structure), and their composite from R–**mod** to **Sets**.

Faithful Functors A.2.3 A functor $F: \mathcal{C} \to \mathcal{D}$ is called *faithful* if the set maps $\mathrm{Hom}_{\mathcal{C}}(C, C') \to \mathrm{Hom}_{\mathcal{D}}(FC, FC')$ are all injections. That is, if f_1 and f_2 are distinct maps from C to C' in $\mathcal{C}$, then $F(f_1) \neq F(f_2)$. Forgetful functors are usually faithful functors, and a category $\mathcal{C}$ with a faithful functor $U: \mathcal{C} \to$ **Sets**

is called a *concrete category*. In a concrete category, morphisms are completely determined by their effect on the underlying sets. *R*–**mod** and **Ab** are examples of concrete categories.

A *subcategory* $\mathcal{B}$ of a category $\mathcal{C}$ is a collection of some of the objects and some of the morphisms, such that the morphisms of $\mathcal{B}$ are closed under composition and include id_B for every object B in $\mathcal{B}$. A subcategory is a category in its own right, and there is an (obvious) *inclusion functor*, which is faithful by definition.

A subcategory $\mathcal{B}$ in which $\mathrm{Hom}_\mathcal{B}(B, B') = \mathrm{Hom}_\mathcal{C}(B, B')$ for every B, B' in $\mathcal{B}$ is called a *full subcategory*. We often refer to it as "the full subcategory on the objects" $\mathrm{obj}(\mathcal{B})$, since this information completely determines $\mathcal{B}$.

A functor $F: \mathcal{C} \to \mathcal{D}$ is *full* if the maps $\mathrm{Hom}_\mathcal{C}(C, C') \to \mathrm{Hom}_\mathcal{D}(FC, FC')$ are all surjections. That is, every $g: F(C) \to F(C')$ in $\mathcal{D}$ is of the form $g = F(f)$ for some $f: C \to C'$. A functor that is both full and faithful is called *fully faithful*. For example, the inclusion of a full subcategory is fully faithful. The Yoneda embedding (see A.3.4) is fully faithful. Another example of a fully faithful functor is "reflection" onto a skeletal subcategory, which we now describe.

Skeletal Subcategories A.2.4 By a *skeletal subcategory* $\mathcal{S}$ of a category $\mathcal{C}$ we mean a full subcategory such that every object of $\mathcal{C}$ is isomorphic to exactly one object of $\mathcal{S}$. For example, the full subcategory of **Sets** on the cardinal numbers $0 = \phi, 1 = \{\phi\}, \ldots$ is skeletal. The category of finitely generated R-modules is not a small category, but it has a small skeletal subcategory.

If we can select an object FC in $\mathcal{S}$ and an isomorphism $\theta_C: C \cong FC$ for each C in $\mathcal{C}$, then F extends to a "reflection" functor as follows: if $f: B \to C$, then $F(f) = \theta_C f \theta_B^{-1}$. Such a reflection functor is fully faithful. We will discuss reflections and reflective subcategories more in A.6.3 below. The set-theoretic issues involved here are discussed in [MacCW, I.6].

Contravariant Functors A.2.5 The functors we have been discussing are sometimes called *covariant* functors to distinguish them from *contravariant* functors. A *contravariant functor* $F: \mathcal{C} \to \mathcal{D}$ is by definition just a covariant functor from $\mathcal{C}^{\mathrm{op}}$ to $\mathcal{D}$. That is, it associates an object $F(C)$ of $\mathcal{D}$ to every object C of $\mathcal{C}$, and a morphism $F(f): F(C_2) \to F(C_1)$ in $\mathcal{D}$ to every $f: C_1 \to C_2$ in $\mathcal{C}$. Moreover, $F(\mathrm{id}_C) = \mathrm{id}_{FC}$ and F reverses composition: $F(gf) = F(f)F(g)$.

The most important example in this book will be the contravariant functor $\mathrm{Hom}_R(-, N)$ from **mod**–R to **Ab** associated with a right R-module N. Its derived functors $\mathrm{Ext}_R^*(-, N)$ are also contravariant (see 2.5.2). Another example is a *presheaf* on a topological space X; this is by definition a contravariant functor from the poset of open subspaces of X to the category **Ab**.

A.3 Natural Transformations

Suppose that F and G are two functors from C to D. A *natural transformation* $\eta: F \Rightarrow G$ is a rule that associates a morphism $\eta_C: F(C) \to G(C)$ in D to every object C of C in such a way that for every morphism $f: C \to C'$ in C the following diagram commutes:

$$
\begin{array}{ccc}
F(C) & \xrightarrow{\;Ff\;} & F(C') \\
\eta\downarrow & & \downarrow\eta \\
G(C) & \xrightarrow{\;Gf\;} & G(C').
\end{array}
$$

This gives a precise meaning to the informal usage, "the map $\eta_C: F(C) \to G(C)$ is *natural* in C." If each η_C is an isomorphism, we say that η is a *natural isomorphism* and write $\eta: F \cong G$.

Examples A.3.1

1. Let $T(A)$ denote the torsion subgroup of an abelian group A. Then T is a functor from **Ab** to itself, and the inclusion $T(A) \subseteq A$ is a natural transformation $T \Rightarrow \mathrm{id}_{\mathbf{Ab}}$.
2. Let $h: M \to M'$ be an R-module homomorphism of right modules. For every left module N there is a natural map $h \otimes N: M \otimes_R N \to M' \otimes_R N$, forming a natural transformation $M \otimes_R \Rightarrow M' \otimes_R$. For every right module N there is a natural map $\eta_N: \mathrm{Hom}_R(M', N) \to \mathrm{Hom}_R(M, N)$ given by $\eta_N(f) = fh$, forming natural transformation $\mathrm{Hom}_R(M', -) \Rightarrow \mathrm{Hom}_R(M, -)$. These natural transformations give rise to maps of Tor and Ext groups; see Chapter 3.
3. In Chapter 2, the definitions of δ-functor and universal δ-functor will revolve around natural transformations.

Equivalence A.3.2 We call a functor $F: C \to D$ an *equivalence of categories* if there is a functor $G: D \to C$ and there are natural isomorphisms $\mathrm{id}_C \cong GF$, $\mathrm{id}_D \cong FG$. For example, the inclusion of a skeletal subcategory is an equivalence (modulo set-theoretic difficulties, which we ignore). The category of based vector spaces (objects = vector spaces with a fixed basis, morphisms =

matrices) is equivalent to the usual category of vector spaces by the forgetful functor. Equivalence of categories is the useful version of "isomorphism" most often encountered in practice. As a case in point, the category of based vector spaces is not isomorphic to the category of vector spaces, in which the basis choices are not explicitly given.

Functor Categories A.3.3 Given a category I and a category $\mathcal{A}$, the functors $F: I \to \mathcal{A}$ form the objects of the *functor category* $\mathcal{A}^I$. The morphisms in $\mathcal{A}^I$ from F to G are the natural transformations $\eta: F \Rightarrow G$, the composition $\zeta \eta$ of η with $\zeta: G \Rightarrow H$ is given by $(\zeta \eta)_i = \zeta_i \eta_i$, and the identity morphism of F is given by $(id_F)_i = id_{F(i)}$. (*Exercise:* show that $\mathcal{A}^I$ is a category when I is a small category.) We list several examples of funtor categories in Chapter 1, section 7 in connection with abelian categories; if $\mathcal{A}$ is an abelian category, then so is $\mathcal{A}^I$ (exercise A.4.3). Here is one example: If G is a group, the **Ab**^{G} is the category of G-modules discussed in Chapter 6.

Example A.3.4 The *Yoneda embedding* is the functor $h: I \to \mathbf{Sets}^{I^{op}}$ given by letting h_i be the functor $h_i(j) = \mathrm{Hom}_I(j, i)$. This is a fully faithful functor. If I is an **Ab**-category (see A.4.1 below), the Yoneda embedding is sometimes thought of as a functor from I to **Ab**$^{I^{op}}$ (which is an abelian category). In particular, the Yoneda embedding allows us to think of any **Ab**-category (or any additive category) as a full subcategory of an abelian category. We discuss this more in Chapter 1, section 6.

A.4 Abelian Categories

The notion of abelian category extracts the crucial properties of abelian groups out of **Ab**, and gives homological algebra much of its power. We refer the reader to [MacCW] or Chapter 1, section 3 of this book for more details.

A.4.1 A category $\mathcal{A}$ is called an **Ab**-*category* if every hom-set $\mathrm{Hom}_{\mathcal{A}}(C, D)$ in $\mathcal{A}$ is given the structure of an abelian group in such a way that composition distributes over addition. For example, given a diagram in $\mathcal{A}$ of the form

$$ A \xrightarrow{f} B \underset{g}{\overset{g'}{\rightrightarrows}} C \xrightarrow{h} D $$

we have $h(g + g')f = hgf + hg'f$ in $\mathrm{Hom}(A, D)$. Taking $A = B = C = D$, we see that each $\mathrm{Hom}(A, A)$ is an associative ring. Therefore, an **Ab**-category with one object is the same thing as a ring. At the other extreme, R–**mod** is an

Ab-category for every ring R, because the sum of R-module homomorphisms is an R-module homomorphism.

We call $\mathcal{A}$ an *additive category* if it is an **Ab**-category with a zero object 0 and a product $A \times B$ for every pair A, B of objects of $\mathcal{A}$. This structure is enough to make finite products the same as finite coproducts, and it is traditional to write $A \oplus B$ for $A \times B$. Again, R–**mod** is an additive category, but so is the smaller category on objects $\{0, R, R^2, R^3, \ldots\}$ with $\mathrm{Hom}(R^n, R^m) =$ all $m \times n$ matrices in R.

Definition A.4.2 An *abelian category* is an additive category $\mathcal{A}$ such that:

1. (AB1) Every map in $\mathcal{A}$ has a kernel and cokernel,
2. (AB2) Every monic in $\mathcal{A}$ is the kernel of its cokernel, and
3. Every epi in $\mathcal{A}$ is the cokernel of its kernel.

Thus monic = kernel and epi = cokernel in an abelian category. Again, R–**mod** is an abelian category (kernel and cokernel have the usual meanings).

Exercise A.4.1 Let $\mathcal{A}$ be an **Ab**-category and $f: B \to C$ a morphism. Show that:

1. f is monic $\Leftrightarrow$ for every nonzero $e: A \to B$, $\quad fe \neq 0$;
2. f is an epi $\Leftrightarrow$ for every nonzero $g: C \to D$, $\quad gf \neq 0$.

Exercise A.4.2 Show that $\mathcal{A}^{\mathrm{op}}$ is an abelian category if $\mathcal{A}$ is an abelian category.

Exercise A.4.3 Given a category I and an abelian cateory $\mathcal{A}$, show that the functor category $\mathcal{A}^I$ is also an abelian category and that the kernel of $\eta: B \to C$ is the functor A, $A(i) = \ker(\eta_i)$.

In an abelian category every map $f: B \to C$ factors as

$$B \xrightarrow{\ e\ } \mathrm{im}(f) \xrightarrow{\ m\ } C$$

with $m = \ker(\mathrm{coker}\ f)$ monic and e epi. Indeed, m is obviously monic; we leave the proof that e is epi as an exercise. The subobject $\mathrm{im}(f)$ of C is called the *image* of f, because in "concrete" abelian categories like R–**mod** (A.2.3) the image is $\mathrm{im}(f) = \{f(b): b \in B\}$ as a subset of C.

A sequence $A \xrightarrow{f} B \xrightarrow{g} C$ of maps in an abelian category is called *exact* (at B) if $\ker(g) = \mathrm{im}(f)$. This implies in particular that the composite

$gf: A \to C$ is zero. Homological algebra might be thought of as the study of the circumstances when sequences are exact in an abelian category.

A.4.3 The following axioms for an abelian category $\mathcal{A}$ were introduced by Grothendieck in [Tohoku]. Axioms (AB1) and (AB2) were described above. The next four are discussed in Chapter 1, section 3; Chapter 2, sections 3 and 6; and in Chapter 3, section 5.

> (AB3) For every set $\{A_i\}$ of objects of $\mathcal{A}$, the coproduct $\coprod A_i$ exists in $\mathcal{A}$. The coproduct is often called the *direct sum* and is often written as $\oplus A_i$. Rather than say that $\mathcal{A}$ satisfies (AB3), we often say that $\mathcal{A}$ is *cocomplete* (see A.5.1).
>
> (AB3*) For every set $\{A_i\}$ of objects of $\mathcal{A}$, the product $\prod A_i$ exists in $\mathcal{A}$. Rather than say that $\mathcal{A}$ satisfies (AB3*), we usually say that $\mathcal{A}$ is *complete* (see A.5.1 below).

Example A.4.4 **Ab** and R–**mod** satisfy both (AB3) and (AB3*), but the abelian category of finite abelian groups satisfies neither and the abelian category of torsion abelian groups satisfies (AB3) but not (AB3*). For purposes of homological algebra, it is often enough to assume that $\prod A_i$ and $\coprod A_i$ exist for countable sets of objects $\{A_i\}$; for example, this suffices to construct the total complexes of a double complex in 1.2.6 or the functor $\varprojlim^1$ of Chapter 3, section 5.

Exercise A.4.4 (Union and intersection) Let $\{A_i\}$ be a family of subobjects of an object A. Show that if $\mathcal{A}$ is cocomplete, then there is a smallest subobject $\sum A_i$ of A containing all of the A_i. Show that if $\mathcal{A}$ is complete, then there is a largest subobject $\cap A_i$ of A contained in all the A_i.

> (AB4) $\mathcal{A}$ is cocomplete, and the direct sum of monics is a monic.
>
> (AB4*) $\mathcal{A}$ is complete, and the product of epis is an epi.

Example A.4.5 **Ab** and R–**mod** satisfy both (AB4) and (AB4*). The abelian category Sheaves(X) of sheaves of abelian groups on a fixed topological space X (described in Chapter 1, section 7) is a complete abelian category that does not satisfy (AB4*).

Exercise A.4.5

1. Let $\mathcal{A}$ be a complete abelian category. Show that $\mathcal{A}$ satisfies (AB4*) if and only if products of exact sequences are exact sequences, that is,

for every family $\{A_i \to B_i \to C_i\}$ of exact sequences in $\mathcal{A}$ the product sequence

$$\prod A_i \longrightarrow \prod B_i \longrightarrow \prod C_i$$

is also an exact sequence in $\mathcal{A}$.

2. By considering $\mathcal{A}^{op}$, show that a cocomplete abelian category satisfies (AB4) if and only if direct sums of exact sequences are exact sequences.

A.4.6 For the last two axioms, we assume familiarity with filtered colimits and inverse limits (see A.5.3 below). These axioms are discussed in Chapter 2, section 6 and Chapter 3, section 5.

(AB5) $\mathcal{A}$ is cocomplete, and filtered colimits of exact sequences are exact. Equivalently, if $\{A_i\}$ is a lattice of subobjects of an object A, and B is any subobject of A, then

$$\sum (A_i \cap B) = B \cap \left(\sum A_i\right).$$

(AB5*) $\mathcal{A}$ is complete, and filtered inverse limits of exact sequences are exact. Equivalently, if $\{A_i\}$ is a lattice of subobjects of A and B is any subobject of A, then

$$\cap (A_i + B) = B + (\cap A_i).$$

Examples A.4.7

1. We show in 2.6.15 that **Ab** and R–**mod** satisfy (AB5). However, they do not satisfy (AB5*), and this gives rise to the obstruction $\varprojlim^1 A_i$ discussed in Chapter 2, section 7. Hence $(R\text{–}\mathbf{mod})^{op}$ cannot be S–**mod** for any ring S.
2. Sheaves(X) satisfies (AB5) but not (AB5*); see A.4.5.

Exercise A.4.6 Show that (AB5) implies (AB4), and (AB5*) implies (AB4*).

Exercise A.4.7 Show that if $\mathcal{A} \neq 0$, then $\mathcal{A}$ cannot satisfy both axiom (AB5) and axiom (AB5*). *Hint:* Consider $\oplus A_i \to \prod A_i$.

A.5 Limits and Colimits (see Chapter 2, section 6)

A.5.1 The *limit* of a functor $F: I \to \mathcal{A}$ (if it exists) is an object L of $\mathcal{A}$, together with maps $\pi_i: L \to F_i$ $(I \in I)$ in $\mathcal{A}$ which are "compatible" in the sense that for every $\alpha: j \to i$ in I the map π_i factors as $F_\alpha \pi_j: L \to F_j \to F_i$, and that satisfies a universal property: for every $A \in \mathcal{A}$ and every system of "compatible" maps $f_i: A \to F_i$ there is a unique $\lambda: A \to L$ so that $f_i = \pi_i \lambda$. This universal property guarantees that any two limits of F are isomorphic. We write $\lim_{i \in I} F_i$ for such a limit. For example, if I is a discrete category, then $\lim_{i \in I} F_i = \prod_{i \in I} F_i$, so products are a special kind of limit.

A category $\mathcal{A}$ is called *complete* if $\lim F_i$ exists for all functors $F: I \to \mathcal{A}$ in which the indexing category I is small. Many familiar categories like **Sets**, **Ab**, *R*-**mod** are complete. Completeness of an abelian category agrees with the notion (AB3*) introduced in A.4.3 by the following exercise, and will be crucial in our discussion of $\varprojlim^1$ in Chapter 3, section 5.

Exercise A.5.1 Show that an abelian category is complete iff it satisfies (AB3*).

Dually, the *colimit* of $F: I \to \mathcal{A}$ (if it exists) is an object $C = \mathrm{colim}_{i \in I} F_i$ of $\mathcal{A}$, together with maps $\iota_i: F_i \to C$ in $\mathcal{A}$ that are "compatible" in the sense that for every $\alpha: j \to i$ in I the map ι_j factors as $\iota_i F_\alpha: F_i \to F_i \to C$, and that satisfies a universal property: for every $A \in \mathcal{A}$ and every system of "compatible" maps $f_i: F_i \to A$ there is a unique $\gamma: C \to A$ so that $f_i = \gamma \iota_i$. Again, the universal property guarantees that the colimit is unique up to isomorphism, and coproducts are a special kind of colimit. Since $F: I \to \mathcal{A}$ is the same as a functor $F^{\mathrm{op}}: I^{\mathrm{op}} \to \mathcal{A}^{\mathrm{op}}$, it is also clear that a colimit in $\mathcal{A}$ is the same thing as a limit in $\mathcal{A}^{\mathrm{op}}$.

A category $\mathcal{A}$ is called *cocomplete* if $\mathrm{colim}\, F_i$ exists for all functors $F: I \to \mathcal{A}$ in which the indexing category I is small. Many familiar categories like **Sets**, **Ab**, *R*-**mod** are also cocomplete. Cocompleteness plays a less visible role in homological algebra, but we shall discuss it and axiom (AB3) briefly in Chapter 2, section 6.

Exercise A.5.2 Show that an abelian category is cocomplete iff it satisfies axiom (AB3).

As a Natural Transformation A.5.2 There is a diagonal functor $\Delta: \mathcal{A} \to \mathcal{A}^I$ that sends $A \in \mathcal{A}$ to the constant functor: $(\Delta A)_i = A$ for all $i \in I$. The compatibility of the maps $\pi_j: \lim(F_i) \to F_j$ is nothing more than the assertion that π is a natural transformation from $\Delta(\lim F_i)$ to F. Similarly, the compatibility of the maps $\iota_j: F_j \to \mathrm{colim}\, F_i$ is nothing more than the assertion that ι is a natural transformation from F to $\Delta(\mathrm{colim}\, F_i)$. We will see that lim and colim are adjoint functors to Δ in exercise A.6.1.

Filtered Categories and Direct Limits A.5.3 A poset I is called *filtered*, or *directed*, if every two elements $i, j \in I$ have an upper bound $k \in I$ ($i \le k$ and $j \le k$). More generally, a small category I is called *filtered* if

1. For every $i, j \in I$ there is a $k \in I$ and arrows $i \to k, j \to k$ in I.
2. For every two arrows $u, v: i \to j$ there is an arrow $w: j \to k$ such that $wu = wv$ in $\operatorname{Hom}(i, k)$.

This extra generality is to include the following example. Let M be an abelian monoid and write I for the "translation" category whose objects are the elements of M, with $\operatorname{Hom}_I(i, j) = \{m \in M : mi = j\}$. I is a filtered category, because the upper bound in (1) is $k = ij = ji$, and in axiom (2) we can take $w = i \in \operatorname{Hom}_I(j, ij)$.

A *filtered colimit* in a category $\mathcal{A}$ is just the colimit of a functor $A: I \to \mathcal{A}$ in which I is a filtered category. We shall give such a colimit the special symbol $\underrightarrow{\operatorname{colim}}(A_i)$, although (filtered) colimits over directed posets are often called *direct limits* and are often written $\varinjlim A_i$. We shall see in Chapter 1, section 6 that filtered colimits in R–**mod** (and other cocomplete abelian categories) are well behaved; for example, they are exact and commute with Tor. This provides an easy proof (3.2.2) that $S^{-1}R$ is a flat R-module, using the translation category of the monoid S.

Example A.5.4 Let I be the (directed) poset of nonnegative integers. A functor $A: I \to \mathcal{A}$ is just a sequence $A_0 \to A_1 \to A_2 \to \cdots$ of objects in $\mathcal{A}$, and the direct limit $\varinjlim_{i \to \infty} A_i$ is our filtered colimit $\underrightarrow{\operatorname{colim}} A_i$. A contravariant functor from I to $\mathcal{A}$ is just a tower $\cdots \to A_2 \to A_1 \to A_0$, and the "inverse limit" is the filtered limit $\varprojlim A_i$ we discuss in Chapter 3, section 5.

A.6 Adjoint Functors (see sections 2.3 and 2.6)

A.6.1 A pair of functors $L: \mathcal{A} \to \mathcal{B}$ and $R: \mathcal{B} \to \mathcal{A}$ are called *adjoint* if there is a set bijection for all A in $\mathcal{A}$ and B in $\mathcal{B}$:

$$\tau = \tau_{AB}: \operatorname{Hom}_{\mathcal{B}}(L(A), B) \xrightarrow{\cong} \operatorname{Hom}_{\mathcal{A}}(A, R(B)),$$

which is "natural" in A and B in the sense that for all $f: A \to A'$ in $\mathcal{A}$ and $g: B \to B'$ in $\mathcal{B}$ the following diagram commutes.

$$\mathrm{Hom}_B(L(A'), B) \xrightarrow{Lf^*} \mathrm{Hom}_B(L(A), B) \xrightarrow{g_*} \mathrm{Hom}_B(L(A), B')$$

$$\downarrow \tau \qquad\qquad\qquad \downarrow \tau \qquad\qquad\qquad \downarrow \tau$$

$$\mathrm{Hom}_A(A', R(B)) \xrightarrow{f^*} \mathrm{Hom}_A(A, R(B)) \xrightarrow{Rg_*} \mathrm{Hom}_A(A, R(B'))$$

That is, τ is a natural isomorphism between the functors $\mathrm{Hom}_B(L, -)$ and $\mathrm{Hom}_A(-, R)$ from $\mathcal{A}^{\mathrm{op}} \times \mathcal{B}$ to **Sets**. We say that L is the *left adjoint* of R, and R is the *right adjoint* of L. We also say that (L, R) is an *adjoint pair*.

Here is a familiar example of a pair of adjoint functors. Let k be a field and L: **Sets** $\to$ (k-vector spaces) the functor sending a set X to the vector space with basis X. ($L(X)$ is the set of formal linear combinations of elements of X). This is left adjoint to the forgetful functor U, because $\mathrm{Hom}_k(L(X), V)$ is the same as $\mathrm{Hom}_{\mathbf{Sets}}(X, U(V))$.

We will see many other examples of adjoint functors in Chapter 2, section 6. The most important for Chapter 3 is the following adjunction between Hom and tensor product. Let R be a ring and B a left R-module. For every abelian group C $\mathrm{Hom}_{\mathbf{Ab}}(B, C)$ is a right R-module: $(fr)(b) = f(rb)$. The resulting functor $\mathrm{Hom}_{\mathbf{Ab}}(B, -)$: **Ab** $\to$ **mod**–R has $L(A) = A \otimes_R B$ as its left adjoint. (See 2.3.8 and 2.6.2.)

Exercise A.6.1 Fix categories I and $\mathcal{A}$. When every functor $F: I \to \mathcal{A}$ has a limit, show that lim: $\mathcal{A}^I \to \mathcal{A}$ is a functor. Show that the universal property of lim F_i is nothing more than the assertion that lim is right adjoint to Δ. Dually, show that the universal property of colim F_i is nothing more than the assertion that colim: $\mathcal{A}^I \to \mathcal{A}$ is left adjoint to Δ.

Theorem A.6.2 *An adjoint pair* (L, R): $\mathcal{A} \to \mathcal{B}$ *determines*

1. *A natural transformation* η: $\mathrm{id}_A \Rightarrow RL$ *(called the* unit *of the adjunction), such that the right adjoint of* f: $L(A) \to B$ *is* $R(f) \circ \eta_A$: $A \to R(B)$.
2. *A natural transformation* ε: $LR \Rightarrow \mathrm{id}_B$ *(called the* counit *of the adjunction), such that the left adjoint of* g: $A \to R(B)$ *is* $\varepsilon_B \circ L(g)$: $L(A) \to B$.

Moreover, both of the following composites are the identity:

$$(*) \quad L(A) \xrightarrow{L(\eta)} LRL(A) \xrightarrow{\varepsilon L} L(A) \quad \text{and} \quad R(B) \xrightarrow{\eta R} RLR(B) \xrightarrow{R(\varepsilon)} R(B).$$

Proof The map η_A: $A \to RL(A)$ is the element of $\mathrm{Hom}(A, RL(A))$ corresponding to $\mathrm{id}_{LA} \in \mathrm{Hom}(L(A), L(A))$. The map ε_B: $LR(B) \to B$ is the element of $\mathrm{Hom}(LR(B), B)$ corresponding to $\mathrm{id}_{RB} \in \mathrm{Hom}(R(B), R(B))$. The rest of the assertions are elementary manipulations using the naturality of τ and are left to the reader as an exercise. The lazy reader may find a proof in [MacCW, IV.1]. ◇

Exercise A.6.2 Suppose given functors $L: \mathcal{A} \to \mathcal{B}$, $R: \mathcal{B} \to \mathcal{A}$ and natural transformations $\eta: \mathrm{id}_{\mathcal{A}} \Rightarrow RL$, $\varepsilon: LR \Rightarrow \mathrm{id}_{\mathcal{B}}$ such that the composites (*) are the identity. Show that (L, R) is an adjoint pair of functions.

Exercise A.6.3 Show that $\varepsilon \circ (LR\varepsilon) = \varepsilon \circ (\varepsilon LR)$ and that $(RL\eta) \circ \eta = (\eta RL) \circ \eta$. That is, show that the following diagrams commute:

$$
\begin{array}{ccc}
LR(LR(B)) & \xrightarrow{\ LR\varepsilon\ } & LR(B) \\
\ \downarrow{\varepsilon_{LR(B)}} & & \ \downarrow{\varepsilon_B} \\
LR(B) & \xrightarrow{\ \cdot\varepsilon\ } & B
\end{array}
\qquad
\begin{array}{ccc}
A & \xrightarrow{\ \eta\ } & RL(A) \\
\ \downarrow{\eta_A} & & \ \downarrow{\eta_{RLA}} \\
RL(A) & \xrightarrow{\ RL\eta\ } & RL(RL(A))
\end{array}
$$

Reflective Subcategories A.6.3 A subcategory $\mathcal{B}$ of $\mathcal{A}$ is called a *reflective subcategory* if the inclusion functor $\iota: \mathcal{B} \subseteq \mathcal{A}$ has a left adjoint $L: \mathcal{A} \to \mathcal{B}$; L is often called the *reflection* of $\mathcal{A}$ onto $\mathcal{B}$. If $\mathcal{B}$ is a full subcategory, then by the above exercise $B \cong R(B)$ for all B in $\mathcal{B}$. The "reflection" onto a skeletal subcategory is a reflection in this sense.

Here are two examples of reflective subcategories. **Ab** is reflective in **Groups**; the reflection is the quotient $L(G) = G/[G, G]$ by the commutator subgroup. In 2.6.5 we will see that for every topological space X the category of sheaves on X is a reflective subcategory of the category of presheaves on X; in this case the reflection functor is called "sheafification."

References

[A]   Adams, J. F. *Stable Homotopy and Generalized Homology*. Chicago: University of Chicago Press, 1974.

[BAI]   Jacobson, N. *Basic Algebra I*. San Francisco: Freeman and Co., 1974.

[BAII] Jacobson, N. *Basic Algebra II*. San Francisco: Freeman and Co., 1980.

[Barr] Barr, M. "Harrison homology, Hochschild homology and triples." *J. Alg.* **8** (1968): 314–323.

[BB]   Barr, M., and J. Beck. "Homology and standard constructions." In *Seminar on Triples and Categorical Homology Theory*. Lecture Notes in Math. No. 80. Berlin, Heidelberg, New York: Springer-Verlag, 1969.

[BBD]   Beilinson, A., J. Bernstein, and P. Deligne, "Faisceaux pervers." *Asterisque* **100** (1982).

[Bour] Bourbaki, N. Groupes et algèbres de Lie. Chapters 2 and 3 of *Éléments de Mathématiques*. Paris: Hermann, 1972.

[Br]   Browder, W. "Torsion in H-spaces." *Annals Math.* **74** (1961): 24–51.

[Brown] Brown, K. *Cohomology of Groups*. Berlin, Heidelberg, New York: Springer-Verlag, 1982.

[BX]   Bourbaki, N. Algèbre homologique. Ch. X of *Algèbre*. Paris: Masson Publ., 1980.

[CE]   Cartan, H., and S. Eilenberg. *Homological Algebra*. Princeton: Princeton University Press, 1956.

[ChE]   Chevalley, C., and S. Eilenberg. "Cohomology theory of Lie groups and Lie algebras." *Trans. AMS* **63** (1948): 85–124.

[Connes] Connes, A. "Cohomologie cyclique et foncteurs Ext^n." *C. R. Acad. Sci.* (Paris) **296** (1983): 953–958.

[Dold] Dold, A. "Homology of symmetric products and other functors of complexes." *Annals Math.* **68** (1958): 54–80.

[DP]   Dold, A., and D. Puppe. "Homologie nicht-additiver funktoren." *Ann. Inst. Fourier* (Grenoble) **11** (1961): 201–312.

[EGA]   Grothendieck, A., and J. Dieudonné. Éléments de Géométrie Algébrique. *Publ. Math. I.H.E.S.*, Part I: 4 (1960); Part II: 8 (1961); Part III: 11 (1961), 17 (1963); Part IV: 20 (1964), 24 (1965), 28 (1966), 32 (1967).

[EZ]   Eilenberg, S., and J. Zilber. "Semisimplicial complexes and singular homology." *Annals Math.* **51** (1950): 499–513.

[Faith] Faith, C. *Algebra II: Ring Theory*. Berlin, Heidelberg, New York: Springer-Verlag, 1976.

432

[Freyd] Freyd, J. P. *Abelian Categories*. New York: Harper & Row, 1964.

[GLC] Grothendieck, A. *Local Cohomology*. Lecture Notes in Math. No. 41. Berlin, Heidelberg, New York: Springer-Verlag, 1967.

[Gode] Godement, R. *Topologie Algébrique et Théorie des Faisceaux*. Paris: Hermann, 1958.

[Good] Goodearl, K. *Von Neumann Regular Rings*. London, Melbourne, San Francisco: Pitman, 1979.

[GS] Gerstenhaber, M., and S. D. Schack. "A Hodge-type decomposition for commutative algebra cohomology." *J. Pure Applied Algebra* **48** (1987): 229–247.

[Gw] Goodwillie, T. "Cyclic homology, derivations and the free loopspace." *Topology* **24** (1985): 187–215.

[G-W] Geller, S., and C. Weibel. "Étale descent for Hochschild and cyclic homology." *Comment. Math. Helv.* **66** (1991): 368–388.

[GZ] Gabriel, P., and M. Zisman. *Calculus of Fractions and Homotopy Theory*. Berlin, Heidelberg, New York: Springer-Verlag, 1967.

[Hart] Hartshorne, R. *Algebraic Geometry*. Berlin, Heidelberg, New York: Springer-Verlag, 1977.

[HartRD] Hartshorne, R. *Residues and Duality*. Lecture Notes in Math. No. 20. Berlin, Heidelberg, New York: Springer-Verlag, 1966.

[HK] Husemoller, D., and C. Kassel. *Cyclic Homology*. In preparation.

[Hopf] Hopf, H. "Fundamentalgruppe und zweite Bettische Gruppe." *Comment. Math. Helv.* **14** (1941–42): 257–309.

[HS] Hilton, P., and U. Stammbach. *A Course in Homological Algebra*. Berlin, Heidelberg, New York: Springer-Verlag, 1971.

[Humph] Humphreys, J. *Introduction to Lie Algebras and Representation Theory*. Grad. Texts in Math. 9. Berlin, Heidelberg, New York: Springer-Verlag, 1972.

[Hus] Husemoller, D. *Fibre Bundles*. 2d Edition. Grad. Texts in Math. 20. Berlin, Heidelberg, New York: Springer-Verlag, 1974.

[JLA] Jacobson, N. *Lie Algebras*. New York: Dover, 1962.

[KapIAB] Kaplansky, I. *Infinite Abelian Groups*. Ann Arbor: University of Michigan Press, 1971.

[KapCR] Kaplansky, I. *Commutative Rings*. Boston: Allyn and Bacon, 1970.

[Lang] Lang, S. *Algebra*. Reading, Mass.: Addison-Wesley, 1965.

[Leray] Leray, J. "Structure de l'anneau d'homologie d'une représentation." *C. R. Acad. Sci.* (Paris) **222** (1946): 1419–1422.

[LMS] Lewis, L. G., J. P. May, and M. Steinberger. *Equivariant Stable Homotopy Theory*. Lecture Notes in Math. No. 1213. Berlin, Heidelberg, New York: Springer-Verlag, 1986.

[Loday] Loday, J.-L. *Cyclic Homology*. Berlin, Heidelberg, New York: Springer-Verlag, 1992.

[LQ] Loday, J.-L., and D. Quillen. "Cyclic homology and the Lie algebra of matrices." *Comment. Math. Helv.* **59** (1984): 565–591.

[MacCW] MacLane, S. *Categories for the Working Mathematician*. Berlin, Heidelberg, New York: Springer-Verlag, 1971.

[MacH] MacLane, S. *Homology*. Berlin, Heidelberg, New York: Springer-Verlag, 1963.

[Magid] Magid, A. *The Separable Galois Theory of Commutative Rings*. New York: Marcel Dekker, 1974.

[Massey] Massey, W. "Exact couples in algebraic topology." *Annals Math.* **56** (1952): 363–396.

[Mat] Matsumura, H. *Commutative Algebra.* New York: Benjamin, 1970.

[May] May, J.P. *Simplicial Objects in Algebraic Topology.* Princeton: Van Nostrand, 1967.

[Milnor] Milnor, J. "On axiomatic homology theory." *Pacific J. Math.* **12** (1962): 337–341.

[Mitch] Mitchell, B. "Rings with several objects." *Adv. Math.* **8** (1972): 1–161.

[MM] Milnor, J., and J. Moore. "On the structure of Hopf algebras." *Annals Math.* **81** (1965): 211–264.

[Osof] Osofsky, B. *Homological Dimensions of Modules.* CBMS Regional Conf. Ser. Math. 12. Providence, R.I.: AMS, 1973.

[Q] Quillen, D. "On the (co-) homology of commutative rings." *Proc. Symp. Pure Math.* XVII. Providence, R.I.: AMS, 1970.

[Rob] Robinson, A. "The extraordinary derived category." *Math. Zeit.* **196** (1987): 231–238.

[Roos] Roos, J. E. "Sur les foncteurs dérivés de lim." *C. R. Acad. Sci.* (Paris) **252** (1961): 3702–3704.

[Rot] Rotman, J. *An Introduction to Homological Algebra.* New York: Academic Press, 1979.

[Schur] Schur, I. "Über die Darstellungen der endlichen Gruppen durch gegebene lineare Substitutionen." *J. Reine Ang. Math.* **127** (1904): 20–50.

[Serre] Serre, J.-P. *Cohomologie Galoisienne.* Lecture Notes in Math. No. 5. Berlin, Heidelberg, New York: Springer-Verlag, 1964.

[Shatz] Shatz, S. *Profinite Groups, Arithmetic, and Geometry.* Annals of Math. Study No. 67. Princeton: Princeton University Press, 1972.

[Smith] Smith, L. "Homological algebra and the Eilenberg-Moore spectral sequence." *Trans. AMS* **129** (1967): 58–93.

[Spal] Spaltenstein, N. "Resolutions of unbounded complexes." *Compositio Math.* **65** (1988): 121–154.

[St] Stallings, J. "On torsion-free groups with infinitely many ends." *Annals Math.* **88** (1968): 312–334.

[Suz] Suzuki, M. *Group Theory I.* Berlin, Heidelberg, New York: Springer-Verlag, 1982.

[Swan] Swan, R. *Algebraic K-Theory.* Lecture Notes in Math. No. 76. Berlin, Heidelberg, New York: Springer-Verlag, 1968.

[Swan 1] Swan, R. "Some relations between K-functors." *J. Algebra* **21** (1972): 113–136.

[SwCd1] Swan, R. "Groups of cohomological dimension one." *J. Algebra* **12** (1969): 585–610.

[Th] Thomason, R. "Algebraic K-theory and Etale Cohomology." *Ann. Scient. Ec. Norm. Sup.* (Paris) **18** (1985): 437–552.

[Tohoku] Grothendieck, A. "Sur quelques points d'algèbre homologique." *Tohoku J. Math.* **9** (1957): 119–221.

[Tsy] Tsygan, B. "Homology of matrix algebras over rings and the Hochschild homology" (in Russian). *Uspekhi Mat. Nauk* **38** (1983): 217–218.

[Verd] Verdier, J.-L. "Catégories dérivées," état 0." In *SGA 4½.* Lecture Notes in Math. No. 569. Berlin, Heidelberg, New York: Springer-Verlag, 1977.

[Weyl] Weyl, H. *The Classical Groups.* Princeton: Princeton University Press, 1946.

[Wil] Wilson, R. "Euclidean Lie algebras are universal central extensions." In Lecture Notes in Math. No. 933. Berlin, Heidelberg, New York: Springer-Verlag, 1982.

Index